THE
QUANTITATIVE ANALYSIS
OF PLANT GROWTH

STUDIES IN ECOLOGY

GENERAL EDITORS

D. J. ANDERSON B.Sc. Ph.D.
Department of Botany
University of New South Wales
Sydney

P. GREIG-SMITH M.A. Sc.D.
School of Plant Biology
University College of North Wales
Bangor

FRANK A. PITELKA Ph.D.
Department of Zoology
University of California, Berkeley

STUDIES IN ECOLOGY · VOLUME 1

THE
QUANTITATIVE ANALYSIS
OF PLANT GROWTH

G. CLIFFORD EVANS

M.A. Ph.D.

Department of Botany,
University of Cambridge

UNIVERSITY OF CALIFORNIA PRESS

Berkeley and Los Angeles 1972

UNIVERSITY OF CALIFORNIA PRESS
Berkeley and Los Angeles, California

ISBN 0 520 09432 8
Library of Congress Catalog
Card Number 77-183156

Printed in Great Britain

To the memory of
DR F. F. BLACKMAN
and
PROFESSOR V. H. BLACKMAN,
whose students, to the
second and third generation,
carried out most of
the work discussed
herein.

PREFACE

The quantitative study of the growth of individual higher plants, considered as whole organisms in their 'normal' physiological state, and against a background of natural or semi-natural environments, has grown up gradually over a long period. The literature is scattered, and reviews are few. To plan the writing of a book was therefore not a matter of arranging to fill a neatly demarcated gap, but involved a number of decisions on scope, content and treatment.

Readership. It would have been simple to address the book to biologists with a background at least equivalent to first-year level in a British university, with all that this implies in the way of access to the latest developments in biology, but it seemed that it would be a pity to do so, for four reasons.

Firstly, we are now in the middle of the International Biological Programme, and this book is intended, in part at least, as a contribution to it. It is hoped that it will be useful to many of those collaborating in this programme; but this immediately raises the question of what common background should be assumed.

Secondly, many years ago I assisted the late Professor E.J.Maskell in running a vacation refresher course for school teachers at Cambridge, and then became aware both of the very wide keenness among teachers of biology from all over the country to keep up with recent developments and the very great difficulties in the way of doing so. These early impressions have been strengthened by experience both as a university teacher, and as examiner and drafter of syllabuses for the Cambridge Local Examination Syndicate. It is hoped that this book will help teachers to introduce their pupils to the study of higher plants as whole organisms, and at the same time that it will serve as one line of entry into the study of quantitative biology.

Thirdly, for those in schools with limited resources, those doing private research, or those working under field or expedition conditions remote from laboratory facilities, the basic methods which will be discussed have the advantage that much can be done by very simple means. Where appropriate, such methods will be described in detail, because often success in an investigation hinges more on the observance of a number of simple precautions than on the use of elaborate techniques and apparatus. There are more than a quarter of a million species of higher plants, and very few of these have ever been the subject of detailed quantitative study. In few scientific fields are there such opportunities for the amateur (in the old sense of the word) to make useful contributions to the sum of ordered knowledge.

Fourthly, I have been fortunate in that much of my university teaching has been in the form of practical courses. In particular, it has long been customary in the University of Cambridge for those studying plants to come into residence for a six-week practical course in field work of all kinds in the middle of the so-called 'Long Vacation'. This has provided an opportunity for a practical class on the quantitative analysis of plant growth. With the exception of the war years, I have been associated with this course during every Long Vacation since 1934, at first assisting Professor Briggs, and then since 1948 taking charge. This intimate contact with students from many different backgrounds, at home and abroad, has made clear what an excellent introduction to quantitative biology this work provides for those students without much previous biological experience.

For these four reasons I have tried to make the main contents of the book available to those with little or no biological training, while not diminishing the completeness and rigour (in the mathematical sense) of the main treatment. This has been attempted partly by the use here and there of analogies which teaching experience suggests are helpful in providing students with readily-grasped mental images to correspond to unfamiliar concepts, and partly by supplying an introductory Part of four chapters on the life of higher plants as whole organisms, stripped as far as possible of jargon and technical terms.

The quantitative study of plant growth impinges on many other branches of science, some of them well provided with textbooks and works of reference. It seemed, however, unwise to assume that the reader would necessarily be familiar with these, and the bibliography includes a classification leading to a few useful texts in closely related fields. The advice of friends and colleagues has been sought, in order to select books which are known to be useful to students, but to make such a list complete would be an impossible task, and I cheerfully accept responsibility for the inevitable deficiencies. To a degree the treatment of particular aspects in the text has been coloured by the coverage of cognate subjects in textbooks and works of reference. In the hope, however, that the book will be found useful in places where extensive reference libraries are not available, the attempt has been made to include in summary form sufficient basic material to make the main thesis complete in itself.

Experimental planning and technique. I have already mentioned that, while the practical methods used are often simple, nevertheless success often depends on careful planning of experiments and observations. An apparently straightforward situation may conceal systems of extreme intricacy, which often baffle the young investigator. At the same time the pressure on space in scientific journals nowadays puts a premium on short papers, and experienced investigators must usually confine themselves to short descriptions of the most important points of technique, without discussing the why and the wherefore. How to tackle the complex situation presented by the growing plant and its surroundings therefore forms the subject of the second Part of the book.

Mathematics: statistics. The quantitative data once obtained, analysis of them can begin, and to this the third Part of the book is devoted. Success here does not require the investigator to be a skilled mathematician. It suffices if he has enough grasp of the principles involved to be able to select the most appropriate analytical methods. The treatment has been developed mainly for that increasing body of biologists who are interested in quantitative investigation, but who are not highly trained in mathematics generally, and in the mathematical aspects of statistics in particular. Much can be done by the use of techniques so simple as hardly to warrant the label of mathematics; even so, the working has been set out in full, so that those interested need not puzzle out short cuts for themselves. These cases are interspersed with others so complex that they cannot be reduced to handleable mathematical terms without the use of simplifying assumptions so gross as to remove the resultant treatment from contact with reality on to the plane of human imagination. In between are those rare cases where a worthwhile biological advance can be made by a treatment which is at the same time rewarding to the mathematician. Generally, the biologist finds that, if he knows enough of mathematical modes of thought and notation to be able to express his problem in suitable terms, and if the problem is soluble at all, a mathematician can fill in the gaps off the cuff.

Equally, if the reader already possesses considerable statistical expertise, and has elaborate computing facilities available, so much the better; but the assumptions on which the book is written are that he has neither, and the worked examples are arranged accordingly. Although in these studies little progress can be made without an acquaintance with the principles underlying basic statistical modes of thought, the minimum requirement here is readily met from one of the many good statistical text books, of which a short selection is given in the bibliography under the classification letter b.

Arrangement of the argument. The trains of reasoning in the first three Parts are interrelated, and some of them are intricate; and further interrelations develop in the fourth Part, where a variety of cognate problems are considered, involving many branches of plant science. But arguments do not necessarily proceed in a linear manner—the lines of reasoning may branch and recombine, and cross-connections may be numerous. It is never easy to present such arguments in the linear style of a European book, in which sentences, paragraphs and chapters succeed one another in inexorable progression. A variety of schemes have been devised to cope with these difficulties, but most of them impede the smooth flow of ideas and make the work difficult to read. I have preferred as a first priority to concentrate on readability, so that generally it is possible to follow out each of the main sequences of ideas without a break. If, later, it is necessary to follow another sequence which has branched from the first partway through, a brief recapitulation of the state of the argument at the branching point has been inserted. Such repetitions must obviously be kept to a minimum, and generally they are confined to the main trains of reasoning.

The question then arises of how to cope with the other cross-references, of which very many are needed. Page references always have the twin disadvantages that the page is an artificial division of the text which rarely corresponds to a natural division of the argument, and that these artificial divisions are first established at the page proof stage, so that the identification and insertion of references must begin then. Very numerous page references in themselves entail a great deal of work for author, publisher and printer, and also make it necessary to guard against the possibility of extensive corruption of the text. These were the reasons which led to the devising of the fully sectional book, where the argument is broken up into natural divisions of varying length comparable in number to the pages, and running section numbers are provided in the lower, outer, corners of the pages. The Figures, Plates, Tables and Equations carry the number of the section in which they stand, so that one common system of reference can be used throughout, and the page numbers become nothing more than a convenience in book production. All the references can then be inserted in the manuscript, thus overcoming the production difficulties just mentioned, and there is the further advantage that indexing can begin as soon as the manuscript is ready, instead of having to wait for page proofs, making possible a fuller index. Such a book has obvious advantages for author, publisher and printer. It is hoped that it will prove to be equally convenient to the reader.

The Index has been used to convey two classes of information deliberately omitted from the text: authorities for plant names, and the latitude and longitude of places. It was felt that this is the most convenient way of combining accessibility with avoidance of repetition.

The perfect index is always out of reach, because its production would require insight into the mental processes of all its users. In an endeavour to minimize the inevitable imperfections there has been close collaboration between author and indexer, but any views from readers on how the index might be improved will be welcomed.

Units. A word is necessary about units. As far as practicable I follow the Système International d'Unitiés, adopted by the Eleventh General Conference of Weights and Measures in 1960, and now gradually coming into general use (*The International System* (*SI*) *Units.* British Standards Publication 3763, 1964). This encourages the use of basic units as far as possible, and otherwise uses a series of derived units formed directly from them. Some of the deviations here are small, and dictated by convenience. For example, the mean rate of increase of dry weight of plants per unit area of leaves will be expressed in $g\ m^{-2}\ week^{-1}$, taking a week as being equal to $60 \times 60 \times 24 \times 7 = 604{,}800$ seconds. It could be said that as a unit $\mu\mathrm{g\ m^{-2}\ s^{-1}}$ would meet the requirements of the system better, while being within one power of 2 in size. But the preferred unit reflects better the actualities of measurement. As we shall see, a week is a very convenient period over which to make an observation: but the observation is then of the mean rate of increase over the period, averaged over good days and bad, night, twilight and midday sun. To express this as mean rate per second would be misleading, in regard both to the practicability of the observation and to its meaning. The chosen unit has also the minor incidental advantage that when thinking of smaller or larger systems, $\mathrm{g\ m^{-2}}$ can be scaled down to $\mu\mathrm{g\ mm^{-2}}$, or scaled up to tonne $\mathrm{km^{-2}}$, without moving the decimal point.

The most awkward question of the proper unit arises for the measurement of solar and terrestrial radiation, usually expressed for purposes of absolute measurement in terms of equivalent heat, for which up to 1960 the absolute unit of quantity had long been the calorie. The SI unit of heat is the joule (J, 10^7 erg; one 15°C calorie $= 4 \cdot 1855$ J) and the use of the various forms of calorie unit is now discouraged. Unfortunately it seems that when the international system was framed, not all the consequences for radiation measurement were foreseen, and the present situation is chaotic. In meteorology the calorie is well entrenched in the literature: the basic meteorological tables are expressed in calories (Smithsonian Meteorological Tables, 6th Edn, 1951), and so were most of the data collected during the International Geophysical Year and the subsequent period of International Geophysical Cooperation. Some important authorities, but not all, have replaced the calorie by the milliwatt-hour (Watt $=$ J s^{-1}; 1 mWh $= 10^{-3} . \mathrm{J.s^{-1}}.3\cdot 6.10^3.\mathrm{s} = 3\cdot 6$ J). Unfortunately, this change of practice does not fit in with the SI system, and with the increasing use of SI units to change from calories to mWh would not be justifiable. It does not preserve the old unit, with the advantages of access to the existing literature: it does not conform to the new international system, instead introducing an unnecessary factor of $3\cdot 6$ by bringing in the number of seconds in an hour and then dividing by a thousand: and in the particular context of field studies, where quantities of radiation must often be considered, it leads to the creation of awkward units such as mWh cm^{-2} day^{-1}. Consistent use of SI units would require the replacement of calories by joules, but until this tangle is sorted out, and there is general international agreement on the absolute unit of quantity to be used in radiation measurement, it seems better to preserve contact with the extensive literature and continue to use calories.

Notation. In the interests of established workers it is obviously desirable to retain as many as possible of the existing conventions of notation, while the interests of beginners require consistency. In a subject having a literature so scattered in space and time, inconsistencies have inevitably developed, creating problems which are discussed in Appendix 4. For convenience of reference this is printed on a distinctive paper.

Ontogenetic drift. An important concept, constantly recurring throughout the chapters which follow, is that of ontogenetic drift. The term has a long history, but there are those who prefer 'ontogenetic progression' or 'ontogenetic change'. My own preference is to continue to use 'drift'. In one or another of its standard usages this term seems to cover just the range of meanings needed. Drift may lack specific direction—'aimless drift'; or it may be highly directional, like that of a stick drifting in a stream; it may be cyclic, like that of a floating object in the Fair Isle Channel coming within a few miles of the island, and carried up the east coast and across the north by the ebb tide, and down the west coast and across the south by the flood, and so on; it may be superimposed on a direction otherwise determined, the sense in which tidal drift is understood in navigation; or more than one of these meanings may be combined. Thus to use 'drift' does not beg any questions or presuppose that underlying mechanisms or directions are known: and yet in all its meanings it retains the notion of being carried forward by something external —in plant ontogeny, the passage of time, with all that that implies. As Motteux wrote nearly three centuries ago when translating Rabelais 'All things tend to their end'. Compared with a word so flexible and rich in shades of meaning, the word 'change' appears to mean too little, and 'progression' too much.

Acknowledgments. In science as in other branches of human endeavour we progress by climbing on the shoulders of our predecessors, and this book owes much to the teachers who, nearly 40 years ago, inspired this attempt to apply the physiological ideas of Dr F.F.Blackman and Professor G.E.Briggs to the study of individual plants in the field, and hence to relate them to the ecological teaching of Professor Sir Harry

Godwin and Dr A.S.Watt, and the field taxonomy of Mr H. Gilbert Carter. In particular, the lectures to third-year students on the physiology of plant growth given by Professor Briggs for more than thirty years have never been published, although parts of this work crop up here and there in the literature, acknowledged by other authors. Sections 13.6–13.10 and 16.6–16.9 are derived directly from his treatments. The pervasive ideas running through this teaching are less easy to acknowledge, although it is possible to pick out the insistent application to growth problems of Blackman's ideas on ontogenetic drift. The way from these early beginnings to the present work has been long and roundabout, sometimes directed for years to related studies such as the measurement of radiation in the field, or the control of artificial environments in the laboratory, but even so with growth studies never very far away in the background. Progress owes much to discussions with innumerable friends at all levels of academic status, which cannot possibly be acknowledged save in general terms.

The genesis of the book goes back to an afternoon in the spring of 1967, spent by Professor A.R.Clapham in stating the case for giving the work priority over other projects. I am happy to acknowledge the correctness of his judgment. Since writing began I have been helped first by Mrs J.M.Evans, who checked the first draft and suggested very many improvements, nearly all of them incorporated in the text, and then by six friends who agreed to act as referees, and who have read and criticized the second draft as it was typed: three ecologists, Professor A.R.Clapham, Dr D.E.Coombe and Dr O.Rackham; and three agriculturists, Mr W.Hadfield, of the Indian Tea Association's research station at Tocklai in Assam, Dr E.J.M.Kirby of the Plant Breeding Institute at Trumpington, and Mr D.G.Morgan of the Cambridge University School of Agricultural Science and Applied Biology. The balance of the book owes much to their unstinting help. Individual portions have been read and criticized by Dr T. ap Rees (for respiration), by Professor P.W.Brian (for growth substances), by Mr B.G.Coulson (from the point of view of a schoolmaster), by Dr A.P.Hughes (Part IV), by Dr J.A.Raven (for photosynthesis) by Professor P.W.Richards (Chapter 31), by Professor J.S.Turner (Chapters 13, 28 and 29), and by Dr T.C.Whitmore (for tropical botany, particularly Chapter 31). Professor M.J.P.Canny has kindly given permission to make use in preparing Appendix 1 of a draft account of the Thoday respirometer written by him some years ago when we were working together, and Professor J.L.Monteith supplied the information in the first paragraph of section 9.5.

A major problem has been finding suitable data for the illustrative examples. In addition to those mentioned below, the following authors have generously given permission for their work to be reproduced: Dr M.C.Anderson (Fig. 6.9), Mr A.E.Canham (Fig. 18.10), Dr K.E. Cockshull (Figs. 23.8, 23.9.1 and 23.9.2, and others published in association with Dr Hughes), Professor Sir Harry Godwin (Fig. 12.4.1), Dr J.A.C.Harrison and Dr I.Isaac (Figs. 30.16, 30.17 and 30.18, and Table 30.17), Professor D.Muller and Dr J. Nielsen (Figs. 12.4.2, 12.9, 31.12.1, 31.12.2, 31.14 and Table 31.16), Dr D.J.Watson (Figs. 13.18.1, 13.19a, and Table 13.19). The Director General of the Meteorological Office gave permission for the reproduction of the figures used to construct Fig. 13.19b, and the Director of the Tocklai Experimental Station of the Indian Tea Association for Fig. 13.18.2. The editors and publishers of the following journals have also kindly given their consent to reproduce the figures and tables indicated:

Annals of Applied Biology (Fig. 30.16), *Annals of Botany* (Figs. 10.6, 13.14, 13.18.1, 13.19(a), 13.20, 23.7, 24.13, 24.17, Tables 13.19, 16.17, 16.18, 24.18), *Forstlige Forsøksvaesen i Danmark* (Figs. 12.9, 31.12.1, 31.12.2, 31.14, Table 31.16), *Journal of Agricultural Science* (Figs. 7.3, 17.4.1, 17.4.2, 17.6, Table 17.4), *Journal of Applied Ecology* (Figs. 20.4.1, 20.4.2), *Journal of Ecology* (Pl. 6.10, Figs. 9.8, 18.5(a), 28.10, 28.11, 30.8, Table 16.10), *Journal of the Linnean Society* (Botany) (Figs. 27.2, 27.3, 27.4, 27.5, 27.6), *Nature* (Figs. 23.8, 23.9.1, 23.9.2), *New Phytologist* (Figs. 3.2.1, 3.4.1, 3.4.2, 4.5, 12.4.1, 16.9.1, 16.13.1, 19.4.2, 19.6, 19.8, 21.2, 21.3, 24.7, 24.8, 26.6, 26.12, 27.8.1, 27.8.2, 27.9, 28.6.1, 28.6.2, 28.8, 28.9, 28.13, 30.8, Tables 16.12, 28.3).

The following publishers have also given permission to reproduce figures from books published by them:

Blackwell Scientific Publications Ltd. (Figs. 6.9, 9.8), Butterworth & Co. Ltd. (Fig. 18.10).

Nevertheless, for the reasons already given, in all too few recent papers are the authors able to include the basic data in full enough form to give the basis of a worked example, fitting into the treatment here proposed. Much use is therefore made of four substantial corpuses of observations, of which only the first has ever been published in full. They are (i) the studies of Professor Kreusler and his coworkers on *Zea mays*, carried out at the Poppelsdorf Agricultural Experimental Station between 1875 and 1878; (ii) the study carried out on the growth of *Helianthus annuus* at Cambridge in 1920 by Professor G.E.Briggs, Dr F.Kidd and Dr C.West, who have kindly entrusted me with the basic data and given permission for it to be used in the preparation of illustrative examples; (iii) the studies on the growth of young plants of *Helianthus* spp. made at Cambridge during the last 20 years by successive classes reading Botany in Part II of the Natural Sciences Tripos: these studies form the basis of many of the worked examples, and I must thank the class members, and also particularly Mr C.T.Sewell and Mr P.W.C.Barham and the teams of assistants under their direction, who did most of the background work, particularly the heavy computation involved in the construction of those Tables comparing the results of many years' experimentation: over the years they also contributed a great deal to the accumulated expertise on the handling of plants which is reflected in the eight chapters of Part II; and (iv) the studies on the growth of *Impatiens parviflora* made at Cambridge and thereabouts since 1948 by a succession of research workers, Dr D.E.Coombe, Dr P.J.Welbank, Dr A.P.Hughes, Dr O.Rackham, Dr J.P.Lewis and Dr T.Hegarty: my thanks are due to all of them, not only for permission to make use of the original data and of their later published work, but also for many helpful and stimulating discussions.

Mr P.Freeman and the team of assistants working with him have also helped with a number of these last studies, as well as some of the work on respiration described in Chapter 12, and the computation of Table App. 1.2.2. Dr D.P.Edwards wrote the programme and constructed from the output of the Titan computer the table of Appendix 2, which is printed by photoreproduction to avoid corruption. Otherwise, personal communications, where they can readily be identified, are acknowledged in the text and in the index. The taxonomy has been checked by Mr P.D.Sell, and cases involving cultivated plants have been examined by Dr P.F.Yeo. Mrs P.A.Chapman supplied the index and put the bibliography into correct form, and Mrs O.A.Robinson produced the cleanest of typescripts from a necessarily complex manuscript. Mr F.T.N.Elborn took the photograph used for Plate 12.8, and Miss S.M.Bishop and Mr P.J.Evans each drew three of the Figures. My thanks are due to them all.

Finally, I owe apologies to a number of authors whose publications are not here as fully discussed as their work merits, or as I would have wished. The implementation of the basic scheme of the book in the form in which it was first drafted has, however, produced a volume much longer than that originally intended, and quite long enough for a first attempt. I hope that readers may enjoy the problems posed by improvement on what follows, and I shall welcome correspondence on relevant examples which have been omitted, especially those which increase the generality of the conclusions; on better illustrations or improvements of points of the argument; and generally on ways of making the book more useful to those working on the numerous, fascinating, tantalizing problems posed by the growth of plants.

Cambridge   G. Clifford Evans
18 October 1971

CONTENTS

of leaves. 11.14 Roots. 11.15 Extraction of root systems. 11.16 Apparatus: ovens. 11.17 Cages, trays and tins. 11.18 Identification of samples. 11.19 Final drying and weighing: balances. 11.20 A moral tale. 11.21 Final check: area and dry weight of leaves. 11.22 Final check: dry weights of leaves and stems. 11.23 Final check: dry weights of tops and roots.

12 MEASUREMENT OF RESPIRATION

12.1 Dry weight changes, metabolic balance and mineral content. 12.2 Dry weight changes, photosynthesis and respiration. 12.3 Conditions of respiration measurements. 12.4 Wounding and handling. 12.5 The rate and the measurement. 12.6 Gas current methods. 12.7 Infra-red gas analysis. 12.8 Less expensive gas current methods. 12.9 Systems not involving gas flow. 12.10 Allowances for the effects of wounding and handling. 12.11 Respiration of the leaves of large plants. 12.12 Respiration of stems and roots of large plants. 12.13 Periods of measurement. 12.14 How accurate? 12.15 Changing conditions and the steady state. 12.16 Intercellular space carbon dioxide concentration.

Part III · Analysis of Data

13 HISTORY AND DEVELOPMENT OF THE MAIN ANALYTICAL CONCEPTS

13.1 Empirical generalizations. 13.2 Analogies. 13.3 The monetary analogy extended. 13.4 Noll and the '*Substanzquotient*'. 13.5 V.H.Blackman and the efficiency index. 13.6 Relative growth rate. 13.7 Briggs, Kidd, West and Kreusler. 13.8 Weber and Haberlandt: '*Assimilationsenergie*'. 13.9 Gregory and apparent assimilation. 13.10 Unit leaf rate and net assimilation rate. 13.11 A revised terminology. 13.12 Problems of calculation. 13.13 The unit of photosynthetic machinery. 13.14 Leaf protein–nitrogen. 13.15 Leaf area ratio. 13.16 Leaf weight ratio and specific leaf area. 13.17 Relative leaf growth rate. 13.18 Leaf area index. 13.19 Leaf area duration. 13.20 Roots and mineral uptake. 13.21 Special problems of large and long-lived plants. 13.22 Reproductive structures.

14 FIRST ANALYSIS OF HARVEST DATA

14.1 State of plants at harvest. 14.2 A particular example. 14.3 Dry weights and leaf areas. 14.4 Comparison of dry weight distribution. 14.5 Leaf and stem weight ratios and plant size. 14.6 Comparisons involving leaf area. 14.7 A further analysis. 14.8 A useful moral.

15 RELATIVE GROWTH RATE

15.1 Computation and units. 15.2 Estimation of fiducial limits. 15.3 An example. 15.4 Variances and variance ratios. 15.5 An alternative treatment.

Part IV · Problems Posed by the Growing Plant

CORRIGENDA

page 25, penultimate line. *For* head-one *read* head on

page 27, legend to Fig. 3.2.1, line 1. *For* interest *read* increase

page 30, legend to Fig. 3.3, line 2. *For* 3 μ *read* 3 μm

page 32, legend to Fig, 3.4.1, line 3. *For* in *read* on; *for* neutrals creens *read* neutral screens

page 39, line 12. *For* This *read* this

page 134, Section 11.7, 4 lines from end. *For* 10.8 *read* 11.8

page 149, line 6. *For* means *read* mean
Section 11.20, penultimate line. *For* kinds *read* kind

page 182, line 7. For *pulcherima* read *pulcherrima*

page 203, line 1. *For* of *read* or

page 218, line 11. *For* Plate 13.18 *read* Plate 6.10

page 262, last line. *For* even *read* ever

page 329, line 17. Replace semi-colon by comma

page 356, Delete penultimate sentence *from* We saw *to* 0.047

page 379, 7 lines from foot. *For* 1971 *read* 1972

page 453, line 4. *For* example *read* examples

page 497, Section 28.18, line 9. *For* flowering *read* flowing

page 521, 6 lines from foot. *For* Fig. 3.2 *read* Fig. 3.2.1

page 537, Section 30.18, line 3. *For* Fig. 30.17c *read* Fig. 30.17d

page 552, line 3. *For* forests *read* forest

page 568. After last word add a comma

page 569, line 2. Delete comma after However

page 645, line 16. *For* (in the press) *read* **47**, 113-127

page 658, 5 lines from foot. For *Physol.* read *Phytol.*

PART I

INTRODUCTION

INTRODUCTION

1.1 Aims

This book is intended to assist ecologists and others who wish to arrive at quantitative solutions of problems involving plants growing in natural or semi-natural conditions. It is hoped that it will be useful not only to those who have had a formal training in the plant sciences, but also to many others interested in plants, and in the parts which plants play in the natural economy of the world. Although the problems considered are intricate, anyone with a general knowledge of scientific method and of the properties of matter can readily grasp the basic principles involved.

1.2 Synopsis

The four chapters of the introductory Part I examine these basic principles with the intention of establishing a consensus of view. The eight chapters of Part II survey the special techniques involved in manipulating plants and their environments for the purposes of experiment, leading up to the acquisition of quantitative data. The twelve chapters of Part III consider the various ways of analysing the data once obtained, and the types of conclusions which can be drawn from them. Finally the eight chapters of Part IV point to ways in which these conclusions can be related to various branches of plant science—well established techniques providing solutions to problems raised by considering the growth of the whole plant, and the study of this growth providing a quantitative framework within which can be fitted studies of many aspects of plant life.

1.3 Definitions of plant growth

We must first reach a common view of the senses in which we are to use the expression 'plant growth', and define the scope of the studies with which we are to deal. While the first reference to the noun 'growth' in the

Oxford English Dictionary is dated 1557, the verb 'to grow' has ancient roots, and before the time of Chaucer had already acquired a number of senses, including 'to have vegetative life. Hence also, to exist as a living plant in a specific habitat, or with specified characteristics', which senses it has retained to the present day (*Oxford English Dictionary*, 1901). More recent usage has added many more senses, and in adapting the word to scientific purposes authors have tended to frame their own definitions. This is both an attractive and a dangerous process: attractive, in that scientific usage requires a precise statement, translatable into terms of quantities which can be measured; and dangerous because we are trying to define natural phenomena in terms of ideas which are products of the human mind. Experience shows that a definition so framed can easily become a Procrustean bed into which nature must be fitted either by stretching it or lopping a bit off.

We can avoid this difficulty by retaining the old, broad, senses of the word, and by not attempting at this stage to specify any one characteristic of the plant to be used as a measure of growth. Suitable measures will then be chosen to fit the circumstances of the intended observations. Chapter 4 is devoted to discussion of problems of mensuration, and of how the various possible measures can be related to each other.

While all the topics to be considered are concerned with 'having vegetative life', this is a subject of many ramifications, which would cover a large part of the contents of a botanical library. Our immediate interests are more circumscribed. Fortunately the old secondary sense of 'to grow' fits the subject matter of plant ecology well—'to have existence as a living plant in a specified habitat, or with specified characteristics'.

Although this definition covers all plants, we shall deal almost exclusively with higher plants. Many of the concepts and trains of reasoning which we shall examine could be applied equally to lower plants, but so far little has been done in this direction. Most investigators have been concerned either with the higher plant components of plant communities, or with agricultural or horticultural situations. We shall however consider in Chapter 30 an instance of fungal parasitism, and its effects on the growth of higher plants.

Our primary concern will be with the living higher plant: the whole plant, living in its natural environment. But even accepting this restriction of sense, nevertheless every branch of plant science impinges to some degree on one or another of our problems—inevitably so, seeing that the

1.3

study of plants has progressed most rapidly during the last century and a half by breaking up the overall problem into components and by studying in detail some aspect of the form, structure or functioning of plants to the exclusion of all others. Meantime the plant growing quietly away in its natural environment embodies all these aspects in its normal life.

1.4 The development of plant science

Science progresses by taking the easy cases first. Three centuries ago, growing plants were an obvious subject of investigation; but a few decades of work made clear that progress was limited—limited, as we now know, by the state of physics and chemistry at the time. The situation was neatly satirized by Voltaire, who was in London from 1726–1729, during which time he took a great interest in the progress of natural science. In *Les oreilles du Comte de Chesterfield et le chapelain Goudman* he wrote (I translate), 'After many observations of nature, made with my five senses, and with lenses and microscopes, I said one day to M. Sidrac "They're pulling our legs; there's no such thing as nature, every-thing is art.... Animals, vegetables, minerals, everything seems to me to be arranged with weight, measure, number, movement. Everything is a spring, a lever, a pulley, a hydraulic machine, a chemical laboratory, from grass to an oak, from a flea to a man, from a grain of sand to the clouds.

"Certainly, there is nothing but art, and nature is an illusion." "You're right" said M. Sidrac.'

It is not clear when this squib was written, but by the time it was published in 1775 the main attention of students of animals and plants had for several decades been turned away from studies of organisms as functioning entities to concentrate on taxonomy, following the genius of Linnaeus.

Meantime the physical sciences advanced, and it was against a very different background of physics and chemistry that the quantitative study of the growth of whole plants was resumed a century ago. But progress was again hindered, this time by a different set of difficulties. No statistical means were yet available to deal with small samples; and plant physiol-ogists were only just beginning to study the individual processes which together make up plant growth. In consequence, as an example, the very voluminous data collected over four seasons of field work by Kreusler and his co-workers, and all published before 1880, were not analysed until 1920. But in those days there was a habit of publishing original data in full; and one can profitably turn back to their extensive tables for data to test some new hypothesis. Several of our examples have been derived in this way (e.g. Fig. 2.7, from the growing season of 1875; Figs. 16.3, 16.4, 16.8, 16.9.2, and 16.13.2, from the season of 1876; Fig. 2.9, data from trees felled in 1887–1888).

The intervening period up to the present day has seen much progress towards understanding the structures, the mechanisms and the processes underlying the integrated whole of vegetative life. At the same time substantial advances have been made in related fields, in statistical methods, in basic meteorological measurements, in microclimatic studies, in soil science, and in many important techniques including those involved in growing plants under artificial conditions. We are thus better equipped than ever before to tackle problems involving plants growing in natural or semi-natural environments. Yet 'the larger the light, the larger the circle of darkness which surrounds it'; the more we know, the more we are aware of the range of problems awaiting solution.

1.5   The inaccessible plant

The plant growing in its natural environment is an object peculiarly inaccessible to scientific investigation. Most of the experimental methods which have proved so successful in the laboratory would interfere destructively with the subject of the investigation, altering either the environment, or the plant, or both [4.6]. Great circumspection is therefore needed in the choice of methods. It is necessary to consider carefully the nature of the biological and physico-chemical systems involved both when planning observations, and when considering the meaning of the results.

At the same time it is essential to study the natural plant in the field. The great accumulation of knowledge on the form and functioning of plants does not provide us with enough data to infer without observation how any particular plant would behave as a functioning whole. This is partly because different aspects have been studied using different plants, and partly because of the great diversity of form and functioning of the organs of higher plants, which makes it difficult and often impossible at the present stage of our knowledge to formulate general statements of universal application. No sooner is this done than exceptions accumulate on every hand. In consequence it is difficult to advance to the concept of the plant as a whole by putting together studies of the individual parts of different species of plants; just as it would be difficult to construct a working engine out of parts each derived from a different make of motor car. There are also gaps in our knowledge which cannot always be filled by laboratory studies alone. As an example, a modest library could be built up devoted entirely to laboratory studies of leaves or parts of leaves; their structure; studies of photosynthesis, respiration, stomatal behaviour and so on in intact leaves; studies of the biochemistry of leaf preparations, and so on. Yet, using all this information, without field observations it would be hard to set limits to the values of the concentration of carbon dioxide in the chloroplasts of a specific plant, growing undisturbed in a natural environment (a problem to which we shall return in Chapter 29).

Our studies must therefore be based on observations made on entire plants growing in the field and interfered with as little as possible; and if the observations cannot be made under wholly natural conditions because of the effects of interference, then we must use plants growing under specified conditions sufficiently 'normal' to allow the plant to continue function as a whole in a manner not grossly dissimilar to growth under natural conditions. From such studies we can hope gradually to build up a framework of quantitative knowledge about the growth of plants as complete entities—plants belonging to particular species, growing in habitats within their natural range.

1.6 Sources of variation?

If a plant has to be grown under natural or semi-natural conditions it is difficult to get it into a particular state at a particular time. Thirty years ago the writer knew a gardener whose habit it was every year to have three or four potted laburnum trees, 2 to 3 m high, in the perfection of full flower for a special occasion on 6 May. In the absence of modern aids this involved much art, and it is not surprising that the same gardener won innumerable Gold Medals and other awards at the Royal Horticultural Society's shows. Such a feat involves a good deal of interference with the natural environment, and normally the investigator accepts that no experiment made under natural or semi-natural conditions can be repeated exactly, while often two successive sets of observations will show wide divergences. It will be part of our task to examine the causes of these natural variations, and to see how far it is possible to make allowances for them, or, indeed, to integrate them into the scheme which we are trying to construct of the life of the plant as a whole.

Much is already known, from investigations in the laboratory, about many of the individual processes involved in the growth of whole plants, and about the form of the structures within which they operate. Far less is known about the scope of these operations in the intact plant, about the natural balance of chemical pathways, the natural concentrations of important metabolites, the natural morphogenetic mechanisms, and so on. A great field of investigation here lies open to the inquisitive.

1.7 A quantitative framework of reference

Many ecological problems can be solved in terms of critical conditions which affect whole communities, or of critical phases in the life cycles of individual plant species. The solutions of other problems may depend on more generalized interactions between the plant and all the other elements of the ecosystem within which it lives, operating over substantial periods. Here a quantitative analysis of growth can be a most valuable tool. In favourable cases it may point to the outlines of a specific solution; but in any case it will provide a quantitative framework of measurement against which hypotheses may be tested.

One of the essential preliminaries to an understanding of the whole plant in its natural environment is the production of such a quantitative framework within which the individual studies of particular processes can be fitted, and indeed within which they must fit. We shall have to consider problems involved in the creation of this quantitative framework, and the relation of it to particular states of the environment and particular processes within the plant, when one must accept as restrictions all the disabilities which go with investigations of whole plants growing in natural environments.

Once a quantitative framework has been set up, it is possible to draw together separate studies of different aspects of plant life, and to fit them into a single picture of the life of the natural plant in its natural surroundings. As we shall see in Part IV, there are many outstanding problems, but work has been going on long enough for much valuable practical experience to have accumulated. Even if for most plants so far investigated the central synthesis is still elusive, nevertheless partial solutions can readily be obtained which suffice to answer many questions arising when studying the behaviour of plants in the field.

1.8 Investigations in the field

The ecologist usually finds it necessary to make his field observations in places remote from all normal laboratory facilities; and indeed, if he is to study some natural communities, remote from most of the activities of civilization. Many of the methods discussed are inherently suitable for such conditions, and we shall consider means by which others can be adapted. Under less severe conditions, these characteristics of the methods of investigation also render them very suitable for studying agricultural and horticultural problems. The necessarily semi-artificial nature of the agricultural and horticultural environment also simplifies investigation, so that in a number of respects progress here has been more rapid than under the more natural conditions of ecological research.

ORGANIZATION AND GROWTH

2.1 Plants and animals

Higher plants as organisms are not easily understood by the human mind. We feel a certain intuitive sympathy with the life of an animal—perhaps it is left over from the nursery: certainly it is a danger when endeavouring to interpret animal behaviour—but we have no such sympathy with the plant. Complex intellectual processes are needed to attain even an imperfect understanding of the life of a plant; and every step of the argument must be checked by observation and experiment, as organisms so alien to us are full of surprises—only too often the response of a plant to some change of circumstances is anything but what a human being would expect.

The differences extend to every department of life: the typical higher plant depends on the sun's radiation for its energy source and on the inorganic world for its basic molecular supply—whereas the higher animal depends on other organisms for both. The necessity of exposing a large surface to the environment, and particularly to radiation, demands from the plant a very different basic structure from the animal: the plant is normally sessile, whereas the typical higher animal must move in order to eat and avoid being eaten. With these differences go great differences in metabolic rate—respiration of a typical higher plant is much slower on any basis than that of a typical higher animal. The differences are no less marked in organography—the organs of higher animals are very numerous, and when all groups are taken together they fall into no simple classification: the organs of the higher plant are readily grouped into four broad categories of roots, stems, leaves, and reproductive structures, nearly always assembled in the same regular architectural plan. With these differences go differences of biochemical specialization—in the typical higher animal, cells are frequently specialized, performing some specific biochemical or biophysical function, and such specialized cells are frequently grouped into a specialized organ; whereas the average cell of the average green leaf can perform a greater range of syntheses than all the organs of a higher animal put together. Yet for all their standardization of basic architectural pattern, the organs of a higher plant make up for what they lack in number of types by an extreme diversity of form, seen at its richest in the wet tropics, and faintly mirrored in our gardens and hothouses. This variability is no doubt connected with the enormous range of size in the higher plants, reaching its maximum in the Angiosperms, whose range of mass from smallest to largest is about a hundred thousand times greater than in any major group of higher animals. Also connected with this are the differences of make-up of the adult form: higher animals have characteristically a set adult form, which can succeed and reproduce only within relatively narrow limits of size and shape: most higher plants have no such restriction—they are capable of successful growth and reproduction over a very wide range of form, size, and relative proportions of parts.

Working against this background of extreme variability we must obviously take particular care in endeavouring to generalize on the basis of scattered observations of a few species. More than half a century ago Anton Kerner von Marilaun devoted over a thousand pages to a summary of current knowledge on the natural history of plants—their architecture, structure, life cycles and relations with their natural environments. Since then, to take one example, the *Journal of Ecology* has grown to occupy more than two metres of shelving dealing with ecological relations alone. Yet we are still children paddling on the shores of a great ocean. No species of plant has ever been studied exhaustively, and for every species whose growth has been the subject of a quantitative investigation there are hundreds that have never been investigated at all.

2.2 Plants and animals: life cycles

Like the higher animal, the characteristic form of the higher plant has two similar sets of chromosomes; its cells are at least diploid, although much more frequently than in the animal each of these similar sets may itself be found to contain two, three, or more times the basic number, making the plant tetraploid, hexaploid, and so on. But unlike the great majority of higher animals, in the reproductive cycle a haploid phase with a single set of chromosomes intervenes between one diploid organism and its next diploid descendant. In those higher plants which reproduce and are dispersed by spores—the ferns, horsetails and clubmosses—this haploid phase, produced by the germination of the spore, is usually free-living, and produces sexual organs. On the fertilization of the egg cell, the young diploid plant develops *in situ*, at first being nourished by the haploid gametophyte, and then anchoring itself to the substratum by a root, producing aerial photosynthetic organs, and becoming fully self-supporting. In the seed plants, on the other hand, the female spores are not shed, and the haploid phase is much reduced and never free-living. Once again fertilization is *in situ*, and a young diploid embryo develops, still enclosed within the tissue of the diploid plant of the previous generation. After a certain degree of development, which varies very widely from one species to another, growth of the embryo checks. When it is shed, the mature seed is a complex structure of which the seed coat is part of the female parental generation, and more or less of the intervening haploid generation may remain between it and the diploid embryo, together with other tissues which may have more than two sets of chromosomes. In the seed plants, reproduction and dispersal take place at different parts of the life cycle, but the effect is the same—after dispersal the seed germinates by the outgrowth of the young root, which anchors the embryo to the substratum before the aerial part is developed. Consequently all typical higher plants are sessile, and must spend their lives in the neighbourhood of the place where the young embryo first put down its roots, as distinct from the typical higher animal, which is free to move.

This has an important corollary for the study of the growth of higher plants—it means that their growth can only profitably be studied in conjunction with studies of the environment, with consequences which will run through our whole work.

2.3 Plant architecture

Once established, the aerial part of the typical higher plant consists of a stem bearing unlike pairs of associated structures, leaves and axillary buds, which have the potentiality of developing into further stems. There are a variety of reasons for regarding the stem as being at the centre of higher plant organization, but here we need to notice only two—the great majority of plants can be increased by stem cuttings, but much more rarely by cuttings of leaves or roots; and all types of tissues found in any part of a particular higher plant are found in some stem, though not necessarily in the stem of the same species. Most young stems, for example, possess stomata and photosynthetic tissue, which in certain species persist for months and even years, but which in woody plants are more usually soon replaced by a corky periderm. At times this raises difficult questions about the role of the photosynthetic tissue of the stem (or, more rarely, of the root [2.5]) in the overall carbon assimilation of the plant. In the annual sunflower, for example, the stem area may be a fifth of the leaf area; in wheat the stems, leaf sheaths, glumes and lemmas of the inflorescences may together equal the area of the laminae of the foliage leaves. But although it is easy to show that the area may not be negligible, to determine what proportion of total photosynthesis takes place in these organs is another matter, about which relatively little is known. In the typical stem not only is stomatal frequency different from the leaf, but stomatal aperture also; and the photosynthetic tissue itself has a different structure.

In herbs the main stem often terminates in a flower, further growth being carried on by lateral branches developed from nearby buds, but in many herbs, and the majority of woody plants, growth of the main stem is indefinite. This is not, however, necessarily true of lateral branches. In many woody plants the laterals are regularly of limited growth, the so-called 'short shoots', familiar examples being the larch (*Larix decidua*) or the Virginia Creeper (*Parthenocissus tricuspidata*). Sometimes the short laterals themselves produce a regular system of even shorter second-order laterals, and the elaboration may proceed as far as five orders of branching, as in the 'Asparagus Fern' (*Asparagus plumosus*). Very little is known about the correlation mechanisms controlling such development, which is of great importance in the architecture of many trees.

2.4   Modes of higher plant growth

In many plants growth of the stem proceeds continuously, an active apical growing zone producing new stem material, with its continuous succession of new leaves and buds, throughout the active life of the stem. In others, growth is cyclical, the growing zone having a succession of active and dormant phases or ceasing to be active for some reason and being replaced by a lateral bud. In each active phase it produces a 'flush' of growth of limited length. In the wet tropics trees exhibiting each mode of growth flourish side by side and there are many variants (see, e.g., Koriba, 1958). If the seasonal variation of climate is comparatively small, the flushes may succeed each other three or four times a year: adjacent trees of the same species may be found out of phase, and in some cases even individual main branches of a tree may develop a rhythm different from their neighbours. If, however, the climate is seasonal there is a tendency for the flushing cycle to become locked to the seasonal one, and this irrespective of whether the unfavourable season is cold or dry. It does not do, however, to take too anthropocentric a view of what are unfavourable seasons to a plant. For example, if the unfavourable season is a hot dry one, it is not infrequently found that the new flush of leaves is produced towards the end of the dry season before the favourable, warm, wet, season is reestablished.

Many native British trees, such as the oak (*Quercus* spp.) and the beech (*Fagus sylvatica*) follow a flushing pattern of growth, and it is noticeable that under favourable growth conditions, with a warm season and a moist and fertile soil, a second flush is produced in the middle of summer, this happening frequently enough to have the common name of 'Lammas shoots'.

On the other hand, without changing over to some cyclical habit the continuous mode of growth can only be adapted to a climate with a markedly unfavourable season by shortening the life cycle until it is comprised within a single favourable season. It is noticeable that many temperate woody perennials have an intermediate mode—they unfold a flush in the spring, but then growth of the apex continues as long as conditions are favourable. With the onset of unfavourable conditions— hot and dry, or cold and dull—the apex becomes dormant again, and usually remains so until the following spring. Common examples of this intermediate mode are the apple (*Malus sylvestris*) and the holly (*Ilex aquifolium*). These two examples also show the marked difference in degree of response exhibited by different species. The apical meristems of the apple frequently continue active after the first flush of growth has expanded, whereas in the holly they so commonly become dormant again after the first spring flush of growth has expanded as to create the impression of a flushing habit of growth similar to the oak and the beech. Under exceptionally favourable conditions the holly has, however, been known to continue active growth for almost the whole of the growing season (Peterken and Lloyd, 1967).

2.4

2.5 Plant architecture: roots

Roots are very different from stems in structure and relations: the root system is usually branched, although it may (especially in Monocotyledons) consist of many simple roots each attached to the base of the stem. It usually bears only one class of appendages, other roots, which arise in the pericycle (often opposite the protoxylem elements, thus appearing in rows rather than spirally) and force their way out through the cortex. These laterals may persist, or be short-lived and deciduous. Roots never possess stomata, and only rather rarely (e.g., in certain orchids) do they have chloroplasts in the cells of the cortex. The root system thus normally derives its whole energy supply, together with its supply of carbon compounds for growth, from the rest of the plant.

2.6 Plant architecture: leaves

Despite the possibility of photosynthesis by stems, roots or reproductive structures, in the great majority of the higher plants it is the leaves which are the principal photosynthetic organs, and it is particularly to the leaves that we shall turn when considering the plant's supply of new reduced-carbon compounds. Later we shall have to explore the structure, development, and functioning of leaves in more detail; here we need to notice only that normally leaf development is profoundly influenced by the environment up to maturity of the tissues, when further expansion becomes impossible. The time of attainment of this state usually differs from one part of the leaf to another, the tip frequently being more mature than the base. Consequently, when a plant has been grown in a changing environment, the individual leaf, at maturity, will present a kind of integration of the effects of the varying environmental factors for a period up to that time; and when leaves are produced sequentially, each may differ in structure from the next. The proportion of the total leaf area which is mature will also change with time, so that even supposing that a given environmental change always produced the same effect on young developing leaf tissue, the effect of such a change on the leaf system as a whole would change with time.

When the leaf tissue is once mature there is normally little morphological change thereafter, anatomical change usually being limited to the thickness and composition of the cell walls; but biochemical and physiological change continues, and there may be changes in stomatal behaviour, in photosynthetic rate, in storage and in the pattern of biochemical activity generally. Accordingly, when dealing with sequentially produced leaves, or consecutive flushes, it is unwise to assume that all mature leaves are behaving alike—the point must be checked experimentally.

2.7 Growth and ontogenetic drift

Such problems of changing structure and function are inseparable from any study of plant growth. They are by no means confined to the leaves but inevitably involve the whole of the plant. Fig. 2.7 shows some typical forms of growth progressions in one particular annual crop plant. There is first an example of an attribute of the plant which increases throughout life; then one which rises to a maximum and remains constant; finally others which rise, reach a maximum and then decline more or less as the plant ages. Simple hypothetical instances can be used to show more generally that as a plant grows its form and function must change. For example, the autotrophic habit of the higher plants requires the maintenance of as large a photosynthetic area as possible, exposed to radiation and accessible to the inward diffusion of carbon dioxide. Let us suppose a plant to possess initially a particular shape, and imagine that growth takes the form of a doubling of all linear dimensions. This would increase the weight of the whole system eight-fold, while a cross-sectional area of any part would increase only four-fold. The mechanical interrelations of the system would therefore change, and so would the diffusive relations, since all the cell layers are now twice as thick. In other words, a mere maintenance of shape has not resulted in a maintenance of functional

---

Fig. 2.7. Some typical plant growth curves: showing throughout the life cycle of an annual plant a number of different attributes plotted against time from sowing, based on samples from two populations of *Zea mays* grown in Germany in 1875. (a) Mean total plant dry weight and (b) mean total ash content, both in grams, for samples from a population sown on 19 May and grown at Halle; (c) mean plant height and (d) mean total leaf area per plant for samples from a population sown on 11 May and grown at Poppelsdorf near Bonn. The smooth curves are intended only to emphasize the broad trends in each case. ((a) and (b) plotted from data given by Märker (1876); (c) and (d) from Kreusler *et al.* (1877a)).

2.7

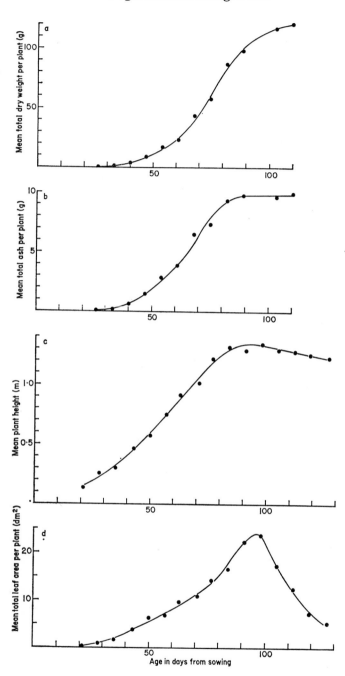

relations. Alternatively, let us imagine growth to take place by a multiplication in number of the original units. Clearly, as this goes on, an additional mechanical basis would be needed to hold them apart, and unless this is very extensive, they would soon begin to shade each other, and the radiation relations would also alter. Thus growth in any higher plant is inseparable from ontogenetic changes in form and functioning, known generally as ontogenetic drift. As examples, in Fig. 2.7 not only do the rates of change of all the various attributes themselves change with time, but each of the individual drifts is so different from the others as to make clear how fundamental are the changes constantly going on in the plant as a whole.

There is a further aspect, related to the fact that different organs of the higher plant have a different length of life—stems may be very long-lived, leaves relatively short-lived; a main root system may be long-lived, its laterals short-lived, and so on. Thus the pattern of ontogenetic drift may easily vary from organ to organ in more complicated ways than the simple progressions shown in Fig. 2.7, and this is connected with the wide range of adult form which we have already noticed [2.1].

2.8   Higher plant organization

Like all living things, the immensely elaborate biochemical and biophysical system of the higher plant is composed mainly of carbon compounds. Many other elements enter into the composition of the various simple and complex entities which together make up the whole; but in plants carbon, hydrogen and oxygen make up much the greatest part. It is accordingly a useful simplification to consider the higher plant as built up of reduced-carbon units, produced by the reduction of carbon dioxide by energy derived ultimately from the sun, and to relate the content of other elements to that of carbon. Because the plant takes in the carbon dioxide molecule by molecule, it is also convenient to regard the individual carbon atom, in any more reduced state than carbon dioxide itself, as the ultimate building brick, and to think of the plant as possessing a stock of these, the size of the stock being an important determinant of the size, form and activity of the plant at any stage. Additions to the stock take place by photosynthesis, and Fig. 2.8 shows in outline form the deployment of these additions. We must imagine that all over the diagram the chemical energy of some of the reduced-carbon units is

2.8

being utilized, more or less efficiently, to do metabolic work, maintaining the existing structure or effecting various biochemical and biophysical changes. Some units will be consumed in this way, being oxidized back to

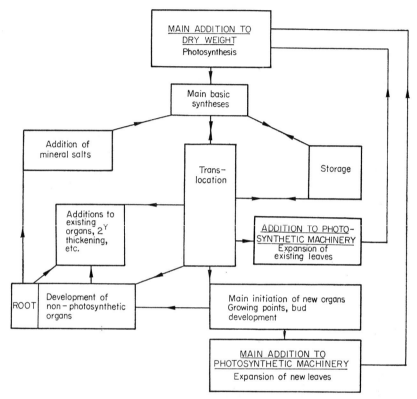

Fig. 2.8. A very simplified block diagram of the main bulk movements of organic materials in the vegetative phase of a higher plant. To simplify so far it is necessary to omit many items and cross connections, and to make no attempt to represent either the movement of inorganic materials within the plant, or the small but very important coordinating movements of plant hormones.

carbon dioxide and lost to the stock in the process of respiration. The remainder will pass through one or another of the various transformations which we shall now consider. First of all, from the immediate products of photosynthesis will be synthesized a range of basic molecular configurations, mostly sugars and amino-acids. Some of these will at once be

removed from the neighbourhood of the photosynthetic sites by translocation, the amino-acids as such or as the corresponding amides, the
sugars mainly as sucrose, the last accounting for the great bulk of translocatory movement. The rate of photosynthesis may at times exceed the
rate of removal, and there may then be temporary storage near the site
of photosynthesis, in the form of plastid starch grains, for example.
Translocation itself may, on the other hand, lead to more permanent
storage in the stem or in some specialized organ. But in due course,
possibly after many months, such stores will also be mobilized and translocated away. Some will then be utilized to form additions to existing
organs, by way of secondary thickening or other changes involving the
creation of new cells, such as the activity of a cork cambium; or by additions to or transformations in existing cells, such as thickening of cell
walls, or the changes accompanying bark formation. Such additions to
existing organs can consume a considerable fraction of the total production of new reduced-carbon units, as may happen in the addition of an
annual ring to a tree [Chapter 31]. Some is translocated straight down the
stem into the root, and then on to regions where new growing roots are
being formed. As the root system grows through the soil, water and mineral salts are absorbed. Some of these mineral salts pass up the stem in the
transpiration stream to the leaves, where they become involved in the
main basic syntheses, particularly of amino acids. Others are absorbed
at various points along their passage through the plant, some passively
bound by the non-mobile anions of the cell wall material, some actively
taken up into the vacuoles of living cells, and some becoming involved in
cellular metabolism in the cytoplasm. In particular nitrate may be reduced and incorporated into amino-acids in the root.

2.9   Higher plant organization and ontogenetic drift

All this activity and consumption of the newly formed units is necessary
to maintain the existing photosynthetic machinery in a fully functioning
state. However, this machinery usually has a limited life, so that for a
plant growing continuously, new photosynthetic organs must be produced in order to maintain photosynthesis at a fixed level, let alone to
increase it. We see from Fig. 2.8 that the main addition of new photosynthetic machinery is not just a simple multiplication of the old, but
involves directly or indirectly the whole metabolic system of the growing
plant.

2.9

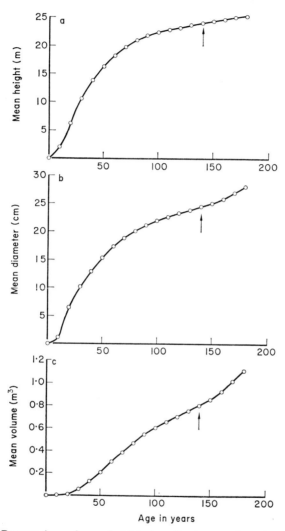

Fig. 2.9. Progressions of mean height, stem diameter, and trunk volume for the mean of two adjacent specimens of *Pinus sylvestris* growing at Erdweis-Thiergarten on the Weitra estate of the Landgraf Fürstenberg in Lower Austria. ↑ Thinning of neighbouring trees. Least significant differences between two means (at the 5 per cent probability level), in the age range from 20 to 170 years, as percentages of the relevant mean values: (a) height, 1 per cent; (b) diameter, 1 per cent; (c) volume, 3 per cent. Calculated from tabulated data for increments of height and diameter, together with form factors, published by Tischendorf (1926).

Very little consideration is needed to show that it is most unlikely that any system so complex can be maintained for long in a steady state, with all forms of consumption exactly balancing a steady production of new reduced-carbon units, and the development of new photosynthetic machinery exactly matching the decay of the old. In reality the plant is much more complex than the diagram has indicated—to avoid undue elaboration many particular items and cross-connections have been ignored. It is almost inevitable that such a system should be on the up-grade, with a constantly increasing production and a plant growing ever larger; or that it should be on the downgrade, with insufficient production to meet all demands, with decay of the photosynthetic machinery exceeding the rate of replacement, leading to a still further restriction of production.

Considered purely from the viewpoint of functioning we should therefore expect marked ontogenetic drifts in all parts of the overall metabolism. Fig. 2.7 showed some examples during the vegetative life of an annual plant. In just the same way longer lived plants show longer term progressions, over and above their annual cycles of change. Fig. 2.9 shows mean stem height, diameter, and volume during 180 years of the life of specimens of *Pinus sylvestris* growing in Austria. We see also how the broad progressions may be modified by environmental influences, here exemplified by thinning of neighbouring trees, even when the plants have already reached an age of 140 years.

2.10 Genetic differences

Finally, we must remember that any or all of the various developmental changes and responses to the environment which we have been considering are liable to be affected by the genetic make-up of the plant itself. Higher plants are not the most suitable organisms for genetical experiments—under carefully selected artificial conditions *Arabidopsis thaliana* will complete its whole life cycle from seed to seed in 35 days, but this is an exceptionally short cycle, and for some plants, in particular for trees, the period runs into many years. Nevertheless, the agricultural and horticultural interest in plant genetics, now extending back over very many generations of some plants, has sufficed to show what a very wide range of effects on the form and functioning of a higher plant can be produced by genetic differences. More recently this has been supplemented by studies made in the experimental taxonomy of natural populations of plants, which have shown the existence in nature of many genetic aggregates of characters within the species. In one species or another these cover almost every possible aspect of plant form, and every possible type of response to environment [5.4]. A small number of general accounts, giving an entry to the extensive literature, are listed in the bibliography under the classification letter a.

Mention has already been made [2.3–2.4] of the great variety of modes of higher plant growth. Later we shall consider a few other examples of genetically determined differences, both between species and between variants within a species, for instance on germination behaviour [5.5, 5.7–5.9], on tolerance of unfavourable conditions [10.5], and on reproductive behaviour [23.1–23.7, 23.10, 23.11]. We shall also examine cases where useful inferences can be drawn from comparisons between closely related species [e.g. 24.3–24.6, 26.3–26.9, 26.12 and see 32.2].

We have spoken separately of the responses of the plant to its environment; of the developmental pattern inherent in the make-up of the plant; and of the influences of the plant's genetic make-up on both of these. But we must remember that in nature the three are not separable; they are three facets, three ways of looking at one complex system, in which the plant reacts at the present time to its present environment in a way partly determined by its genetic make-up, and partly by its present form and stage of development, both of which in their turn are the products of many past reactions to past environments.

GROWTH AND ENVIRONMENT

3.1. The influence of past history

We have already seen that the plant body grows gradually, tissues maturing progressively and being added to those matured earlier; and that as they grow and mature these tissues are affected by the current environment in various ways. The plant body at any moment is therefore an epitome of the effects of past environments. It is this plant body which is reacting to the present environment, a fact to be remembered when studying plants which have grown in a natural, changing, environment. Only rarely can experiments at different times be made on plants which embody the same structural history: and in consequence it is easy to carry out at different times two apparently identical experiments, whose results do not agree. To understand in detail the growth of a plant in a natural environment at any particular time, it would thus appear to be necessary to record in detail and to interpret the results of the past environments which have contributed to its make-up. But in a climate such as that in Great Britain where substantial fluctuations of hot and cold, wet and dry, can succeed each other irregularly throughout the growing season, this is a counsel of despair. We must accept the variability of the effects of past history on the make-up of the experimental plants as contributing to the overall variability of the experimental result; and we must beware of generalizing from the results of work confined to a single season, which may have impressed its idiosyncrasies on the pattern of the observations (see, e.g., the discussion in 16.15 of the data presented in Figs. 4.5, 16.9.1, and 16.13; or Fig. 26.4.2).

Variations in the genetic make-up of the experimental plants will greatly complicate these problems, as they make it possible for two similar plants growing side by side to react in different ways to a given change in their common environment. In the early stages at least of any scheme of investigation into plant growth, it will be wise to use material as uniform as possible genetically, and sometimes clones can be used for this purpose. Granted relative genetic uniformity, it should then be possible to raise side by side two plants which respond similarly to environmental changes, so that a common past will have the effect of producing closely comparable plants. If we now wish to investigate the effects of a specific environmental change, one plant can be placed in condition *A*, the other in condition *B*, and their subsequent growth and development can be studied with the reasonable assurance that at least the plants were comparable when the comparison began. Using modern techniques for controlling plant environment we can go further, and raise large numbers of plants under standard conditions, producing a supply of standard plant material, which can then be used as a basis for a whole series of experiments radiating from a common centre. This was done, for example, by Hughes (1959a), who took as standard a plant of a specific dry weight, raised from seed in a carefully specified and controlled environment, and then compared the growth of this with other plants raised in six different environments, each of which differed from the standard one in a particular, specified way [17.3–17.10].

Unfortunately, this aid of the standard environment loses its usefulness, the more the experimental plants differ from each other in genetic make-up, and hence in reaction to the common environment. It would be a great convenience if it were possible to propose some specific environment, which could be agreed on as a standard one for use in comparisons between the growth of different species, but this is not possible. Different species of plants differ in their responses to their environment over a wide range, and in many different ways. There is therefore no *a priori* reason to suppose that conditions which produce optimum growth for one species will produce optimum growth in another, even if the two species are found in nature both growing in the same climatic region; nor are the conditions for optimum growth of a particular species necessarily the same at all parts of the life cycle. When one considers the very large range of climates occupied by plants, and also the great range of special techniques which gardeners have had to devise in order to keep a range of plants in cultivation, it is easy to see that the simple solution of comparing one species with another by growing them side by side in a standard environment is not a practicable one.

Clearly the problems raised by the incorporation in the plant body of a series of responses to past environmental changes are too complex to meet head-one; but as we shall see later, in favourable cases they can be circumvented by an indirect approach.

3.2 Non-linearity of response

Other general aspects of the effects of environment on growth must be considered at this stage, these being the form of the response of the plant to changes in individual environmental factors, and the nature of the interactions between the effects of one factor and those of another.

In the first place, the relationships between the different environmental factors and the reactions of the plant are rarely linear. Occasionally, one does find a very close approach to linearity over a specific range, as, for example, in the relationship observed by Hughes (1965a) between rate of increase of the total dry weight of a plant per unit leaf area and total daily visible radiation [Fig. 3.2.1a]. Here the approximation to linearity is very close over a range from 1·5 to 20 cal cm^{-2} day^{-1}: but the total for a bright day near midsummer would be more than 200 cal cm^{-2}, so that the observed linear range is less than one-tenth of that found in natural environments in the summer. Above the linear range the graph curves over: the slope becomes less, and finally, considerably before the maximum possible daily total has been reached, the rate of increase of dry weight per unit leaf area has become independent of the radiation total [Fig. 3.2.1b]. This type of relationship, with approximation to linearity near the origin, gradually changing to independence at high values, is commonly observed between a number of aspects of plant growth and various factors of the environment.

Also commonly observed is a more complex relationship which may take a variety of forms, having in common a graph with a point of inflection, sometimes referred to as sigmoid. Fig. 3.2.2 shows an example of the relationship between total daily radiation and the ratio of the total leaf area of the plant to the total leaf dry weight, or the specific leaf area—in effect, the degree of expansion of the leaf tissue. We see that between the full light of a bright summer's day and dense shade the leaf area per gram of leaf tissue in this particular plant can increase roughly three-fold. However, more than half this expansion takes place over less than one-tenth of the range, between 10 and 30 cal cm^{-2} day^{-1}, the rate of change falling off on either side of these limits. A straight line is a good approximation to such a curve near the point of inflection, but it becomes a worse and worse approximation as the range is extended.

3.2

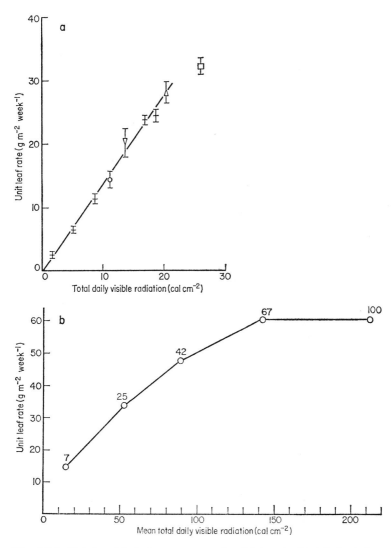

Fig. 3.2.1. Relationship between mean rate of interest of total plant dry weight per unit leaf area (unit leaf rate) and mean total daily visible radiation for *Impatiens parviflora*. (a) Plants aged between 17 and 44 days, growing under constant controlled conditions in the laboratory. From Hughes (1965a). (b) Plants aged initially 33 days, growing under neutral screens in the field between 27 May and 3 June 1957. 100 = full exposure, transmissions of the four screens shown in percent. From Hughes and Evans (1962).

3.2

In this particular plant, Fig. 3.2.2 shows another point: that between daily light totals of 10–20 cal cm^{-2} day^{-1} specific leaf area is falling approximately linearly, while Fig. 3.2.1a showed that the rate of increase of dry weight per unit leaf area is increasing linearly. Therefore a study

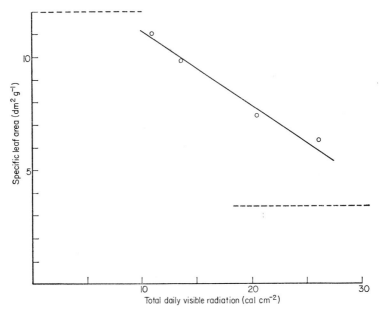

Fig. 3.2.2. Mean specific leaf areas of plants of *Impatiens parviflora* of 0·15 g total plant dry weight as a function of the daily total of visible radiation. All plants grown under constant, controlled conditions with 16 hours of illumination per day, a temperature of 15°C and a water vapour pressure deficit of 2 mmHg. Upper broken line, maximum specific leaf area observed during a long series of experiments under controlled conditions, including some with daily radiation totals down to 1·5 cal cm^{-2}. Lower broken line, minimum specific leaf area observed during field experiments in the open, with mean daily totals of visible radiation exceeding 200 cal cm^{-2}.

of some aspect of growth such as the rate of increase of dry weight per unit leaf dry weight (which can be produced by multiplying these two quantities together) would show that it changed relatively little with total daily visible radiation, a fall of one component compensating for a rise in the other. Indeed, in this plant the effect persists beyond the linear portions of the first two relationships, and results in the rate of increase of dry weight per unit leaf dry weight being almost independent

3.2

of total visible radiation around a daily light total of about 50 cal cm^{-2}. Such compensatory effects are not uncommon.

If non-linearity were the only difficulty, it would still place a formidable barrier in the way of using the statistical methods of correlation and regression for the interpretation of field observations on the growth of plants in terms of environmental changes. Correlation coefficients test the departure of a relationship from linearity: if we expect a relationship to be non-linear, it must be transformed into a linear form before such methods can be used with effect. This involves knowing beforehand what the relationship may be expected to be. Regression equations can readily be adapted to a non-linear form, but every complication adds to the number of arbitrary constants, and to the effort involved in computation. However, non-linearity is not the only difficulty, as we shall see.

3.3 Mathematical transformations

Curved relationships can on occasion be converted into linear ones by a suitable transformation of the basis of presentation: we shall meet one such in Figs. 5.11 and 13.6.1, where for a limited period a linear relationship existed between the logarithm of total plant dry weight and time. We must always remember, however, that linear relationships demonstrated in this way may be of limited duration; and the method of transformation used may be equally capable of converting a linear relationship into a curved one. A convenient example is afforded by the relationships which we have just examined in Fig. 3.2.1. Blackman and Rutter (1946) suggested that for the bluebell, *Endymion nonscriptus*, growing in the field, there was a linear relationship between the rate of increase of total plant dry weight per unit leaf area (which we shall call unit leaf rate) and the logarithm of the relative light intensity in which the plants were growing, taking full daylight as 100. Subsequently, using artificial screens, Blackman and Wilson (1951a, 1951b) extended this finding to a number of other species. Yet in Fig. 3.2.1a we observed a linear relationship between unit leaf rate in *Impatiens parviflora* and the total daily light quantity itself, in absolute units.

These two findings are not incompatible, as is shown by Fig. 3.3, where again for *I. parviflora* unit leaf rates observed under varying degrees of woodland shade by Coombe (1966) (see Figs. 28.11 and 30.8), combined with some of Hughes's observations under artificial screens in the field [Fig. 3.2.1b], are together plotted against the logarithm of the total daily short-wave radiation in absolute units. The relationship is now sigmoid, as in Fig. 3.2.2, but in the opposite sense. Where, in Fig. 3.2.1a (and Fig. 28.13a) there was a straight line, Fig. 3.3 shows a curve. Where, in Fig. 3.2.1b, there was a curved relationship between 70 and 200 cal

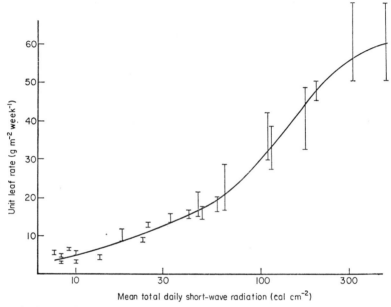

Fig. 3.3. Unit leaf rate of *Impatiens parviflora* (with 19:1 probability limits) in relation to the logarithm of the mean daily short-wave radiation ($\lambda < 3 \mu$). Observations of Coombe (1966) in Madingley Wood near Cambridge [Figs. 28.11 and 30.8] combined with those of Evans and Hughes (1961) on plants grown under neutral screens in the field [Figs. 3.2.1.b and 3.4.1, Expts. I and II (ii)]. The curved line is intended to illustrate only the probable form of the broad relationship. For the relation of visible to total short-wave radiation see 28.11, 30.6–30.8.

cm^{-2} day^{-1}, is now a roughly linear region surrounding a point of inflection. Allowing for the uncertainties of observation, shown in the figure as 19:1 probability limits, a linear expression would be an adequate description of this region, provided that we wished to work wholly within it, and for certain purposes this approximation to linearity serves a useful purpose. On the other hand it cannot be extrapolated, and the figure shows clearly that the approximately linear region is part of a much more complex relationship, as we saw in the case of Fig. 3.2.2.

3.3

The particular case of Fig. 3.3 gives an opportunity to consider also a more general point. There is reason to believe that many species of higher plants would show a relationship of the broad type shown by Fig. 3.3, if they could be grown under a suitable range of conditions. But for a particular species the detailed relationships of the different parts of the curve may vary, and the curve itself may be displaced along either axis. The work of several authors (e.g., Blackman and Wilson, 1951; Blackman and Black, 1959) has shown that for many species the region of decreasing slope above the point of inflection may lie above the light level associated with full daylight. Provided that one does not wish to work at very low daily light totals it may then be convenient to use a second-order equation to approximate to the region about the point of inflection and just below it. Such expressions cannot, however, conveniently be used to characterize the low light end of the relationship, and we shall see later in Chapters 28 and 29 several examples of the utility, in this region, of the linear relationship of Fig. 3.2.1a.

3.4 Interaction

One example of interaction, where a change in one particular aspect of the environment brought about contrary changes in two different aspects of plant growth, has been discussed in section 3.2. The complementary form of interaction, where two or more aspects of the environment all affect one particular aspect of growth, is very common. The aspect of growth affected may be the rate of some process, such as photosynthesis, or the state of some organ, such as degree of expansion of a leaf. If the interaction took the form of the simple 'limiting factors' notion of F. F. Blackman (1905), the situation would once again be relatively simple. Changing one factor would either bring about a linear addition to the rate, or have no effect, and the operation of such a system could readily be disentangled by statistical means. However, this is not what is found in practice. If factor (1) of the environment be fixed, a graph of the relationship of some aspect of growth with factor (2) is curved; if now factor (1) be changed in a stepwise fashion, every change will produce a new curve, the whole result being a family of curves where the effects of the individual factors are neither linear nor additive. Fig. 3.4.1 shows such a situation for some experiments on artificial shading reported by Evans and Hughes (1961). Rate of increase of dry weight per unit leaf area is here plotted against total daily short-wave radiation for three separate experiments, carried out at different times of year and in different weather conditions. We notice that although at low radiation totals the curves all follow a similar course, they diverge into a family of curves at higher totals.

Leaf expansion is a particularly complex case of cumulative effect of a number of different factors on some aspect of plant structure. We shall see later [19.4, 19.8] that the ratio of leaf area to leaf dry weight (specific

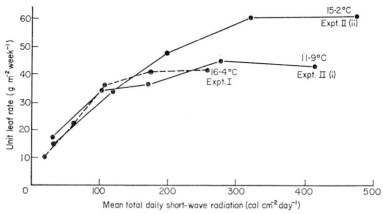

Fig. 3.4.1. Relationships between mean rate of increase of dry weight per unit leaf area (unit leaf rate), and mean total daily short-wave radiation for three experiments in *Impatiens parviflora* growing under neutrals creens in the field. Experiment I, 19–26 September 1956; Experiment II (i), 17–27 May 1957; II (ii), 27 May–3 June 1957. Mean daylengths 12¼, 16 and 16¼ hours respectively. Mean temperatures from continuous records for each period. For further details of these experiments see 19.6, 21.2 and 24.7. From Evans and Hughes (1961).

leaf area, here used as an index of leaf expansion) is a complicated function of plant size, and also that its value is affected by many aspects of the environment [28.8, 28.9]. Fig. 3.4.2 illustrates some of these from data given by Hughes (1965c). Here the effect of plant size has been allowed for by making all the comparisons at a fixed dry weight of 100 mg (for a discussion of the validity of this, see 28.7). Changes in daily totals of visible radiation produce large effects, when all other aspects of the environment are kept constant. Change of rooting medium produces two curves of similar form, but displaced laterally relative to each other.

3·4

A change of temperature can be seen to produce further displacements. We shall see later [28.8, 28.9] that alterations in the spectral composition of the visible radiation produce yet further divergences. What is more,

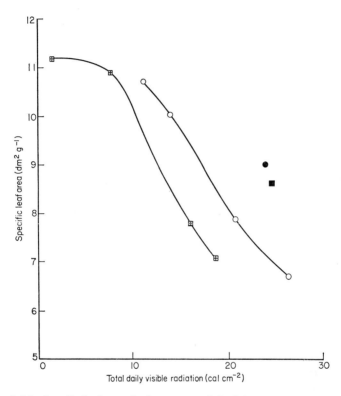

Fig. 3.4.2. Specific leaf area (leaf area per unit leaf dry weight) of *Impatiens parviflora* at a total plant dry weight of 100 mg, after growth from seed under constant, controlled conditions, as related to (i) the daily total of visible radiation (cal cm^{-2}); (ii) the rooting medium, loam and sand (○, ●) or vermiculite, sand and gravel with added nutrient solution (⊞, ■); (iii) temperature, 15°C (⊞, ○) or 20°C (■, ●). From Hughes (1965c) [see also Figs. 28.8. and 28.9.].

taken together, the overall consequences of these varied effects cannot readily be expressed in a linear and additive form.

These interactions add greatly to the difficulties of analysing field experiments by statistical means.

3.4

3.5 The natural environment further considered

So far the natural environment has been considered in the way which
statistical analysis demands—in other words, as if it consisted of a number of discrete and measurable 'factors', each of which makes some
specific contribution to the total effect of the environment on the growth
of a plant—a specific contribution which it is possible to separate from the
effects of other 'factors', and to characterize and measure. For natural
environments, this is true only to a limited degree. Environmental 'factors'
are products of the human mind, convenient abstractions devised in
order to make it possible to analyse and discuss complex problems: and
like all such convenient abstractions they are separated by a gulf from
things as they are. These abstractions are necessary if progress is to be
made, but the existence of the gulf, and the added complexities which lie
beyond it, must be borne in mind [9.1–9.4]. Consider temperature as a
'factor'. No doubt the temperature of the below-ground parts of most
plants approximates closely to that of the surrounding soil: it changes
only slowly and is relatively easy to measure because of the heat capacity
of the system as a whole. The temperature of the above-ground parts, and
particularly of the leaves, is another matter. Here the heat capacity is so
small in relation to the rates of energy exchange that the sudden appearance of a sunfleck in a wood, for example, can change leaf temperature at
a rate which initially may be as high as 1°C per second; and may bring
about an overall temperature change of leaf temperature of 10°C. Air
temperature round the leaf is only one of a number of aspects of the total
environment involved, conveniently summarized in the 'heat balance
equation'; and in consequence leaf temperature is a complex function of
long- and short-wave radiation, of air movement and the other 'factors'
involved in advective heat exchange, of all the 'factors' influencing
transpiration, and last (and usually least) of the heat component of
cellular metabolism. Indeed, even the concept of a 'leaf temperature'
itself is difficult to sustain; partly because, as can be shown using an
infra-red image converter, different parts of a leaf may well be at markedly
different temperatures, especially when exposed to full sunlight; and
partly because, under the conditions of very rapid temperature change
which we have mentioned, it is obvious that the various tissues lying
under a particular part of the leaf surface cannot be isothermal. What is

more, both the advective and radiative heat exchanges of the leaf react back on the air temperature, and to a lesser degree on the temperature of surrounding objects. It is thus apparent that any physical measurement or recording of air temperature gives us only a tenuous clue to the temperature of the aerial parts of the plant (and see 9.3).

We have here a system of built-in correlations. In periods of high radiation the relations between leaf temperature and air temperature will differ systematically from those in periods of low radiation; and at a fixed air temperature, the whole of this system will be different if the humidity were high, from what it would be if the humidity were low.

To such interrelations due to the totality of plant plus environment must be added other correlations inherent in the natural environmental system itself. In nature the values we measure as the various environmental 'factors' do not occur at random and independently: for example, a time of year with long days is also usually one of high daily total of visible radiation; periods of high net radiation gain are necessarily warmer than those of low gain, and so on.

The presence of these necessary correlations between the effective values of one environmental 'factor' and those of another puts a final stumbling block in the way of the direct interpretation of field observations; for if a strong association exists in nature, it is not to be expected that it can readily be broken by statistical means, unless one is able to multiply observations, and to continue for prolonged periods in search of those infrequent occasions when the natural correlations are broken of themselves.

3.6 The labyrinth

To recapitulate: the growth of plants can be understood only against the background of the environment in which they are growing, and it is not possible to propose any single environment as a standard of comparison for different plants. We have glanced briefly at some of the difficulties of characterizing this environment—of abstracting from its totality specific 'factors' which can be measured; and we have seen that the planning of experiments of the classical statistical kind designed to test the influences on growth of specific factors of the natural environment is hampered by natural correlations between 'factors', by interrelating plant responses, and by unforeseeable fluctuations. We have noted that the relations between an aspect of plant growth and a 'factor' of the environment are only rarely linear. We have also noted that the effect of a second 'factor' is rarely to be *added* to the effect of a first—in fact that the two effects interact, producing a combined result which is not a simple function of either separately. Finally, we have seen that the plant itself is a constantly changing system, and that consequently all these relationships with the environment change with it, and are bound up with its ontogenetic drift. It follows in a climate such as that of western Europe that a plant growing in a particular place at a particular date in one year will have had a different experience of climate from that of another plant, identical in genetic constitution, growing there on the same date in another year. This will have produced a different plant—a different summation of past experience (e.g. Fig. 26.4.2). At this date the plant will be acted upon by a complex of environmental factors which may not be easy to disentangle; and the effects of these factors on the functioning of the plant may be neither linear nor additive. Furthermore ontogenetic drift will ensure that when the plant is confronted with an identical external situation later in its life, it will react in a different way.

These interlocking complications can be summarized in a few phrases (see also 7.1; 17.1):
genetic differences;
non-linearity of plant response;
interaction of factors;
reaction of the plant on its environment;
natural environmental correlations;
ontogenetic drift;
persisting effects of past environments;
but each has many aspects, and together they form a formidable labyrinth. To attempt to elucidate so complex a situation by collecting field data and by statistical analysis alone is like trying to find a way through the maze by trial and error.

3.7 Ariadne's thread

If meaningful results are to be obtained, it is important to try to avoid meeting so complex a situation head on. Any steps which can be taken to reduce the number of variables which must be controlled and recorded will ease both the experimental programme itself, and the interpretation of the results. Otherwise, if no information is available about one or more of the variables, no allowance can be made for its systematic effects in analysing the results, and these systematic effects have to be thrown into the general random experimental error. At the best this will lead to a reduction in significance of the observations, and at the worst may make it impossible to make meaningful comparisons between different conditions. However, much can be done in planning the experimental programme to eliminate one or other of these sources of variation. Later on in Chapter 7 we shall give more attention to the details of this, and here it will suffice to give one or two simple examples. The effects of past climate on the state of the plants at the beginning of the experimental period can be systematized if the experiment allows of them being grown up to this stage under carefully controlled environmental conditions. Such a method has the added advantage that a particular plant form can be re-created at will for the repetition of particular observations. But for many experiments this approach is not possible, and here care has to be taken to raise a population as uniform as possible by careful attention to uniformity, rather than control, of conditions during the pre-experimental period of growth. Ontogenetic drift can be dealt with in a similar way— once again if it is possible to raise the experimental plant under controlled environmental conditions, a series of experiments can all be begun at the same stage of the plant's ontogeny. Otherwise, we are faced with a complicated situation which will require us to assess what ontogenetic stage the plant has reached at the beginning of the experiment, and this too is a subject to which we must give further attention later [17.2; Chapters 18–24, *passim*].

Natural environmental correlations can also be broken by the use of controlled environmental conditions, and indeed in some circumstances such experimentation affords the only means of interpreting observations made on comparable plants under natural conditions. Careful planning is also needed to deal with problems arising from non-linearity of plant response. Once again it may not be possible to elucidate the form of these non-linear relationships by considering the results of any particular experiment, and it may be necessary to use background information derived from experimentation under controlled environmental conditions to supply this.

Controlled environmental conditions can be a great help in sorting out the plant-environment relationships, but always subject to one proviso. We must be sure that the plants which we are growing and observing under controlled conditions fall within the naturally occurring range of plant form. If all that we have succeeded in doing by using controlled environments is to create a travesty of the natural plant, information derived from studying it will be of little help. It is therefore always necessary to pursue observations under natural conditions in parallel with any made in controlled environments.

All this applies to plants of uniform genetic constitution—for example, a pure line of an agricultural crop. But if our experimental programme requires us to deal with a natural population, in which there is also genetic variability, further dimensions of complication are added to those which we have already mentioned. It would be outside our scope to deal with these in detail, but we shall have to bear the consequent problems in mind as our discussions proceed.

3.7

4

PROBLEMS OF MENSURATION

4.1 Qualitative and quantitative

An ecologist is usually confronted by an ecosystem comprising both plants and animals. Among the most important differences between higher plants and higher animals, from the standpoint of a quantitative study [2.1], are

(a) the relative immobility of the plant, and

(b) the relatively wide range of adult form within which a particular species of plant can live and reproduce. Also

(c) that a consequence of immobility is that the plant and its environment must be studied together. It is not profitable to attempt to separate the two, the more so as the plant responds to environmental changes by changes in both form and function; and

(d) This has the result that the adult form of a higher plant sums up its successive responses to environmental changes from germination onward.

In considering the relations between a plant and its environment the ecologist must consider the whole life cycle, because the appearance and persistence of a particular plant in a particular community cannot *a priori* be ascribed to any particular phase, in the absence of information gathered in field studies. It may happen that such studies succeed in uncovering the existence of a critical phase or phases, when some particular characteristic of the plant and some particular state of the environment interact to determine absolutely success or failure in a particular habitat.

There are times, however, where no such simple solution can be detected, and it seems likely that success or failure is determined rather by relative performance over substantial parts of the life cycle. Here, quantitative growth studies may be essential to elucidate the problem.

Similar considerations apply to many agricultural problems involving plant growth. The success or failure of a crop from the farmer's point of view may often be determined by relatively small differences in yield, and here again quantitative growth studies may be helpful in shedding light on the underlying causes of such differences.

The historical development of the study of plant growth affords a good example of how much more difficult it is to carry out quantitative than qualitative studies. The general outline of the development of higher plants was well understood by the end of last century, and yet, by the time that Solereder's compendious work on the comparative anatomy of the dicotyledons began to appear in 1899, the quantitative study of plant growth was only just beginning. This delay in development has been quite as much due to the difficulty of analysing the measurements once made as to the difficulty in making the measurements themselves [1.4; 13.4–13.9]. For example, in the years 1875–1878, as part of a programme of agricultural research organized by the Prussian government, Kreusler and a number of co-workers had made a long series of quantitative observations on the growth of a number of different varieties of maize, but although these were all published by 1880, their analysis had to wait for another 40 years [Figs. 2.7, 13.7, 16.3, 16.4, and others].

Part of the difficulty has lain in plant variability, which can be affected by many causes. For convenience these can be formed into three interlocking groups—differences in genetic constitution; differences in environmental experience; and the summation of both of these in the plant's past history leading to differences in structure and functioning of the existing plant body, within which at any one moment the first two causes are operating. However great are the endeavours to reduce variability from these causes, some will inevitably remain. In consequence it would not in any case have been possible to make much progress in this field until the development of the statistics of small samples.

It is in many cases difficult, if not impossible, to make a useful series of observations of a particular characteristic without destroying the plant on which they are made [4.3–4.4]. This means that most estimates of growth must be based on at least two plants, one measured at the beginning of the growth period, and the other at the end. The problems due to variability are thus aggravated.

4.2 What to measure?

A great choice of measurement is open to us when confronted by a higher plant. We may measure numbers—numbers of parts in the various generally recognized main morphological categories of stems, leaves, roots and reproductive structures, or numbers of cells in a particular tissue or organ: we may measure linear dimensions, heights of stems, lengths of roots: or areas and volumes—leaf area, stem volume, dimensions of cells: we may measure fresh weights or dry weights of parts: or we may measure angles between parts or record particular shapes. We may also require to record the relationships of the different parts to each other—how large is the plant before it begins to branch?—or whereabouts on the plant does the first flower appear? Alternatively, at the sub-cellular level we may wish to make analyses for particular molecules or ions, or to determine the rate of some process such as respiration.

Some of these measurements can easily be made without seriously disturbing a plant growing in soil—overall height, or number of leaves; others less easily—leaf area, sometimes impracticable, as in a carrot; others not at all—root measurements, dry weights of parts. In any case it would be unwise to try to measure everything automatically, however much labour is available to assist in the experiment, however excellent the recording facilities, however powerful the computer, because the nature of the measurements may react back upon the results, as is always liable to happen when dealing with a living organism. For example, measuring the lengths and breadths of living leaves almost inevitably means handling them, and this puts up the respiration rate above what it would be if the leaf had been left undisturbed [12.4]. Thus, the more frequent the leaf measurements, the bigger the respiratory loss, which inevitably has effects upon other measurements. Again, if a plant is to be harvested, and its parts divided into categories and dried, the more measurements are taken between harvest and drying the longer the time interval and the larger the respiratory loss.

If labour and facilities are limited, it is essential so to plan the observations that they yield the optimum return in relevant information. For example, when dealing with a population of plants of high variability it may be profitable to measure fewer quantities, but to use more plants in order to obtain more accurate estimates of error. We have therefore to ask ourselves—what are likely to be the most important quantities to measure? Which leads to another question—is the ecological situation we are dealing with one where some specific dimension, such as extent of root system, or overall height, is likely to outweigh in importance all others, as may happen, for example, in the colonization of fertile but bare ground? Or are we more likely to be concerned with the growth of the plant as a whole? If the latter, we must remember that we are dealing with a situation at least as complex as that summarized in Fig. 2.8. Thinking back to that argument [2.8] we remember deciding that it is convenient to consider the basic building brick as the reduced-carbon unit: the individual carbon atom with its associated atoms at an oxidation level below CO_2.

4.3   Reduced-carbon units and energy pools

During the past history of the plant's growth, many of these units have been built irreversibly into the framework within which the life of the plant goes on—for example, most cellulose cell walls and nearly all lignified ones. They are as essential to the functioning of the organism as a whole as the bones of a mammal, but they are no longer available to take part in the continual metabolic changes of the protoplasm itself. At the other extreme, other units form the metabolic structures themselves —enzymes, structural proteins, membranes—and are intimately associated with the extent and rate of metabolism. Yet others are in an intermediate position—the compounds which they form are labile and capable of transformation into either of the first categories; or, alternatively, they can be oxidized back to the CO_2 level, and their stored chemical energy can be made available, at least in part, to do a variety of metabolic work. Such are transitory stores of starch grains, or vacuolar sugars, or the sucrose of the translocation stream. The additions to the system made by photosynthesis are added in the first instance to the labile energy pool; and then they are either used in supplying energy; or converted into more metabolic machinery; or into more structure.

It is apparent that as between different plants, or within the same plant or the same organ at different phases of the life history, the proportions of these different categories differ very widely. In a wheat grain just beginning to germinate, for example, the great bulk of the dry weight is formed by the energy pool; the same is true of a freshly formed storage organ, such as a carrot or potato, which has an energy pool large enough to supply all the respiratory needs of the winter months and leave enough over to construct a fresh growing phase in the spring. In a vigorously growing young annual the metabolic machinery may form the largest part of the total; in a tree the nonavailable structural material is largely predominant. In some circumstances, as in an autumn leaf, the metabolic machinery itself may be broken down, converted into units of the energy pool, and translocated away. What is then shed consists only of the irretrievable units of the leaf structure itself. The structural material records the organism's past; the metabolic machinery determines the rate and pattern of its present activity; the energy pool may set limits to its future activity, in conjunction with its future income.

How then are we to estimate the plant's total stock of reduced-carbon units [2.8], and how are we to determine to which category they belong? There is no doubt that the easiest measurement to make is that of total dry weight. There are instances, such as the germination of a seed with fatty endosperm, where this would be a most misleading index, either of the total size of the stock of reduced-carbon units at a particular stage of germination, or of the proportions of the different categories (see, e.g. App. 1.12). There are other cases where it can yield much information, particularly when the plant is growing in such a way that the proportions of the different categories are changing but little (e.g., the example discussed in 12.1). In any case from a practical point of view of experimental planning it is likely that we shall wish to measure total dry weight. If later we wish to consider what this means in terms of carbon, we have the stable dried material available. By a suitable combustion technique the carbon of weighed samples can be converted into carbon dioxide, which can be collected and measured, so that overall we can convert dry weight into carbon units with confidence. Furthermore, if we carry out a combustion in a bomb-calorimeter, the heat produced in the combustion can be measured and an estimate made of the average energy level of the reduced carbon units. Other samples of the dried material can be analysed for atoms other than carbon—nitrogen and the various mineral constituents; or for particular organic compounds such as sugars, starch, or cell wall materials. Of course we shall have limited the possibilities of organic analysis by drying the material, but useful information can still be gained.

It seems, then, that at some stage in most of our series of observations we shall almost certainly, as a matter of convenience, have to harvest the plant and dry it, however much or little subsequent analysis is to be done. With the rarest exceptions drying will kill it, and its subsequent growth cannot be followed. Could we, then, not use fresh weight as a measure of dry weight? Unfortunately experience shows that this is usually not practicable. The average plant contains several times as large a weight of water as it does of dry matter. At the same time the relations of dry weight to fresh weight are very complex [Chapter 24], and it is a very difficult, and at times impossible, task to infer the dry matter content from the fresh weight. There is therefore no escape from actually destroying the plant, either by drying it and weighing it dry, or by using some other destructive analytical technique.

4.4 Harvesting and sampling

The death of the plant to be measured would not matter if we could be sure of replacing the killed plant by an identical one. This could grow on and be measured and harvested later, giving information on growth in the interval. But as we have seen, the numerous possible causes of variations between one plant and another minimize the chances of having two identical plants, sometimes making the chances impossibly small. When this happens we cannot study the growth of individual plants, except for the limited range of information which can be obtained by measurement of the plant without a harvest. We must instead study the growth of the members of a population, from which we shall withdraw at intervals samples each of several plants for harvesting. We shall then have estimates not just of the mean growth of the population but also of its variability at each harvest. From these estimates of variability we can estimate the errors of our growth measurements. These errors may prove to be tolerable in the context within which we are working. If so, well and good; if not, it may prove practicable in subsequent experiments to make use of the information gained on variability to plan to reduce the mean errors by increasing the sample sizes, by some system of preliminary grading, or the like. But some error there is bound to be, and this will have important repercussions on our experimental planning. The necessity of working with populations of plants increases the labour involved, but it also brings other complications in its train, of which the most difficult to avoid are those associated with variability of the natural environment. Under field conditions it may be very difficult to ensure that all members of a population of plants are growing under identical environmental conditions; and if they are not, not only are further serious causes of variability imported into the experiment, but the question also arises of whether a larger programme of measurements of the natural environment itself should be planned, in order to assess this variability. These are serious difficulties which increase rapidly as the size of the population increases. It will be apparent how great is the importance of trying to keep the variability of the experimental plants to a minimum, so that the necessary level of accuracy can be reached on the smallest possible sample.

This level of accuracy depends on the difference between two estimates, each made on a variable population, and it is a function both of the two variabilities, and on the size of the difference in relation to them. For a given level of variability, obviously the larger the difference, the greater the accuracy; and conversely, for a given size of difference, the lower the variability, the greater the accuracy.

It follows that it is not worth planning experiments involving sampling from a population unless differences can be expected larger than a certain minimum size, and this in turn influences the length of time which must elapse between harvests. Under very favourable conditions, the total dry weight of an annual sunflower may increase by a third in a single day—in such a case significant results can easily be obtained by harvests at daily intervals, but this is unusual. For most plants daily harvests are not profitable.

There are also objections to leaving the plants to grow many times larger during an experimental period, as uncertainties appear in the computation of certain measures of growth [16.10, 16.12]. Usually, as a rough guide we should aim at a doubling of the plant's size between two harvests, and generally it will not cause any difficulties if the increase is half or double this amount. If it were practicable to arrange for exact doubling during each period between harvests, 15 such periods would see the plant size increase by a factor of $2^{15} \simeq 30,000$, and 20 periods by a factor of $2^{20} \simeq 1,000,000$. These are the right orders of magnitude for the increase in size during the whole growth period of a vigorous temperate annual, and if we took the growing season of such a plant as about half the year, we see that a harvest on average between once a week and once a fortnight would suffice. In practice, most experiments fall within these limits of 1–14 days, and this being so, it is not surprising that the favourite period is 7 days, as fitting so well into the working rhythm of what the Muslim calls 'the people of the Book'. It is interesting in this connection that although Kreusler's first year's work on maize in 1875 included several 6- and 8-day periods, for the remaining three years 1876–1878 he kept to a rigid 7-day rhythm.

4.4

4.5 The harvest and the interval

When a plant is harvested precise information of many kinds about it can be obtained; but there will then elapse a period of one, several, perhaps seven or fourteen days before we obtain our next precise set of figures. The growth of the plant, on the other hand, is a continuous process—granted that it has a daily rhythm, that the bulk of the increase in dry weight happens during the light hours of the day, nevertheless the

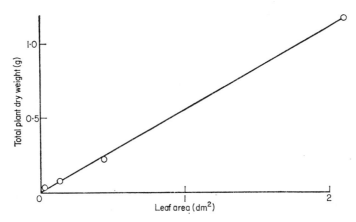

Fig. 4.5. Part of the data of Briggs, Kidd and West for the growth of *Helianthus annuus* at the Cambridge University Botanic Garden in 1920, showing the relationship between mean total plant dry weight and mean total leaf area for the first four harvests, at weekly intervals (data for the remaining harvests in Figs. 16.9.1. and 16.13.). From Evans and Hughes (1962).

processes of growth are going on all the time. Indeed, in some circumstances it is found that the bulk of extension growth takes place during the hours of darkness.

It will be necessary to draw inferences from the harvest data about what has been going on during the periods between harvests. Sometimes, with a sequence of harvests, it is possible to show that one regular pattern of plant behaviour runs through the whole sequence, as in Fig. 4.5 giving some of the data collected by Briggs, Kidd and West on the growth of sunflowers in 1920. This plot of total plant dry weight against leaf area approximates very closely to a straight line through the origin: so closely that we should be justified in inferring that this relationship continued to hold throughout the three inter-harvest intervals, and basing our calculations accordingly.

On the other hand, frequently there is no long sequence of harvests; or, if there is one, it may not yield any such regular result. We shall consider in Part III examples of handling such situations, but we must always remember the distinction between the particular size and morphological state of the plants at harvest, which have been measured; and the physiological changes between harvests, which have been inferred.

4.6 Advantages of harvesting at intervals

While the system of harvests at intervals has drawbacks as a means of investigating the growth of plants, it has also compensating advantages of a very decisive kind, for it makes it possible for us to plan growth studies under almost natural conditions.

A plant growing under wholly natural conditions is an extremely difficult subject for physiological investigation, because almost any attempt at investigation involves interference of some kind either with the plant or with its environment [1.5]. For example, none of the standard laboratory methods of measuring carbon assimilation can be applied to a plant growing under *wholly* natural conditions, and the question is, how much interference can be tolerated? However, in many studies of plant growth we do wish to study the development of the plant under natural conditions, or at least as close an approximation to these as possible.

Using the system we have been considering, this can easily be done as far as the aerial environment is concerned, provided that the plants being investigated are small enough to grow satisfactorily in pots or other containers, which can be placed in the appropriate environment and left without interference during the period between one harvest and another.

To use a natural root environment poses much greater problems, and it is frequently impossible because of the complex nature of the soil and the virtual impossibility of subjecting a natural soil to any process of investigation which will leave all its properties unaltered—the whole physical, chemical and biological complex being extremely sensitive to interference of any kind [10.1]. Apart from using an artificial soil as close to the natural one as can be contrived, all that can usually be done is to take advantage of favourable circumstances when they occur—for example, samples can be harvested at intervals from a sufficiently extensive natural population of seedlings, in just the way we have been considering. On the other hand, these difficulties are much less serious in experiments on agricultural or horticultural crops, and they may be absent altogether. Here we have in any case to deal with a partly artificial soil, and there may be no great difficulty in arranging for the necessary crop of plants to be growing in exactly the soil which we wish to investigate.

An essentially similar method has been used for studying the growth of trees under natural conditions. This involves making exhaustive measurements of selected specimens in an extensive woodland, and then harvesting other comparable trees, growing under comparable conditions but far enough away for the harvesting itself not to affect the selected specimens [31.11, and Newbould 1967, chapters 2 and 4]. From measurements of these harvested specimens the dry weight, etc. of the selected growing specimens is inferred. Later the selected specimens are measured again, and if necessary other comparable trees are harvested, and so on. Substantial errors are unavoidable with such a technique, and the periods between measurements are usually counted in years, but nevertheless much useful information can be gained in this way about plants growing under conditions indistinguishable from natural ones (Chapter 31). In suitable cases it can be used to provide data on rates of growth long before the observations themselves were made. For example, the specimens of *Pinus sylvestris* which were the subjects of Fig. 2.9 were seedlings in the earliest years of the eighteenth century, being harvested in 1887–1888 after 18 completed decades of growth. Indeed, even when there is no more information than the average ring width, it has been possible to make inferences on growing conditions far back into the past (Grock, 1937; Ferguson, 1968).

4.7 Measurements of respiration

The methods of investigation involving harvests may lead to morphological and anatomical measurements of plant parts and may supply dried material which can be subjected to inorganic or organic analysis. These analyses in turn may lead to estimates of the total stock of reduced-carbon units; of the average energy level associated with these; of the proportions of compounds particularly associated with the energy pool, or with cell wall architecture. But all this gives little information on the vital core of the whole system itself—the metabolic machinery. It is true that by taking trouble we could learn much about the structure of the machinery from the harvested plants—we could cut sections and count the cells and the chloroplasts—from electron micrographs we might estimate the numbers of mitochondria—and so on. But all this information, if gathered together, would still not tell us how fast the metabolic machine was running.

It seems that one of the most useful accessory methods of investigation readily available is to measure the respiration of the plant during the dark period of its daily cycle. The practical precautions which have to be observed when doing so will be considered in Chapter 12. Here it suffices to say that for the above-ground parts of the plant the effects on respiration of unavoidable interference with the natural environment are much less serious than they are in the case of carbon assimilation or transpiration, sensitive as these latter are both to natural radiation and to air movement [12.2]; and whereas small changes in atmospheric carbon dioxide concentration may have a marked effect on the rate of photosynthesis, comparable absolute changes in oxygen concentration around the mean value in normal air have inappreciable effects on the rate of respiration. However, these considerations apply to the above-ground parts only; and in natural or semi-natural conditions those parts growing below ground may be exceedingly inaccessible.

Nevertheless, measurements of respiration rates of above-ground parts in the dark may still yield much useful information. It is clear from recent biochemical and physiological investigations that the rate of plant respiration is set at anything but haphazard—that over the temperature range within which the plant normally grows the metabolic machinery is very well regulated in such a way as to produce adequate, but not superfluous, supplies of high-energy compounds taking part in all those transformations in the plant requiring metabolic work to be done—syntheses of all kinds, active transport, and so on (for a general account see Beevers, 1961, chapters 7 and 10). Inferences about the activity of the metabolic machinery of an organ can therefore be drawn from its dark respiration rate at a particular time.

Such measurements have a further value. They make possible a step in the direction of inferring the net rate of carbon assimilation by the plant during the light phase of the daily cycle from the data which we have of rate of dry weight gain over periods involving at least one light–dark cycle [28.5, 29.3, 29.4]. Unless we wish to study plants under natural conditions, it is possible, by growing them in pots in a suitable inert and sterile medium, to measure the dark respiration of the whole plant, both above- and below-ground [12.5, 12.13, App. 1.18–20]. From the ratio of the rates for the roots and tops under these artificial conditions, we may be able to obtain information which may help to elucidate, in this respect at least, the perennially difficult problems of root behaviour under natural conditions. Furthermore, if the respiration measurements include simultaneous studies of oxygen uptake and carbon dioxide evolution, light can be thrown on the extent of dark fixation of carbon dioxide in those plants where the process goes on at a measurable rate.

4.8 Photosynthesis

The problems presented by photosynthesis, including photorespiration and other problems connected with the detailed operations of the photosynthetic mechanism, are much more difficult than those of respiration measurement [4.7, 12.2, 32.6, 36.7]. Many different types of measurement can be made in the laboratory which would be quite impracticable in the field, especially if they are to be made on plants growing under more or less natural conditions, and if it is desired to relate them to the average rates at which processes are going on within the plant over substantial parts of a growing season [32.7]. In this chapter we have so far considered types of measurement which can be made with these last two purposes in mind. In the twenty chapters of the next two Parts we shall consider how these measurements can be arranged and how the data collected can be handled, in order amongst other things to establish a framework of quantitative description of the changes going on in the plant as a whole, and in its parts. This framework is then available as a standard of reference when, in the final chapters, we come to consider related problems in many branches of plant science. We shall find in Chapter 29 that it is possible to establish with reasonable accuracy the mean rate of net carbon dioxide uptake in the light, over periods of the order of a week, by the leaves of plants growing undisturbed under natural conditions of the aerial environment. This mean rate, corresponding to mean apparent assimilation rate, can then form the basis of further inferences, for example relating to the mean carbon dioxide concentration in the intercellular spaces of the leaves in the light [29.6], and to the declining efficiency of the photosynthetic mechanism as the plant ages [Fig. 29.6]. Many different kinds of observations, and many precautions, are needed to establish such conclusions with reasonable accuracy, and as yet this has rarely been done. However, the possibility having now been established, it is to be hoped that soon comparative figures will be available for a range of species under a range of natural conditions. But such work carries us only as far as the estimation of apparent assimilation. The problems posed by the known properties of the photosynthetic mechanism itself, including those of photorespiration, lie beyond, and it seems likely that some time must elapse before field methods are available to the ecologist for the investigation of problems of this kind [32.6, 36.7, 36.9].

PART II

EXPERIMENTAL TECHNIQUES

THE EXPERIMENTAL MATERIAL

5.1 Uniformity and variability

Our preliminary discussions have established that in studying plant growth we shall be unlikely to gain the information which we need unless we harvest plants from time to time, thereby destroying them, so that future observations must be made on other plants. For a growth experiment we therefore need a population of plants, from which samples can be withdrawn. If at any particular time the plants were known to be identical, it would suffice to harvest a single one. But they never are. Winston Churchill (1923), in *The World Crisis, 1911–1915* notes that Fisher used an aphorism 'Uniformity is death'. Although the First Sea Lord had a particular context in mind (a ship-building programme), he was clearly enunciating a piece of epigrammatic wisdom quite worthy to rank with the sayings of the Seven Sages. One of the most characteristic features of living systems is that they never are uniform, and all that one can hope to do in dealing with them is to keep the inevitable variability within tolerable limits.

5.2 Preliminary investigations of variability

When beginning work with a species not previously studied, or even with a new variety of a well-studied species, it is usually necessary to make preliminary investigations, both of the average magnitude of particular attributes of the populations in which we are interested, and also of the variability of these attributes within the populations concerned. In living systems this variability can take many different forms, but for convenience these are usually considered as approximations to a relatively small number of different types of distribution (for a simple treatment see Bailey, 1964, chapter 2). It is necessary first to identify the appropriate form of distribution, and in cases of doubt or difficulty it is always worth consulting a statistician before embarking on any extended treatment of the data. Variability is often found to approximate to the 'normal distribution', a continuous distribution resulting in theory from the combination of the chances of a very large number of independent events, all of which affect the magnitude of the quantity being measured. This particular distribution can be characterized by a mean and a standard deviation. The perfect theoretical forms of all these distributions hold only in infinite populations, but of course in practice one must always deal with a finite population. Even supposing that the rules governing variability in this finite population take a standard form, and apply rigidly, nevertheless the characteristics will be only an approximation to the theoretical state of affairs, although in large populations the deviations may be so small as to be undetectable in practice. In small samples from such populations, on the other hand, the deviations from mean attributes of the population as a whole may be considerable, and it is usual to consider means based on samples as estimates of the mean for the whole population, and in the case of normal distributions to distinguish between the true standard deviation of an assumed infinite population (σ) and the estimate of this quantity based on a small sample (s). During this century the relations between the two have been the subject of much research. We need not consider in detail the resulting standard statistical methods, which are covered by many textbooks (see the classified bibliography). Certain general results must, however, engage our attention from time to time, and in planning programmes of observation it is important to consider the element of uncertainty introduced into our conclusions by the use of small samples. From a sample of n observations of some quantity x, we can calculate an estimate, $\bar{x}$, of the mean of the assumed very large population from which the sample was taken, and also an estimate of the standard deviation of a single observation, s. Then our estimate of the standard deviation of the mean, $\bar{x}$, of n observations is $s\sqrt{(1/n)}$; and this is a measure of the variability which we should notice in our estimate of the mean, if we took many repeated samples, each of n observations, from the same very large population. The uncertainty can therefore be decreased by increasing n, but to halve the standard deviation of the mean requires n to be multiplied by 4, so that a limit is soon reached beyond which the gain in accuracy by using very large samples does not justify the extra work involved.

The next question concerns the value of the mean. Assuming that the particular set of observations being considered is a sample from a

5.2

hypothetically infinite population of possible observations with a normal distribution, is it possible to relate the sample mean to the mean of the whole population from which the sample was drawn (the 'true mean'), which although unknown is assumed to have a fixed value? To pursue this matter in detail leads into the complexities of fiducial theory; in the present context it suffices to pose a further question. Suppose that we use a particular method to set limits on either side of our sample mean, and then state that the 'true mean' lies within these limits, how likely are we to be right? Or, more precisely, supposing that this same particular method was used to work out limits on a very large number of occasions using a different set of data each time, in what proportion of cases could it be expected that the 'true mean' would be found to lie within the appropriate limits? (It is important to remember that the probability is concerned with the likelihood that our estimate is correct, not with the value of the 'true mean' itself, which is a fixed quantity.) To have a 19:1 chance of being right one calculates 'fiducial limits' by multiplying s by the appropriate value of 'Student's' t for the probability level 0·05 (for a short account of fiducial limits and their meaning and use see Bailey, 1964, section 3.2 and Chapter 4). This value of t, near 2·0 for very large samples, increases as the size of the sample decreases (and therefore the inaccuracy of the estimate based on it increases). If the number in the sample was 20, the value of t would be 2·09; if 10, 2·26; if 5, 2·78; if 3, 4·30. In practice samples usually lie within these limits, because if we dropped our sample size down to 2, t would increase to over 12; we should have trebled the inaccuracy of the estimate by a saving of only a third of the work, and this is usually not worth while. By increasing the number of observations from 3 to 5 we have increased the labour by two-thirds, and $s.(t/\sqrt{n})$ has decreased from 2·48s to 1·24s, halving the uncertainty of the mean. On the other hand, when the sample size is increased from 10 to 20 the work is doubled, but $s.(t/\sqrt{n})$ decreases from 0·72s to 0·47s, a reduction of only a third. Above a sample size of 20, the gain in accuracy from the use of larger samples often becomes negligible compared to the extra labour involved, although for certain purposes statisticians prefer a number around 30, if this is practicable.

In planning the preliminary investigation, the labour involved in making the desired measurements must therefore be balanced against the accuracy desired from the estimates to be based on them.

5.3 Determination of sample numbers

At the next stage of planning we shall need to consider the desired degree of accuracy of the sample mean. An example will illustrate the process of reasoning. Suppose that at the end of a particular treatment A of an experiment we should expect that x, the quantity in which we are interested, would have a mean value of 1·0, and that the standard deviation of a single observation has been estimated as 0·2, based on a very large preliminary sample survey. Then t being near 2·0, we should expect single observations to show deviations as large as 0·4 on one occasion in 20. That is to say, on one occasion in 40 we should expect an individual value as large as 1·4, and on one occasion in 40 an individual value as small as 0·6. Suppose that after treatment B of the same experiment (with which A is being compared), we should expect the same quantity x to have changed by 0·5, the standard deviation remaining unchanged at 0·2.

The question we now have to ask is, 'How large a sample should we need to take from each of the two populations, A and B, in order to demonstrate that the expected difference between them was significant?' In our particular context, this means that the difference, $x_B - x_A = \pm 0·5$, must be significantly different from 0. If s be the estimated standard deviation of a single observation, the standard deviation of a mean of n observations is $s/\sqrt{n}$; and for the difference between two means, each of n observations, from two populations having the same standard deviation, it is $(s\sqrt{2})/\sqrt{n}$. If a difference of 0·5 is to be found to be significantly different from 0, its standard deviation multiplied by the appropriate value of t must be smaller than 0·5.

We have then $0·2(\sqrt{2}/\sqrt{n}) \times t \leqslant 0·5$, which is not satisfied by n = 3, but satisfied if n = 4 or more. We should therefore need a sample of at least 4 from each population in order to demonstrate that they were significantly different.

If, on the other hand, we asked, 'How large a sample should we need in order to measure the expected difference, 0·5, with a probable accuracy, as expressed by "fiducial limits", of 10 per cent?', then $0·2(\sqrt{2}/\sqrt{n}) \times t \leqslant 0·05$. n will clearly now be large enough for us to be able to make a rough estimate by writing t as 2, when we have

$$\sqrt{\frac{n}{2}} = \frac{0·2 \times 2}{0·05} = 8,$$

$n/2 = 64$, $n = 128$. This would usually rank as an unacceptably large sample, and would lead to the conclusion that the desired degree of accuracy in an experiment of this kind could be attained only by taking steps to reduce the variability; for, if we can reduce the standard deviation from $0\cdot2$ to $0\cdot1$, then the desired accuracy could be met by a mean of a sample of 32, and if s could be reduced to $0\cdot05$, by a mean of a sample of 8.

Thus, in the early planning stages of an experimental programme, it is necessary to have some idea both of the magnitude of the effects to be distinguished, and of the variability to be expected within the populations concerned; and also to consider what statistical procedure will be appropriate for dealing with the results. It may be necessary to deal with very much more complex situations, but here also, texts are available on the statistical aspects of the planning of experiments, adapted to the various types of system commonly encountered (see the classified bibliography). Again, in cases of doubt or difficulty in deciding on the most appropriate treatment, much trouble can often be saved in the long run by consulting a statistician in the planning stages. But good statistical planning is only part of our problem. It is necessary to consider the characteristics of the biological systems involved, so that, where appropriate within the ambit of the problem, sources of variability can be controlled, and as far as possible, reduced.

5.4   Sources of initial variability—genetic

When working with known varieties of horticultural and agricultural plants, genetic variability usually gives relatively little trouble, although it must be remembered that many named varieties still contain within themselves considerable genetic variability. But the ecologist is more usually confronted by a situation where little, and sometimes nothing, is known of the genetic structure of the natural populations with which he must work. Indeed the study of this structure may well be part of his problem. It is, however, outside the scope of our present discussions, and reference should be made to standard works on the subject, an introduction to which is given in the classified bibliography.

In a country such as Great Britain, where there have been numerous field workers for a long time, it is quite possible that the species in question will already have been studied. It may then be possible to take advantage of favourable circumstances—a species may be known to be relatively invariable; or it may be known to be an apomict, without a process of sexual reproduction, so that genetically uniform seed is easily collected. Alternatively, the species may be known to possess some regular outbreeding mechanism, in which case the natural population can be expected to possess considerable variability, which may be difficult or impossible to reduce by a breeding programme within a reasonable time. This is particularly true of plants such as trees, where many years may elapse before sexual maturity, and where matters may be complicated by irregular or infrequent seed production. This last problem is by no means confined to temperate trees, but is also found in a significant number of species of the wet tropics. In such cases, to raise a population of young plants from seed at a particular time may be out of the question; and in other cases, of very great variability, it may be undesirable. If experiments are to be made over short periods, there is then no escape from using some means of vegetative propagation, appropriate to the species in question. It must then be remembered that in reducing the genetic variability of a species for the sake of experimental clarity we are confining our conclusions to those forms actually studied. It may be worth while to select two or more forms covering as much as possible of the natural range of variability of the species, so that our conclusions can be extended by comparisons between them.

5.5   Size and weight of seed or propagule

Seeds of different species differ widely from each other both in structure and in the relative proportions of the different parts. In particular, the initial stock of reduced-carbon units [4.3] may be divided in very different ways between the embryo and the food reserves. Such differences may also exist within a particular species, over and above any differences due to the actual size of individual seeds. Briggs, Kidd and West investigated this during their work on *Helianthus annuus* in 1920, using the variety *uniflorus* Sutton's Giant Yellow. They divided the seeds into five classes on the basis of initial air-dry seed weight [Table 5.5, line 1]. Equal numbers of seeds of each size class were sown in the open field, and 7 days after germination the leaf areas of samples of each size class were measured [Table 5.5, line 2]. Thereafter samples of plants grown from seeds of each size class were harvested at weekly intervals, and the mean dry weight and leaf area of the plants of each class were expressed as a percentage of the largest on that occasion. In lines 3 and 4 of Table 5.5 we have the means of these percentages for harvests between 2 and 8 weeks from germination, all scaled up to 100 for the largest score in each case.

Table 5.5. Relative seed weights compared with relative plant sizes in five groups, for seedlings and young plants of *Helianthus annuus* grown by Briggs, Kidd and West at the Botanic Garden, Cambridge, in 1920. In each line the largest value = 100. Harvests every 7 days. Least significant differences: line 2, $p = 0.05$, 15; $p = 0.01$, 21; lines 3 and 4 (except column 4), $p = 0.05$, 8; $p = 0.01$, 11; $p = 0.001$, 15; lines 5 and 6 (except column 4), $p = 0.05$, 13; $p = 0.01$, 17.

Line		Grade of seed				
		1	2	3	4	5
	Air-dry seed before sowing					
1.	Mean seed weight	100	93	86	79	72
	7 days after germination					
2.	Mean leaf area	96	77	100	76	88
	Mean of harvests 2–8					
3.	Mean total dry weight	100	91	98	84	82
4.	Mean total leaf area	100	90	96	85	82
	Mean of harvests 6–8					
5.	Mean total dry weight	100	85	100	87	86
6.	Mean total leaf area	100	84	96	90	82

We see that there is no more than a broad correspondence between initial seed weight and mean plant weight. The correspondence is much closer between leaf area at 7 days and plant weight later, suggesting that a large seed gives rise to a large plant only if it contains a large embryo which will early develop a large leaf area. In this respect the distribution among the original seeds is at least bimodal, because in line 2 the mean leaf area of group 2 is significantly below that of both groups 1 and 3. The data as a whole suggest even further heterogeneity, because although the variance of lines 3 and 4 appears to be homogeneous for all groups except those of column 4, the variance associated with plants grown from the fourth grade of seeds is very significantly higher

5.5

($p \ll 0.001$) that that of the remainder, suggesting that this size class includes seeds which will give rise to plants of more than one different population. It is interesting that under the fluctuating conditions of growth in the field these relatively small influences associated with initial seed weight should have persisted throughout a period during which the dry weight of the plant increased more than a thousand-fold. If we look at the means for the harvests between 6 and 8 weeks after germination (lines 5 and 6 of Table 5.5) we see the same pattern as for the period as a whole.

Finally, there is a suggestion that even at 7 days the immediate influence of initial seed weight on the development of the young seedling is not exhausted. The two groups of heaviest seed produce larger plants throughout than the 7-day leaf area would lead one to expect, while the group of lightest seed produces smaller ones.

Thus we see that if one wishes to reduce variability the initial grading procedure should provide both for a grading of dry seeds and for at least one subsequent grading of seedlings.

Similar procedures are needed for other forms of propagules—bulbs, corms, tubers, rhizomes, cuttings, etc.

5.6 Mineral content

It may be necessary to take further precautions if the experiments are concerned in any way with soil conditions or with mineral nutrition. An annual plant may grow until its total dry weight is between 10^4 and 10^6 times that of the seed. Except in the very earliest phases of this growth, the seed's content of minerals obviously makes a negligible contribution to the mineral content of the mature plant. The same is not usually true of plants growing from bulbs, corms, tubers, etc. where often the mature plant may represent less than a hundred times the dry weight of the perennating organ, and where the increase in mineral content may be less still. Here if experiments are being carried out on the effects of decreased nutrient status of the soil on plant growth, mineral nutrients already present in the propagule may be of great importance. At times these stored nutrients may much reduce the impact of mineral deficiency in the first growing season, and the effects may not be fully developed until the second growing season, when growth begins from a propagule itself already deficient.

It follows that in any studies involving mineral nutrition it is usually wise to analyse the seed for initial minerals. It is essential to analyse other types of propagules, where the increase in total minerals may be much less.

In some plant communities, such as a wet tropical forest on a deep, nutrient-poor, sandy soil, the great bulk of the available mineral nutrients may be found incorporated in the vegetation at any one time, and thus under suitable conditions mineral nutrients in short supply may be conserved for long periods, being gradually released by decay and promptly taken up by other plants. On the other hand, the persistence of some plant with a high nutrient requirement in particular places within a general area of nutrient-poor soil is not necessarily to be explained in this way, because accidental accessions of some nutrients may become bound in the soil, being only very gradually released. For example, Pigott and Taylor (1964) note that '... the occurrence of *Urtica dioica* on the site of settlements which have long since been abandoned is readily explained in terms of the persistence of a high concentration of phosphorus, as organically-bound phosphorus is very stable in soil and phosphate ions are strongly adsorbed. The total phosphorus concentration in such sites is often as high (150–200 mg per 100 g dry soil) as in soils from the vicinity of modern farms.'

Similarly, it should not be assumed that all the mineral nutrients present in the propagule are free to move into growing tissues, as a certain proportion, particularly of divalent cations such as calcium, is likely to be bound to the non-mobile anions of the Donnan phase of the permanent cell walls (Briggs *et al.*, 1961, chapters 4 and 5; Nobel, 1970). From this position these ions can be released only when replaced by other ions, until the breakdown of the wall structure itself.

The results of initial analyses have therefore to be interpreted with care as far as subsequent growth studies go: but for our present purposes of ensuring the minimum variability of our initial material, it suffices to use them as an additional check on comparative uniformity, especially in the case of large seeds or of vegetative propagules derived from more than one source.

5.7   Germination requirements

Those studying standard horticultural or agricultural crops enjoy an advantage over the field ecologist, in that crops grown from seed have been selected, consciously or unconsciously, over hundreds or even thousands of generations for ease and certainty of germination. Of the remaining quarter of a million or so species of seed plants, germination has only been studied in detail for perhaps 1 per cent of species. The results of these studies are not uniformly encouraging. The seeds of many plants of the wet tropics never become dormant—their percentage of germination is highest when they are shed, and it diminishes thereafter, often so rapidly that the seeds, if stored, are all dead in a matter of weeks. Such an early death of the seed is less common in temperate regions, but here also some species have a short viability, e.g., elm, willow and poplar. In many temperate plants dormancy is much affected by the weather at the time that the seed is maturing. If the weather has been mild and damp, the seed of many species will germinate at once or, if dried out, will become dormant; whereas in hot, dry seasons, seed collected from plants in the same natural communities would already be dormant when shed, and would require some specific stimulus to bring about germination. Some seeds are shed with an immature embryo, which slowly develops during the first months or years after the shedding of the seed. Such seed cannot be brought to germination by any means until the embryo is mature.

Once the seed has become dormant, its length of life and the conditions which will bring about germination vary very widely. Many seeds deteriorate rapidly in storage, and a large percentage die within a few years. Others have a storage life of many tens of years, and a few are known to have survived for much longer periods, the classic example being the seeds of the North American lotus, *Nelumbo lutea* from the British Museum, which were acquired by Sloane for his herbarium about 1727, and germinated by Robert Brown in 1848. (For an account of this example, and a review of other instances of longevity see Barton (1961), chapter 1.) But among seed plants as a whole such instances are rare, and appear to be confined to a small number of families, prominent among them the Leguminosae. As a general rule the seed used should be as fresh as practicable, partly because of possible germination difficulties, and partly because in some species deterioration of the seed in storage leads to the production of abnormal plants (for examples see Barton (1961), chapter 14). Storage conditions are also very important, many of the troubles experienced in trying to germinate seeds being due to the use of dead or dying material. Once again, no universal rules can be laid down, and when beginning work on a species not previously studied it is desirable to find by experiment the best conditions for storing the seed.

5·7

Many seeds do not survive well in paper packets in a desk drawer. The use of anhydrous calcium chloride in a desiccator, or a deep-freeze cabinet at -10 to $-20°C$ can be useful aids to long-term storage in the laboratory, although neither is by any means universally applicable.

Some seeds contain water soluble inhibitors, which must be washed from the seed coat before germination will take place. Others, including a number of species from markedly seasonal climates, require a period of frost, or at least of temperatures below $5°C$, as a pretreatment; and the seeds frequently have to be moist at the time. This is the basis of the regular horticultural procedure of "stratification', where seeds are exposed during a winter in open trays, covered by a very thin layer of soil.

When all these requirements have been met, there still remain the questions of the breaking of dormancy and of actual germination, and here the requirements may again vary very widely. When once they have passed through the necessary stages of development and pretreatment, some seeds need only be moistened for germination to begin. Others may need dark, or light; radiation of specific wavelengths; warmth, or cold; mechanical rupture of the seed coat, and so on. Some seeds are known to germinate more readily after passage through the gut of animals; some difficult tropical members of the Leguminosae (such as *Mucuna* spp), normally apparently inert, can be brought to germination by three weeks' immersion in a 50 per cent mixture of concentrated hydrochloric acid and water—the conditions under which they germinate in nature being a mystery. A variety of treatments involving plant hormones or other organic chemicals are effective in certain cases (for reviews see Crocker and Barton, 1953; Toole *et al.*, 1956; Koller *et al.*, 1962).

Thus the seeds of a proposed experimental species may not germinate when wanted, and if their properties are unknown, it is necessary to carry out preliminary investigations to see if a suitable means can be found to bring about germination (for discussion of the various available conditions see Bünning, 1948, Part II; Crocker and Barton, 1953; Koller *et al.*, 1962; Mayer and Poljakoff-Mayber, 1963; Sussman and Halverson, 1966, chapter 7; Wareing, 1969; and below, 5.8).

5.8   Determination of conditions for germination

As in many other branches of plant studies, the world distribution of botanists in time has resulted in a very high proportion of the work done on seed germination being concentrated on plants of the North Temperate zone. The small sample of the world's flora already studied is therefore not typical, and in particular there is reason to suppose that the flora of Western Europe contains an unusually large proportion of cases of complicated germination mechanisms. At the present time detailed investigations are in progress at the Jodrell Laboratory of the Royal Botanic Gardens at Kew on the germination requirements of a large number of species selected to cover a wide variety of families and of geographical regions. This work has not yet gone far enough for the publication of general conclusions, but Dr P. A. Thompson has supplied the following interim information and advice on what seems to be the most profitable sequence of trials to use when beginning work on a species not previously studied.

1   Always include a period of light in each daily cycle in the first instance.
2   Try a range of different temperatures, spread at around 5°C intervals between 5°C and 30°C.
3   Try stratification (5·7): many seeds as shed require a period of after-ripening (1–3 months), which often proceeds satisfactorily in moist storage at 2–5°C.
4   Try fluctuating temperatures, with a shift of 10°C or more.
5   Try treatment with gibberellins–Gibberellic acid at up to 1000 mg l^{-1}, or Gibberellin A$_4$ at up to 100 mg l^{-1}.

Finally it must be remembered that the investigations will not necessarily succeed. There are a number of plants whose seeds have never been brought to germination artificially, although they are known to germinate in nature (e.g., *Asperula odorata*, A. R. Clapham, personal communication*), while in other cases, although much of a particular stock of the seed will germinate given a long enough time, the chances of any particular seed germinating during a fixed period of a few days are low.

5.9   Difficult cases

A good example of the complications which may be encountered is furnished by Pigott's work on the British limes, *Tilia cordata* and *T. platyphyllos*. In northern England on average less than 1 per cent of the indehiscent, nut-like fruits of *T. cordata* contain viable seed (a rather

------

* This little problem has now been solved, and the seed brought to germination by a very prolonged cold treatment (K. Taylor, personal communication).

higher proportion may occur in *T. platyphyllos*), and then rarely more than one or two seeds each; a very large initial collection is therefore necessary. To bring about normal germination requires a pretreatment of one or two cold periods; thus in natural conditions, after the first winter not more than 20 per cent of the seeds will usually germinate; and most of the remainder do not germinate until the end of the second winter. Much time can be consumed in working out the details of such a case, and the possibility that similar complications may be encountered must always be borne in mind when starting work on a species whose germination requirements are unknown.

Finally, even when all the conditions mentioned had been met, germination was found to be far from simultaneous. After any of the effective pretreatments, a given seed of either *T. cordata* or *T. platyphyllos* is likely to germinate at any time within about a month (C. D. Pigott, personal communication). For such species, any experimental plan requiring 100 seedlings all germinated within 24 hours might thus demand an initial collection of about a million fruits.

In a similar, but less extreme, situation Faheemuddin (1969) found that the germination of stratified seeds of *Iris pseudacorus* could be speeded up and also increased from about 30 per cent to about 70 per cent by removing the seed coat, but that only about 10 per cent of these seedlings developed normally. In the remainder the coleoptile failed to emerge and the seedling died. However, normal development could be induced in nearly all of these by supplying an aqueous extract of the ground seed coats. In suitable cases such a method might be useful in producing a substantial crop of seedlings from seeds which normally germinate only sparingly.

Means of dealing with populations of seedlings whose germination was not effectively simultaneous are considered in sections 5.10, 5.11; 15.1, 15.2; and 20.4.

5.10   Time of germination

Let us suppose that the difficulties just mentioned have been overcome and that a technique for bringing about germination at the desired time has been worked out. Even so, it is most unlikely that all the seeds sown will germinate simultaneously, and the process is likely to be spread over several days at least, if not over a few weeks. When one considers that under really favourable conditions a young plant of *Helianthus annuus* can grow by 30 per cent in a single day (although granted that most plants do not grow as fast as this) one can see that germination spread over several days can cause considerable variability, even if all the other factors affecting growth were constant.

This situation can be dealt with in one or another of three ways. One is to specify certain tolerable limits of time within which germination must take place. These limits will depend on the rate of growth, and on the tolerable limits of variability. The total number of seeds which must be sown to produce the necessary crop within the specified time is then worked out from the known pattern of germination. This number is sown, and all those germinating outside the specified time are eliminated. It will be obvious that in certain cases (e.g., *Tilia* spp., 5.9) this will call for a very large sowing indeed.

If the seeds are to be sown in pots, or otherwise in conditions where labour is involved in preparation, this first simple procedure may be unsuitable, as causing a great waste of effort. The second possibility is then to sow several seeds in each pot, once again eliminating the individual seedlings which emerge outside the specified time. The number needed in relation to the number of pots can easily be worked out from the distribution of germination. Suppose that there is one chance in five that a given seed will germinate within the specified time. Then, by planting 4 seeds per pot, we should have 4 one-in-five chances of a pot with a suitable seedling, with 5 seeds 5 one-in-five chances, and so on. But as always with lotteries, the mere possession of a number of tickets equal to the odds against drawing a lucky number does not guarantee a prize. The chances fall into a binomial distribution $(\frac{1}{5} + \frac{4}{5})^n$, where n is the number of seeds sown, and the terms of the expansion correspond to the chances of n, n-1, n-2 ... 0 seeds germinating within the specified time. We must obviously concentrate on the last term, $(\frac{4}{5})^n$, representing the chance of no seed germinating within the specified time, and therefore the wastage of pots. Table 5.10 sets out the percentage wastage against number of seeds. We see that at first there is a large improvement, but

Table 5.10. Mean percentage of wasted pots as a function of the number of seeds per pot, on the assumption that on average only one seed in five will germinate within the required time interval.

Number of seeds per pot	1	2	3	4	5	6	10
Percentage of wasted pots	80	64	51	41	33	26	11

5.10

that at each step the improvement brought about by adding one more seed gets less. At 5 seeds we could expect to succeed in two-thirds of the pots, at 10 seeds in nearly 90 per cent. The practical decision on how many seeds to sow will be based on the balance between two considerations. One is the labour of preparing the extra, wasted, pots, and the other that of collecting, preparing, and grading the extra, wasted, seeds. Thus, to secure 8 pots of correctly germinated seeds with 5 seeds per pot we should on average need an extra 4 pots, making 12 pots and 60 seeds in all. To secure 8 pots with 10 seeds per pot we should need only one extra pot, but 90 seeds in all—the question being whether it is easier to collect and grade 30 extra seeds, or prepare 3 extra pots.

The answer may be forced upon us by shortage of seed, or a limit may be set by the positions in which the seeds can be arranged in the pot—it might not be convenient to have as many as 10, for example.

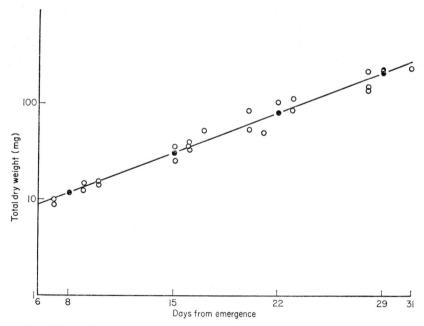

Fig. 5.11. Relationship between total dry weight, plotted on a logarithmic scale, and age from emergence of the cotyledons, for young plants of *Impatiens parviflora* cultured under constant conditions in growth cabinets. Four harvests at weekly intervals. ○ Points for individual plants; ● estimated mean figures for each harvest, taking account of the average ages of all plants. From Lewis (1963).

5.11 Allowances for variations in time of germination

On occasion the two methods of dealing with variability in time of germination which we have just considered may not be applicable: for example, the experimental design may be such as not to allow for rejects, or this reason may be combined with either of those mentioned in the preceding paragraph to force us to accept a substantial variation in time of germination. The third possibility is then to note the time of germination of each plant, so that when the results come to be analysed, allowance can be made for the variable periods which have elapsed between germination and the (fixed) date of harvest. Fig. 5.11 shows an example of the use of this method of allowing for unavoidable variability in date of germination, the logarithm of the dry weight of individual plants being plotted against the number of days from germination, for four successive harvests. It will be seen that a large part of the variability within these harvest samples has been accounted for by this procedure. We shall consider later [20.4] some of the statistical aspects of the method, and also means of dealing with non-linear relationships uncovered in this way.

THE PRELIMINARY PHASE

6.1 Natural conditions

So far we have been concerned with the initial variability of material during the preparatory stages. The observations themselves may not begin until the plants have been growing for some weeks, and during this period we have to take care not to add unnecessarily to the variability. How much can be done to reduce variability during this preliminary phase depends, of course, upon the nature of the planned observations themselves. If the plants are to grow under wholly natural conditions, rooted in a natural soil, then one can only observe, and not interfere, although such observations at times may turn out mainly negative. For example, a research student wishing to make observations on the germination and establishment of two species of grasses, both of which are present in the ground flora of a wood near Cambridge, sowed on the woodland floor 10,000 graded seeds. The seeds all disappeared before germination, and there is no doubt that they had been eaten or carried away by animals, mainly mice, of which there are many in the wood.

6.2 The reduction of biotic pressure

Thus even if the plants are to grow under conditions as nearly natural as possible, it may be necessary to reduce biotic pressure in order to reduce the area and scale of the observations to manageable proportions. It must be remembered, however, that such measures often bring side effects in their train. It may be necessary to tolerate these in order to make the experimental programme possible, but account must be taken of them in interpreting the results. Thus, in the open, fences erected to keep out grazing mammals affect the pattern of air turbulence down wind, the usual allowance being fifty times the height of the fence, and might therefore affect measurements made on the air above the plants, while netting cages to keep out birds may have considerable effects on wind speed, particularly in high winds. It must also be remembered that the drip from galvanized wire is toxic to some plants, so that it may be advisable to paint it, or to use string net. The use of metaldehyde to attract and kill slugs can usually be arranged outside the root run of the plants, so that side effects can readily be eliminated. Sucking, boring or tunnelling insects are best dealt with by a systemic insecticide, but here it is wise to carry out preliminary experiments to make sure that there are no observable adverse effects of the selected insecticide on the growth of the plants. Even the presence of such an insecticide does not necessarily prevent damage by chewing and gnawing insects, such as cockroaches and earwigs, and it may be advisable to reduce the population of these by trapping.

In any case, plans should always be made in the preliminary phase for the regular inspection of the plants, so that any damage can be noted and allowance made for it.

6.3 Semi-natural soils

The difficulties and uncertainties attached to attempts to carry out observations under natural conditions often make necessary observations under semi-natural conditions, particularly if the time available is limited and the observer cannot wait for the natural occurrence of some particularly favourable combination of circumstances.

Such experiments most frequently involve manipulation of the soil, either by eliminating root competition from a natural soil (which inevitably involves also a whole series of changes in the activities of the soil micro-flora and micro-fauna [7.3, 10.1, 10.2]); by growing the plant in an agricultural or horticultural soil of some kind; or by growing the plants in pots containing one or another of these soils, or some artificial soil substitute such as vermiculite.

We are not here concerned with problems of mineral nutrition (see classified bibliography) but with the techniques appropriate to the preliminary phase of growth of our experimental plants, where we are trying to keep variability down to a minimum. Our main concern must therefore be with uniformity, and with the avoidance of any conditions likely to bring about damage to the plants. Gradients or pockets of variability are a common feature of natural and semi-natural soils, and the consequences may be inseparable from their use as an experimental substratum. Soils with several distinct horizons, e.g., podsols, present exceptional difficulties. For use in pots or other containers, however, unless stratification is to be part of the technique, the soil should be very thoroughly mixed, together with any fertilizers which are to be added. Equally, unless waterlogging is part of the problem being investigated, care should be taken to ensure that the soil is free draining. It should be remembered that a soil containing many fine particles may have a natural capillary rise of many centimetres. Thus a soil, such as a very fine and uniform sand, which may be free draining in nature, where the natural water table may be several metres down, may turn out to be completely waterlogged in a pot. In such cases it may be necessary to mix in with the soil a proportion of coarse sand or fine gravel to improve drainage and aeration; and the same may be necessary with clay or silt soils which tend to become excessively compact. If the experimental conditions require a water table below the rooting zone, it may be better to grow the plant in a vertical drain-pipe than in a normal pot [see 6.5].

6.4 Vermiculite

If an artificial substratum is to be used, vermiculite has great attractions, with its combination of a very large adsorbent area with relatively large air spaces, thereby combining ample aeration with a large water-retaining capacity, which may be very advantageous in experiments where plants have to be left unattended for periods of several days. Plants grown in vermiculite should normally be watered with a dilute culture solution, for which there are several standard formulae (see the section of the classified bibliography on mineral nutrition). There are, however, snags which must be watched for.

Vermiculite in the mineralogical sense belongs to the large Clay Group of minerals, and is probably derived by a weathering process from the widely distributed mica, biotite. The typical formula of the mineral vermiculite is given as

$$(Mg, Ca)_{0 \cdot 7}(Mg, Fe^{+3}, Al)_{6 \cdot 0}[(Al, Si)_8 O_{20}](OH)_4 . 8H_2O$$

When a solid fragment is heated rapidly to 300°C or more, it decomposes, steam is generated, and the mineral exfoliates, producing the granules of platelets, which constitute vermiculite in the commercial sense. These platelets have the largest cation-exchange capacity of any clay mineral.

6.4

If dilute culture solution is allowed to percolate through a bed of vermiculite, prepared as above, the liquid draining out may be found to be alkaline. It is therefore worth shaking a sample from any new batch with culture solution, and if there is a marked increase of pH, back-titrating the whole to determine the amount of acid which should be added to bring the liquid back to neutrality. The bulk of vermiculite is then stirred with water containing the calculated quantity of acid. Plastic dustbins are convenient for this purpose, and nitric acid is advantageous unless the intended experiments preclude this addition of nitrate. The vermiculite so treated is then washed with water, and subsequently with excess of culture solution, if it is desired to establish a standard cation balance before the experiment begins.

A further difficulty may be encountered, due to fluoride toxicity (Coombe, 1966, p. 161). It is very rare for analyses of vermiculite, in the strict mineralogical sense, to show any fluorine at all, but it is commonly found in nature with interleavings of other 'micaceous' minerals, which may contain fluorine. Different species of plants differ very widely in their tolerance of fluorine, and the reaction of a particular species can be determined only by experiment. If there is any reason to suspect that the plant is sensitive to fluoride toxicity it is always worth testing each new batch of vermiculite for exchangeable fluoride before using it for experimental purposes.

Finally, when using vermiculite in pots, it is necessary to check the aeration of the root system. The height of a small pot may be within the possible capillary rise of finely divided vermiculite, so that after watering the pot may stand waterlogged. If this is found to be the case, waterlogging can be avoided either by using a coarser grade of vermiculite, or by mixing the finer grade with a sufficiency of coarse silver sand and washed gravel or crushed flint [6.3]. For certain purposes such a mixture may in fact be preferable, as facilitating the washing out of the root system.

6.5   Root run

Apart from the regulation of mineral nutrients and aeration, root run may also be important. Some potted plants appear to go on making normal growth after the roots have reached the outside of the pot, provided that ample water and mineral nutrients are available. Others either

6.5

suffer a check to growth, or the plant form itself may be altered by a restricted root run. If plants are to be grown in pots it is necessary to carry out preliminary observations to see whether growth is affected by the pot size intended, as compared with growth in a larger pot. It should also be remembered that the root systems of some plants naturally spread downwards, rather than outwards, so that the normal pot shape is very wasteful both of soil and space. It may turn out that such a plant is more economically grown in a drain-pipe. A convenient recent introduction to the building trade has been the Hepsleve pipe, which is cylindrical with a separate, watertight, plastic collar. This collar has lugs overlapping the end of the pipe, which makes it a simple matter to seal off with a per-forated plastic end piece. The pipes can be obtained in various widths from 4 in (10 cm) upwards, and lengths from 4 ft to 6 ft (1·2 to 1·8 m).

6.6 Water supply

Finally, having seen to soil composition and aeration, mineral nutrients and root run, there is the all-important question of water supply. If the semi-artificial conditions with which we are dealing involve natural conditions above ground, the plants are subjected to natural rainfall and evaporation from the soil, as well as transpiration. If the plants are in pots in the field the pots themselves should be plunged so that the soil surface is level with the ground; evaporation will be higher from a pot standing above the ground, because of increased wind movement, and the soil temperature may tend to fluctuate less in a plunged pot.

The problem is how much to water. If too little water is given, water stress will be set up and growth will be checked. If too much water is given, mineral nutrients may be leached out of the bottom of the pot, or washed down in the soil below the rooting zone. This last difficulty does not apply, of course, to pots which are being watered with nutrient solution.

Various partly theoretical, partly empirical formulae have been worked out to connect plant transpiration to various quantities nor-mally measured at a meteorological station (Thornthwaite, 1948; Leeper, 1950; Hamon, 1961; Penman, 1963; Papadakis, 1966; and for a short account see Slatyer, 1967, pp. 62–64). The difficulty about using any of them is that although they often give a good approximation to total water loss over a considerable period, say a month, and thus are useful to farmers for regulating irrigation, they are liable to considerable errors in either direction over short periods. But particularly in the case of potted plants, where the water reservoir is small, it is essential that we should know from day to day how fast water is being lost.

Undoubtedly the best method is to have a few spare pots, prepared in the same way as the experimental ones, and to measure water loss from them at intervals. This may be done in one of two ways. If a suitable balance is available, the pots may simply be weighed, and the loss in weight made up. This method has the advantage that, if desired, the soil may be kept permanently somewhat below field capacity. If the necessary root run demands a pot size which, when filled with the chosen substrate, is too heavy to weigh on the only available balance, it can be lightened by mixing in a proportion of granules of expanded polystyrene. The method is particularly applicable when the plants are growing in vermiculite. It is advisable to check by a pilot experiment that no undesirable effects are produced. Alternatively, if weighing is not possible, a measured quantity of water can be sprayed uniformly all over the surface of the soil, using what is expected to be an excess, and the volume which trickles through can be collected and measured. This then gives us, by difference, the amount of water which should be added to bring the other pots up to field capacity. It will be noticed that both these techniques measure water loss from both pot and soil; and they also measure the net figure for loss minus rainfall, which is what we need. Furthermore, if rainfall is very heavy, and brings the soil above field capacity, so that some water runs out of the bottom of some of the pots, this also is allowed for. However, it is as well to be aware of such an occurrence, which may have affected growth in some way (such as by leaching nitrate out of the soil), so that it is wise to have a rain gauge among the instruments for measuring factors of the aerial environment, which we shall consider later.

Very heavy rain may cause a good deal of splashing of earth on to the stem and lower leaves of the plant. If this is likely to be troublesome, and if the experimental design allows, it can be largely prevented by covering the surface of the soil with a shallow layer of small gravel, say about 1–2 cm thick.

6.7   The aerial environment

The preliminary period may serve one or both of two purposes, either to provide an aerial environment as equal as possible for all plants, in

6.7

preparation for an experiment which is to begin later; or in addition to raise the plants in some very specific environment. The latter will involve techniques of measurement and control identical to those which we shall use during the experimental period itself, and we shall deal with them later under this head. The former is less exacting, but worth careful attention.

The object should be to ensure that all plants have an equal exposure to radiation from sun and sky, and that they are exposed to equal temperatures, humidities and air movements. If these conditions can be met, and if at the beginning of the preliminary period the plants are all comparable, then carbon assimilation and transpiration during the period should also be the same; plant temperature and the effects of radiation and water stress on growth should be the same; and we may reasonably hope for comparable plants at the end of the preliminary period. If the conditions cannot be met, we shall almost certainly be faced with a much more variable population when the experiment proper begins.

6.8 Plant spacing

Several methods are available to meet the requirement for equal exposure, depending on the purpose and arrangement of the experiments. If the experimental plants are to grow in the open without competition from other plants, either above or below ground, then they can be spaced far enough apart for the root runs not to overlap, and also far enough apart for the plants not to shade each other. The avoidance of shading by other plants requires some thought. If the plants are set out in north–south rows, then the spacing in the row needed to avoid shading of one plant by another can easily be worked out from the expected maximum height and width of the plants and the minimum altitude of the sun at solar noon during the experimental period. If the plants in adjacent rows were also opposite each other in an east–west direction, then the east–west spacing would have to be much greater, because of the lower altitude of the sun at 0600 and 1800 hours. On the other hand, if plants in alternate rows are moved on half a space, we may be able to save a good deal in experimental area. Fig. 6.8 shows four sample patterns of shadow movement for Cambridge, at monthly intervals between the summer solstice and the equinoxes and also shows how plant spacing can be

6.8

arranged so that plants avoid each other's shadows. It is rarely necessary to allow for shadows when the sun is very close to the horizon. Experimental sites which at first sight seem to have an unobstructed sky nevertheless on examination frequently prove to have some tree, building, hedge, fold of ground, or other obstruction within 5° or 10° of the horizon.

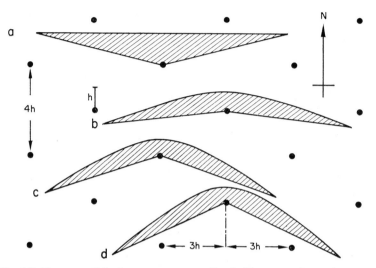

Fig. 6.8. Patterns of shadow movement at Cambridge: a, at the equinoxes; b, on 22 April and 22 August; c, on 22 May and 22 July; d, at the summer solstice. The shaded areas shown are those traversed by the shadows of narrow vertical rods of height h during periods when the altitude of the sun exceeds 10° above the horizon. In planning an experiment these minimum areas would be increased to allow for the expected width of the plant. The particular spacing shown, of 4h within rows and 3h between rows, would avoid mutual shading of direct sunlight on any day during the summer half of the year, provided that the width of the plants did not exceed the height.

If one is in doubt about the effects of such obstructions on sunshine early in the morning and late in the evening, the matter can be settled by a hemispherical photograph with solar track diagram, as in Evans and Coombe (1959) and see also Plate 6.10. Also, in tall growing plants the lower leaves frequently become senescent and drop before the plant has reached full height, so that a length of stem near ground level becomes bare, thus much reducing the shading effects in early morning and late evening. The patterns of shadow movement shown in Fig. 6.8 cover periods

6.8

when the sun is at least 10° above the horizon, and show the area which would be shaded at some time during these periods by a narrow vertical rod of height h (the height of the plant being considered), standing over the base of the stem. These minimum shadow patterns must obviously be widened to allow for the expected width of the plant, but we see that at the spacing shown there would be no mutual shading at any time during the summer half of the year, even if the spread of the plant's leaves at the stem apex equalled its height. This spacing is of 4h between plants in a row, and 3h between rows, but both these dimensions could be reduced if the spread of the leaves were less, or if the experiment were confined to a short period.

6.9 Edge effects

The second possibility is that the experimental conditions are intended to include root competition below ground, or shading by adjacent plants above ground, or both. The spacing will then be set by the experimental conditions intended, which may include particular conditions of light interception, patterns of leaf orientation, etc. Here the important thing is to avoid an edge effect, whereby the outermost experimental plants are exposed to different conditions of temperature, humidity, air movement, radiation, or root competition from the rest. It is then necessary to surround the experimental plants on all sides by guard plants. Ideally they should be planted in the middle of a large area of similar plants, so that a regular atmospheric concentration and wind profile can build up above them, bringing about uniform conditions. If this is not possible, it may be necessary to be content with several, or even with just two or three, rows of guard plants. Conditions will then not be ideal, but it may be necessary to tolerate this. If desired, radiation conditions half way between two plants in the outermost row of experimental plants can be compared with those between two plants in the middle of the block by hemispherical photographs: the principle of comparison is illustrated, for example, by the photographs in Plate 6.10 and of Anderson (1966). The latter illustrates both comparisons between photographs taken at different heights in a plant stand (her Table 1 and Plate 2 Figs. 1 and 2), and (more similar to our problem here) between photographs taken at the same height, but in different positions (her Plate 1 Fig. 1, and Fig. 1, reproduced here as Fig. 6.9). She notes 'Samples 5 and 9, plotted in Fig. 1

[Fig. 6.9], were 120 cm apart. The mean distance between trees was about 4 m. Although the direct site factors are higher in the summer, when with the greater altitude of the sun, there is more opportunity for it to shine through gaps in the canopy, there is no close proportionality between the direct site factor for successive months between different samples. It cannot be assumed that variations in the total or direct light through the season at one point parallel those at another.'

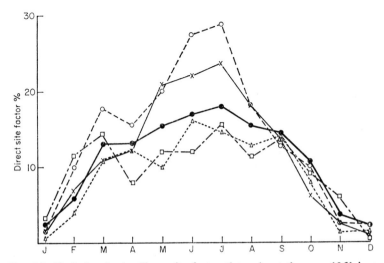

Fig. 6.9. Variation in the direct site factor throughout the year 1961 in a stand of *Pinus sylvestris* at four individual sample points, together with the mean of ten samples. From Anderson (1966).

●———●	Mean	×———×	Sample 5
△------△	Sample 9	○----○	Sample 6
□—·—□	Sample 10		

If the plants are in pots, water loss can also be monitored by a comparison of plants in the outer row and in the middle, by one of the methods of 6.6.

6.10 The glasshouse

The third possibility is that the plants are being raised in a glasshouse. This is a most difficult site as far as uniformity goes (see, for example, the hemispherical photographs in Coombe and Hadfield (1962) reproduced

(a)

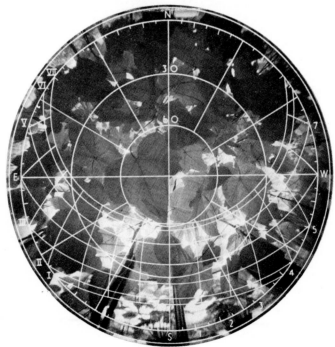

(b)

Plate 6.10

in Plate 6.10; and Coombe and Evans (1960)); and furthermore, space is frequently limited, and guard plants can be used only to a very limited degree, or not at all. If possible, plants should be set out in the middle of the glasshouse on a bench or on the ground, thus avoiding as much as possible temperature gradients (which may be very steep near the glass) and making possible access on all sides. The plants should then be moved about each day, so that all spend equal times on the outside and inside of the block, on the north side and the south side, and (if the central position is not possible) on the side nearer the glass and the side farther away from it. In this way, experience of the very uneven conditions of most glasshouses can be averaged out, each plant having in theory a similar experience. Needless to say, this does not always work out in practice, especially in a climate such as the British one where hours of bright sunshine are very irregular and erratic in their appearance, so that some variability will almost always accumulate in crops grown in a glasshouse. This is, however, inherent in the method, and is best dealt with by raising an ample supply of plants so that variability can be brought down again to a reasonable level at the stage of grading and setting up the experiment.

Plate 6.10. Hemispherical photographs taken at bench level under crops of *Musanga cecropioides* in a glasshouse at Cambridge on 23rd April 1961, with superimposed solar tracks for the solstices (I and VII) and at monthly intervals in between: spacing of the even-aged potted plants arranged to give (a) a leaf area index of 1·0; (b) a leaf area index of 4·0. In (b) one row of plants stood south of the camera, and three rows to the north. From Coombe and Hadfield (1962).

PRINCIPLES OF EXPERIMENTAL DESIGN

7.1 General considerations

Two distinct series of considerations have to be borne in mind in planning an experiment in plant growth: the first are concerned with the nature of the plant, the nature of the environment, and their interactions, all of which have determining influences on the methods of investigating particular problems, and on the types of experimental design most appropriate for the purpose. Secondly, bearing in mind the inevitable variability of material which we have just been discussing; and bearing in mind also that investigations of plant growth tend to be very laborious, and that we shall almost inevitably be working with limited resources, there are the statistical questions of experimental design, to enable us to extract the maximum of relevant information from the available data. With this latter aspect we shall not concern ourselves here, as good handbooks are readily available (for examples see the classified bibliography).

The main outline of the problems posed jointly by the plant and its environment has been discussed in Chapter 3. Recapitulating our conclusions [3.6, 17.1]:

(a) genetic factors may influence the growth of the plant in a specific environment;

(b) the response of a plant to change in a particular aspect of the environment is usually not linear;

(c) frequently the responses of the plant to changes in two different aspects of the environment interact, so that the overall effect is not a simple addition of individual responses;

(d) parts of the plant itself may react on its surroundings, producing environmental changes which may immediately affect other parts of the same plant, or produce effects remoter in space or time;

(e) under natural conditions, changes in different aspects of the environment are frequently correlated;

(f)   the plant is subject to inherent ontogenetic drift—its responses to specific changes in the environment vary throughout the life of the plant; (g)   when grown in a fluctuating environment, those parts of the plant which matured at various stages in the past embody the past reactions of the plant (b) and (c) under the influence of (f), to past states of the environment.

When working under natural conditions, (g) ensures that it is not possible to repeat an experiment at a later date using material identical with that used in an earlier experiment, unless one is working in a climate where the weather in one period can be guaranteed to be identical with that in another.

At the same time, if we confine ourselves to observations of the plant growing under natural conditions and the environment in which it is growing, the complex web of relationships produced by the simultaneous operation of (b), (c), (d), (e), (f) and (g) will always be difficult, and frequently impossible to disentangle by statistical means alone. It must therefore be disentangled by experiment. In the previous chapters we have considered means by which experimental material of minimum variability can be produced, taking account in detail of the consequences of (a), (f) and (g).

7.2   Use of controlled environments

We have now to consider types of experimental design which will enable us to investigate certain strands of the web, with the minimum of interference from the others. Ideally, one would wish to proceed by the classical means of physical investigation, holding constant all the aspects of the plant's environment except that under investigation. This involves the use of controlled environments—growth rooms, cabinets, phytotrons. We see that here we have got rid of all, or nearly all, the complications due to (e), while still needing to take account of (a), (b), (c), (f), (g) and possibly (d). Up to the point when the experiment begins, we can grow the plants in a specified environment, taking care of (g), and making it possible to return to a specific plant form at will. We can use stepwise changes in the aspects of the environment under investigation, thus mapping out the responses of the plant to particular environmental changes, and working out in detail (b) and (c), and also clearing up any outstanding ambiguities due to (e). By beginning such experimental observations at different stages in the plant's life history we can also investigate and map out the ontogenetic drift of response, (f). The use of controlled environments is thus a most useful tool in unravelling the web, and in accumulating information on specific plant/environment relationships which can be used to interpret observations made under less closely controlled conditions.

7.3 Their limitations

There are limitations on the use of even this valuable tool. Suppose that it were possible to isolate a single 'factor' of the environment and investigate that. Suppose the factor to be irradiance: mere intensity of short-wave radiation, leaving unchanged day length, spectral composition, and spatial distribution of sources. This has powerful influences on plant form. Investigations of the effects of changing irradiance therefore have two aspects, corresponding to (f) and (g). We must not merely investigate the effects of changing irradiance throughout the life of the plant, (f); we must also repeat this on the various types of plants produced by different past experience: plants which have been grown all their lives in a fixed irradiance before the experiments begin; plants which have been grown first in high light, and then transferred to low; and those which have been grown first in low light, and then transferred to high. All these changes produce different plants, as can be seen in Plate 7.3, reproduced from Hughes's experiments on the growth of *Impatiens parviflora*. We must therefore make extensive checks of the form and functioning of these different artificially-grown plants against comparable plants grown under more natural conditions, to see how far the former fall within the naturally occurring range [3.6], and hence how far the information gained under artificial conditions is relevant to the interpretation of natural situations.

Next there is the question of daily march of irradiance. So far we have considered a 'flat-topped' day of constant irradiance (see Fig. 7.3). What happens if, without altering the day length, we give the same total in the form of a smooth curve of first rising and then falling intensity, as happens in nature on a day of cloudless blue sky? Or in the form of a constant low intensity interspersed with periods of high, as happens on a day of cloud and occasional blue sky? All these, combined with the effects of ontogenetic drift of response, and the effects of past history, would in themselves constitute a very formidable programme of research.

7.3

Full daylight   67%   67% after 25%

42%   25%   25% after 67%

7%

Plate 7.3. Plants of *Impatiens parviflora* grown individually in pots at Cambridge. The plants were grown initially in a glasshouse, selected and graded. Comparable samples were plunged in the open field, one set being in fully daylight and the others covered by neutral screens with transmissions of 67, 42, 25 and 7%. After 20 days samples were transferred between the 67% and 25% screens. Photographs taken 7 days later. For experimental details see Evans and Hughes (1961). From Hughes (1966).

And it is a programme devoted to a single 'factor' only; one which, on the terms which we have given, is fairly easy to define; and one which, within limits, lends itself to artificial manipulation. This is not true of many aspects of the natural environment of the plant, and particularly of many aspects of the natural soil [10.1 *et seq.*]. It is out of the question to extend such an investigation to a whole range of factors *and their*

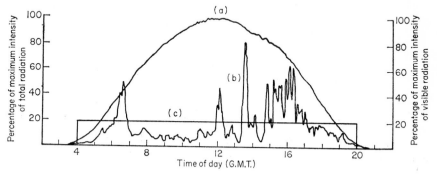

Fig. 7.3. Tracings from Callendar total radiation recorder charts for (a) 19 June 1957, a very bright day (upper tracing); (b) 19 June 1956, a mainly dull day (lower tracing), for comparison with (c) the illumination of a chamber arranged to given an intensity of visible light equivalent to 20 per cent of the full midday light in the open for a 16-hour day. The left-hand ordinates apply to (a) and (b), but not to (c). The right-hand ordinates apply to (c), and also within a close approximation to (a), but not to (b). From Evans (1959).

interactions; once again we should be trying to find our way through the labyrinth by trial and error.

We conclude that, useful though observations made in controlled environments may be, they cannot present the whole solution to the problems of plant growth in natural environments. These problems must be studied in the natural environments themselves. Particular observations under controlled environmental conditions can then be used to elucidate particular problems thrown up by the field investigations.

7.4 Natural environments and ontogenetic drift

Although there may be difficulties in the way of the use of wholly artificial controlled environments, nevertheless they do enable us to study the web of growth/environment relationships at almost any desired point;

and to return to a particular plant form, if desired, for further experimentation. In natural environments we do not have these conveniences. At any moment the full set of complexities due to the structure of the natural environment and the nature of the plant's responses—(b), (c), (d), and (e) of our list [7.1]—are in operation, and nothing can be done about this without turning the natural environment into a partially artificial one. We have, however, a measure of control over (a), the genetical factor, (f) ontogenetic drift, and (g), the historic factor, the influence of past environments on plant form.

The simplest type of observation which we could make would be to produce a population of young seedlings, as uniform as possible, by raising them at first by the methods aiming at reducing variability, which we have considered in Chapters 5 and 6. These would then be set out in the natural environment in which we are interested, and samples would be harvested at intervals. In this way we could follow the ontogenetic drift of plant response to the environment—or could we?

7.5 Some simple observations

In Table 7.5 we have data collected at the Botanic Garden, Cambridge on the growth of young plants of *Helianthus annuus* towards the end of July in each of 9 years between 1952 and 1960. The plants were grown in pots in a uniform loam, plunged up to the rim in open ground, well supplied with mineral nutrients and watered when necessary. The root systems had not filled the pots, so that differences from year to year can be ascribed almost wholly to differences in the aerial environment, both before and during the experimental period. We see that in different years plants of similar ages differ widely in size, both measured as total dry weight and total leaf area. We also notice that the proportional makeup of the plants varies from year to year: there is a wide range of ratios of leaf area to total dry weight (leaf area ratio, LAR), of leaf area to leaf dry weight (specific leaf area, SLA), and of leaf dry weight to total dry weight (leaf weight ratio, LWR) (of course in each case the last two of these ratios multiplied together give the first, SLA × LWR = LAR): but as we might expect, this range of ratios is much smaller than the range of plant size itself. Each of these ratios is an index of plant form, its value being determined partly by the responses of the plant to its environment, partly by ontogenetic drift. Let us now consider how these ratios change

Table 7.5. Data on plants of *Helianthus annuus* from 2–4 weeks old, all of the same strain, grown in pots plunged in open ground at the Botanic Garden, Cambridge, during July in the years 1952–1960. Two harvests each of nine plants (except 1958, when the number was six) at an interval of 7 days, of plants matched at the first harvest. (a) Age, mean dry weight and mean leaf area per plant; (b) Mean ratios of leaf area to total plant dry weight (Leaf Area Ratio, LAR), of leaf weight to total plant dry weight (Leaf Weight Ratio, LWR) and of leaf area to leaf dry weight (Specific Leaf Area, SLA) at each harvest, together with the percentage change in the mean value of each ratio between harvests. Least significant differences between two mean ratios: LAR, 0.22 dm^2 g^{-1}; LWR, 0.07; SLA, 0.21 dm^2 g^{-1}. Significances in the third section of the table, 0.05;* 0.01;† 0.001.‡

(a) Age, mean dry weight and mean leaf area per plant

	First harvest			Second harvest		
Year	Age days	W g	L_A dm^2	Age days	W g	L_A dm^2
1952	21	1·36	1·96	28	2·66	3·17
1953	14	0·296	0·488	21	1·11	1·15
1954	16	1·20	1·34	23	2·10	2·28
1955	21	3·74	6·36	28	12·75	19·22
1956	22	1·61	1·89	29	3·10	2·65
1957	21	1·18	1·79	28	2·59	3·47
1958	21	1·77	2·49	28	3·82	4·34
1959	19	0·279	0·397	26	0·456	0·432
1960	18	1·09	1·79	25	3·62	5·40

(b) Mean leaf area ratio, leaf weight ratio and specific leaf area

	First harvest (I)			Second harvest (II)			100 (II–I)/I		
Year	LAR dm^2 g^{-1}	LWR	SLA dm^2 g^{-1}	LAR dm^2 g^{-1}	LWR	SLA dm^2 g^{-1}	LAR	LWR	SLA
1952	1·50	0·53	2·85	1·19	0·47	2·52	−21†	−10	−12†
1953	1·69	0·45	3·79	1·04	0·46	2·25	−37‡	+4	−41‡
1954	1·13	0·45	2·49	1·12	0·41	2·74	−1	−9	+10*
1955	1·72	0·55	3·13	1·51	0·48	3·13	−12	−12	0
1956	1·18	0·43	2·73	0·86	0·35	2·45	−27†	−19*	−10†
1957	1·52	0·57	2·68	1·34	0·53	2·51	−12	−6	−6
1958	1·42	0·55	2·58	1·13	0·44	2·60	−20*	−21†	+1
1959	1·34	0·52	2·61	0·98	0·44	2·24	−27†	−15	−14†
1960	1·66	0·58	2·85	1·49	0·56	2·64	−10	−3	−7*

during a single week, the third or fourth week of growth. We see that leaf area ratio fell during this week in all 9 years, the fall averaging 19 per cent; but the variability was very large, ranging from a fall of 1 per cent in 1954 to one of 37 per cent in 1953. If we now analyse these particular differences by considering how they were made up from the product of the leaf weight ratio (a measure of the distribution of dry material between the leaves and the rest of the plant) and specific leaf area (a measure of leaf expansion), we see that in 1953 specific leaf area fell by 41 per cent, while leaf weight ratio rose by 4 per cent. On the other hand, in 1954 specific leaf area rose by 10 per cent and leaf weight ratio fell by 9 per cent. Now, to what extent are these changes due to inherent ontogenetic drift? and what proportion of them is to be ascribed to the weather of the week concerned? and what proportion to variations in plant form at the beginning of the week, brought about by the plant's previous climatic experience? That the latter must be of considerable importance is shown, not only by considering the large changes in plant form brought about by the the weather during the specific week considered, but also by considering the large differences in size and form at the beginning of the experiment.

These results are typical of field experimentation in a fluctuating and unpredictable climate. They illustrate both the difficulty of drawing conclusions simply by observing growth under natural conditions; and also the wisdom of not drawing sweeping conclusions from a single season's observations (see also 15.3–15.7 and 26.4).

7.6 Further experimental designs

There is therefore a severe limit to the amount of useful information which we can gain simply by observation of plants over a growing season in the field. We must if possible plan more complex experiments, and these can be broadly of two kinds. Firstly, we can investigate the combined effects of ontogenetic drift and past climatic experience by exposing two sets of plants of different ages to an identical natural climate. Secondly, we can investigate the effect of some particular difference of environment, say shade, or soil conditions, by exposing to the two conditions two sets of plants which are initially as nearly identical as possible. In this way differences due to differences in past experience are eliminated, and the two sets of plants are initially as close together ontogenetically as possible.

7.6

The first of these methods has considerable attraction as a means of investigating ontogenetic drift, but once again it is necessary to generalize with care, and in particular to remember that such an experiment covers only a short range of ontogenetic difference; that past climatic experience affects the initial size and form of both sets of plants; and hence it also affects the starting points for comparison on the overall drift. These points may be illustrated by reference to Table 7.6, which gives data for

Table 7.6. Data on young plants of *Helianthus annuus*, all of the same strain as Table 7.5, grown in pots plunged in open ground at the Botanic Garden, Cambridge, in two years. Two harvests, at an interval of 7 days in each year, dates of final harvests 26 July 1966 and 2 August 1967. Two groups of plants in each year, aged respectively 28 and 42 days from germination at the final harvest. Nine plants in each sample. For each age group and each harvest, mean total dry weight and mean total leaf area per plant, mean ratio of leaf area to total plant dry weight (LAR), mean ratio of leaf area to leaf dry weight (SLA), and mean ratio of leaf dry weight to total plant dry weight (LWR).

Year	1966				1967			
Harvest	1	2	1	2	1	2	1	2
Age from germination (days)	21	28	35	42	21	28	35	42
Mean total dry weight (g)	1·57	4·51	9·66	23·4	45·4	131	233	431
Mean total leaf area (dm²)	2·56	7·17	14·3	27·6	65·1	187	257	409
LAR (dm² g⁻¹)	1·63	1·59	1·48	1·18	1·43	1·43	1·10	0·95
SLA (dm² g⁻¹)	3·24	3·03	3·32	2·86	2·65	3·11	2·46	2·58
LWR	0·50	0·52	0·45	0·41	0·56	0·46	0·45	0·37

two sets of seedlings of *Helianthus annuus* grown in the open in Cambridge for just such an experiment in two successive years, 1966 and 1967. We see that although the plants at the beginning of the experimental period in the two years were identical in age, in 1967 they were around twenty times the size of those in 1966, and hence the comparison of the two age groups took place at a very different part of the ontogenetic drift, and it was based on plants having very different past climatic experiences—cold and dull in 1966, warm and sunny in 1967. This, combined with the effects of ontogenetic drift itself, has led to considerable differences in plant structure between the plants of both ages at the beginning of the experimental period in the two years, as illustrated by leaf area ratio, leaf weight ratio, and specific leaf area. When comparing the results of the experimental week in the two years there is the same difficulty again of not knowing how much of the difference between the results of the two years to attribute to differences in the environmental conditions during the week, how much to differences in initial structure and physiological activity, and how much to differences in the two parts of the overall ontogenetic drifts being compared. In planning experiments of this kind it is therefore desirable to consider whether the batches of plants should be raised in the first instance in a glasshouse, or some other wholly or partly artificial conditions, so that there can be some control of development during the preliminary phase, avoiding such gross differences as those shown in Table 7.6.

7.7 Problems of adaptation to environment

It might be thought that the second, alternative method of investigation, in which we start with a uniform set of plants and distribute them between two different conditions, would be free of this particular complication, and so it may be; but the one complication is simply replaced by another. Plants can respond to changes in environmental conditions in a number of different ways—to shading by changes in leafiness, to changes in mineral nutrition by changes in the balance between the roots and the rest of the plant, and so on. It is clearly desirable that at the beginning of an experimental period a plant should already be adapted to the conditions under which the experiment is to take place; for if not, we shall be faced with the necessity of disentangling the responses of the adapted plant from the adaptive changes which will be taking place during some, or all, of the experimental period. If we are comparing two conditions, we therefore need two sets of adapted plants, one adapted to each of the two conditions. But in the nature of the case these two sets of plants cannot be identical; and this contradicts our initial premise that the attraction of this particular type of experimental design is that they *are* identical. There is no way round this particular snag, and when we are using this method of investigation we shall simply have to decide, from our background knowledge of the magnitude of the two effects which have to be

7.7

balanced against each other, either to prefer identical plants, and have one or both sets not adapted; or to prefer adapted plants, and have them not identical; or to adopt some compromise arrangement.

7.8 A practical example: (i) general

So far this discussion has been largely hypothetical, in order to cover the very wide range of possible situations encountered in ecological field work in general terms. In conclusion we will consider the planning of a particular series of observations, choosing those described by Coombe (1966) on the growth of the woodland annual *Impatiens parviflora* in Madingley Wood, near Cambridge, throughout a growing season from March to September. At the time of writing this is one of the most comprehensive autecological studies in which extensive use has been made of the type of technique which we have been considering. The object was to study the growth and form of the plant from early spring through summer to autumn, as a function of factors of the aerial environment, particularly temperature and irradiance, so as to obtain information on the particular factors limiting growth at different times of year, and on their mode of operation. There was every reason to suppose that as the canopy of this deciduous wood expanded, growth would be limited by irradiance, and arrangements were therefore made to compare two sites, one deep in the wood, where expansion of the canopy would have maximum effect, and one in a clearing, where it would have much less. As the observations were concerned with the aerial environment, the plants were to be grown in pots, in vermiculite watered regularly with culture solution, so that there should at all times be an ample supply of water and mineral nutrients.

7.9 A practical example: (ii) ontogenetic drift

The first question was—should a large population of plants be grown throughout the season in the two situations, samples being harvested at intervals, or should plants of a limited age range be used throughout? The first alternative was rejected, for five reasons. (i) The aim was to follow the effects of specific aspects of the environment, particularly with regard to the development and functioning of new foliage. It was not necessary that the plants should be getting older as the season advanced. (ii) As the plants got older, the proportions of expanding, mature, and ageing foliage would change as would the proportions of main stem and side branches. Both these would superimpose on any other effects a marked ontogenetic drift. (iii) Although up to about 7 weeks old the plants were known to have a very effective leaf mosaic [Fig. 18.5], with very short upper internodes, so that all the leaves were subjected to an essentially similar light climate, above this age self shading would begin to set in, creating a variety of lighting conditions for one single plant. (iv) The very numerous harvests needed would require initially a very large population, and in the dense shade of the wood it would be practically impossible to set this population out in uniform conditions. (v) Towards the end of the season, the older and large plants would need large pots. Repotting was inadvisable, because of disturbance to growth, and the larger pot size needed would aggravate the problem of uniformity of environment posed by (iv).

It was therefore decided that the experiments should be confined to plants less than 49 days old, and that there should be successive sowings, providing plants within this age group throughout the growing season. This particular species is frequently in flower around 7 weeks from germination, ripe seed being shed around 10–11 weeks, so that the period selected would cover a substantial fraction of the minimum life cycle.

7.10 A practical example: (iii) Frequency of harvests

The next question was frequency of harvests. The relevant period stretched roughly from the vernal to the autumnal equinoxes, about 26 weeks, so that weekly harvests would be laborious, and would involve a very large number of plants. However, in the partial or dense shade growth rates are never very high, so that plants could safely be left for $1\frac{1}{2}$ weeks without danger that the leaf area at the end of the period would greatly exceed twice that at the beginning (see 16.10, 16.12 for the reasons for this restriction), It was therefore decided to have alternating growth periods of 10 and 11 days, so that harvests always fell on a Monday or a Thursday, and the necessary harvesting processes could be fitted conveniently into the weekly laboratory routine. This procedure reduced the number of harvests needed over the season from 27 to 18 with very little loss of relevant information.

7.10

7.11 A practical example: (iv) Number of harvests per batch of plants

The next question was 'How long should harvesting continue with any one batch of plants?' The point is this. To study growth during a period we need two sets of comparable plants, one to harvest at the beginning and one at the end of the period. If a particular batch lasts for only a single period, this means two harvests per period, one at the beginning and one at the end. If the batch lasts for two periods, only three harvests in all are needed, as the middle one serves to define both the end of the first period, and the beginning of the second. Similarly, for three periods only four harvests are needed, so that in order to save labour there is an incentive to make one batch of plants last as long as possible. There are, then, three possible alternative plans, as shown in Table 7.11. Before

Table 7.11. Alternative harvest plans, providing (i) a preliminary period of cultivation, selection and grading, and adaptation to the chosen environment; (ii) harvests at intervals of 10 and 11 days alternately; (iii) the last harvest at 49 days.

Plan	Each batch to last	Preliminary period	Day of harvest			
			1	2	3	4
a	1 period	39 days	39	49		
b	2 periods	28 days	28	39	49	
c	3 periods	18 days	18	28	39	49

harvesting can begin, two processes are needed; firstly grading and selection of a batch of plants as uniform as possible from a large population (see the procedures of Chapters 5 and 6), so as to reduce variability to as low a level as possible. The object of this is not just to save labour in harvesting, but more important, to keep to a minimum the area occupied by the plants in the wood, so that the environmental conditions during the experiment can be as uniform as possible. It is therefore essential, and cannot be omitted. Secondly, the plants must be adapted to the conditions under which the experiment is to take place. In some circumstances the whole procedure could be carried out in the experimental conditions themselves, so that as soon as selection and grading is complete, harvesting can begin. Here this is not practicable: a large enough uniform area for growing on the young seedlings cannot easily be provided in the wood, and damage of one sort and another to very young seedlings is frequent enough to make the attempt not worth while—to ensure a large enough population at the end of the grading and selecting procedures, so many pots would have to be set out initially that the gain through having plants continuously adapted would be nullified by the increase in the area occupied, and consequently in the variability of the woodland climate that they would encounter. This being so, it is better, and certainly more convenient, to grow the seedlings initially under uniform conditions in a glasshouse, using light shade so that the conditions in the initial growth phase fall roughly between those in the clearing and those in dense shade. This means that a period of adaptation to woodland conditions must be allowed. Previous studies show that in young plants of this age adaptation (which here is practically confined to the foliage) is complete in about a week: an allowance of 10 or 11 days should then be on the safe side. Subtracting this from the preliminary periods shown in Table 7.11, we are left with 28 days, 18 days, and 7 days respectively to cover the initial cultivation, grading and selection before the young plants go out to the wood. Of these periods, 7 days is inadequate, 18 days is sufficient, 28 days more than enough. We therefore select from the alternatives in Table 7.11 plan *b*, involving two growth periods and three harvests.

7.12 A practical example: (v) Harvest plans

The selected plan is then developed as shown in Table 7.12.1.

It will be seen that a feature of this arrangement is that a batch of 3n plants is taken out at the time of every other harvest: and that the alternate harvests are of n and 2n plants. The next obvious step is to arrange that the harvest plans for the two sites should operate alternately as in Table 7.12.2, where Site A represents the plan we have already seen in Fig. 7.12.1, and Site B is moved on one period by starting with a batch of only 2n plants, giving a single interval for the first batch, so that the second batch is moved out one interval earlier, and harvested one interval earlier, than the second batch on Site A. By this means we have been able to meet all the various requirements, and at the same time produce an overall plan which gives a steady flow of work, with the same number of plants to take out to the wood at the time of each harvest, and the same number to harvest.

Table 7.12.1. Development of plan *b* of Table 7.11 for a single site, showing the time relations of operations for each sequential batch of plants; numbers of plants taken to the site and numbers harvested on particular days; and numbers of plants on site between harvests; all on the basis of n plants per batch per harvest. S, sowing; T, transfer to wood; $_1$A, $_2$A, $_1$B, etc., harvests.

Harvest number	—	—	1	2	3	4	5
Time (days)	0	18	28	39	49	60	70
Batch A	S———T———$_1$A———$_2$A———$_3$A						
Interval (days)		18	10	11	10		
Batch B			S———————T———$_1$B———$_2$B———$_3$B				
				18	10	11	10
Batch C, etc.					S———————T———$_1$C		
Number of plants taken to site		3n	—	3n	—	3n	—
Number of plants on site		3n	2n	4n	2n	4n	
Number of plants harvested			n	n	2n	n	2n
Time (days)		18	28	39	49	60	70

Table 7.12.2. Combination of harvest plans as in Table 7.12.1 for two sites, to give an even flow of work.

Time (days)	18	28	39	49	60	70	81	91	102
Harvest Number	—	1	2	3	4	5	6	7	8
Number of plants taken on site:									
Site A	3n	—	3n	—	3n	—	3n	—	3n
Site B	2n	3n	—	3n	—	3n	—	3n	—
Numbers of plants harvested:									
Site A	—	n	n	2n	n	2n	n	2n	n
Site B	—	n	2n	n	2n	n	2n	n	2n

7.12

7.13 A practical example: (vi) Plant numbers

The number of plants to be harvested as a sample is set by the variability of the material and the accuracy desired. In this particular case, preliminary studies had shown that a sample of n = 8 made possible estimates of the principal measures of growth with standard errors of only a few percent of the means, giving quite a sufficient accuracy. In case of some accidental damage, however, it is always wise to have a few spare plants, which have already been subjected to the grading and selecting process, so that they are known to be comparable with the others, and can be used by the method of 8.10 as complete substitutes in the event of damage, thus maintaining a steady sample size (but see also 15.7). Suppose that for this purpose we increased n from 8 to 10, and took out 30 plants every three weeks instead of 24, this would mean that the maximum number on site at any one time is 4n = 40; alternatively, if we used method (iii) of 8.8 and provided n graded spare plants in each batch, n would remain at 8, but we should take 4n = 32 plants to the wood every three weeks, and the maximum number on site at any one time would be 6n = 48. In either case these can be accommodated in a rectangular grid of 7 × 7, and the former arrangement would leave some spare spaces for measuring instruments. These particular plants were grown in pots 9 cm in diameter, so that at a pinch a square of 70 cm side would have sufficed. In fact environmental heterogeneity in dense shade on the woodland floor did not compel the space occupied to be quite so small, and the pots were spaced about 6 cm apart, making a square of just over 1 m side. This extra allowance of space was advantageous for two reasons; firstly because in order to even out what environmental irregularities there were, the plants were moved about the grid at regular intervals during the periods between harvests, and the extra allowance of space made it possible to do this without risk of damage. Secondly, the ground area per plant was just over 2·2 dm². The leaf area per plant exceeded 2 dm² on one occasion only, early in June, and apart from this no special arrangements to avoid mutual shading were needed, as the leaf area index was always well below 1·0, and there was an effective leaf mosaic, close to one horizontal level [Fig. 18.5].

Although the plan we have just considered was an extremely convenient one for the purpose for which it was devised, it is not put forward as a model, but rather as an example of the ways in which general planning considerations can be applied to a particular case: other plans would no doubt be more appropriate to other circumstances, and other plants possessing different characteristics (see also 15.7). The observations resulting from this particular plan will be considered later [28.10, 28.11, 30.8].

SELECTION OF
EXPERIMENTAL PLANTS

The object of these procedures is to leave the final distribution of variability among the experimental plants on the random basis required by the statistical techniques necessary for analysing the results, while if possible making a final attempt to reduce variability before the experiment itself begins. They may take a variety of forms depending upon the planning of the experiment, which we have just discussed.

8.1 The simplest case

In the simplest case it is not intended to reduce variability at this stage. This might happen in an agricultural or horticultural field experiment on the plants in an area of crop, or in an experiment involving untouched plants growing under natural conditions. In either of these cases the variability from one plant to another might well be amongst the objects of investigation. Experience shows that in a situation of this sort human judgment is not to be relied on to make a random selection. It is worth using some standard experimental design of which examples will be found in standard texts on design of experiments such as those marked in the classified bibliography. These experimental designs have the added advantage that they can be arranged if necessary to allow for systematic sources of variation, such as unsuspected gradients of fertility in the soil, variations in crop density between drill rows, and so on.

8.2 Selection for uniformity

Usually it is necessary to select a number of plants all as uniform as possible, from a larger number which have been grown on up to the time of selection, using the methods which we have already discussed; that is to say, attention has already been given to initial genetic uniformity; to grading of size of seed or propagule; to uniformity of starting date, whether it be seed germination or some other appropriate measure for other types of propagule; and to environmental uniformity during the preliminary growth phase.

8.3   Double selection

If the plants belong to some variety or species which is known to have a very low genetic variability, it may suffice to follow up the precautions of the preliminary phase by selection just before the experiment proper begins. If not, it is desirable to have two selection points, the first perhaps half way through the preliminary growth period, or at some convenient time at least a week before the final selection. The reason for this is that the plant's genetic constitution may affect its growth in many different ways, connected with the rate of carbon assimilation, the rate of unfolding of new foliage, plant architecture, and so on. These genetic differences may affect the overall rate of growth to differing degrees in different phases of the plant's life, so that it is quite possible to have two closely related plants, one of which grows relatively more rapidly early in its vegetative growth, and the other relatively more rapidly later on. If we have only two points of comparison, the starting date and a fixed size n days later, we might then choose as identical (having the same size) two plants which were at that time growing at different rates. It might at first sight seem that the chances of this happening are rather low, but we shall encounter a number of examples as our discussions develop, and we have to remember that this precaution is most necessary in populations which for one reason and another (usually lack of initial genetic uniformity) are in any case very variable, so that in selecting plants as uniform as possible it is advantageous to have the extra criterion of equality of growth rate, as well as equality of size.

This first sorting out need not necessarily be as exacting as the final selection. It may suffice to divide the plants by eye into a number of size categories, say 5 or 7, and give the members of each category a distinctive label. Plants which at the final selection are found to have moved from one category into another can then be eliminated. If time allows for a pilot survey before the main experimental programme begins, this matter of a second selection, and the form it should take, is worth settling at this stage, as much time can be saved in handling large samples, if use is made of experience already gained on small ones.

Table 8.3 illustrates this effect, from some data on a population of plants of *Helianthus debilis*, from seed not selected to give a genetically uniform strain. When they were 40 days old, three large, three medium

Table 8.3. Data on plants of *Helianthus debilis* grown in pots plunged in open ground at the Botanic Garden, Cambridge, in June and July 1961.

	Harvest at age 47 days	
Grade at age 40 days	Dry weight (g)	Leaf area (dm²)
Large	17·98	22·5
	16·86	15·1
	14·53	16·5
Medium	10·84	13·4
	10·59	13·9
	5·96	7·4
Small	9·36	13·5
	9·23	10·4
	8·96	13·4

and three small plants were selected. They remained undisturbed, growing in open ground for a week, and were then harvested. What was the initial 'medium' sample now includes one plant which, although apparently quite healthy, was considerably smaller than any in the 'small' sample; whatever may have been the reason, clearly its rate of growth during the week between grading and harvest differed widely from that of the other 'medium' plants, and had there been a second selection it would have been eliminated.

8.4 Characterization by measurement

We have now to consider the selection procedures themselves. We require plants as uniform as possible—in size, in leaf area, in rate of growth, in rate of carbon assimilation, and so on: but we cannot harvest, damage or accidentally injure the plants. What measures are open to us?

From the point of view of the addition of new stocks of reduced-carbon units, the leaves are the most important part of the plant: equality of leaf area is much more important than equality of stem size [Table

8.4

5.5]. On the other hand, there are a number of instances among young dicotyledonous plants, having a plant architecture based on a stem supporting the leaves, where at a given plant size stem weight is a comparatively constant fraction of total dry weight. Here changes in conditions of growth (including soil conditions) have much larger effects on proportions of root and leaf than on proportions of stem. In such a case stem size can be a very useful index of plant size as a whole, as little can be done to measure the root directly. In any case, we are aiming at uniformity, and it is worth measuring the stem if it is practicable to do so.

8.5   Stem measurements

Stem measurements should include both height and diameter near the ground, measured at some standard place, such as the hypocotyl or the lowest internode. The measurement should be made in such a way as

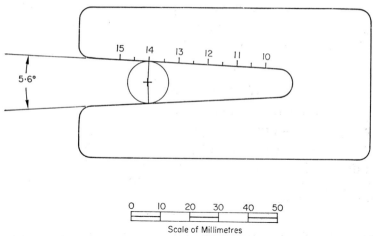

Fig. 8.5. A simple cardboard gauge, for rapid estimation of stem diameter with minimal risk of damage, particularly to herbaceous stems. In the example shown, an angle of 5·6° gives a ten-fold magnification, convenient for small stems, and the gauge reads in millimetres.

not to risk damage to the stem itself, such as can easily be inflicted accidentally by a steel calliper gauge. If many plants are to be graded, it is worth making up a simple gauge from cardboard or some spongy material such as balsa wood, to a pattern such as that illustrated in Fig. 8.5. The range and magnification of the gauge can then conveniently be suited to the expected range of stem thicknesses, and the categories into which these are to fall. Numbers, positions and stage of development (length, or number of nodes) of side branches will also probably be worth recording.

8.6 Leaf measurements

In some cases it may be necessary to measure the length and breadth of every leaf, but it is worth taking steps to avoid this if possible, both on account of the labour involved, and also because of the risk of accidental damage to the plant by much handling (and see also the remarks in 4.2 and 12.4 on the effects of handling on leaf respiration). The number of fully expanded and the number of expanding leaves can be recorded separately, setting some arbitrary standard for the smallest young leaf to be counted in this way. It may also be necessary to record separately leaves on side branches, which may differ in several ways from those on the main stem. In some cases these numbers, together with the maximum span across the foliage, will characterize the foliage sufficiently for grading purposes. If this proves to be insufficient, it may be necessary in addition to measure the lengths and breadths of certain selected leaves.

8.7 Reproductive structures

If flowers, or visible flower buds are present, their numbers, states of development and positions should also be recorded.

There are thus three classes of information on which to base the selection: degree of morphological development, as evidenced by numbers of leaves and internodes, numbers and positions of side branches, with their leaves and internodes, and buds or flowers if any; size of stem; and some measure of leaf area. Ideally, plants should agree in all these respects; but if a preference has to be given to one criterion, leaf area is usually to be preferred [Table 5.5, 8.4].

8.8 Methods of selection

The use to be made of these measurements will depend on the experimental design. They can be used as a means of estimating the actual size and weight of a particular plant at the time of measurement [8.10]. Alternatively they can form the basis for selection of the experimental plants, in which case we can proceed broadly in one of three ways.

8.8

(i)   We can use the measurements to reject all plants lying outside some standard of uniformity, producing one single population of experimental plants. The assignment of individuals to immediate harvest or to some particular experimental treatment is then done at random, say by drawing lots, shaking dice, or using a table of random numbers. From the statistical point of view, this will in the end give us the most flexible possible set of data.

(ii)   If the initial population from which selection is to be made is very variable, so that method (i) would produce too few plants for the experimental plan; or alternatively, if the allowable limits were widened to include the correct number, the variability would be too great if treated as a single population, we can use our measurements to select from the initial stock of plants not one uniform population but several, say three, each of which populations conforms to our desired standard of uniformity. We then employ the restriction that the initial harvest sample, and all other samples used in the experiment, must each contain equal numbers of our three uniform populations, as in Table 8.3. Reference to standard statistical texts will show that to make the best use of data of this kind the standard deviations of the different populations should be effectively identical, or bear some standard relation to each other. Such criteria are, however, usually met by the kind of plant populations we have been discussing, and there are times when by this means significant effects can be detected which would otherwise have been swamped by the variability of the experimental plants.

(iii)   The third method goes a stage further, and divides the initial population into groups, each of which contains a number of uniform plants equal to the number of harvests after the experimental treatments, plus one for the initial harvest, and one for a spare. Taking as a simple example the Coombe woodland experiment which we have already discussed [7.8–7.13], there are here three successive harvests. Each group of plants would therefore contain either three plants (just enough) or four (one spare). In the particular experimental design there considered, eight plants are to be harvested in each sample. We therefore require eight groups each of three or four plants. Within each group the plants will, as before, be assigned at random to a particular harvest, and the spares are there as complete substitutes in the event of any accident during the period of observation.

By this method, which is often very valuable even when populations are not inherently highly variable, we can greatly reduce the random element in the overall variability, giving the maximum accuracy for estimating differences between means of samples having different treatments, which is usually the object of the experiment.

Like method (ii), method (iii) is easiest to apply if one can rely on homogeneity of variance. If every plant can be harvested, measured, weighed and recorded separately, no assumptions need be made, and the necessary tests for homogeneity of variance can be applied at the beginning of the analysis of the results. At times the test has the added value that its results reveal the existence of an unsuspected situation (e.g., 5.5, 11.20). If in the main experiment the recording of every plant separately would make too much work, the method can still be used by checking for homogeneity of variance in a separate, pilot experiment. In planning this pilot experiment it should be remembered that the object is to investigate variability, and that therefore the numbers used should be as large as practicable, and be chosen with reference to tables of variance ratio distribution.

8.9 A practical example

By attention to all these details of management and selection it is possible to achieve remarkable accuracies, even when working with populations of natural species which have not been subjected to agricultural or horticultural selection. Table 8.9 gives some figures to illustrate this from part of the data from Coombe's experiment on the growth of *Impatiens parviflora* in Madingley Wood near Cambridge (Coombe, 1966; 7.8–7.13, 8.8).

When the second and third plant of each grade of this particular batch of plants was harvested, both the total dry weights and the leaf areas fell into the correct order in every case. The same had happened also at harvest 1, except for the dry weight of plant 6. Considering the mean rate of increase of dry weight per unit leaf area during the periods between harvests (unit leaf rate, $\mathbf{E}$; 13.7, and Chapter 16), we see that the variability was small, as compared with the large variability of plant size in the initial population from which the seven grades were chosen. For the first interval the mean unit leaf rate was $3 \cdot 1$ g m^{-2} week^{-1}, with a standard error of $0 \cdot 19$, around 6 per cent, the corresponding figures for the second interval being $3 \cdot 9$ and $0 \cdot 39$, 10 per cent. This is a remarkable accuracy with which to measure the unit leaf rate of a woodland plant in a natural aerial environment, and growing so close to the compensation point. Fig. 3.4.1, for example, shows that the unit leaf rate of the same plant in full daylight is about 50–60 g m^{-2} week^{-1}.

Table 8.9. The growth of graded plants of *Impatiens parviflora* in deep shade in Madingley Wood near Cambridge, between 15 August and 5 September 1963. Experimental plan as in Table 7.12.1. Initial selection of seven grades, with three plants in each grade; one plant of each grade assigned at random to each of three harvests. For each grade: total dry weight (W) and total leaf area (L_A) per plant for each harvest; and also the mean rate of increase of dry weight per unit leaf area ($\bar{E}_1$, see 16.12) during the period between the first and second (1–2) and between the second and third harvests (2–3) (data of D. E. Coombe, personal communication).

	15 August Harvest 1		26 August Harvest 2		5 September Harvest 3		$\bar{E}_1$	
Initial grade	W (mg)	L_A (cm^2)	W (mg)	L_A (cm^2)	W (mg)	L_A (cm^2)	g m^{-2} wk^{-1} (1–2)	(2–3)
1	56·5	36·6	78·0	55·0	—	—	3·0	—
2	45·8	30·8	63·4	45·6	107·4	67·6	2·9	5·5
3	43·3	26·8	59·0	43·5	92·1	60·4	2·9	4·5
4	41·0	26·1	52·8	38·1	80·0	52·2	2·3	4·2
5	32·5	23·2	49·9	34·5	65·4	47·4	3·9	2·6
6	33·4	21·8	48·2	32·9	63·5	46·9	3·4	2·7
7	29·7	19·6	43·3	30·7	63·3	44·2	3·5	3·7

The flexibility of a plan based on this method of grading is illustrated by the fact that this accuracy has been achieved in spite of experimental difficulties very familiar to all those accustomed to field work of this kind [15.6, 15.7]. It was originally intended [7.13] that there should be eight plants per harvest: because of the accidental failure of part of this particular batch in the preliminary phase, the number had to be reduced to seven, and it was not possible to provide any graded spare plants. Consequently, when at a routine inspection the grade 1 plant for the final harvest was found to have been damaged, it could not be replaced, and the number at harvest 3 was reduced to 6.

Before leaving this example the question arises of whether there is evidence here of an increase in variability with increasing time after the initial grading, such as that shown for *Helianthus debilis* in Table 8.3. Table 8.9 showed that there were no gross effects, the order of the grades being preserved after a lapse of 3 weeks. Simple comparisons of variability within grades being ruled out by growth between one harvest and the next, the obvious test is to compare the variance of the unit leaf rates for the interval between harvests 2 and 3 with that for harvests 1 and 2. For the data of Table 8.9 the variance ratio is 4·52, just high enough to reach significance at the 5 per cent level (4·39), but nowhere near the 1 per cent level (8·75). It is therefore worth looking at the whole body of data between the closing of the canopy in late May and early September. We can then compare four periods between harvests 1 and 2 for which there are at least six values of unit leaf rate (26 degrees of freedom) with three periods between harvest 2 and 3 (18 degrees of freedom). The variance ratio is now 1·01, so near unity that it is hardly worth looking up the value, 2·02, needed for the 5 per cent level of significance. This seems to be a case where, after a careful initial grading, there is a negligible increase in the variability of the graded population during the following 3 weeks. What has happened in the particular case of Table 8.9 is that the variance of $\bar{E}_1$ for the period between harvests 1 and 2 was somewhat lower than the mean for the season as a whole, and the corresponding variance for harvests 2 and 3 somewhat higher, two small differences happening to add up to a 'significant' one, and showing how much care is needed in interpreting statistical indications.

8.10 An alternative procedure

The example of 8.9 was in several ways an unusually favourable one for the use of selection: the plants were all small: they were of a species apparently of remarkable genetic uniformity (at least as concerns its British populations): they were all available in small pots in a glasshouse, so that grading and selection could be done under near ideal conditions. There is an alternative way of proceeding, first described by Goodall (1945), which is always worth considering when planning an experimental programme on plant growth. In principle this uses the information which we have collected [8.2, 8.4–8.7] on the size, shape, and make-up and history of individual plants, not as an adjunct to the process of dividing them into grades, but as a means of estimating what their dry weight and leaf area would have been, had they been harvested.

8.10

We measure all the available plants, decide on the size range within which we propose to work, and divide the plants into a number of samples equal to the number of harvests, plus some spare plants, as in 8.8, method (iii). If there is to be no second selection, we proceed at once with the first harvest: otherwise, at the time of the second selection we pay particular attention to the measurement of the plants for harvests 1 and 2. After the first harvest we plot the measurements of the harvest 1 plants against the appropriate harvest data for the same plants, and work out regressions which can then be used to estimate the dry weights and leaf areas, at the time of the first harvest, of the plants intended for the second harvest. In this way leaf measurements can be connected with leaf area and with leaf dry weight: stem measurements with stem dry weight: and estimated total top dry weight with root dry weight. At each stage the errors involved in the estimations can be determined from the deviations of the observed points from the appropriate regressions. At harvest 2, the sets of plants for both harvest 2 and harvest 3 are measured, and the same procedure gone through again; and so on.

This method has the great advantage that it is not necessary to pair the plants or divide them into size classes. Use can be made at the next harvest of any plant whose measurements fall within the range of those being measured at the current harvest. In a particular case this method may be found to involve a good deal more work than the methods described in 8.8, but under favourable conditions it is capable of great accuracy (e.g., compare Fig. 28.13, showing results obtained by this method, with Fig. 28.11, where method (iii) of 8.8 was used on plants of the same species growing under comparable conditions). We shall see later that the relationships on which the regressions are based are sometimes very regular, even for plants grown in fluctuating environments (e.g., Figs. 18.2.2, 18.6, and 19.10); but the method is also useful for critical work of high accuracy undertaken under controlled environmental conditions.

8.10

MEASUREMENT AND CONTROL OF
THE AERIAL ENVIRONMENT

9.1 Environmental 'factors'

We began our discussions by pointing out explicitly how greatly higher plants as organisms differ from man and the higher animals, and the consequences of these differences have been implicit in our subsequent considerations. Nowhere is it more important to bear these differences in mind than when thinking of the environment in which the organisms live. Not only are higher plants sessile, effectively confined to the immediate neighbourhood of the place where the young plant puts down its roots, but the environment affects them in ways different from the ways in which it affects higher animals. It is accordingly most unwise to apply to plants the anthropocentric notions of environmental 'factors' which have been worked out by human beings for their own convenience [3.5], and which are, to a degree, convenient when thinking about organisms closely related to human beings.

9.2 Rain

These considerations apply with even greater force when one considers the particular methods of measuring environmental factors which have been worked out by meteorologists for the convenient study of meteorological phenomena. Rainfall, for example, is a straightforward meteorological quantity, and a good deal of time and thought has gone into the devising of rain gauges which will register rain and nothing but rain. However, mere absence of rain does not mean that there is no atmospheric source of water for plants. For example, there are parts of the coastal strip of Peru and Chile, where rainfall as measured meteorologically is excessively rare; but every year, in winter, mists roll in from the sea, and for a period of several months enshroud parts of the coastal plain having a suitable topography. The mist settles on the irregular surface of the ground, which is soon covered with dense, hygrophytic vegetation. Anyone visiting the region would not doubt that it is wet: water drips off every leaf: the plant forms are those appropriate to an extremely moist environment: yet it practically never rains. In due course the mists roll away: the vegetation dries up, and is eroded away by the desert winds: all trace disappears until the next season.

This is an extreme instance, a cautionary tale. In less extreme form it applies to the relations between the lives of plants and many standard meteorological measurements—the connection is there, but it is often extremely indirect. Rainfall again furnishes a good example. Rain is subject to large topographical variations; unless the nearest meteorological station is very close indeed to the experimental area, it is worth setting up one's own rain gauge or gauges, if the area being studied is extensive.

9.3 Air temperature

Again, air temperatures are usually measured in a Stevenson's screen, set up in the open with attention to a number of factors of siting. This is not necessarily the same thing as air temperature in the immediate neighbourhood of the plants being studied, although in favourable cases the two may bear a recognizable relation to each other. Furthermore it is most unlikely that the cells of a leaf, for example, would be at exactly the same temperature as the surrounding air [3.5]. The relationship between the two will depend on the net absorption of radiation, the heat equivalent of transpiration and metabolism, and the advective heat exchange, which will be a function both of the temperature difference between the leaf and the air, and of air movement. This combined function of all these different variables is anything but straightforward: yet there is no doubt that the important aspects of temperature in relation to metabolism in the leaf is leaf temperature, not air temperature. These differences can be shown to have important practical consequences. Hadfield (1968), for example, was able to explain by an extensive combination of field and laboratory studies the observation that tea bushes in Assam give a higher yield in moderate shade than in full sun, because the effect of this is to bring the temperature of the upper leaves down to near air temperature. In full sun the upper leaves which he examined were so much hotter than the surrounding air that net photosynthesis by these leaves was very greatly depressed, being at times reduced to zero.

9.4 Aspects of the aerial environment

Temperature, in fact, provides an excellent example of how difficult it is, when dealing with plants, to isolate an environmental 'factor' from the totality of the plant's surroundings, and this is why, in our discussions hitherto, we have either spoken of a 'factor' in inverted commas, or more usually, avoided the use of the word altogether, and spoken instead of some aspect of the environment, which does not carry the same implication that it may be possible to consider the 'factor' in isolation.

All this is not to say that meteorological measurements are useless in connection with plant studies. On the contrary, they form the foundation on which sound work on the relationships between plants and their environment must be built. What we have been concerned to stress is that they are not more than the foundation, and much care has to be devoted to the erection of a suitable superstructure upon them.

9.5 Meteorological measurements

Standard meteorological measuring instruments, and the methods of using them, are set out in the *Handbook of Meteorological Instruments*, and in other references quoted in the classified bibliography. The relation of climatological measurements to the International Biological Programme was surveyed at the Second General Assembly (Paris, 1966; Section PP: Report of Instruments Sub-Committee on climatological measurements).* In this context the report listed as essential equipment for a climatological station a large Stevenson screen with stand, 2 maximum, 2 minimum and 3 ordinary thermometers, 1 thermohygrograph, 1 whirling hygrometer, 1 rain gauge and measuring jar, 1 cup anemometer, and 1 solarimeter with integrator; and as desirable equipment, in addition to the above, 1 Assmann psychrometer, 3 earth thermometers (2 for 10 cm and 1 for 30 cm) and 1 sunshine recorder.

A selection of special techniques may be needed to bridge the gap between the climatic records of such a station and those aspects of the aerial environment impinging directly on a specific ecological problem. These special techniques are very numerous (see the classified bibliography). Ecologists will find useful parallels to many types of field investigation in the British Ecological Society's Symposium Volume *The Measurement of Environmental Factors in Terrestrial Ecology* (1968),

* and see Monteith (1972).

which also contains an appendix on available apparatus. Many of these techniques are not suited to continuous recording—they provide only 'spot' readings at a particular time. If these spots readings can be related to the standard meteorological observations recorded at the same time at a neighbouring station, it is possible in certain cases to weight the spot readings in such a way as to produce useful long-term average values (e.g., Evans, 1956, pp. 410–419).

9.6 Light as an ecological factor

Of all the aspects of the aerial environment, however, there is no doubt that much the most complex are those associated with radiation. The measurement of radiation for ecological purposes poses multi-dimensional problems which also require special techniques for their solution. Many of these are dealt with in the British Ecological Society's Symposium Volume on *Light as an Ecological Factor* (1966), which contains papers covering the whole range of measuring techniques from those appropriate to a national meteorological network to those which can be used under expedition conditions in parts of the world difficult of access and lacking all normal laboratory facilities.

The study of radiation in relation to plants growing in the field is closely linked with the study of the temperatures of the above-ground parts, and of water loss from them, the link being provided by the heat balance [9.3]. For investigations of this kind it is necessary to take account of long-wave ('thermal') radiations from the plant and from its surroundings, as well as of the short-wave radiation from the sun or from some artificial source. Hitherto the bulk of heat balance studies of vegetation have been made on large uniform stands of crop plants (see e.g. Penman, 1963), but it has recently been shown that under favourable conditions heat balance methods can yield much useful information on conditions inside a natural plant community (e.g. Evans, 1966a).

9.7 Radiation inside plant communities

The small scale of many natural plant communities poses particular difficulties in studying the radiation climate as it affects individual plant parts below the top layer of leaves, or whole plants growing below the main canopy. Consequently most progress has been made in the study of conditions in the lower layers of communities on the largest scale, particularly of woodlands and forests. Even here, the subject is one of such intricacy that great care is needed in the selection and use of suitable methods of investigation, and separate sections of the classified bibliography have been prepared, relating to particular aspects of the overall problem.

9.8   Comparison of environments:
(i) temperature and humidity

It is convenient at this point to notice certain uncommon but very useful methods of presenting comparisons between two different environments. In 1959 Coombe (1960) made a glasshouse study of the growth of *Trema guineensis*, a shrub or small tree common in the regrowth of Nigerian high forest. The observations were made near the equinox, so that the natural daylength was similar to that near the equator; total short-wave radiation and hours of bright sunshine are quoted. The principal interest in our present context attaches to the comparison made between temperature and humidity in the glasshouse and observations made earlier in the high forest zone of Nigeria, at the same time of year. This comparison, is reproduced in Fig. 9.8. It will be seen that the basic plot is of relative humidity against temperature, hour by hour, the hourly readings being joined up to give the daily march of both temperature and humidity. To this plot are added two other sets of lines, those of constant saturation deficit (Buxton and Lewis, 1934) and those of constant absolute humidity (Evans, 1939). It is thus possible in a single graph to follow the daily march of temperature, and also the daily march of three aspects of humidity, so that several simultaneous comparisons can be made. We see that in Nigeria the night temperature was about 1·5°C warmer, and in the middle of the day about 3°C warmer. On the other hand, although the mean relative humidity over the day was similar for both, the saturation deficit in the glasshouse was considerably lower during the day, and considerably higher during the night. A very interesting difference between the two climates is shown by the values for absolute humidity, which in Nigeria was remarkably constant, whereas in the glasshouse there was a very marked fall in absolute humidity between 1600 hours and 0600 hours, during which time the saturation deficit actually rose from 4 to 5·5 mmHg. This reflects partly ventilation exchanges with the outside air, whose absolute humidity would be likely to fall during the night, and partly condensation on the inside of the glass.

9.8

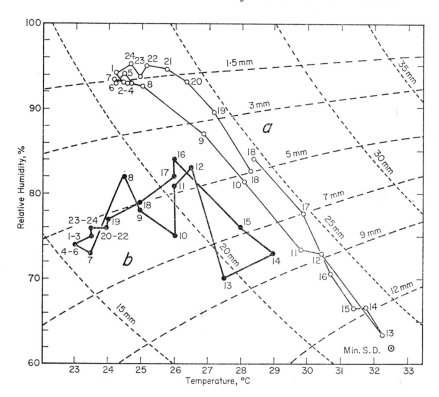

Fig. 9.8. Temperature and relative humidity (*a*, above, right) in the canopy of tropical rain forest in Nigeria, 8–9 March 1935 (after Evans (1939), Fig. 4) and (*b*, below, left) in the third section of the tropical experimental glass-house at Cambridge, 17–24 March 1959. The data from Nigeria are hourly readings over one 24-hour period; the Cambridge data are the means for each hour of the day over the experimental period of a week. – – – – constant saturation deficit; - - - - - - constant absolute humidity. Note that absolute humidity in the forest was nearly constant, but that it fell considerably at night in the glasshouse. The absolute minimum (instantaneous) saturation deficit in the glasshouse (Min. S.D. ⊙) is also shown. From Coombe (1960).

9.9   Comparison of environments:
(ii) Hemispherical photography

A second interesting comparison of two sets of glasshouse conditions is given by Coombe and Hadfield (1962), reproduced here in Plate 6.10. This shows hemispherical photographs of two experiments similar to the last named, but this time carried out on *Musanga cecropioides*, a common tree of second growth in Nigerian high forest. These experiments were made at two different leaf area indices (ratio of leaf area to ground area [13.18]), of 1·0 and 4·0, and the hemispherical photographs show the two resultant seedling canopies. From the solar track diagrams the track of the sun on the day in question can be inferred—it is very close to track no. V. We see the degree of self shading offered by the plants, particularly in Photograph 1, where at a leaf area index of 1·0, all the open sky corresponds to self shading. We can also see that, although the glasshouse occupied an open site, nevertheless obstructions by trees and buildings at a distance cut out any possible sunshine for half an hour after sunrise and before sunset each day.

Both these are instances of condensed comparisons which it would be difficult to present so concisely in any other way.

9.10   Fully controlled environments

An extensive literature has now grown up on this subject (for an entry, see the classified bibliography). Although a number of phytotrons have been built, many workers still have no access to phytotron facilities, and, indeed, for many purposes it is more convenient to have limited controlled environment apparatus near to the main work site, rather than to have to transport material to more extensive facilities at a long distance. Hence the attraction of growth rooms, or controlled environment cabinets, which are appearing on the market in increasing numbers and a variety of patterns. Not all these are suitable for all types of ecological work—for example, some have the main ventilation coming up through the floor. This arrangement, made to improve the uniformity of the lighting, is very far removed from natural conditions, and gives rise to complications if the floor is needed for apparatus of one kind or another. Those wishing to design their own equipment, or to assess the appropriateness of particular commercial designs for their own experimental requirements, will find a useful account of general principles of cabinet design in Evans (1959), 'The design of equipment for producing accurate control of artificial aerial environments at low cost'. This gathers together the results of twenty years' experience of work in controlled environments; and as the basic physics and plant science have not changed since then, it retains its relevance at the present day. This paper lays particular stress on reliability and serviceability of equipment, essential for studies where

9.10

a breakdown may ruin weeks or months of work. It also considers capital and running costs of equipment, and the comparative costs of raising plants in wholly artificial environments as compared with others. Technical improvements in lighting are fully dealt with by Canham (1966; and see 28.10).

9.11 The standardized plant

There is one point about the possession of fully controlled environmental equipment which is worth bearing in mind in many studies of the physiology of higher plants, as well as for ecological purposes, and that is the possibility of using the equipment to generate a standard plant for experimental purposes. All those who have had extended experience of the preparation of higher plant material for physiological research will agree that one of the major difficulties is to reproduce, for a later experiment, plant material identical with that used in an earlier experiment. Those who are fortunate enough not to have experienced the frustrations which can result, but who have followed our discussions so far, will not be surprised that this should be so. If one can find genetically uniform seed, the fully controlled artificial environment solves this problem, with the added advantage of being able to generate a particular plant form at will, at any time of year.

9.12 Partially controlled environments

Essentially these involve making use of certain properties of the natural environment relatively unchanged, while controlling or modifying others. The simplest example is the glasshouse, including in this generic term such simple structures as warm and cold frames.

9.13 The light climate in glasshouses

The object here is to make use of natural solar radiation, which produces over most of the land surface of the globe radiant intensities much higher than can be achieved by artificial means over any considerable area. There is no particular difficulty in producing artificially higher intensities over small areas; or in producing the same daily total of radiation as on an average summer day in a climate like the British one. The difficulty lies in reproducing the intensity of full midsummer noon sunlight, which can be attained nearly enough in a well-designed glasshouse. The light climate there has the added advantage, for certain experimental purposes, of possessing good approximations both to the natural spectral energy distribution in the visible region of the spectrum, and also to the natural daily march of irradiance. Both these latter are exceedingly troublesome to reproduce artificially, and the three restrictions taken together provide one reason why most phytotrons have glasshouses as part of their organization, and why they are sited in climates where regular insolation can be relied upon.

9.14   Uniformity of conditions

For experimental purposes it is important that there should be an extended area of relatively uniform conditions, if possible at all times of year, and this means that the glasshouse must be most carefully sited with respect to obstructions of all kinds, including other glasshouses. The orientation must also be carefully planned, in relation to the structure of the house itself. While a single house of suitable design can be satisfactory with the long axis running east and west, two or more parallel houses are better arranged north and south. Hemispherical photographs furnished with solar track diagrams can be a great help in siting glasshouses, and assessing the effect of obstructions (Evans and Coombe, 1959; Coombe and Evans, 1960). The latter reference includes a glasshouse photograph, and others are shown in Plate 6.10, although there the photographs were taken for another purpose, and most of the distant view is obscured by plants. It is also a help to know the distribution of hours of bright sunshine over the day, which is not necessarily symmetrical about noon, and which may influence a decision about the importance of a particular obstacle.

Fig. 9.8 showed an example of a glasshouse climate used in an endeavour to reproduce the temperature and humidity of the wet tropics. The other main overall effects of the glasshouse on the aerial environment are on air movement, and, partly as a corollary of this, on atmospheric carbon dioxide concentration. This last facet of the glasshouse climate is frequently neglected, but is worth remembering. It can be a difficult matter to assess, *a priori*, the relative effects of reduced air movement, carbon assimilation by the glasshouse crop, and respiration by a compost-

9.14

rich potting soil. If possible, measurements should be made (for an account of a number of possible methods see 12.6–12.9).

Even when all these precautions have been taken, any high degree of uniformity of conditions is unattainable in a structure like a conventional glasshouse, for reasons inherent in the properties of glass. Firstly, glass is largely transparent to solar radiation (99 per cent of the solar radiation reaching the earth's surface has wavelengths below 3 μ), and largely opaque to the long-wave radiation emitted by bodies on the earth's surface. Glasshouses have therefore normally to be kept cool by ventilation, or by shielding them from the sun's rays. But if our object is to expose the experimental plants to full solar radiation (or at least, as much as the glazed structure will transmit) the latter method cannot be used. Full sunlight from a clear sky, penetrating at normal incidence through one square metre of window glass, suffices to heat about 40 m^3 of air per minute through 1°C. Therefore if the temperature rise inside the glasshouse is to be kept to 1°C, the ventilation system must renew 40 m^3 of air per minute for every square metre of glass, projected on a plane normal to the sun's rays; or, if the average depth of the glasshouse is 4 m, we require 600 air changes per hour. No conventional ventilation system will do this, so that when the sun shines during daylight hours the glasshouse heats up, convection currents are established, and conditions at bench level cease to be uniform. Secondly, glass is a good conductor of heat, and, when the temperature outside falls below that inside, the glass surfaces become colder than the rest of the house: once again convection currents are set up, and in addition, condensation on the glass may lower the absolute humidity [9.8].

One consequence is that really uniform conditions cannot be produced in a glasshouse, and the precautions already noticed [6.10] must always be observed, so that even if the experience of particular plants at a particular moment is different, averaged over a period the experience of all the experimental plants will be roughly the same.

A second consequence, in the absence of a really powerful ventilation system, is that blinds must often be used to prevent plants in the glasshouse from becoming too hot. In high humidity and still air the leaves of temperate plants can reach a lethal temperature when exposed to full sun, and this has been known to happen in a Cambridgeshire wood under suitable climatic conditions (*Mercurialis perennis* in a large sunfleck in Hayley Wood, after rain; O. Rackham, personal communication). The high background temperature needed for tropical plants makes the risk of damage greater. When the sun is near the horizon its direct light can enter a glasshouse under blinds of the conventional pattern, and particularly near midsummer damage can ensue in the early morning or late evening when no-one is about.

9.15 Screens

An even simpler method of partially controlling the aerial environment is by artificial screens of one kind of another, such as perforated metal, plastic mesh, or bamboo slats. If uniform conditions are wanted under these screens, and if we are to avoid a pattern of light and shade on the ground underneath, the widths and spacing of the obstructions in relation to their height above the ground must be such that the various penumbras (or regions from which only part of the sun's disc is visible) overlap to give a uniform irradiance on the ground when the sun is shining. If the screen is very dense, there may be a great deal of interference with air movement, with consequential undesired effects on temperature and humidity. It may then be desirable to make a pilot study of conditions under the screen, and if necessary to record them during the experiment. Sometimes a gap must be left between the ground and the base of the screen. If so, this must be shielded by a suitable baffle at a distance, to prevent access of direct sunlight when the sun is near the horizon. Other precautions concern rain and dewfall. It is wise to check the amount and distribution of precipitation under the screen (Ovington and MacRae, 1960), and also to ensure that there is no toxic drip from the screening material. Perforated zinc or galvanized metal, for example, must always be protected by a thin layer of paint, and this must be done with great care so that the perforations remain uniform.

A good general account of the effects of artificial screens on the aerial environment is given by Keith Lucas (1968), in connection with work on the growth of British species of *Primula*.

10

MEASUREMENT AND CONTROL OF THE ROOT ENVIRONMENT

10.1 An analogy

A natural soil represents one of the most complex and inaccessible of all biological systems, and in consequence many of the problems of root growth under natural conditions have not yet been solved, and show no signs of being so. An analogy may be helpful. Air travel has made many people familiar with the appearance of a great city from a height of 10,000 m. Imagine several hundred Londons constructed one on top of another till they reached this height, with a population exceeding the present total of the human race, provided at all levels with elaborate intercommunications, water supplies, sewers, factories, power stations, pipelines, housing, warehouses, libraries, museums, art galleries, hospitals, schools, parliaments, local administrations, law courts—the whole complex of modern society condensed into a single block ten kilometres deep. Now imagine a giant twenty kilometres high coming along with a spade a couple of kilometres across over his shoulder. He thrusts it into the top of the three dimensional city, and digs out a piece of the complex with a volume of several cubic kilometres, which he chops into smaller pieces. Some he dissolves in acid, and determines the principal chemical constituents; some he shakes in water and analyses what dissolves; some he suspends and allows to settle, so that rough estimates can be made of the proportions of such things as bricks and paving stones. And so on, and so forth. But how much does he learn of what was going on in that city before the spade entered?

10.2 The soil atmosphere

This analogy is not as far-fetched as it may seem at first sight. A most important part of the structure of most natural soils is the system of air spaces, which are of all sorts and sizes, from mole runs and mouse holes, worm tunnels and channels left by decaying roots, to cracks produced by drying, gaps between crumbs of soil, and finally down to the fine capillaries between some of the mineral particles. Into this labyrinth diffuses oxygen from the outside air, out of it diffuses carbon dioxide. But the soil is unlikely to be for long in a steady state—either it is gaining water from rain [9.2], or losing it by drainage and uptake by plants [6.6], so that the volume of the air spaces is continually increasing and decreasing, within limits set by the soil, the climate and the vegetation. All this maintains within the soil a range of concentrations of oxygen and carbon dioxide in the soil atmosphere, which favours certain microorganisms and inhibits others, and the activities of these microorganisms together with the larger soil-living forms and the plant roots in their turn add up to the respiration of the soil as a whole. But the moment the soil is interfered with in any way, these elaborate interrelationships are disturbed, the composition of the soil atmosphere changes and the balance of the various types of microorganism begins to change. With these changes go changes in the whole organic complex, living and dead. The soil, in fact, becomes something other than what it was when the interference began. It may then be a more or less suitable place for the growth of a particular species of plant, but it is no longer identical with the soil from which it was derived.

It has long been known that such changes in the soil, brought about by natural or artificial means, may have profound effects on the distribution of certain plants (e.g. *Euphorbia lathyrus* in Wiltshire, Babington, 1843); and soils which in nature are compact and ill-aerated, such as fen peats, show these effects to a marked degree (e.g. Haslam, 1960; Clymo, 1964). Often, however, under natural conditions it is difficult to disentangle the effects of soil changes from simultaneous changes in other aspects of the environment. For example, consequences of soil disturbances may be among the factors determining the changes in species composition during the colonization of gaps in tropical rain forest, but here wind damage or the fall of over-mature trees bring many other consequences in their train. Coppicing may produce similar effects.

10.3 Agricultural and horticultural soils

Such problems of interference, although present, are much less serious for agricultural and horticultural soils, where human interference is part of the normal biological pattern [6.3]. If the plants are growing

10.3

in the field there may then be no problems: but if our experimental design involves, in addition to normal field conditions, such treatments as small-scale cultivations, or growing plants in pots or in the pans of soil balances, the problem is to ensure that the small-scale operation produces the same effects of aeration, compaction, and so on as the normal large-scale cultivation methods used in the area. More precisely, the state of the soil in the small-scale operation should lie within the limits of the soil states produced by the large-scale cultivations. These may differ considerably, not only from one part of a field or experimental garden to another, but also over short distances, because of compaction due to tractor wheels, and so on. The difficulty is to prove that the soil of the small-scale experiment lies within the desired limits. If this cannot be done it is wise to make surveys of easily measurable plant attributes at intervals during the growing season, to see whether differences in the plants are discernible which might be due to unintended differences in the soil. The measurements might include dates of germination, numbers of seedlings per unit area or length of row, rates of establishment, rate of growth in height, leaf colour, and so on. Such measurements may detect a discrepancy at an early stage, or else possible growth differences due to differences in soil may be inextricably involved in the overall experimental results. Probably the best insurance against unjustifiable conclusions is to adopt a thoroughly sceptical attitude to the possibility of producing two identical soils after different treatments, even when dealing with semi-artificial systems such as agricultural and horticultural soils.

10.4 Energy supply

As with most natural biological systems, the energy supply maintaining the population of plants, animals and microorganisms in the soil is ultimately derived almost wholly from photosynthesis by the plants which are or have been associated with it, so that the soil and its vegetation form one great complex. Chemical energy, mainly in the form of carbon compounds, is fed into the soil by active translocation down the stems of plants into the roots, by the fall on to the soil surface of leaves, reproductive structures, and plant and animal debris generally, and by animal and fungal activity and water movement which carry this material into the deeper layers of the soil. In a mature, well aerated soil forming part of a mature ecosystem we have in the long term a steady state in which gains and losses balance, the great bulk of the accessions of carbon compounds being ultimately oxidized to carbon dioxide and given off in soil respiration [31.9]. But this long-term steady state may contain fluctuations—in time, in the form of seasonal changes—or in space, corresponding to spatial differences in cycles of vegetational change, such as gaps in forest caused by the fall of over-mature trees [10.2, 31.18].

Soil respiration provides a possible means of integrating the total activity of the whole soil ecosystem at any one time, but the interpretation of data is hedged about with many difficulties, particularly connected with changes in the amount of carbon dioxide stored in the soil, and the non-attainment of a steady state under natural conditions [31.9]. When considering the measurement of the respiration of whole plants and of separate organs we shall see that these difficulties can bedevil the interpretation of measured rates of carbon dioxide evolution from much simpler systems than soil [12.5, 12.15]. In consequence, little progress has yet been made in relating soil respiration to the activities of the various types of organism making up the ecosystem of natural soils. In fact, soil problems generally are very numerous and very fascinating, very important and very elusive, and here we can only examine a few of the more important series of problems connected with the soil which particularly affect plant growth (see also 31.7–31.9, and the classified bibliography).

10.5 Plant competition

In most natural plant communities one of the most important influences on the growth of individual plants is competitition—either between plants of different species, or between individuals of the same species. Much less is known about the mechanism of plant competition than about the corresponding phenomena of natural selection among animals. One important reason for this is that plant competition is likely to involve at least an element of root competition, plant roots growing in natural soil being very difficult subjects of investigation. Even if they were easier to deal with there would still persist the difficulty of disentangling effects of competition above and below the soil, which many studies of competition have failed to do.

When plants are brought from their natural habitats into a garden, and grown in the absence of competition in a fertile soil, it is a matter of

10.5

common observation that they often make larger plants than they would do in nature. One thinks of seedlings of the beech (*Fagus sylvatica*) on the Chiltern escarpment, of *Carex humilis* on shallow, dry, calcareous soils, of the bluebell (*Endymion nonscriptus*) in woods, of *Viola rupestris* on the sugar limestone of Upper Teesdale, of *Arabis turrita* half-way up a decaying, shaded wall in Cambridge. In all these cases the plants are growing under less than optimal conditions; they would all grow more rapidly, attain a greater size and produce more seed in a deep, fertile garden soil. Yet the details of all these cases are different: perhaps *A. turrita* is the most obvious—it is confined to a situation inaccessible to the rest of the flora of Cambridgeshire, and thus avoids competition. Slow and precarious as is the growth of a young beech seedling on the steep chalk slopes of the Chilterns, field observation shows the growth of oak and ash to be even worse, so that in the end beech is the only dominant in conditions where its growth is far from optimal. Other cases require more quantitative investigation for their elucidation, as in the classic case of the bluebell (Blackman and Rutter, 1946, 1947, 1948, 1949, 1950). Yet others are still obscure—why should the *locus classicus* for *V. rupestris* have been confined for more than a century to certain areas of a sugar limestone outcrop, while apparently identical contiguous areas with the same exposure are without the plant?

Confronted with a situation of such complexity, involving so many possible mechanisms, some authors would prefer a different terminology from the well-established word 'competition' (for a discussion see, e.g., Harper, 1961). This attempt seems to be premature. 'Competition' in its common usages is a broad enough term, so long as we do not seek to narrow its meaning. More cases are needed where competitive situations have been fully worked out in quantitative terms, providing knowledge of the detailed operation of the mechanisms involved. To have relevance to the natural situations these investigations will often have to be made under carefully selected ranges of conditions well away from the optimum conditions for the growth of the species concerned.

10.6 An experimental design

Fig. 10.6 shows one ingenious means of separating below-ground from above-ground competition. A large number of pans were prepared with a uniform soil, and plants from a clone of *Agropyron repens* were established

10.6

in a ring round the outside, the centre being occupied by a soil-filled pot. When the *Agropyron* was well established, a set of pans could be selected showing uniform growth of the grass. These were then divided at random into two groups. In one, the central pot was withdrawn, the hole filled with soil, and one of a batch of uniform seedlings of *Impatiens parviflora*, knocked out of a thumb pot, was put in centrally. A situation

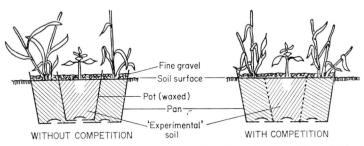

Fine gravel
Soil surface
Pot (waxed)
Pan
'Experimental'
WITHOUT COMPETITION   soil   WITH COMPETITION

Fig. 10.6. Arrangement of pots for growing *Impatiens parviflora* with and without root competition with *Agropyron repens*. When the experiment began the outer hatched part of the soil contained the root systems of established plants of *A. repens*, and the central clear part the root system of the seedling of *I. parviflora*, just knocked out of a thumb pot. From Welbank (1961).

was then created with two undisturbed root systems, separated from each other by an unoccupied volume of soil, within which competition must inevitably develop. The mineral nutrient content of the unoccupied soil could be varied at will. As a control, in the other group, the central pot was withdrawn and replaced by a waxed pot, forming an impermeable barrier to roots, or to the passage of water or dissolved substances. This was filled with the same soil, carrying in the middle another of the same batch of uniform seedlings. Thus by the use of suitable combinations of treatments it was possible to investigate the competition of the root systems of the two species, while having above-ground conditions as closely comparable as possible.

10.7 Root competition

At the crudest level, root competition may take broadly three forms— either the competitor withdraws from the soil (i) water and/or (ii) mineral nutrients, which are then not available to the plant with which we are concerned, or alternatively, (iii) it may produce some toxic substance, which, when released into the soil, has an adverse effect on the other plant. To distinguish this latter effect from the other two can be a matter of great experimental difficulty if experiments are confined to competition within a soil, because competition for water and mineral nutrients is bound to be going on at the same time. None of these effects are necessarily linear, nor are they strictly additive. In this complex situation it may be possible to prove or disprove the existence of the further factor of toxic exudation, or it may be impossible. Welbank (1961, p. 133) discusses this matter in relation to the criteria which may be applied.

10.8 The soil solution

It can also be difficult in practice to settle the next point about the effects of soil conditions on plant growth, which is the question of determining the actual concentrations of ions and molecules in the soil solution immediately in contact with the growing roots. The soil will normally be below field capacity (effectively the maximum stable water content of a free-draining soil), so that samples of the actual soil solution are not easy to extract; and what is more, the soil solution itself is by no means homogeneous. Some of it (in some soils a considerable amount) is held in more or less close association with the soil minerals and the organic soil colloids, and this part can be difficult if not impossible to extract unchanged. Yet this part can be most important for the plant, whose root hairs grow in close association with the film of water surrounding the soil particles, so that the situation is further complicated by partition of solutes between adsorption on the soil particles, the local soil solution itself, and the Donnan phase of the outermost layers of the walls of the root hairs [5.6]. To this situation must be added mass flow into the root hair, under the tension gradient generated by the transpiring plant.

Little progress has been made hitherto in characterizing such a situation for a natural soil. Yet these studies cannot be abandoned, because the plant must be studied in relation to its environment, of which the soil is always an important, and may be a decisive part.

Fortunately many problems of mineral nutrition, of the effects of hydrogen ion concentration and of toxic ions, do not depend for their solution on a knowledge of the exact conditions immediately surrounding plant roots. A convenient entry to the literature concerning ecological problems is provided by the British Ecological Society's Symposium

10.8

Volume *Ecological Aspects of the Mineral Nutrition of Plants* (1969; and also see the classified bibliography).

10.9 Artificial rooting media

The use of artificial rooting media provides another possible line of attack, both on problems connected with the effects on growth of specific changes in the root environment, and on problems concerned with distinguishing between the effects on the plant of particular conditions of the aerial and root environments. An example of this latter approach is given by Coombe's observations on the growth of *Impatiens parviflora* in Madingley Wood [7.8–7.13]. Here the approach is the opposite of the Welbank experiments—by using a wholly artificial soil (vermiculite and culture solution), the roots are amply provided with everything needed for growth: and we then sink the pots containing these roots systems in the soil under various different canopies in the wood, and compare their effects. We can thus distinguish the overwhelming importance of the unfolding of the deciduous canopy during May and the early part of June. It would now be possible to go a stage further and compare the growth of such plants with those rooted in the natural woodland soil; but there is a large gap to bridge, involving simultaneous changes in the availability of water and mineral nutrients, in soil structure, and in root competition. It would certainly be necessary to plan intermediate experiments to bridge this gap [30.13, 30.14]; and in doing so to consider the climatic idiosyncrasies of individual years which, as we saw when considering the growth of young sunflowers in a sequence of years [7.5], can have very profound influences in themselves (e.g. Fig. 26.4.2).

The other type of indirect approach involves the use of artificial rooting media of one kind and another, specially contrived to ensure particular conditions of supply or deficiency of water, or mineral nutrients, or both [28.18]. Such experiments can take a very great variety of forms, and Hewitt (1966) and Slatyer (1967) give entry into the extensive literature of this subject.

10.10 And other simplifications

At the very end of this road lies the total elimination of a solid soil, the plant being ensured support by some other means, and the roots being supplied with water and mineral nutrients together in the form of a culture

10.10

solution. In the simplest form the culture solution may be supplied in bulk. Different species vary very widely in their reactions to this treatment, as the practice of hydroponics shows. Some plants will thrive, indeed in the Tropics some floating aquatics may thrive so well as to be a serious nuisance in artificial waters, e.g., *Salvinia natans* in tanks in Ceylon, and *Eichhornia crassipes* and *Pistia stratiotes* in rivers, canals, and lakes in Africa. In temperate regions *Lemna minor* has proved useful as an experimental plant: but these are all examples of plants which in their natural habitat float on water. At the other end of the scale, other plants cannot be grown at all in this way, usually because of problems of aeration. In any case in bulk culture the problem of the concentration of particular ions still remains. Because of damage to the root system there is a limit to the vigour of stirring, so that as the root surface is approached, laminar flow becomes slower and slower, in effect creating a volume round the root through which ions must move by diffusion. Even if all the other properties of the system are known (itself a large 'if') uncertainty about the length of this diffusion path creates uncertainty about the actual concentration of the particular ion immediately outside the absorbing cell. Again, plants vary widely in the rate of water movement which the root system can withstand without damage; e.g., *Ricinus communis* will make excellent growth in a culture solution subjected to violent agitation by a powerful aerating system, and is thus another useful plant for physiological experimentation, but many plants will not make normal growth under these conditions.

These difficulties of providing a known concentration of ions immediately outside the absorbing cells of the root, together with adequate aeration (both stemming from the same diffusion problem) can sometimes be largely overcome by using some arrangement whereby the solution flows over the root system. Then, however, the root system must be supported. The supporting medium, however chemically inert, must be either polar or non-polar. If it is non-polar, and therefore hydrophobic, it will not adsorb ions, but at the same time it is necessarily non-wettable. This may make very difficult the uniform supply of culture solution to the root system which is the only wettable part of the whole. If the supporting medium is polar, and wettable, uniform supply of culture solution is easy to arrange, but the medium will inevitably adsorb either anions, or cations, or both. This once again complicates the question of nutrient concentration. The matter has been explored in detail by Lewis (1963) and Hegarty (1968). However, in the special case of the apple tree, where the main root system is woody (and can therefore be made self-supporting), the absorbing rootlets being short, deciduous laterals, all these problems have been solved at one blow at East Malling by enclosing the root system in a damp chamber and spraying it with culture solution (Roach *et al.*, 1957; and for the history of this development, with several Figures, see Hewitt, 1966).

Provided that these various complications are borne in mind, much useful work can be done using culture solution techniques for elucidating problems posed both by major nutrients and by micronutrients. The standard text is Hewitt (1966), *Sand and Water Culture Methods used in the Study of Plant Nutrition.*

With this very short and partial survey of some major problems of importance to our own particular studies we must leave the fascinating subject of the root environment, referring readers to the works listed in the classified bibliography for more details, and for an entry into the extensive literature.

10.10

11

HARVEST

We come now to the harvest. How ought the plants to be handled in order to give the best possibility of understandable answers to the questions we are asking?

11.1 Time of harvest

Firstly, we are studying the growth of the plant: that is to say, the changes which have taken place in the plant over a period of time. How is the period of time determined? It is unusual to be able to make useful studies of plant growth over periods of less than a day, and usually several days—seven, ten, fourteen—have to elapse between one harvest and the next [4.4]. But the growth of the plant does not proceed in a steady progression with time. In any natural environment there is a daily rhythm—photosynthesis waxes and wanes, the products of photosynthesis are translocated through the plant, the meristems divide and their products mature, at varying rates throughout the daily period of growth. It is important, therefore, that our period between harvests should consist of an integral number of days, as exactly as is practicable. But what is the moment from which the harvest is to be counted? The moment when the plant is removed from its natural environment? The moment when all enzyme activity ceases, in consequence of heat or drying of the tissues? These are difficult questions, and we do not need to answer them provided that we can ensure that the harvesting procedure is always spread in the same way over the same period of time. It is then immaterial what precise instant we count from, because on each occasion the interval between a particular phase of harvest and the same phase of the next harvest is an exact number of days. This requires a certain amount of forethought. If we have, say, four consecutive harvests, separated by three periods in each of which the plants double in size, then at the last harvest the plants will be eight times as large as those at the first harvest and the harvest itself can be expected to take a good deal longer.

11.2 Measurements of the living plant

We have already considered [4.2] the various possible measurements. Some of these can be made without disturbing the plants—numbers of leaves, number and position of side branches, flower buds, flowers, etc. These measurements and records should obviously all be cleared out of the way before the plants are touched. They are thus excluded from the harvest period [11.1].

11.3 Division of the plant

Next, the plant must be divided into categories—the main photosynthetic system, the leaves; the main architectural supporting system, the stem; the roots; the reproductive structures. This poses a series of choices, broadly on the theme of whether to follow the functional pattern of the plant, or to follow strictly morphological criteria? Is the petiole part of the stem, or of the leaf? Is the hypocotyl part of the stem, or of the root?

11.4 (i) Leaves

It is here suggested that we ought to be guided by functional criteria, rather than using strict morphological ones, which are equally products of the human mind. Thus in a leaf like that of the sunflower, the lamina contains the great bulk of the photosynthetic tissue; the petiole is like the stem in having an important function of support, in having relatively little photosynthetic tissue, and relatively much strengthening tissue, in relation to its total weight. We therefore divide the leaf lamina at the top of the petiole, and count the petioles as part of the stem, although in some plants with partially decurrent laminae a judgment is necessary as to where the division should be made. Similarly, we only begin to count a leaf as such when it has begun to photosynthesize. To apply this criterion one needs some knowledge of the stage of leaf development at which photosynthesis begins, which varies from species to species. Young leaves, which have not yet begun photosynthesis, are counted as part of the stem.

We may wish to obtain the fresh weight of the various organs. In this case the leaves should all be severed as expeditiously as possible, and at

11.4

once be put into a tin or weighing bottle [11.17], which is closed up and weighed. The leaves will not change appreciably in area while this is being done. They would be much more likely to change in fresh weight while the area was being determined.

11.5 (ii) Stems

Similar considerations apply to the harvesting of the stems. With the leaves once out of the way, we may wish to measure the length of the various internodes. We shall also remove all buds, flowers, etc. and deal with them separately—they are put in a separate closed container ready for the fresh weight to be determined. Now we have to decide where the stem ends and the root begins. Here a little flexibility is necessary. If we applied our functional criterion rigidly, the stem would end and the root begin at ground level. But the lowest part of the stem and the top of the root, in those plants which pass through the soil surface as one single structure, is usually one of the most massive parts of the plant. Accidents of seed depth at germination may place the first aerial leaves (the cotyledons, in the case of epigeal germination) at varying heights above the ground for plants otherwise identical. If we use the soil surface to decide where the stem ended and the roots began for such plants, we should transfer a short but heavy piece of tissue from stem to root or vice versa, introducing an undesirable element of arbitrary variability. Thus, if there is some morphological guide, it is well to choose it, and adapting our general functional criteria for this purpose, to count everything below the cotyledonary node (for epigeal cotyledons) or the node of the first leaf (for hypogeal) as part of the root. So long as the cotyledons or first leaves, or their scars, remain, this criterion is an easy one to apply. There are also favourable cases, of which *Helianthus annuus* is one, where the distinction is a permanent one, the stem of this species being rough and hairy while the hypocotyl is glabrous. The division can then be made neatly long after all traces of the cotyledons and their scars have disappeared. In some cases there is a similar permanent criterion, but of colour rather than surface roughness. Even on our functional basis this criterion is less arbitrary than it appears at first sight. The assimilatory foliage is all above the cotyledonary node; and from there down to the first branch of the root system we are dealing with a region whose principal functions are two; (i) conduction up and down between the root sys-

11.5

tem below and the foliage region above: and (ii) mechanical connection between these two. There is therefore no obvious reason why this region should be divided arbitrarily at ground level, while there are good reasons for dividing it at the cotyledonary node (or at that of the first leaf) and including the whole of this region with the root. Secondly, if in the course of development this region develops appendages, as sometimes happens, they are much more likely to be roots than anything else, which makes the distinction even more convenient.

Many plants do not have such a convenient distinction as the annual sunflower. Some, including many monocotyledons, produce very many adventitious roots from the base of the stem: as the plant become older, these may emerge higher and higher, producing 'prop', or 'stilt' roots, which are also characteristic of a number of dicotyledonous plants of the wet tropics, particularly trees growing in swamps. In all these cases it seems best to shave the root off as nearly level with the surface of the stem as is practicable.

Having divided the stem from the root, we brush off any adhering earth, cut the stem into convenient sized pieces, place them in a tin, and determine the fresh weight. At once, if there is plenty of labour available, we prepare them for drying. The object here must be to heat and dry the material as rapidly as possible, so that there are the minimum losses through respiration and the minimum breakdown by autolysis. The respiration rate of much plant tissue roughly doubles for every 10°C rise in temperature up to say 40–50°C, so that the slow heating of a solid mass of tissue may cause appreciable respiration losses. Both rapid heating up to a lethal temperature, and subsequent drying, are facilitated by splitting down thick pieces of stem, and arranging all the pieces loosely, so that there is ample air circulation. The pieces are put in a perforated metal or wire mesh cage, for all but the smallest stem samples, which can be dried in their tin or weighing bottle. Before drying, petioles should be severed from the stems—otherwise they are liable to act like certain dehiscence mechanisms. They curl up and sometimes distribute the contents of the various cages over the bottom of the oven.

11.6 First check

Harvesting any considerable number of plants is a great labour for a single experimenter; there is also the question of respiratory losses during harvest, which we have noticed, and which are likely to be greater than in the intact plant because of wounding and handling [12.4]. It is therefore advantageous to have a number of helpers, thus cutting down the time taken: but as soon as a task passes from one individual to several, muddle is liable to creep in, and avoidable sources of error multiply. If there are several persons engaged in the harvest, steps must be taken to keep the number of accidental errors to a minimum, and the first of these comes at this stage. Before starting to dry the stems we plot leaf fresh weight against stem fresh weight: things can be so organized that this plot is kept up as the weighings go on. There is usually no reason to expect the plants to be in any particular, standard, state as far as water content goes [4.3], but if they have been growing side by side in the same conditions, they can be expected to be all in the same state, so that the suggested comparison can usefully be made.

The results may be as in Fig. 11.6, which shows such a plot for stems and leaves of plants of *Helianthus debilis* of two age groups, grown in the open and harvested simultaneously 35 and 49 days after sowing, on 26 July 1967 [7.6]. Further details of these experiments will be considered in 11.21. These plants having been graded into size classes, it will be seen that the points fall into groups, within each of which, as expected, there is a scatter. The scattered groups themselves also fall into a broad relationship with each other, indicated by the curved line, which has been drawn in by eye, simply as a convenience in detecting points which appear to lie outside the general pattern of scatter. The implication in Fig. 11.6 is that the points for both age groups all form part of one single relationship. This need not necessarily be so, as we shall see later (Chapter 26), and often each age group must be considered separately.

If a point appears not to belong to the general pattern of scatter, the corresponding material can be examined before it is too late to identify and rectify the sources of error, if any. Experience shows that every possible error will crop up sooner or later—pieces dropped out of tins, or fallen into other tins, tin numbers switched, whole stems or leaves put in the wrong tin, and so on quite apart from misweighings and all sorts of arithmetic errors (see, e.g., the legend to Fig. 23.5.2). Usually if one has the fresh material to hand, a quick inquest will decide, firstly, whether there was just a misweighing or some other simple error not affecting the material itself; or if not, whether the material itself is all present and correct. If so, then the departure from expectation can be accepted

11.6

without the nagging suspicion that the apparent difference may not be real.

In the particular example shown in Fig. 11.6, there were three groups of three students, taking part in a Long Vacation practical course at the end of their second year of study, none having had previous

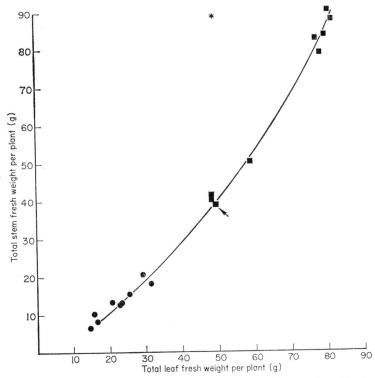

Fig. 11.6 Total stem fresh weight plotted against total leaf fresh weight for individual plants of *Helianthus debilis* of two age groups 35 (●) and 49 (■) days from sowing, grown in the open at the Botanic Garden, Cambridge, and harvested simultaneously on 26 July 1967. For * and ↑, see text.

experience of this particular type of experimentation. Each group of students harvested one plant from each of the graded size classes of plants, so that the general scatter includes throughout any element due to differences in technique between one group and another. Fig. 11.6 shows that with care this element need not be large. Indeed, this particular set of data is remarkably uniform, and even with the most careful manipu-

11.6

lation by trained workers much more scatter will often be found, following inevitably from the complex of causes already considered [3.6, 7.1].

Even with this high standard of experimentation, when the data were first plotted the point indicated by an arrow in Fig. 11.6 appeared at the point shown by the asterisk—there had been a misweighing by exactly 50 g; and a second displacement, too large to be shown on the figure, resulted from a gross weight of 229·1 g having been written down as 291·2 g. Such large errors are particularly easy to detect and rectify, but the procedure we have considered also serves for the detection of much smaller accidental errors, which might otherwise pass unnoticed until the opportunity of rectifying them had passed.

11.7   Leaf area

As soon as the leaf fresh weights have been obtained, the next essential datum is leaf area. Obtaining this is one of the most essential processes, and at the same time may be one of the most tedious. It is accordingly not surprising that very many attempts have been made to produce some quick, simple, and reliable means of determining leaf area during harvest. Many of these are elaborate, and some are specialized for particular tasks, or particular plants; other have a more general applicability, and we shall notice a few of the simplest and most generally useful later [11.8–11.12].

It should always be remembered, when determining leaf area, that the leaves are losing water and that they may be shrinking during the whole period that they are exposed to the atmosphere between cutting and the completion of the area determining process. There is also the question of respiratory losses [12.4]. The process should therefore be as quick as possible, and until the area has been recorded, leaves should never be left lying about—during any intervals they should be kept together in a tin or polythene bag or box to minimize water loss. The extent of the shrinkage varies a great deal from one species to another, being a function both of the extensibility of the leaf cell walls, and of the leaf architecture, especially as concerns mechanical tissues.

In those cases where a record of the leaf is to be preserved, it is often worth preserving also the regular order of the leaves up the stem, so that a complete record of the leafage is available later if wanted. Comparatively little extra organization at the time of harvesting the leaves is needed to achieve this, and it may be valuable to have the record for subsequent examination, perhaps to settle some problem which had not been thought of when the experiment was planned.

Sections 11.8–11.11 deal mainly with simpler cases of flat leaves of relatively simple outline, which are fortunately also common; but for many leaves the determination of the area is more difficult—the leaf area may be much divided, as in many umbellifers, or multipinnate as in the Mimosaceae; the leaf surface may be much wrinkled, or if smooth may be curved in three dimensions or rolled, as in some grasses; or the leaf itself may be cylindrical, needle-like, or some even more awkward shape. With careful adjustment the method described in Section 11.11 can be used for much divided leaves, provided that the lobes do not over-lap when the leaf is flattened; if they do the leaf must be cut up. Curved leaves may also have to be cut into pieces, and occasionally it is possible to flatten wrinkled leaves very considerably by cutting along the main veins. But there are no methods that will deal with all the awkward cases, and if it is necessary to work with plants whose leaf areas are difficult to determine it may be necessary to make a pilot investigation to relate leaf area to some form of measurement, or to determine ways of selecting a small representative sample of leaves whose area can be measured and then related to that of the plant as a whole. Here the method of relating area to leaf water content [10.8, 24.17–24.19] may prove useful. Should it prove necessary to cut the leaves or handle them roughly in order to flatten them, they should be killed and dried as soon as possible there-after, to minimize respiratory loss [12.4].

11.8   The simplest methods

If the leaf shape is reasonably regular, the leaves can be outlined on squared paper, and the squares counted. If the margins are toothed, the outline is best drawn to pass through the mid-line of the toothing: this usually gives rise to less error than the attempt to outline the whole leaf with great accuracy. Similarly, for counting the squares, unless the leaves are very small it is unnecessary to count all the squares—it suffices to aim at an accuracy of about 1 per cent, so that various short cuts can be used. For example, for leaves of more than about 100 cm^2 area, and a more or less regular rounded shape, it suffices to count all the complete centimetre squares within the leaf outline, and then to check over those centimetre squares which are cut through by the leaf outline, counting all those where the leaf occupies more than half the square as 1, and all those of less than half the square, as 0.

If the leaves are large, or if there are very many of them, the determination of the areas by square counting may be altogether too tedious. The leaves can then be outlined on sheets of white paper of a standard size, the number of sheets used being counted. Later the leaf shapes are cut out, and both they and the discarded portions of paper outside the outlines are weighed. The total area of paper involved being known, the leaf area is obtained by proportion. This method (which was used by Kreusler *et al.*, 1877a, b) is particularly useful in places where high quality uniform paper is difficult to obtain; otherwise it supplies a useful check on possible losses, and on changes in the water content of the paper.

If the leaves are small, or not numerous, they can conveniently be printed on cheap document-copying paper, which may be quicker, and for small and thin leaves more accurate, than outlining. The leaves are kept flat by glass plates, and after developing the leaf shapes are either cut out, as above, or planimetered [11.9].

In certain cases it may be possible to estimate leaf area from absolute leaf water content (Hughes, Cockshull and Heath, 1970, and 24.17–24.19) with sufficient accuracy. Preliminary investigations are necessary in order to establish whether a simple relationship holds in any particular case. The method is likely to be particularly useful when dealing with very large numbers of leaves, or with leaves of difficult shapes.

11.9   The planimeter

When dealing with the outlines of leaves of moderate size it is quicker to use a planimeter; there is the added advantage that the record of leaves is preserved intact. For following leaf outlines, the pattern having a lens marked with a small central circle will be found more convenient than the more familiar pattern with a metal point as index.

Important points to bear in mind are to use a smooth, uniform, matt paper, such as a good quality drawing paper, free from creases or folds; to arrange the leaf outlines or photographs so that the wheel never runs over an edge of paper; to make two (and if necessary more) determinations of each area until there is agreement within some predetermined limit of error; and when using an instrument with variable scale, always to check the scale setting before beginning. The accuracy of the instrument is greatest when the arms are most nearly at right angles; therefore, when laying out a sequence of leaves whose areas are to be determined in

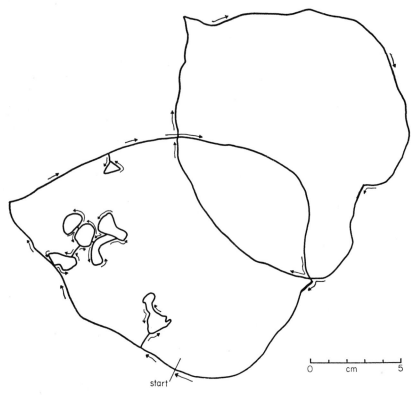

Fig. 11.9. Outlines of two leaves of *Helianthus annuus*, overlapping for convenience of planimetering. Arrows show the direction taken by the planimeter index, beginning and ending at the arbitrary point marked 'start'. The first leaf happening to have a number of holes, straight lines are drawn joining these to each other and to the margin of the leaf, and the index follows the course indicated by the arrows.

one operation, they should be arranged in an arc of a circle round the fixed pin of the instrument, rather than in a straight line.

    If several leaves have to be measured at once, it saves time to overlap the outlines, as shown in Fig. 11.9, where arrows indicate the route followed by the index. Holes in leaves can be allowed for by joining up the hole to the leaf margin by a single line, following it from the margin

11.9

Plate 11.10. A photograph of standard areas of black paper and leaves of *Helianthus annuus* arranged for planimetry, the leaves joined by lines to form a single area.

to the hole, going round this backwards and returning down the same line, as shown in the same figure.

11.10   Photography

In some plants these methods may be awkward because the leaves are very large—a mature plant of *Helianthus annuus*, for example, will have a leaf area of many thousands of square centimetres, and some leaves too large to measure with the single span of an ordinary planimeter. Here a convenient method is to photograph the leaves, together with standard squares of black paper, as shown in Plate 11.10. They are laid out on a white background, and kept flat with glass plates. A whole plate (17 cm × 22 cm) print is then made from the photograph, the leaves are joined by lines into a single area and planimetered, and the standard squares are measured separately and the areas added up. The squares can be measured more accurately with a travelling microscope or vernier callipers than with the planimeter, but by doing both the planimeter calibration can be checked. The fact that the standard squares have been distributed regularly in the four corners makes an automatic allowance for any errors of magnification due to faulty alignment either in the initial photograph or in the enlarger. The photographs can be taken with any camera which will focus down to about 1·5 m (or an even greater distance, if the leaves are very large and numerous), and as this method involves the minimum of special apparatus, and produces a record of a reasonable size it is a useful one for dealing with very large leaf areas.

11.11   Automatic integration

If many determinations have to be made, or if the leaves are small and very numerous, or are finely divided (e.g., the work of Schwabe, 1951, on the bracken, *Pteridium aquilinum*), it is worth having some device which avoids the necessity of planimetering the areas. This is not wholly a matter of saving time. The errors of the planimeter are associated with the length of margin which must be traversed. Measuring a single, large, entire leaf like a sunflower, where the ratio of margin to area is minimal, the area can with practice be measured with an accuracy of rather better than 1 per cent. But as the length of margin for a given area increases, so do the errors mount up.

On balance it seems that the most convenient device, combining the generation of a permanent record with ease of operation and the minimum of special equipment and consequent expense, is that described by Hughes (1964a, following Schwabe, 1951, where the apparatus is figured). This involves a glass table, fitted below with fluorescent lamps and a diffusing screen to give uniform illumination. The camera, a standard 35 mm loaded with document film, is held in a fixed clamp above, giving a fixed reduction. The leaves are arranged on the table, held down by glass plates, and photographed at a fixed exposure together with a small identifying label. At intervals in the sequence photographs of standard areas are inserted which are roughly equal in total area to the batches of leaves being photographed, and similarly distributed over the glass table. These areas are all the same size, adjustment being made by counting out a suitable number. At the end of the sequence is a blank exposure. The film strip is processed so as to give black and white constrast. It is then inserted in a standard 35-mm enlarger, with the lens removed and replaced by a photoelectric cell, suitably mounted to cut out stray light. The current output of the photocell with the blank exposure in position measures the background transmission of the fully exposed film, plus any stray light. This reading should be near zero, but an allowance can be made if it is not. The strip is then run through, the current output of the cell being recorded for each frame: if no constant voltage device is available, it is advisable to avoid times of day when changes of mains voltage can be expected. If this is not possible, a reading of standard areas is inserted between each one of leaf areas, and in this way the effects of voltage changes can be detected and allowed for.

The use of standard squares of roughly the same area as the leaves and similarly disposed on the glass table, provides an automatic compensation mechanism for dealing with the projectional errors inherent in a method of this kind. At the same time it provides an automatic calibration for the lamp—photocell system, and one which is easy to apply as often as needed.

A photographic method has the great merit of preserving a permanent record. This may be needed for two purposes—either to repeat the area estimation later, if subsequent work casts doubt on the original determination; or to make any further measurements on the leaves, which may form part of the experimental programme, or which may be suggested by subsequent reflection. The impossibility of doing either of these is a

11.11

grave defect of measuring machines which do not preserve a permanent record, and they are to be preferred only if the special circumstances of the investigation make them necessary.

11.12 Large quantities of leaves

Further problems arise if there are very many plants, or if one is harvesting a very large plant, such as a tree. It may then be necessary to build special equipment, but the total leaf area can often be estimated with sufficient accuracy by taking samples from the whole mass of leaves, measuring them and determining their dry weight, while the remainder of the mass of leaves is simply dried and weighed, the area being determined by proportion. In all such cases of large quantities of leaves it is worth making a pilot investigation to see whether leaf area is not more closely related to leaf water content than to dry weight (11.8, and see 24.17–24.19). In either case it is most important that the samples shall be representative. If we use a dry weight basis, what we are doing, in effect, is determining the specific leaf area—the leaf area produced by unit leaf dry weight. This is an aspect of plant form which has a very marked ontogenetic drift, and which is very sensitive to a great many different environmental influences (Chapter 19). It is therefore particularly important that the sets of leaves used as samples should contain the same proportions of young, middle-aged and old leaves, that the leaves should have been similarly disposed on the plant, and been subject to the same range of environmental conditions, as the bulk. If we are faced with a very great many plants, it is therefore worth including whole plants in each sample, so as to have the whole ontogenetic range; and if the plants differ from each other, say in size or exposure to varied environmental conditions, to include representative plants in each sample, in much the same proportions as they occur in the bulk.

Similarly, when harvesting the foliage of a large tree, there should be included in each sample leaves from different exposures, inside and at the margins of the crown, and also leaves of different ages, representative positions in flushes, and so on.

It may be worth measuring and weighing several, say six or eight samples, so as to make an estimate of the variability of a single sample, and hence of the reliability of the estimate of total leaf area.

11.13 Drying of leaves

If the method used for leaf area allows of a quick estimate, which will not be the case with a photographic method, leaf area should be plotted against leaf fresh weight, in just the same way as for stem weight and leaf weight [11.6], before starting to dry the leaf material. Once again, discrepancies may be detected and eliminated at this early stage.

Otherwise, we should proceed to dry the leaves at once, as soon as the leaf area record has been produced, in order to minimize respiratory losses. Again, unless the leaves are very few and small, it is worth carrying out the initial killing and drying in cages. The leaves should be arranged with ample air spaces between them, not so many in a cage that they will pack down into a sodden mass as soon as they become hot enough for the cell membranes to be destroyed and the leaves to lose their turgidity. If so, not only is drying slowed down, but cell sap may exude from the mass, causing a loss in dry matter. This is liable to be a much more serious problem with leaves than with stems, where there is normally much more mechanical tissue. In really difficult cases it may save time and give a better result in the end, if the leaves are spread out at first into a single layer, until they have become crisp, when they can be arranged several deep in cages for further drying.

11.14 Roots

So far we have been concentrating our attention on dealing with the tops of the plants, leaving the roots undisturbed, apart from the trauma caused by cutting off the top. Unless a great deal of labour is available, this is the wisest plan, in order to get the material of the tops killed and dried as soon as possible. Respiratory losses from the root system will be going on all this time, at a rate which may be substantially higher than in the intact plant because of the effects of cutting [12.4]. However, in estimating growth during a period we shall be working by difference between beginning and end, and provided the two harvest programmes are the same, and the two sets of root systems have been kept for the same short length of time, the errors introduced are likely to be negligible.

The errors will obviously increase with the time which elapses between severance and killing of the root system. This time should be kept to a

11.14

minimum, and the root systems should not be left overnight if it is possible to avoid doing so. If labour is very limited, and it proves impossible to harvest the root systems the same day, the extent of the respiratory losses should be estimated, in order to see whether an allowance for it should be made. If the plants are in pots, it may be worth storing the root systems in their pots in a refrigerator, to reduce the respiration rate.

11.15 Extraction of root systems

Conditions of root system, root structure, and soil vary so widely in nature that no general rules can be laid down as to the best methods of extracting roots, and particular cases must be solved as far as possible by experiment. If initially any of the above conditions is unknown, it is advisable to make a preliminary investigation before the experimentation proper begins, because many unsuspected difficulties may be encountered.

There is usually no particular problem in extracting the root system of plants which have been grown in artificial substrates. They are usually washed out and indeed, ease of extraction was probably one of the factors considered when the substrate was arranged. Similar considerations apply to some semi-artificial soils made up with this end in view, for example if a substantial proportion of sand has been included. Extraction may be less easy in a normal agricultural or horticultural soil, and it may be very difficult, and sometimes impossible, in a natural soil. Stirring in water followed by sedimentation may be found useful, but unless the roots are very tough care has to be taken to avoid excessive fragmentation, and the method is liable to be complicated by the adhesion of some of the fine roots to heavy soil particles. In difficult cases the digging out of the root system may not merely involve a great deal of labour, but there may also be a great deal of ambiguity about what was part of the root system, what belonged to other plants, and what was part of the soil debris.

In such cases a given amount of labour is often best spent on the very careful extraction of a limited number of root systems, and determination of what proportion of the dry weight of the whole root system is in the easily extracted core of the system, the main root and the large branches, and what in the remaining, finer, absorbing system. It not infrequently turns out that the great bulk, perhaps 80–90 per cent of the total, is in the readily extracted main root system, and that by careful extraction of sample root systems, the remainder can be estimated proportionally with an accuracy of, say, 20 per cent. Then if the whole root system is, say, about a third of the whole plant dry weight, and the main core of the system forms 90 per cent of this, the extensive absorbing system, so difficult to extract, forms less than 4 per cent of the total dry weight; and if we can estimate it within 20 per cent by proportion from the main core, our final uncertainty is less than 1 per cent, and not worth a great deal of labour to fine down any further.

Another method which has been applied to this difficult problem is to sample the soil for root systems by putting down borings, extracting soil cores, and washing out the roots (Welbank and Williams, 1968, and references quoted therein). These methods have limited application, being mainly confined to agricultural situations where the number of species is very limited, and to soils free from stones and other obstructions such as tree roots. It would be a fortunate and rather special case to which such a method could be applied in a natural soil: our root system would have to possess some readily recognizable characteristic, such as the tough, bright yellow roots of *Urtica dioica*, a state of affairs which is not very common; and the soil would have to be one which could be bored out. Under these special circumstances, the method can be very useful as giving evidence of spatial distribution of root systems, without disturbing large areas of soil.

The fact that root systems have usually to be washed out, ultimately, in order to remove the fine adhering particles of soil, means that it is often not worth determining their fresh weight; if this should be wanted (see, e.g., Chapter 24) the root system must be evenly blotted and then put in a tin or other container, but it is not easy to obtain consistent results for fresh weight by this means. If the matter is important, the accuracy of the technique adopted can be tested by repeated wetting up, blotting, and weighing of a few root systems.

The procedure for drying roots is the same as that for stems, thicker roots being split down if necessary, and the initial drying being done in cages unless the whole root system is very small. It sometimes happens that in spite of every care, unnoticed particles of soil remain trapped within the root system. When the roots shrink during drying, these usually become detached, and they can be shaken out or picked out when the roots are transferred from their cages to tins or weighing bottles for the final drying. If there is any reason to suspect that soil

11.15

may have been retained in this way, it is very important to remove it at this stage, because a few small soil particles may cause considerable errors on account of their high specific gravity.

11.16 Apparatus: ovens

Whatever measurements are to be made during harvesting, it is then nearly always necessary to kill the plant parts and dry them sufficiently to bring enzyme activity to a halt, thus stabilizing the material. This must be done as quickly as possible. When the material is once stable, it can be further dried at leisure for the determination of dry weights. Sometimes the two processes are combined, but they are in principle quite distinct, and can be widely separated in space and time. Under expedition conditions, when good drying facilities and sensitive balances may be lacking, the writer has prepared material as far as the rough dry stage, enclosed each sample in a labelled bag of greaseproof paper, and packed these in large tins together with a volatile insecticide and bags of dry silica gel. The tins were sealed up and the material would then withstand travel under damp and difficult conditions for many weeks on its way back to the laboratory for the later processes of final drying, weighing, and so on. This procedure involves no complications, apart from the necessity of ensuring that the material remains dry during the interval.

It follows that it is not necessary to use the same apparatus and materials for the two processes, although in a well equipped laboratory it is convenient to do so for small quantities of material, the ideal for both processes being a relatively large volume of well ventilated oven space at around 80°C.

Under expedition conditions there are some things which can not be achieved and are not worth attempting, and the rapid drying and subsequent preservation of large quantities of plant material may be amongst them. Small quantities, however, can usually be dealt with. A convenient small oven can be contrived from a cubical tin (approximately 20 cm side), laid on its side and furnished with two loose, perforated shelves. It has ventilation holes at the top and bottom of the two sides, and a hole in the middle of the top for a thermometer. The push-on lid has a handle bolted to the middle; for carriage this is unbolted and re-fixed on the inside. In use it stands on a small portable cooking stove (bottled gas, if available; but if kerosene, then one with a wick, so that it can be left unattended). When travelling the oven earns its keep as a packing case for small pieces of apparatus.

Under less exacting conditions, but where resources are limited, it is worth concentrating available funds on a good ventilated oven for the final drying, and a good rapid balance for the final weight. At a pinch this oven need not be a large one, because once the plant material has been stabilized by the initial drying, it can if necessary be dealt with in batches. On the other hand, the initial killing and rough drying oven must be large enough to take all the plant material from one harvest at once, and allow ample space for spreading it out in thin layers in cages. A tall metal clothes drying cabinet can readily be modified for this purpose by fitting sufficient metal cased electric heaters at the bottom to raise the temperature to between 80–90°C. A relatively crude thermostat operating on a temperature swing of 1–2°C is quite adequate, provided that its setting is stable. Temperature can then be controlled by 3 or 4 thermometers inserted through holes in a vertical row in one side of the cabinet. Perforated shelves are fitted at intervals of approximately 30 cm. Galvanized expanded metal riblath is suitable for these, as the rib provides sufficient stiffness to avoid any necessity for further strengthening. The most readily dried material, small trays of leaves and so on, are put at the bottom of the oven. If there turns out to be a shortage of space later, they will soon be dry and can be removed, the rest moved down, and fresh, wet, material such as roots put in at the top.

11.17 Cages, trays and tins

There should be available an ample number of cages and trays of various sizes made of wire gauze (for small ones) or expanded metal (for larger ones), with small, new, wired on labels on which the sample code and tin number can be written. Space can then be saved by using appropriate sizes, large enough to give good exposure to the hot air.

The samples will probably first have been harvested into tins for fresh weight and leaf area determinations. Any convenient flat tins with overlapping hinged or push-on lids are suitable, but those with push-in lids are not, because of the difficulty of emptying them. The writer has for years found floor polish tins very satisfactory because when the tin is open, the lid can be put on the bottom, keeping tin and lid together.

The tins should be free of labels, paint or lacquer, and each should bear its own number, punched on; and punched on the lid also if this is detachable. Before the experiment begins the tins are all dried, weighed, and put in a desiccator, the weights being entered in a special book. At the end of the experiment, after the tins are emptied, they are again dried, weighed, and put back in the desiccator ready for the next harvest. The keeping of a special book of tin weights, where successive weighings of the same tin can be seen side by side, provides an automatic check on weighing errors, and also on the cleanliness of the tin. Any exuded sap which has dried on should be cleaned off: any remaining paint, lacquer, or glue will also cause the weight of the empty tin to change with time. So, ultimately, will handling with bare fingers, and the empty tins should be cleaned with an organic solvent from time to time to remove human grease.

At this stage it is not necessary to use a glass laboratory desiccator; any polythene, polyurethane, or polypropylene container with a tight-fitting lid is suitable, modified as a desiccator with a tray of calcium chloride or silica gel at the bottom, covered by a sheet of galvanized expanded metal. A dustbin with a close-fitting lid can be used for a large numbers of tins.

For delicate work, or severe conditions, it is worth having the tins cadmium-plated: this is particularly useful for preventing rusting under expedition conditions. It is less likely to be worthwhile for laboratory studies, where for delicate work one would in any case use weighing bottles. Trouble can be saved by never touching these with bare hands.

11.18 Identification of samples

From harvest through to final dry weight, it is most convenient to identify each individual sample by its tin number or numbers. Before the experiment begins a book is prepared in which the experimental plan is set out, with a line for each individual sample, and columns for tin number, empty tin weight, gross and net fresh weights, leaf area (if applicable), and gross and net dry weights. Before beginning to harvest, say, the leaves of a plant, a tin is selected, and its number written in the appropriate space. This number identifies the sample thereafter. When it is unloaded from the tin into a cage, the tin number is written on the label and the empty tin goes back into the desiccator; after killing and rough drying the sample goes back into the same tin for final drying and weighing. Thus if sap has exuded, or hairs fallen off into the tin, the whole comes back together again for the final weighing. The weight of the empty tin at the end then acts as a check on the whole process.

11.19   Final drying and weighing: balances

Final drying must be done in an open tin or weighing bottle which on removal from the oven is at once closed, because dry plant material is very hygroscopic, and when really dry it must be kept in a closed container Tins and weighing bottles are also hygroscopic, but much less so. It suffices to keep them in a desiccator when they are not in the oven or on the balance.

In other respects the final drying is less straightforward than one might think. 100°C is often quoted as a drying oven temperature. If plant material is maintained for long periods in an oven at this temperature, one of three things happens: (a) its weight falls to a steady value, and remains there; if so, well and good: (b) its weight falls, and then rises again: (c) its weight falls, at first rapidly, and then goes on falling, more slowly. This last is probably due to the slow evaporation of sparingly volatile substances. It can continue for a long time, and may then be followed by a rise, as shown in Fig. 11.19. In either case the slow increase in weight is probably produced by a process analogous to caramelization, accompanied by oxidation. Anyone who has caramelized a tin of sweetened condensed milk by placing the sealed tin in boiling water and maintaining the temperature for a few hours knows how rapidly these changes can take place at 100°C in a suitable system of biological materials. In consequence some early experimenters (e.g., Hackenberg, 1909) dried their samples in an oxygen-free atmosphere. This precaution is rarely taken nowadays, but it may be needed in particularly difficult cases. Experiment shows that both processes (b) and (c) are very much slower at 80°C, so much so that they can usually be neglected. As there is otherwise only a slight and often inappreciable difference in the final dry weight of material dried at 80°C as compared with 100°C, 80°C is generally preferable. Sometimes 70°C may be found to be better, and in cases of doubt the best procedure must be settled by experiment.

After the first rough drying is over, 24 hours in a ventilated oven at 80°C brings most plant material close to its final dry weight. The tins are removed from the oven, closed, put in a dessicator, weighed when cold and put back in the oven for a further 24 hours. If this second weight agrees with the record of the first one, it can be taken as final; if not, a further 24 hours drying can be allowed. This procedure has the advantage that the two weighings act as checks on each other. For fine work,

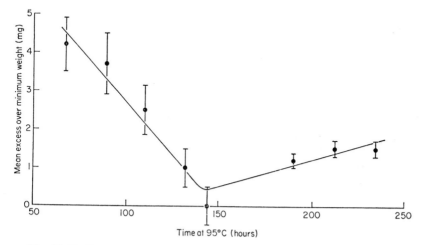

Fig. 11.19. Long-term changes in dry weight of ten samples of the leaves of *Icacina trichantha* Oliv., in relation to time exposed to the air in an oven at 95°C; mean dry weight of samples at 234 hours, 1·0648 g; mean excesses over minimum weight, with 0·05 probability limits. The phase of rapid loss of water was over by 21 hours, when the mean excess over the minimum weight was 11 mg.

supposing that a number of weighing bottles have to be weighed in sequence, it will be found that the weight recorded depends on the position of the particular bottle in the sequence. Desiccators do not dry the contained air instantaneously, and every time one is opened and a bottle removed, some relatively damp laboratory air enters. If one wishes to check that the dry weight of the contained plant material is steady, the same sequence of bottles should be used for both first and second weighings, and also for the dry weights of the bottles themselves. By this means variations in bottle weight can be kept to a minimum, but any particular technique of drying, cooling, handling and weighing will have its variations in weight, the magnitude of which can be checked by using a set of empty bottles, and putting them through the sequence of operations several times. One can thus determine the minimum weight of plant material, below which variations in bottle weight would cause significant uncertainty. Smaller pieces are taken out of their bottles for weighing, as hygroscopic increase of weight would be a less serious source of error. This may in any case be necessary on very sensitive balances. If a correction must be made, a calibration of weight increase against time of exposure to the air can readily be constructed for the particular material and circumstances.

Much time can be saved by using a modern single pan, air damped, semi-automatic balance, and if funds are limited this is one of the most worthwhile items of equipment to obtain. For use with seedlings, and very small pieces of plant material generally, the Cahn electrobalance, reading to less than 1 μg, is a most valuable tool. Considering its sensitivity it is very stable and easy to use, not requiring a balance slab.

11.20   A moral tale

Many years ago the late Mr Udny Yule told the writer how interested he was to find, in making a study of the census returns, that there appeared to be significantly more people in the country aged 60 than were aged 59 or 61; significantly more people aged 70 than were aged 69 or 71, and so on. However, he did not follow appearances, but inferred that there were significant numbers of people who, for whatever reason, rounded their ages off to the nearest 10.

The moral is plain. An elaborate analysis of data containing errors can lead to erroneous conclusions. Two measures are therefore necessary. Before beginning the analysis we must ensure that any accidental errors have been detected and eliminated. We have already taken some steps in this direction in examining fresh weights and leaf areas before drying [11.6]. We must now carry this process a stage further. Secondly, even after this has been done, we must never lose sight of the primary data while building a superstructure upon them, but always be prepared to return to them for clues to explain any anomalies which we observe.

This can be conveniently illustrated by stating the bare bones of a problem which will be examined in detail later [28.3], concerning the growth of plants under low intensities of radiation. At this stage we need not concern ourselves either with the quantities being compared, or with the treatments, but simply say that the question at issue is—is A in Table 11.20 larger than B, or vice versa?

11.20

In two of the six comparisons between the means the probability of the difference being observed by chance is so high that six comparisons would be very likely to include these two. Of the other four differences, two are negative, and two are positive. No more elaborate statistical tests need be applied to show that there is no evidence of any systematic difference between the means values of A and B. There is, however, a

Table 11.20. A comparison of differences between two treatments, A and B. Means of six observations in each case, with fiducial limits.

A	B	Difference	Probability by chance
3.52 ± 0.72	3.68 ± 0.32	+0.16	0.6
2.21 ± 0.13	1.77 ± 0.73	−0.44	0.2
4.07 ± 0.81	3.94 ± 0.44	−0.13	0.7
2.26 ± 0.60	2.45 ± 0.30	+0.19	0.5
4.76 ± 0.25	4.74 ± 0.34	−0.02	0.9
1.46 ± 0.13	1.27 ± 0.28	−0.19	0.1

systematic difference of another kind. In all four instances the standard deviation of the lower value (whether A or B) is at least twice as large as that of the higher one; in three instances factors of 2·0, 2·2, 2·2, and in the fourth 5·6. Examination of the original data shows that in each case this is due to one exceptionally low value. Now these plants grew under low irradiance; they possessed very little mechanical tissue and were easily damaged by handling, although the effects were not necessarily immediately visible. There seems little doubt that this is what happened. The inclusion of the odd damaged plant not only increases the variability (because it belongs to a different kinds of population, and should not be included) but also lowers the mean value.

11.21   Final check: area and dry weight of leaves

When the observations have advanced so far, and when the usual checks on transcriptions and on arithmetic have been made, it is necessary to have a last look for inconsistencies in the basic data before beginning to combine and analyse them in any way. We may take as an example of the procedure the experiment carried out at Cambridge by the third-year botany class in the course of the normal series of field experiments during the long vacation term of 1967, and already mentioned [11.6], as an example of the results achieved by a group without previous experience of this kind of investigation. These observations were made on *Helianthus*

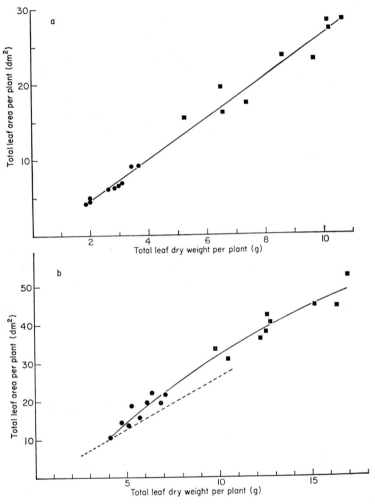

Fig. 11.21. Total leaf area plotted against total leaf dry weight for individual plants of *Helianthus debilis* of two age groups, sown on 7 June (■) and 21 June (●) 1967; (a), at the first harvest on 26 July; (b), at the second harvest on 2 August 1967. The straight line in (a) is represented as a broken line in (b). For details of the experiment, see 11.21; for other related data see Figs. 11.6, 11.22 and 11.23.

11.21

debilis, an annual sunflower showing early branching, whose growth was being compared with the almost unbranched *H. annuus*. The plants were of two ages, having been sown on 7 June and 21 June, so that they were respectively 35 and 49 days old at the time of the first harvest on 26 July. The plants were growing in pots, plunged in open ground at the Botanic Garden. Thirty-six plants of each age group were available, showing considerable variability. There were therefore selected three groups each of seven plants, large, medium and small, the members of each group being as uniform as possible. Three of each group were assigned by lot to be harvested at once, three to be harvested a week later, and the remaining one was kept as a spare in case of any accident to an experimental plant during the week.

The randomly selected plants were harvested, following the procedure already considered, and figures were obtained for the leaf areas of individual plants, and for the dry weights of leaves, stems and roots.

We first plot in Fig. 11.21 the figures for leaf area against the corresponding figures for leaf dry weight. Here we are dealing only with factors which have affected the expansion of the leaf material, and may expect to get a consistent progression within a particular age group at a particular harvest. In this instance at each harvest both age groups can be regarded as parts of a single progression, but as already noticed when considering some of the fresh weights of the same material [11.6], this is not always found to be so, and more generally we should have to consider separately each age group at each harvest. As for Fig. 11.6, the lines are put in by eye: they are intended at this stage only as an aid in detecting points outside the general scatter, and elaborate computation of a line of best fit would not be justified. In fact, if the straight line shown in Fig. 11.21a be plotted in Fig. 11.21b, it can be seen that the progression has in this case changed during the week between the two harvests. Possible reasons for such changes will be considered in Chapters 19 and 27: meantime the consistency observed in Fig. 11.21a and 11.21b has acted as a check both on the figures for leaf area, and on those for leaf dry weight.

11.22 Final check: dry weights of leaves and stems

We next compare the figures for leaf dry weight, just checked, with those for stem dry weight, as set out in Fig. 11.22. Here we see a somewhat less consistent picture, and more scatter, than for the data for the leaves alone, particularly for the largest plants. One point stands out as above

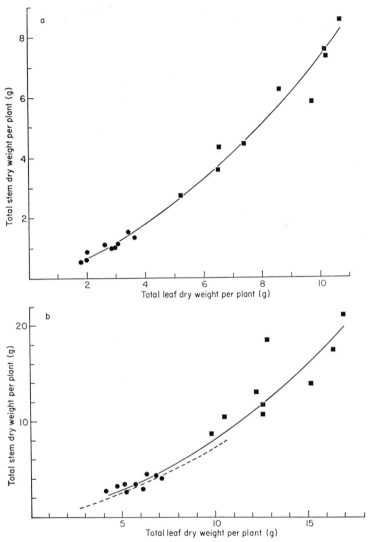

Fig. 11.22. Total stem dry weight plotted against total leaf dry weight for individual plants of *Helianthus debilis* of two age groups, sown on 7 June (■) and 21 June (●) 1967; (a), at the first harvest on 26 July; (b), at the second harvest on 2 August 1967. The solid line in (a) is shown as a broken line in (b). For details of the experiment, see 11.21; for other related data see Figs. 11.6, 11.21 and 11.23.

11.22

the general scatter, that for a stem dry weight of 18·6 g and a leaf dry weight of 12·8 g. We check the transcriptions and arithmetic on which both of these figures are based. We examine the dried stem material, and check the number of nodes and petioles against the corresponding number of leaves. There is no mistake. This must be a plant with an unusually large amount of stem. The seed did not belong to a carefully selected line, and such variations in plant architecture are not unexpected. We must, however, follow this plant through, and if it proves to be anomalous in other ways, eliminate it. We conclude that, with this one exception, the data are free from anomalies so far.

11.23   Final check: dry weights of tops and roots

As the next step in the checking process, leaf and stem dry weights are added together to give the total top dry weight, and root dry weights are plotted against this, as shown in Fig. 11.23. This gives a picture of very much greater variability, some plants having nearly twice as much root as others for a given dry weight of top. This is a general scatter, found equally in the smallest and the largest plants. There are two possible explanations—are the plants themselves very variable in this respect? or are different groups of students very variable in the proportions of the total root system which they have extracted? The tins of root systems are examined, and it is apparent that there are large differences in the sizes of the main core of the root system—the differences are not due to variations in the amounts of fine roots recovered. The variability is clearly due to the plants themselves. But at the same time we see that the root systems as a whole do not exceed 20 per cent of the dry weights of the tops, so that the extremes of variability of root dry weights fall within a range of less than 10 per cent of the total dry weight of the plants. The collection of such information on variations in plant form may be a valuable feature of a field experiment of this kind, carried out, as we have said, using seed which does not come from a particular variety, and which has not been subjected to a long process of selection for uniformity.

We also look at the point representing the plant which we previously found anomalous—it is the third largest, and we see that, although the root dry weight is high, it is not relatively higher than that of a number of other plants.

Further analysis of the data can now begin.

11.23

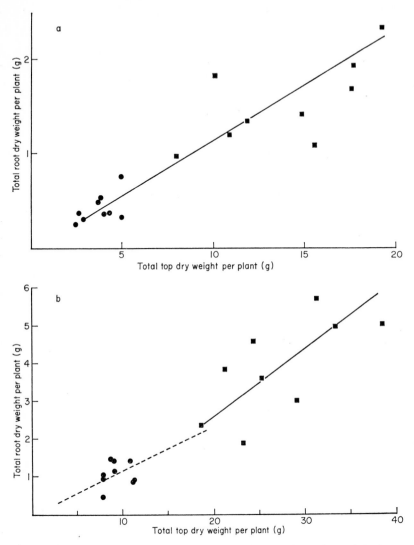

Fig. 11.23. Total root dry weight plotted against total top dry weight for individual plants of *Helianthus debilis* of two age groups, sown on 7 June (■) and 21 June (●) 1967; (a), at the first harvest on 26 July; (b), at the second harvest on 2 August 1967. The solid line in (a) is shown as a broken line in (b). For details of the experiment, see 11.21; for other related data, see Figs. 11.6, 11.21 and 11.22.

11.23

12

MEASUREMENT OF RESPIRATION

12.1 Dry weight changes, metabolic balance and mineral content

In Section 4.3, and in 13.11 we discuss the four main components which together determine daily dry weight gain by healthy plants: net gain by assimilation in the light; net loss by respiration in the dark; mineral uptake; and what we have called 'metabolic balance'. The last term takes care of changes in the average dry weight of a reduced carbon unit [2.8, 4.3] from causes other than the incorporation of mineral elements. These are almost wholly concerned with changes in hydrogen and oxygen. Condensation and hydrolysis may be important in the formation and germination of certain seeds and storage organs, particularly those containing much condensed carbohydrate such as starch, and condensation may also be important when much woody tissue is being laid down. Here the elements are incorporated or given up in the same proportion as in water, so that there is no change in the average level of oxidation or reduction; but these last processes may also be important when they are proceeding on a large scale, as in the formation and germination of seeds containing much fat or oil (App. 1.12). In all these cases the weight of the reduced-carbon units affected is changed substantially; in cellulose, for example, they have 10 per cent less dry weight than in glucose; in fats and oils the reduction relative to glucose is much larger, around 45 per cent. But it is necessary to take account of this only when the proportions of the different compounds are changing markedly, and during the main phase of vegetative growth of many plants this is not so. Supposing that during a particular interval between harvests the proportion of the dry weight of a plant in the form of stem increased by 20 per cent; and supposing that half this was cellulose; then if, on balance, this extra cellulose replaced glucose molecules in the average composition at the beginning of the period, the loss in weight would be 1 per cent. If it replaced sucrose, the loss in weight would be little over half this; if starch, the change would be negligible.

The interfering effects of changing mineral content are usually small, because although there may be substantial changes in the relative mineral content of the plant as a whole, they are usually spread over long periods. Fig. 2.7a and 2.7b afford an example; during the 5 weeks from 26 to 61 days when the mean dry weight of the maize plants increased 43-fold, ash content varied only from 16·95 to 16·20 per cent. Thereafter, as the Figure shows, the accumulation of dry material began to exceed that of minerals; between 61 and 82 days, the ash content fell by 5·6 per cent, an average of 1·9 per cent per week, while the dry weight increased 3·7-fold; in the following 3 weeks by 2·5 per cent, an average of 0·8 per cent; in the following 2 weeks by 0·25 per cent. On the other hand, during the first 26 days, the ash content rose from the 1·65 per cent of the seed to 16·95 per cent, an average increase of 4 per cent per week. During this period of germination and seedling establishment large metabolic changes are going on, while the overall dry weight change was small. The mean weight of the young plant at 26 days was 0·55 g, against 0·36 g for the seed. At such periods it is clearly important to allow for changes in mineral content (and also metabolic balance), while during the main period of rapid relative dry weight increase by these plants it would not be necessary to do so.

12.2 Dry weight changes, photosynthesis and respiration

Net loss of dry weight by respiration of the whole plant is much more important, being usually at least 10 per cent of the dry weight gain by photosynthesis, and it may well be more. In 12.14 are mentioned instances (examined in more detail in Chapters 29 and 31) where the proportion lay between 50 and 75 per cent. The proportion is markedly affected both by the physiological state of the plant, and by external conditions, notably temperature. A 10°C rise of temperature in the temperate range causes roughly a doubling of the respiration rate in many cases, and here temperature may play an important part in determining the overall net gain in dry weight over a period. Studies of respiration rates can thus be a great help in understanding the carbohydrate economy of the plant as a whole; and they are also essential in evaluating the processes going on during the light phase, in relation to the daily cycle of change (Chapter 29).

If we wish to study an individual plant's carbon assimilation under natural conditions in the light, the growth method offers the only approach which does not alter the natural aerial environment [4.6]. All other methods (even the ingenious controlled system of Bosian, 1965) involve interference with natural air movement and the natural radiation climate, and hence with leaf temperature; and they may involve changes in air temperature, humidity and carbon dioxide concentration as well. It follows that the consequential changes are complex, and that it is usually difficult to make an adequate allowance for them.

12.3 Conditions of respiration measurements

The situation as regards respiration in the dark is much less complex than that which confronts us when we wish to measure photosynthesis. In the nature of the case the problems of the radiation climate are now removed; with these go most of the problems of leaf temperature; and the difficulties connected with stomatal movement and the water balance of the plant as a whole are also minimized (but see the end of section 12.6). Air movement has little effect on respiration rate until the wind velocity becomes high enough to cause mechanical deformation of the leaf, and in general, in the absence of large water deficits, the respiration rates of the whole intact plant and its parts are likely to be affected mainly by a single environmental factor—temperature.

The control and measurement of temperature of the aerial parts of a small plant present few problems. When surrounded by damp air in the dark the plant parts will be very close to air temperature, which can be regulated by a thermostat (thick stems and other massive organs present special problems to which we shall return in 12.4 and 12.12). On the other hand, pots of soil or vermiculite are slow to equilibrate, and should remain at a fixed temperature for a period of the order of a day before measurements are made; such slow temperature changes are easy to follow, and should be checked.

Our main concerns must therefore be firstly, to arrange to handle the plants in such a way that our own interference does not alter the respiration rates of the parts, and secondly, to arrange a system where the readings actually correspond to the carbon dioxide produced by the plant.

12.4 Wounding and handling

Respiration rates of all plant organs are liable to be altered substantially by wounding, or in the case of thin parts such as leaves by handling or any other process bringing about deformation (Audus, 1935, 1939). The respiration rate of massive organs such as tubers may also be affected by handling if they are flaccid and thus liable to mechanical deformation; e.g., Barker (1935) observed an increase of respiration rate of 30 per cent immediately after gently squeezing a potato which after 15 months of storage had lost approximately half of its initial fresh weight, ten days elapsing before the rate returned to its previous value.

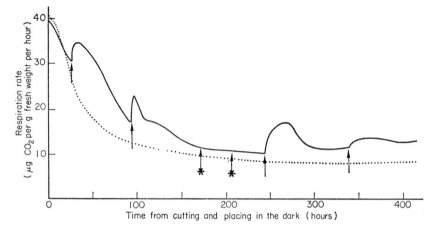

Fig. 12.4.1. Effects of handling on the respiration rates of leaves of *Prunus laurocerasus*, in relation to time from cutting and placing in the dark. Respiration rates of control material (⋯⋯⋯) maintained at a constant temperature of 20·8°C compared with rates for experimental material (———) which was handled at 28, 95, 245 and 341 hours (↑), and also placed in the light and subjected to temperature change for 10 min without being handled (⁎̶), but otherwise also maintained at 20·8°C. From Godwin (1935).

The effect of gentle handling is illustrated in Fig. 12.4.1, showing Godwin's observations on the respiration of detached leaves of *Prunus laurocerasus*. On four occasions, at 28, 95, 245 and 341 hours, the experimental leaves were transferred from one respiration chamber to another, and Godwin (1935), states, 'It should be noted that the handling in the first two cases (28 and 95 hours) consisted in no more than holding the leaves between finger and thumb, singly or together, and bending or rubbing them lightly past one another as is inevitable in transferring leaves from one small vessel to another'. Yet as Fig. 12.4.1 shows this

12.4

gentle handling at 28 hours caused an increase in respiration rate which at its maximum exceeded 75 per cent, and which lasted for more than two days. In more extended studies Audus (1939) investigated the leaves of fourteen other species of temperate regions (six evergreen woody plants, seven herbaceous perennials and one bamboo) from twelve families, and found positive effects of mechanical stimulation on the respiration rate in every case, ranging from 26 per cent in *Rhododendron fargesii* at 45 hours to 183 per cent in *Yucca gloriosa* at 100 hours. The magnitude of the effect was found to vary with the length of time which had elapsed between cutting the leaf and placing it in the dark, and the mechanical stimulation. Several tropical species have also been tested, and all have given similar effects (P. Freeman, personal communication).

The minimum amount of wounding required to sever plant parts can also have profound effects on respiration rates. For example Welbank (1957) measured the respiration of plants of *Impatiens parviflora*, intact except for the disturbance associated with floating the root system out from a potful of a light, sandy soil, and found a mean rate over three experiments of 1·29 mg per g dry weight per hour, over a 12-hour period. Comparable plants were selected from the same batches, the leaves were severed, and simultaneous measurements were made on all the parts of these plants taken together. They gave a mean value over the same 12-hour periods of 2·23 mg per g dry weight per hour, an increase of 73 per cent.

The severance of stems can have equally profound effects. Müller (1924) studied the respiration rates of small branches (1·2–3 cm diameter) of three tree species, comparing short lengths of branch, around 10 cm, with longer lengths, around 20 cm. If the short pieces had a length of 2x cm and the longer pieces a length of y cm, and if we assume that x cm at each end of the long pieces was respiring in a similar manner to the short pieces, it is then possible to work out both the respiration rate of the middle part, y − 2x cm, of the long pieces, and the increases over this rate associated with the cutting of the ends. Fig. 12.4.2 shows an example from his observations on *Picea abies*. Here, for four successive periods during the first 24 hours after cutting, the respiration rates of the intact centre portions of the long pieces work out at 10·6, 12·0, 9·9 and 11·2 μg CO_2 per g fresh weight per hour. Assuming that the respiration rates of the several pieces, had they remained intact, would have been effectively constant at the mean value of 10·9, the percentage increases in respiration rates for the long and short branches are shown in Fig. 12.4.2, in relation to time from cutting. We see that a marked effect is observable at 3 hours, and that it continues to increase up to 24 hours, by which time the respiration rate of the shorter pieces has risen to nearly two and a half times the intact value. We need pursue this question no further here, except to note that in a number of series of Müller's observations there is

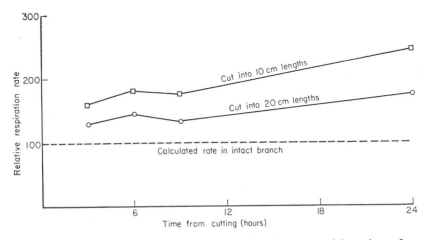

Fig. 12.4.2. Effects of cutting on the respiration rates of branches of *Picea abies*: relative measured rates for branches 1·6 to 2·2 cm diameter cut into 10 cm lengths compared with those for branches 1·8 to 2·5 cm diameter cut into 20 cm lengths and with the calculated rate in intact material (10·9 μg CO_2 evolved per g fresh weight per hour $= 100$), in relation to time from cutting. For details see text. Calculated from data given by Müller (1924).

evidence of a continually increasing respiration rate, not only of the cut ends, but also of the centre sections of the longer pieces during the first 24 hours after cutting. This indicates an even more complex situation than that just discussed, and it would be necessary to make observations on pieces longer than 20 cm before attempting to infer the respiration rate of the intact branch.

　　After cutting and also stripping off the bark of small branches of *Populus vernirubens*, Zelawski and Fuchs (1961) observed much larger increases in the respiration rate of the wood, of up to six-fold within 12 hours, but here the treatment was drastic indeed.

12.4

Yet another complication may arise indirectly from the cutting of massive plant parts lacking a specialized aerenchyma. Little evidence is available on conditions in the interior of intact organs of this kind, but seventy years ago Devaux (1899) suggested that in the centre of sprouting branches the oxygen concentrations might be so low as to cause anaerobiosis. The observations of Scott (1949) indicate that there may be marked concentration gradients both of oxygen and of carbon dioxide, and that on occasion the carbon dioxide concentration near the centre may reach levels between 15 and 20 per cent by volume, where marked inhibition of the rate of respiration can be expected. It should be noticed that, owing to the high oxygen concentration in normal air, and the relatively lower diffusivity of carbon dioxide, this effect is almost certain to become marked before the respiratory pathways are seriously affected by reduced concentration of oxygen. Once again, if it is at all possible we must work with intact plants. If, an account of size or for other reasons, this is impossible, we must endeavour to avoid disturbing these concentration gradients. Such disturbances may arise directly on cutting, by changes in the diffusive pathways, or indirectly by changes in respiration rate similar in character to those shown in Fig. 12.4.2.

To make measurements from which to assess with accuracy the respiration rates of intact plants and their parts, the greatest care is needed both in the design of experimental methods and in the manipulation of the plant material itself. Such well-known techniques as the Warburg respirometer can only be used with confidence if preliminary investigations have shown that for the particular material to be used the effects discussed in this section are negligible; and when investigating plants growing in the field it is obviously desirable to make measurements on intact plants which have been interfered with as little as possible. In later sections we shall consider means of doing so when normal laboratory facilities are not available.

12.5   The rate and the measurement

Our second concern, that the measurement should correspond to the carbon dioxide produced by the plant, involves in the first place ensuring that the respiration of other organisms is not included in the measurement. This makes it impossible to measure the respiration of an intact root system in a soil containing an active microflora and microfauna, although it can be measured in an inorganic system such as vermiculite. Here it is necessary to avoid the growth, in the surface layers of the vermiculite, of algae whose respiration would be added to that of the plant. By using extensive precautions it is feasible to grow plants under wholly axenic conditions, but for a variety of reasons this may be undesirable. Algal growth can then be minimized by shielding the surface of the vermiculite from light during the whole growth of the plant, by using black plastic pots and a black plastic cover, arranged with suitable baffles to keep out light while allowing aeration. There will still remain the problem of bacterial respiration. Most root systems exude small amounts of various organic compounds, which may support bacterial growth in an apparently purely mineral medium. The fairest way to cope with this situation seems to be to regard it as a convenience. Any organic compounds exuded by the root system represent a loss of reduced-carbon units to the plants. If these were simply adsorbed by the vermiculite, or washed out by the culture solution used in watering, as would happen under axenic conditions, their measurement would present very great difficulties on account of the low concentrations involved. If they are mainly respired by bacteria (which, when growing on a purely mineral medium, have no other source of carbon) the carbon dioxide produced can, without appreciable error, be added to respiration by the plant in order to arrive at the total loss of carbon by both processes taken together.

If the measurement is to correspond to the carbon dioxide being produced by the plant at the time, we must also avoid situations in which carbon dioxide is either being stored in some part of the system (leading to low readings), or released from store (leading to high readings). Such situations are always liable to arise, in systems involving diffusion, when conditions of concentration or rate of movement are changing [12.15]. They are best avoided by arranging for the system as a whole to be as nearly in a steady state as possible, so that there are no important changes in rates of movement through the various parts of the system, and we do not need to estimate the actual amount of storage, because it is effectively constant—an amount comes out at one end equal to that which went in at the other. We cannot necessarily arrange that the plant material itself is in a steady state, and changes in respiration rate may be one of the subjects of investigation. Here we are concerned rather with the method of measurement, which is under our control, and which must

12.5

be contrived so that unavoidable changes in respiration rate of the plant material have minimal effects on the interpretation of the results. (Changing conditions will be dealt with in 12.15, 12.16.)

The point can be illustrated by constructing a much simplified diagram (Fig. 12.5). We concentrate all the respiratory systems into a single volume A, which is surrounded by a conducting space B, separated from

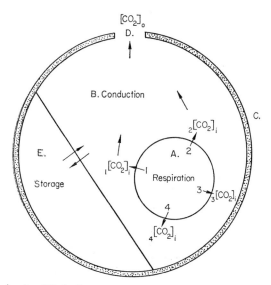

Fig. 12.5. A simplified diagram of the main elements involved when considering movement of carbon dioxide out of a respiring plant organ. The respiring systems A1, 2, 3, 4 each give out carbon dioxide, producing concentrations $_1[CO_2]_i$, $_2[CO_2]_i$, etc. at the nearest points in the conducting system B (mostly intercellular spaces). As the carbon dioxide diffuses through the system B it exchanges with carbon dioxide stored in system E (mainly aqueous phases in cell walls). There is also storage within the volume A (protoplasts and vacuoles). The whole is enclosed in an impermeable wall C, with a variable aperture D. The concentration of carbon dioxide in the air just outside D is $[CO_2]_o < [CO_2]_i$, and this concentration difference maintains an outward diffusive movement.

the external environment by an impermeable wall C, with a single variable aperture D. Also inside the impermeable wall is a storage space E, which communicates with the conducting space B. The conclusion reached during the preceding paragraph can now be restated. The aim of the experimental method is to keep the concentration of carbon dioxide just outside

12.5

the aperture D ($[CO_2]_o$) as constant as possible. If the respiration rate and the resistance to diffusion through the aperture D are both constant, when a steady state is established there will be a constant rate of flow through the conducting space B, and a constant concentration at any one point in this space, including points just outside the respiratory volume A, where the concentrations are $_1[CO_2]_i$, $_2[CO_2]_i$, etc. There will then be exchange between the storage space E and the conducting space B, but no net gain or loss; and the rate of production of carbon dioxide by A will equal the rate of outward diffusion through D. If the rate of respiration changes, the various values of $[CO_2]_i$ may change, and the steady state will be disturbed. Alternatively, if the rate of respiration remains constant and the resistance to diffusion through D changes, the steady state will again be disturbed. Either or both of these changes may lead to difficulties in interpretation of the results but the difficulties will be much worse if at the same time $[CO_2]_o$ is changing due to the way in which our experimental system is arranged. Changes in resistance to outward diffusion, and in storage of carbon dioxide are most likely to be met in leaves. A method of checking on such changes will be discussed in 12.15.

12.6 Gas current methods

When dealing with gas current methods for measuring respiration, the desirability of a steady state system leads to a preference for methods where the concentration of carbon dioxide just outside the plant material is maintained constant. This is not so for closed circuit systems in which the carbon dioxide produced in respiration is allowed to accumulate, the changing concentration being followed by some such method as infra-red gas analysis, which does not involve absorption of carbon dioxide from the gas stream, as in Fig. 12.6a. Such a system does not permit of the establishment of a steady state; it involves measuring the rate of increase of carbon dioxide in the gas phase, which may not be the same thing as its increase in the system as a whole; and unless circulatory mixing is extremely good in all parts of the system, interpretation of the measurements may be made difficult by exchanges with dead space. Particular difficulties are posed by systems such as that of Decker (1954), depending on the time taken for a given concentration change, and consequently these can only be used under special conditions, when it is known that there are no complications due to storage.

12.6

When using a method depending on measurement of carbon dioxide concentration without absorption, it is therefore preferable to have an open system such as Fig. 12.6b, in which the carbon dioxide produced in respiration is exhausted to the air, and $[CO_2]_o$ is maintained constant by using a supply of fresh air, or of bottled gas. The method is equally suitable if the carbon dioxide is to be absorbed in the course of estimation, as in Fig. 12.6c. Measurement is now of the amount of carbon dioxide produced in a given time, and the flow meter is used to check on the maintenance of steady conditions. It can be dispensed with if a constant rate of air supply can be ensured by other means.

Such open, continuous flow systems can be contrived in a very great variety of ways, depending upon the resources available. As a minimum they must contain three elements—an air supply, a plant chamber, and a measuring device. For one reason or another it may be necessary to include one or more of five other elements; a suction pump; a pump working at or above atmospheric pressure; humidification equipment; scrubbers for removing carbon dioxide or water vapour; and a flow meter, which will be necessary if the measuring device is sensitive to concentration, as in an infra-red gas analyser. If the measuring device is sensitive to absolute amounts of carbon dioxide (Fig. 12.6c), then small fluctuations in rate are immaterial, and the rate itself can be adjusted to a value both convenient for the measuring device, and giving a suitable rate of air change in the chamber containing the plant.

If the method of estimation involves the absorption of all or nearly all the carbon dioxide from the gas stream, it is, however, possible to use a closed circuit system and still maintain a steady state in the plant material, as in Fig. 12.6d. For some purposes, especially for field work in places remote from normal laboratory facilities, such a system has advantages [12.8].

In designing a system, attention must be paid to the distribution of pressure within it, and to the avoidance of leaks in critical places. Generally speaking, the nearer to atmospheric pressure the system works, the better, but leaks of laboratory air into the system should always be avoided, as plant respiration is sensitive to a variety of vapours even in very low concentrations (Burg, 1962). In particular, leakage of ethylene from gas supply pipes can easily ruin an experiment in a room where no smell of gas is detectable.

If the plant is to be totally enclosed, a difference between the pressures

12.6

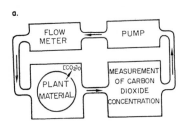

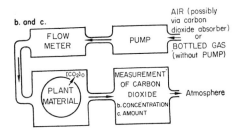

(a) A simple closed circuit system without a carbon dioxide absorber. $[CO_2]_o$, the concentration of carbon dioxide in the air just outside the respiring plant material, changes as the experiment proceeds.

(b) and (c) Variants on open circuit systems, used when the pressure inside the measuring device must be very close to atmospheric pressure. (b) Measuring device not involving absorption of carbon dioxide; $[CO_2]_o$ maintained relatively constant as the experiment proceeds. (c) Measuring device with a carbon dioxide absorber; as (b), but the measurement is now of the amount of carbon dioxide absorbed in a given time, and the flow meter is used to ensure maintenance of constant conditions.

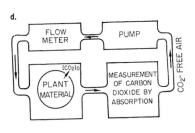

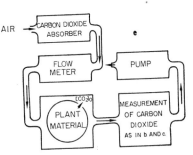

(d) A simple closed circuit system with a carbon dioxide absorber. $[CO_2]_o$ is maintained relatively constant because the air leaving the absorber has a relatively constant, very low level of carbon dioxide concentration; use of flow meter as in (c).

(e) Another open circuit system variant, similar to the systems of 12.6b and 12.6c, avoiding possible contamination of the air supply by the pump, used when it is not necessary that the measuring device should be very close to atmospheric pressure.

Fig. 12.6. Simplified block diagrams of some gas current systems used for respiration measurements.

12.6

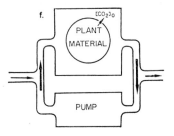

(f) An arrangement replacing the plant material chamber in any of the preceding systems, and permitting a rapid circulation of air over the plant material to be combined with a slow flow through the system as a whole.

Fig. 12.6—*continued*

in the plant chamber and the atmosphere is of no consequence provided that the pressure inside does not fluctuate when the apparatus has once been set up. On the other hand, if a portion of the plant—root, top, branch or leaf—is to be enclosed in a chamber while the rest remains outside, pressure in the chamber must be as close to atmospheric as possible. This not only avoids strain on the seal, but also prevents mass flow of gas through the intercellular spaces of the plant. Any sealing compounds used should also be inert, not giving off any toxic vapours. A useful series of such compounds can be made up from saturated hydrocarbons (white petroleum jelly and medicinal liquid paraffin) mixed in proportions to suit the working temperature.

Finally, when designing a system, the state of water vapour throughout must not be forgotten. It is necessary to ensure that the gas stream passing over the experimental material is not so dry as to cause excessive transpiration leading to loss of turgor, or worse, desiccation. Some of the systems we have discussed normally work close to water saturation (e.g., the version of the closed system of Fig. 12.6d described in 12.8); in others (e.g., those depending on a supply of compressed air) the gas stream would be very dry if not humidified, and here a check is necessary in the preliminary work. On the other hand, some measuring systems, such as an infra-red gas analyser, work best with dry air, and here a dryer may have to be inserted between the plant chamber and the measuring device. Leblond and Carlier (1965) describe such an open-circuit system including control of relative humidity in the plant chamber.

12.7   Infra-red gas analysis

If ample resources are available, the simplest and most effective system to meet all these requirements is to use an infra-red gas analyser fitted with a recorder as the measuring device, and with a flow meter in the gas stream; the details of several possible arrangements are discussed in Eckardt (1965). It is best to use a cylinder of compressed air as a source of supply (as in Fig. 12.6b), if necessary humidifying the air before it enters the plant chamber and drying it after leaving. This ensures an air supply of constant composition, avoids the necessity of a pump, and leaves the exhaust of the system at atmospheric pressure, which is convenient for the operation of the gas analyser. We shall examine data from such a system later [12.13, 12.15, 12.16]. The apparatus requires power supplies and so on, and can only be conveniently operated in or adjacent to a laboratory. It can, however, readily be combined with field observations in remoter places, by substituting gas sampling tubes for the gas analyser and recorder, which are left behind in the laboratory. The sample tubes are then brought back to the laboratory, and the concentrations of carbon dioxide determined by connecting them with the analyser in a closed circuit incorporating a peristaltic pump. For this purpose it is necessary to determine separately the volumes of the sampling tube and of the rest of the closed circuit.

12.8   Less expensive gas current methods

The less expensive methods of determining the concentration of carbon dioxide in air are very numerous, and a suitable one can generally be found to suit all pockets and resources. Under difficult expedition conditions, when resources are minimal, and everything must be both simple and robust, the simplest of all is probably that used by Evans (1939) in Nigeria, where aspiration, volume measurement, carbon dioxide absorption, and titration are all carried out in a single glass jar. By care and attention to detail the analytical method can be made sufficiently accurate. Its defects are the length of time taken to make a single observation, and the impossibility of making serial observations without replication of the apparatus. Unless the jar is filled very quickly with the gas to be analysed, there is also the added complication of exchange with the water with which the jar is filled initially, which has to be allowed for. Under difficult conditions, these restrictions must be accepted, but if more convenient methods can be used the highly simplified one is not to be recommended.

12.8

Plate 12.8. A gas-current field apparatus for respiration measurements, arranged as in Fig. 12.8 except that the special changeover taps F and I are replaced by four simple taps.

If a gas current can be arranged, many methods are available involving absorption of carbon dioxide from the stream and subsequent determination of the amount absorbed, by titration, conductivity, or other means. In the field the creation and measurement of a steady gas stream creates problems when working in situations remote from normal laboratory services including power supplies. A convenient method of providing steady suction, combined with volume measurement, is to use two Mariotte bottles arranged on two levels, as in the diagram in Fig. 12.8, so that the upper one drains into the lower. In this way a constant head device is created, without any water being lost from the system, which is regenerated by changing over the two bottles. To minimize variations in rate due to temperature changes the bottles are lagged, and built into wooden cases, as illustrated in Plate 12.8. This shows how the tubuli of the bottles are protected, together with the two fixed spirit levels, provided to assist in levelling and hence to ensure the maintenance of a constant working pressure head. To monitor the rate, or to measure the aspirated volume, it is best to add a flow meter, as in the diagrammatic circuit of Fig. 12.6d; but although this adds both to the convenience and the accuracy of the determinations, it is not absolutely necessary. If on an expedition the flow meter is out of action, the aspirated volume can be estimated on the hour-glass principle, or, if fractions are needed, by a gauge on the bottle.

An apparatus of this kind can be used with an open circuit of the type of Fig. 12.6c, but it is probably at its most effective if it is used not merely to conserve water, but also CO_2-free air, by an arrangement such as that of Fig. 12.8 which is a form of the circuit of Fig. 12.6d. Here we start with the system full of boiled distilled (and therefore CO_2-free) water and CO_2-free air. As the water from the upper bottle flows into the lower, the air displaced from the lower bottle flows back to the upper, through a system of two changeover taps as at (i). These are arranged so that the CO_2-free air from the lower bottle reaching the first tap passes next through a chamber containing the respiring plant material G, and then through an absorbing and estimating system H, arranged to measure the absolute quantity of carbon dioxide absorbed. It leaves the absorber as CO_2-free air again, and passes on through the second tap I to the upper bottle. When elapsed time or a gauge indicates that the water level in the upper bottle has almost reached the bottom of the air inlet tube the two taps are turned off as at (ii), isolating the plant chamber and

12.8

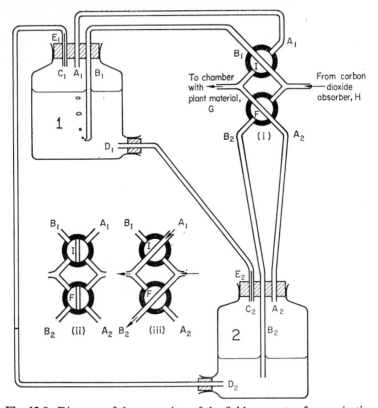

Fig. 12.8. Diagram of the operation of the field apparatus for respiration measurements illustrated in Plate 12.8, and used to replace the pump and, if necessary, the flow meter in a circuit of the type shown in Fig. 12.6d. In this operating position water flows from the lagged Mariotte bottle 1 to its twin 2 through the tube D_1C_2, at a constant rate determined by the pressure difference between the bottoms of tubes B_1 and C_2, and the capillary resistance E_2. This water displaces CO_2-free air, which flows from A_2 to the tap F, and hence to the plant chamber G, the carbon dioxide absorber H, and (now CO_2-free again) through the tap I to B_1, and so into bottle 1. The tube C_1D_2 is sealed by the water in the bottom of bottle 2, and the connection between A_1 and B_2 is closed by the two taps. When the water level in bottle 1 nears the bottom of tube B_1, the taps F and I are turned to position (ii), temporarily isolating the plant chamber and absorber, the positions of bottles 1 and 2 are reversed, and the taps are turned to position (iii). The functional connections are now D_2C_1 and A_1B_2, and flow through G and H is resumed in the same direction and at the same rate as before. Note that the diagram distorts the symmetrical arrangement of the connecting tubes, all of which are flexible, C_1D_2 being the same length as C_2D_1, B_1B_1 as B_2B_2, and A_1A_1 as A_2A_2. The two special change-over taps F and I greatly simplify the operation of the apparatus, and minimize accidental mistakes, but they are not essential. If not available they can be replaced by four normal taps, as shown in Plate 12.8.

absorber, the positions of the bottles are exchanged, and the two taps are turned on in the alternative position, as at (iii). The apparatus is completely symmetrical, as far as the bottles and supply tubes are concerned, and once assembled the system remains closed until new plant material is to be put in, while the tap arrangement ensures that the air flow through the plant chamber and absorber is always in the same direction. With a system arranged in this way there is no question of the air being too dry, or of exchange of carbon dioxide between air and water in the bottles, both being CO_2-free. At the same time measurements can be continued all day in the field without needing a fresh supply of either air or water.

It will be seen that the measurement here is of total carbon dioxide produced during a period of time, which is the easiest form of measurement to make in the field, remote from normal laboratory facilities. At the same time provision is made against the major sources of error. The supply of CO_2-free air to the plant material at a close approximation to a steady rate allows for the establishment of a steady state as far as diffusion relations are concerned. Leaks are always liable to be a nuisance in apparatus of this kind which may have to withstand rough transport, but here the effects of leaks can be reduced by working the system slightly above atmospheric pressure. They are then only important in the plant chamber, absorber and the connection between the two. Small losses of CO_2-free air from the rest of the system are of no consequence. The effects of possible leaks in the important part of the system are most easily minimized by arranging to keep the whole of the gas phase as close as possible to atmospheric pressure, but slightly above it: to this end absorbers are to be preferred which do not require a large pressure head for their operation, and it is better to regulate the flow rate by a capillary resistance in the water stream (E in Fig. 12.8) than by a resistance in the gas stream. In this way the bulk of the pressure change around the system occurs in the liquid phase, where leaks are easily detected.

If the whole of the flexible tube system consists of rubber or plastic, exchange of carbon dioxide across the large area in contact with the air may be a possible source of error. This is easily checked for a particular apparatus under particular conditions by running the apparatus for a time with no plant material, and seeing whether significant quantities of carbon dioxide are detectable by the absorbers. If so the exchange must be reduced to an amount inside the allowable limits of experimental

12.8

error by using a less permeable flexible tube, or sections of rigid tube, say stainless steel, with flexible joints.

Apart from problems of leakage, the principal difficulty likely to be encountered with a closed circulating system of this kind is the gradual accumulation in the gas phase of gases other than carbon dioxide, or volatile substances produced by the respiring plant material, in particular ethylene, which could affect the respiration rate in subsequent measurements ([12.9]; Mapson, 1969). It is therefore advisable to recharge the bottles with a fresh supply of CO_2-free air before starting a new series of observations, and to reboil or renew the water from time to time.

12.9   Systems not involving gas flow

Without using a gas current, the carbon dioxide produced within a closed system can be absorbed, and the amount absorbed can be determined. Alternatively the oxygen consumed can be measured, either by changes in the pressure in the system at constant volume (e.g., the Warburg apparatus) or in the volume of the system at constant pressure (e.g., the Thoday apparatus, described in Appendix I).

The dilemma connected with determination of the amount of carbon dioxide absorbed by analysing the absorber is that if this is highly efficient (e.g., a concentrated solution of strong alkali), changes brought about by the absorption of small amounts of carbon dioxide are difficult to measure with any accuracy; whereas if the absorber is such as to make the changes easily measurable (e.g., a dilute alkali) the absorption is not very efficient, and the efficiency may change with time. For example, in the simple method used by Evans (1939) and mentioned above [12.8], where absorption was in a dilute alkali, even after 25 minutes of shaking there was 1 chance in 20 that 5 per cent of the total carbon dioxide content of the air remained unabsorbed, and a second absorption, shaking and titration was always used, making the method very tedious. In an apparatus such as that used by Müller and Nielsen (1965, and Fig. 12.9), one can be sure that absorption of carbon dioxide was not complete, but it is difficult without a detailed study of the apparatus itself to know how large the error is likely to be.

Methods involving changes of pressure at constant volume are unsuitable for our purposes for three inter-linked reasons. One is that the steady state condition which we have decided to be desirable [12.5] for measurements of this kind on whole plants or whole organs is not attainable in such a system. The second is that in such a system any liquid phase must be constantly agitated in order to maintain it in a condition as close as possible to equilibrium with the changing gas phase. But the mechanical disturbance associated with shaking is most undesirable

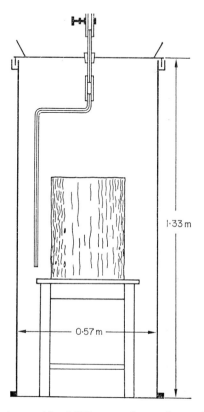

Fig. 12.9. Apparatus used by Müller to estimate the respiration of trunks and branches of trees from la forêt de l'Anguédédou (Ivory Coast), showing a section of trunk in position in the centre of the container, constructed of tin plate covered with paraffin wax, and having a volume of 0·33 m³. From Müller and Nielsen (1965). For results, see 31.12.

when dealing with whole organs of higher plants, whose respiration rate may be increased by such treatment [12.4]. The third reason is that the mechanics involved in shaking make the apparatus very difficult to scale up in size, so that it can be used with intact plants only if these are very small. Otherwise the plant material must be cut into small pieces, with

12.9

the likelihood of the consequences which we have seen [12.4]. If for some reason the method must be used, it is essential to check for both these last two sources of error, and preferably to use an independent method by which the effects of the first possible source of error can also be checked.

Similar objections do not apply to apparatus depending on changes in volume at constant pressure. A concentrated and efficient carbon dioxide absorber can be used, which will rapidly bring the carbon dioxide concentration in the gas phase down to a very low level and maintain it there. The volume change will then correspond to uptake of oxygen by the plant. If the whole system remains at constant pressure, this uptake of oxygen will alter the proportions of oxygen to nitrogen in the gas phase, but both concentrations being high, it is easy to arrange that the proportional changes are negligibly small, so that we are in effect able to maintain a steady state condition. Otherwise, by taking suitable precautions in the proportions and general design of the apparatus, there is no difficulty in attaining an accuracy comparable to that which can be achieved when working at constant volume. The apparatus of which a full account is given in Appendix 1, has the added advantage that it is relatively easy to scale up to a size where work on whole plants in pots is possible. Finally, as elaborate arrangements for stirring or agitation are not necessary, it is a method which can readily be improvised from material available in most laboratories, an important consideration if funds or facilities are limited.

In conclusion, it should be mentioned again that although most plant tissues will respire in the same way in both open flowing air and in closed systems, there are instances of tissues themselves giving off vapours, particularly ethylene, which affect their respiration rate if allowed to accumulate in a closed system (Burg, 1962; Mapson, 1969). The most common examples are ripening fruits, and senescent organs generally. A watch has to be kept for this effect. If it is suspected, it can be checked by trial observations over longer periods than normal, and if necessary an open circuit gas current method can be used.

12.10   Allowances for the effects of wounding and handling

By one or another of these means we can therefore determine the respiration rate of samples of our plant material, preferably as whole plants (but see 12.13), to minimize the interfering effects of wounding and handling. If the size of the whole plants makes this impracticable, it may be possible to assess the magnitude of these effects, and possibly to make an allowance for them, by trial experiments on smaller whole plants of the same species. These would involve respiration measurements on undisturbed whole plants, followed by cutting up and redetermination of the respiration rates of the severed parts. On occasion such experiments show that the wounding effect is very large, as in the observations by Welbank (1957) which we have already discussed [12.4], and the assessment of its magnitude becomes of great importance if the measurements of detached parts are to be used to assess respiration rates in intact plants. If the effects of cutting are not only large but also variable, it may be worth constructing large respiration chambers (e.g. App. 1.18), or else proceeding by one of the methods which we shall now consider.

12.11   Respiration of the leaves of large plants

If the plant is very much larger than anything which could be enclosed in a respiration chamber, the scope of the method of comparison is very much restricted. It may still be possible to use it for leaves, provided that in their structure and functioning the leaves of the large plant—possibly a mature tree—do not differ too widely from the leaves of a seedling small enough for the whole plant to be enclosed in a respiration chamber. This would have to be checked in individual cases. Otherwise it may still be possible to make comparisons between wounded and unwounded plants. For example, when working on a small tree it may be possible exceptionally, and as a check on the more convenient method of detaching leaves, to enclose an individual attached leaf, or twig bearing several leaves, in a chamber. This would allow a respiration measurement to be made on a portion of the intact plant, which could subsequently be detached to determine the wound effect.

For very large trees, such as the emergents in tropical rain forest, the leafage of the intact plant may be inaccessible. Here it may be possible to make comparisons between the standard procedure for respiratory measurement, whatever that is to be, and something on a larger scale. For example, if the standard procedure involves measurements on leaf laminas, detached and placed in a respiration chamber [Fig. 12.9], as in the work of Müller and Nielsen [31.13, 31.16], it may be possible to use as a basis of comparison respiration measurements on a twig bearing several leaves. These are subsequently detached, and the respiration of the severed parts compared with the whole. Alternatively, if the standard procedure involves a twig bearing several leaves, the comparison might be made on a branchlet bearing several such twigs, and so on.

In some plants even an individual leaf could not possibly be put in a respiration chamber—for example a palm of the genus *Raphia*, where the rachis may be 10–15 m long, and the individual leaflets $2\frac{1}{2}$ m; and even more awkward possibilities exist. In such instances it may be possible to seal a portion of an intact leaf into a respiration chamber, taking care to avoid mechanical deformation as far as possible, and comparing the results with the respiration rates of whatever cut pieces are to be used in the standard respiratory measurements. In such comparisons it must be remembered that we may now be faced with the additional difficulty of intercommunicating air spaces. Any pressure difference between the sealed-on chamber and the outside air will cause mass flow of gas through the intercellular spaces, causing an error in the respiration measurements. Equally, it is inadvisable to have a darkened chamber on one part of the leaf, while the rest is illuminated, as otherwise there may be diffusion of carbon dioxide from the relatively high concentration in the intercellular spaces of the darkened part, to the relatively low concentration in the illuminated part. It might turn out in a particular instance that the errors caused by either or both of these sources are negligible, but this can only be settled with reference to the particular system being studied. In exceptionally difficult cases the search for some standard of comparison for assessing the magnitude of the wounding effect may have to be abandoned, but by one or another of the methods which we have discussed some estimate can usually be made.

12.12   Respiration of stems and roots of large plants

Difficulties arise here even more frequently than for leaves, because portions of trunks are not so readily detached, and extensive injury is inevitable. This would argue against prolonged lapses of time between cutting, sealing, and the beginning of the respiration measurements. On the other hand, thick stems take a long time to come to thermal equilibrium. It is worth giving some consideration to the question of stem temperature before the experiments begin. Thermal gradients in stems can sometimes

12.12

be reduced by previous shielding from direct sunlight, and by making use of previous measurements, a working temperature can be selected which is close to mean stem temperature. If other reasons, such as build-up of wound respiration [Fig. 12.4.2], make it desirable to start observations soon after cutting, temperature changes can thus be reduced to a minimum. Similar considerations apply to massive fruits.

Some stems have extensive intercellular space systems communicating either with those of the leaves, or the roots, or both, and playing an important part in the gaseous exchanges of the plant as a whole. When such systems are cut up a very fallacious picture is produced of the respiratory pattern of the intact plant under natural conditions. Fortunately they are the exceptions, and are most common among water, swamp and marsh plants, and some others whose roots grow in soils having a very low oxygen content. Otherwise, most stem tissues are ventilated by diffusive contact with the outside air close by, either through stomata for young stems, or through lenticels in older stems, where cork usually develops (rarely, the epidermis is retained for long periods, as happens, e.g., in the mistletoe, *Viscum album*). In theory, it should be possible to enclose part of such a stem in a respiration chamber, and determine the output of carbon dioxide by a portion of an intact stem, as for leaves. Whether this basis of comparison can be set up in practice depends on the individual stem, its structure (including that of its intercellular spaces), appendages, and general geometry. If this can be done, similar precautions are needed as for respiration chambers attached to portions of leaves.

If a respiration chamber cannot be attached to the intact stem, we are forced to work wholly with wounded material. The cut ends of stems should always be sealed over with an inert sealing compound [12.6] and metal foil, so as to cause minimal disturbance to the normal concentration gradients of oxygen and carbon dioxide corresponding to the diffusive pattern of the intact plant. For the same reason both cutting and sealing should be done as quickly as possible. Even in these difficult cases there is still one basis of comparison open to us—we can compare the respiration rates of pieces of differing lengths, each of which will have two cut surfaces, separated by different bulks of stem tissue not immediately damaged by the cutting process. In this way it may be possible to distinguish the respiration of the damaged ends from that of the intact middle, and also to estimate how far from the immediately cut ends the

12.12

effects of the damage extend, as in the examples discussed in 12.4. Both the respiration rate of the damaged ends and the extent of the damage may be found to vary with time over considerable periods of the order of days, and in a seasonal climate all these effects may vary from one time of year to another (Müller, 1924).

12.13   Periods of measurement

We have seen [12.5] the desirability of establishing steady state systems in order to simplify the interpretation of measurements of respiration. As the pieces of plant material, pots of soil, and parts of the apparatus become larger, more time has to be allowed for the establishment of a steady state; even if there is reason to believe that the rate of respiration is steady, the measurements will at first drift before settling to a steady value [App. 1.19]. This may give rise to a dilemma when dealing with the respiration of cut pieces of stem because it may be desirable to make observations as soon as possible after cutting [12.12]. In such cases it is likely to be advantageous to use a gas current method, possibly with an auxiliary circulation as in Fig. 12.6f, in order to establish a steady state around the plant material in the shortest possible time.

When dealing with leaves alone, a steady state can be established very rapidly, in a period of the order of minutes from setting up the apparatus, if a gas current method and a well designed leaf chamber are used. We then observe a drift in respiration rate as in Fig. 12.13a, based on observations using an infra-red gas analyser. Here there was no question of handling or wounding, and the general disturbance to the leaf was minimal. The plant of *Trema guineensis* was raised from seed in a growth cabinet at a constant temperature of 30°C. The leaf used was the terminal one of a flush of growth, and towards the end of a normal light period the leaf chamber was assembled round it, the leaf being manipulated solely by handling the stem to which it was attached. The gas stream passed through the leaf chamber had a closely similar temperature,

Fig. 12.13. Respiration rates of attached single leaves of two tropical plants, growing in pots in a controlled environment cabinet at a constant temperature of 30°C, in relation to time from darkening. (a) *Trema guineensis*; (b) *Coleus blumei*. Measurements made using an infra-red gas analyser and a circuit as Fig. 12.6b; air supplied from a cylinder and humidified before entering the leaf chamber; flow rate 0·5 litre min^{-1}.

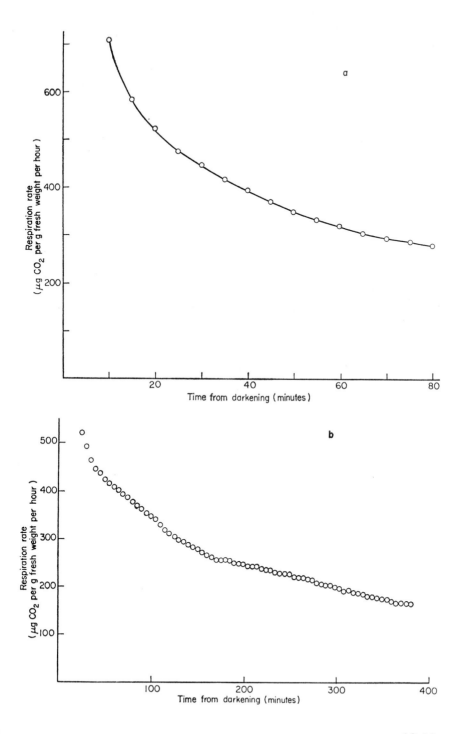

12.13

relative humidity and carbon dioxide concentration to the normal growth cabinet air. The leaf chamber was then darkened at the beginning of the normal artificial night, and observations on respiration made at the same temperature as before, using an open circuit and an infra-red gas analyser, as in Fig. 12.6b. The respiration rate steadily declined throughout the dark period, having at 50 minutes fallen to half the rate measured at 10 minutes from darkening. A somewhat less rapid fall is shown in Fig. 12.13b for another tropical plant, *Coleus blumei*. Here the observations did not begin until 25 minutes after darkening, but half the rate first measured was reached after a further 140 minutes. Such falling rates of respiration are normally observed on darkening leaves which have previously been assimilating in the light, although the decrease may take place more slowly than in the examples of Fig. 12.13. The control leaves in Godwin's observations on *Prunus laurocerasus* [Fig. 12.4.1] showed a falling rate of respiration, but half the initial rate was not reached until 40 hours (it should be noted that owing to the method of measurement the initial rate was the mean over the first hour). Similar declines are seen also in all the species investigated by Audus (1939, and see 12.4), where within 24 hours the rate of respiration fell to half the mean rate during the first hour, in about half of the fourteen species investigated. Clearly, to have meaning in the context of the life of the plant in the field, respiration measurements should be made at a time and over a period close to the length of the natural night, *and should begin as soon as the plant is darkened*.

But as we have seen, this may be impracticable, if we are trying both to work with an intact plant in a pot, and to measure the respiration of the whole plant at once, because several hours may elapse before a steady state can be established for the pots. For example Rackham (1965) found periods of 6–10 hours, using both a gas current method and a Thoday respirometer [App. 1.19].

If this situation is encountered it would be best to make separate assessments of the respiration rates of the roots (where the rate may be expected to be relatively constant, and the time for the establishment of a steady state is long) and the top of the plant, or the leaves alone (where the rate may be expected to drift markedly during the length of a normal night, and the time for establishment of a steady state is short). For potted plants in a wholly mineral medium this is most easily done by a seal round the stem just above the surface of the soil [App. 1.20]. The tops of small plants growing in the field can be studied in a similar

12.13

way; for larger plants the respiration rates of the leaves, or of groups of leaves with the twigs to which they are attached, are better measured separately from those of the rest of the plant.

12.14 How accurate?

If the total quantity of respiration to be measured and allowed for were always a relatively small fraction of unit leaf rate, it would not have been necessary for us to spend so much time on considering all these possible complications which may arise in different types of respiration measurement. Supposing, for example, that photosynthesis of the green parts of the plant during the daylight hours is ten times the respiration of the whole plant during 24 hours, it is not necessary for us to estimate the respiration loss with great accuracy. But there are important cases where the total quantity of respiration is large, as in the instance in Chapter 31, dealing with the observations of Müller and Nielsen on tropical high forest. Here roughly three-quarters of the whole photosynthetic gain was consumed by respiration of the plants themselves, and assessment of this respiratory loss formed an important part of the scheme of research. Chapter 29 affords another instance, where it was necessary to make an allowance for respiration by the whole plant in the dark, and by the non-photosynthetic parts in the light. Over ten sets of observations, covering a wide range of light climates, the allowance for respiration lay between 27 and 53 per cent of the gain in dry weight by the plant as a whole. In such cases we should aim at an accuracy comparable with that of the growth measurements themselves. It may then be necessary to estimate the amount of carbon dioxide stored in the leaf during respiration in the dark [12.5], and to make an allowance for it.

12.15 Changing conditions and the steady state

If there is available a rapidly responding method for carbon dioxide measurement, such as an infra-red gas analyser, it is possible to make use of deliberate changes in the rate of evolution of carbon dioxide from an assimilatory organ such as a leaf to assess the corresponding changes in the amount stored [12.5], and hence to detect changes in the resistances to outward diffusion. The leaf chamber has a transparent top, and is enclosed in a dark box with a source of constant illumination of low intensity which can be switched on from outside.

The first estimate can be made as soon as the rate of evolution of carbon dioxide has settled down to a steady rate, or a slow but steady change, thus avoiding the initial phase of rapidly falling rate of carbon dioxide evolution [12.13], if any. The illumination is then switched on for an accurately determined period, long enough for the establishment of a new steady state, and then switched off again. An experiment similar in technique to that used for Fig. 12.13, on a leaf of *Euphorbia pulcherima*, gave the record shown in Fig. 12.15, there being in each case a lag before the measured concentration of carbon dioxide reached its new value. The rate of flow through the system being known (in this case $0 \cdot 5$ litre min^{-1}), the area of the two lag phases (determined from planimetering the record) can be converted into amounts of carbon dioxide, in this case 12·3 and 22·0 μg. Part of this represents the amount corresponding to the change of concentration throughout the volume of the apparatus, between the leaf and the recorder (for Fig. 12.15, 0·73 litre, corresponding to 7·0 and 7·2 μg). The remainder would represent the change in the amount of stored carbon dioxide [12.5], if there was no change in the rates of photosynthesis or respiration during the two lag phases. We should expect the rate of photosynthesis to change substantially on occasions when there is a large resistance to the outward diffusion of carbon dioxide (represented diagrammatically by the aperture D in Fig. 12.5), and when consequently the internal carbon dioxide concentration may have been relatively large during respiration in the dark. It will be shown in 12.16 that this was the case for the experiment of Fig. 12.15, there being an initial internal concentration of a little over 0·4 per cent by volume. During the lag phase, after the light was switched on, this concentration fell to approximately 0·1 per cent, and one would accordingly expect the rate of photosynthesis to have fallen during the same period. This would account for the difference of 9·5 μg between the apparent amount of carbon dioxide coming out of store after illumination and the amount going into store after darkening. In such circumstances it is necessary to use the lag following darkening to estimate the amount stored. The problems due to changes in respiration rate are minor compared to those due to changes in rate of photosynthesis. It cannot be assumed that after a period of illumination the rate of respiration before illumination will necessarily be regained at once; there is likely to be a temporary increase, as seen in Fig. 12.15. On the other hand, for short periods of illumination of low intensity this increase would be expected to be transient, the pre-illumination rate being

12.15

regained after a period of the order of tens of minutes. The observed drift of concentration of carbon dioxide after a steady state has been reestablished is therefore extrapolated back to the time the light was

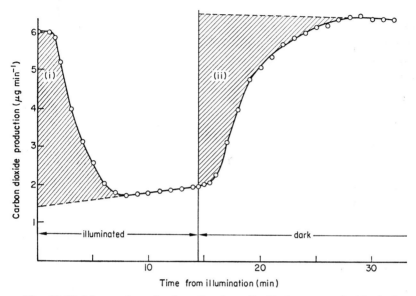

Fig. 12.15. Measured production of carbon dioxide by an attached leaf of *Euphorbia pulcherrima*. Plant grown in a controlled environment cabinet at 30°C and with a daylength of 15·5 hours. Just before the end of a normal 'day' the leaf chamber was assembled round the leaf, which was handled as little as possible and then by the petiole only. A gas current of bottled air having a carbon dioxide concentration of 325 p.p.m. by volume was humidified and passed through the chamber at 0·5 litre min⁻¹, and then on through a drier and infra-red gas analyser. The whole controlled environment cabinet was darkened at the normal time, and a single de luxe warm white-fluorescent tube was switched on 2 hours later (=zero time in the figure), and switched off again 14·5 min later. The rate of evolution of carbon dioxide shown is that being measured at a given time. Counting the whole volume of the analysis tube and half the volume of the leaf chamber the swept volume was approximately 0·73 litre, corresponding to a time lag of approximately 1·5 min.

switched off, in order to determine the area to be integrated, as shown in the figure. In fact, in this instance the correction so introduced is negligible, but it cannot be assumed that it will always be so.

These considerations lead to the conclusion that when the rate of evolution of carbon dioxide increases from 2·0 to 6·4 μg min^{-1}, 14·8 μg were absorbed into store, corresponding to 2·3 min of the respiration rate in the dark. The amount stored was therefore small in relation to the total carbon dioxide evolved during a night's respiration, the quantity in which we are interested.

A test of this kind can be applied at intervals during the dark period, and if necessary, corrections can be made for changes in the amount of carbon dioxide storage. In the particular case we have been considering, the amount of carbon dioxide stored in the leaf declined by 6 μg during a further 6 hours of respiration, approximately one part in 400 of the carbon dioxide produced in respiration during the period. The change is therefore negligible, and no correction need be applied, as it would be inside the limits of experimental error. The arguments relating to the steady state advanced in the earlier sections of this chapter are thus shown to be justified in this instance.

12.16   Intercellular space carbon dioxide concentration

The figures considered in 12.15 provide us with a rough means of estimating the mean carbon dioxide concentration inside the respiring leaf. Assuming that the change of storage is proportional to the increases of the various internal concentrations above the external one, and therefore to the rate of production of carbon dioxide [Fig. 12.5], we see that if the leaf were brought to its compensation point, where there was no net evolution of carbon dioxide, and the concentration just inside the aperture D equalled the external one, then the change in the amount of carbon dioxide stored would have been $14·8 . 6·4/4·4 = 21·6\ \mu$g $\equiv 12.2\ \mu$l. This particular leaf had a volume of 2·7 ml.

There are substantial reasons for believing that the half-time for equilibration of carbon dioxide between the vacuole and the rest of the cell is likely to be less than 1 sec, and that the mean solubility of carbon dioxide in the aqueous phases of the leaf is near 1 (Raven, 1970), so that for our purposes we can assume that the whole volume of the leaf is available for storage.

This gives a mean concentration difference in the darkened leaf of about 4500 p.p.m., a substantial increase above the normal 'atmospheric' concentration in this experiment of 325 p.p.m., and leads us to believe

12.16

that in these leaves in the dark there was considerable resistance to the outward diffusion of carbon dioxide. We shall consider more precise methods of investigating the gaseous exchanges of leaves in Chapter 29.

This estimate ignores the possibility that some of the carbon dioxide produced in respiration may be absorbed reversibly into biochemical combination as the intercellular space carbon dioxide concentration rises, and subsequently released when it falls again. However, the available evidence suggests that the half-time for this process would be much longer than for physical storage. As the relationship of rate of carbon dioxide output with time shown in Fig. 12.15 (ii) is almost exactly exponential over the whole middle part of its course, where the test can most accurately be applied, it seems that here we are dealing with one process only, physical storage. There are also further considerations. Unless the stomata open towards the end of the night, and unless the reversal of the biochemical storage mechanism is rapid enough to lead to substantial losses of carbon dioxide before photosynthetic intake of carbon dioxide overtakes respiratory loss, we ought to exclude biochemical storage from this part of our discussions for the same reasons that we exclude photorespiration—that we are here concerned with loss of carbon dioxide by the plant as a whole, corresponding as it must to loss from the plant's total stock of reduced-carbon units. A mechanism involving biochemical storage of respiratory carbon dioxide and its subsequent reassimilation avoids such a loss, and must be considered in a wider context than the simple measurements of carbon dioxide evolution which have been the concern of this chapter.

12.16

PART III

ANALYSIS OF DATA

13

HISTORY AND DEVELOPMENT OF
THE MAIN ANALYTICAL CONCEPTS

13.1 Empirical generalizations

An important part of the whole process of scientific advance lies in the discovery of empirical generalizations such as Boyle's Law or Ohm's Law which epitomize masses of individual observations, and which at the same time, within recognized limits, can be used for predictions. The early investigators of the quantitative relationships involved in the growth of plants were thus following the classical pathway when they tried to find simple empirical relationships which would in similar fashion sum up many individual measurements of attributes of growing plants. Bearing in mind what we have already said about the complex organization of the higher plant it is not surprising that these attempts should have failed, but at the time it was important that they should have been made as a stage in the overall advance of knowledge. The ideas underlying some of these attempts proved to be of such limited application that they are now almost forgotten. An example is afforded by the attempts of Brailsford Robertson (1908a, 1908b) to demonstrate that plant growth follows the time course of an autocatalytic chemical reaction in a closed system. He had an initial success in fitting an autocatalytic curve to the latter part of the growth curve of an organ of limited growth (a fruit of *Cucurbita pepo*), but even here there were large deviations in the early part of the fruit's development. However, subsequently this line of approach led to the development of more and more complex formulae with increasing numbers of arbitrary constants, without ever reaching the point where the quantitative relationships of plant growth could be predicted. Thus it became clear that the desired simplicity would not be reached by this route, and the attempt is now chiefly memorable for an interesting paper by Enriques (1909) on the use of formulae involving arbitrary constants for what he called 'die kunstliche Nachahmung'—the artificial simulation —of the quantitative relationships involved in natural processes: and for some fine scientific polemic of an acidity rarely met with nowadays (Brailsford Robertson, 1910).

Other attempts were more fruitful. Although too simple in the form in which they were first put forward, the ideas underlying them proved capable of development so as to express in relatively simple form certain facets of plant growth. They thus led into the next phase of development which has extended up to the present day, during which attention has been concentrated on the analysis of the complex problem into simpler and more readily handled components, and the study of these components individually and in combination. Sufficient complications have been uncovered in these studies to deter further attempts to find simple relationships which could be used for predicting the quantitative relationships of plant growth, and it is only with the development of control engineering, with its simulation of elaborate systems using powerful hybrid computers, that possibilities in this direction again begin to open up (for an example of some of the complexities involved in this approach, see Curry and Chen, 1970).

13.2 Analogies

A number of writers have tried to make the processes involved in plant growth less strange to the human mind by the use of familiar analogies. Brailsford Robertson tried to use the analogy of an autocatalytic chemical reaction, and failed. Chodat (1911, p. 136, quoted in 13.5) used the analogy of growth as a geometric progression, the size of the plant at times T, $2T$, $3T$, etc., corresponding to the successive terms. V. H. Blackman (1919) improved upon this by pointing out that growth is a continuous process, so that what is wanted is not a series of discrete terms but a continuous exponential expression. These related analogies fail because the exponent is not found by experience to be constant for more than very limited periods, if at all: the series is not regularly exponential over any considerable number of terms.

What may be termed the 'monetary analogy' (of an invested sum of money growing by accrued interest) has proved to be much closer to actuality. It was used by Hackenberg (1909) and by Kidd and West (1919). If the rate of interest is constant this growth would become a geometric progression, but the notion of constancy is not inherent in the use of the analogy. The idea was improved upon by Gregory, who distinguished

13.2

between productive investment (in the leaves, where the additions to the stock of capital took place) and non-productive investment (in the rest of the plant). As we are accustomed to think in monetary terms it is clear, without an algebraic argument, that with a fixed distribution between productive and non-productive investment the relative rate of growth of the capital as a whole will be higher, if the rate of interest on the productive investment is higher; and that at a fixed rate of interest, the rate of growth of the capital as a whole will be higher, the larger the proportion invested in the productive securities. Converted back from our analogy into plant terms, this means that the rate of growth of the plant relative to its size at any time will be higher if either (a) the rate of increase of dry weight per unit leaf area is higher; or (b) the ratio of leaf area to total dry weight is higher; or both. The monetary analogy has helped us to distinguish between the complementary effects of a morphological change—variation in leaf area per unit total dry weight—and a physiological one—variations in rate of production of new dry matter per unit leaf area; and it has helped us to see that the overall effect on the relative rate of growth of the plant will be the product of these two changes.

The monetary analogy can be carried farther. Companies announce dividend rates, not on the basis of the current price of a share, but on its nominal value; and frequently dividend rates stay relatively unchanged during considerable fluctuations of the market, so that even with a fixed dividend rate the actual return on new capital invested may show large changes. The rate of interest per unit of nominal value is what now corresponds to the rate of increase of dry weight per unit leaf area, mainly an index of physiological activity. The market value of the share, on the other hand, corresponds to the leaf area produced by the plant per unit of leaf dry weight—it is an anatomical index of leaf expansion, and is found in many plants to be very sensitive to a variety of factors, both external and internal. Finally we have the fraction of the total capital productively invested, corresponding to the leaf dry weight as a fraction of the total dry weight. This last may also change within wide limits during the life of a plant, but often in a regular progression following the developing plant form.

So far the analogy has been reasonably precise and quantitative; the growth of the plant relative to its size will be determined by multiplying these three quantities together; and if, say, any two stay fixed and the third increases by 10 per cent, then the overall rate of return on total capital (relative growth rate) will also go up by 10 per cent.

13.2

13.3 The monetary analogy extended

The monetary analogy can be carried a good deal further in helping us to understand the growth of a large and long-lived plant such as a tree, if for the moment we abandon a strict quantitative comparison. Let us imagine an enterprising gentleman of the late seventeenth century. His father has been a successful lawyer and has made a fortune (the seed); this he invests in a substantial landed estate (the seedling leaves). This brings him in a large income, but he lives in a modest way (as yet the proportion of stem and root is small) and the bulk of his income is invested in neighbouring farms (more leaves) which in turn increase his income, and the family becomes in time exceedingly flourishing. However, as time goes on two changes are noticeable in the family fortunes. In the first place, as the estate spreads, it abuts more and more on equally, or even more wealthy, neighbours who refuse to sell and the estate being entailed, it cannot be sold and the proceeds reinvested elsewhere, so that finally it ceases to increase in acreage (the tree has reached its full leafage). Secondly, as the family wealth increases, the style of living of successive generations becomes more magnificent, until finally the family inhabit a Stately Home (the trunk, branches and main root system), the building of which, and the filling of it with *objets d'art et de vertu*, have absorbed a large part of the income of successive generations. We are now in the early nineteenth century, and the family splendour is near its zenith (the tree is mature), but because of the entail, and the lack of alternative investment, trade being unthinkable, its annual income has for the time being reached a maximum, and can increase no further. Then there is a lucky chance. Gambling ruins a neighbouring landholder, who is forced to sell up, creating a further opportunity to expand the estate (a neighbouring tree is blown down by a gale, cf. Fig. 2.9, at ↑). Time goes on, a maximum income is reached once more, but the inescapable calls on this income become continually greater. The rating assessment of the Stately Home is continually being increased, and the rate poundage goes on rising (as the trunk increases in diameter, so does the average dry weight of a full annual ring). Wage rates and the cost of materials go up, and with them the cost of maintaining the building (the material needed to maintain effective bark over an increasing area also increases). Yet the long continued habit of magnificence is not readily abandoned, and

13.3

retrenchment is deemed an impossibility (the tree has no means of self-pruning). A point is reached when the fixed income will no longer meet the outgoings, and farms have to be sold (the tree begins to die back, and fewer, smaller leaves are formed). The income itself then begins to fall, and the end is in sight. The rate payments are kept up as long as possible (the tree goes on adding new wood), but there is not enough money for maintaining the building. The weather gets in, and it begins to fall into ruin (the protection of the bark fails, and wood-rotting fungi gain entry). But the family is not destined to fade away gradually. The twentieth century is advancing, and it perishes in a final blaze of death duties (the partly rotten stump is struck by lightning and bursts into flames. All is consumed).

Of course, the first principle of using analogies is not to push them too far: but the fact that such an analogy can shed light on the growth of a plant is a further illustration of the great differences already mentioned between the adult forms of higher plants and higher animals [2.7].

13.4   Noll and the '*Substanzquotient*'

Modern concepts on the growth in dry weight of whole plants can be traced in unbroken sequence to a paper by Noll in 1906, unfortunately preserved only in title in the proceedings of the society to which it was read. There is, however, an almost contemporary reference, apparently written in the following year, by Noll's pupil Hackenberg (1909). I translate: 'Last year (1906) Professor Noll made a communication to the *Niederrheinischen Gesellschaft für Natur- und Heilkunde* at Bonn concerning the quantity which he had named the plant's *Substanzquotient*. The *Substanzquotient* was to be obtained by determining the quantity of dry substance of a plant at equal intervals and relating each weight thus obtained to the previous one, by dividing the former by the latter. Thus the dry substance quotient gives a measure of the *Assimilationsenergie* of a plant at different periods of its life, in that it relates the assimilatory income to the existing, and increasing, working capital'. (For *Assimilationsenergie*, see 13.8.)

The same period also saw the publication of other researches by Noll's students all making use of the same idea: Gressler (1907) on the growth of a number of species of *Helianthus* [partly abstracted by V. H. Blackman (1919)]; Gericke (1908) on the effect of removal of the cotyledons on the growth of *H. annuus* [reviewed in English by Kidd and West (1919)]; and Kiltz (1909) on *Nicotiana tabacum*.

It thus seems from Hackenberg's account that Noll's *Substanzquotient* was simply the ratio of the total dry weight of the plant at the end of an arbitrary period to the corresponding weight at the beginning; and that it was not originally defined in relation to a specific time interval, although Hackenberg himself used a constant interval of 7 days (Table 13.4), and apparently all Noll's other pupils did the same.

Table 13.4. Ratios of total plant dry weight at the end of successive 7-day periods to the dry weight at the beginning of the same period (Substance Quotient of Noll, 13.4) throughout the growing season for both sexes of *Cannabis sativa* and *C. sativa* cv. *gigantea* [App. 3.2], grown at Bonn in 1906. Abstracted from Hackenberg (1909).

| | *Cannabis sativa* | | *C. sativa* cv. *gigantea* | |
Week	Male	Female	Male	Female
1	0·97		0·93	
2	2·0		1·9	
3	2·3		2·4	
4	2·1		2·4	
5	1·9		2·6	
6	2·3		2·7	
7	2·5		2·4	
8	1·7	1·9	2·1	
9	1·5	2·2	2·0	
10	1·3	1·6	2·3	
11	1·2	1·7	1·5	
12	1·1	1·4	1·3	1·5
13	1·1	1·3	1·3	1·5
14	0·94	1·2	1·3	1·4
15	—	1·2	1·3	1·05
16	—	1·4	1·1	1·05
17	—	1·2	0·92	1·02
18	—	—	—	1·03

The quotient is therefore a function both of the rate at which the plant is growing, and of the length of time which has elapsed between the two determinations of *W*. What is more, determinations of the substance

13.4

quotient made over different periods of time will not bear any very simple relationship to each other. For these reasons Noll's quotient has been abandoned (and largely forgotten), and in effect replaced by the notion of relative growth rate, to which we shall return.

However, it seems to the writer that Noll's idea is well worth preservation, and not simply for historical reasons. Table 13.4 gives Hackenberg's weekly figures for both the male and female plants of two varieties of hemp (the drug problem was a different thing in those days). Considering the growing season as a whole he found that in 14 weeks the male plants of *Cannabis sativa* had increased to a total dry weight of about 1,000 times the seed weight, by which time the plant was chlorotic and dying, and the dry weight had ceased to increase. On the other hand, female plants growing under the same conditions and already six times as large as the males at 14 weeks, further doubled in dry weight ($1 \cdot 2 \times 1 \cdot 4 \times 1 \cdot 2 = 2 \cdot 0$) between the end of the 14th and 17th weeks, finally reaching a total dry weight about 12,000 times the seed weight. In other words, taking as a time interval the whole growing season the substance quotients of male and female plants were approximately 1,000 and 12,000. This substance quotient over the whole growing season seems to be a useful figure when considering the life of plants growing in highly seasonal climates. It is particularly useful for annuals. By removing the arbitrary time element of days or weeks, and replacing it by the recognizable biological unit of the growing season, we lose the possibility of using the substance quotient to analyse events over short periods, which was Noll's original intention, but there remains an index of considerable biological significance, setting an upper limit to the overall resources of the plant. This index can vary within wide limits, and although the range of values for Hackenberg's hemp are typical of many annuals, some attain values two orders of magnitude higher. Shamsi (1970) gives figures indicating values around a million times the seed weight for *Lythrum salicaria*, and of over half a million for *Epilobium hirsutum*. Plant architecture will then determine how large a fraction of this total can be mobilized to form seeds —around 10 per cent for *L. salicaria*, with over 100,000 seeds, and rather more (13·6 per cent in one particular measured instance) for *E. hirsutum* with about 80,000 seeds. Higher percentages are often observed [23.4]. Both this 'mobilization efficiency' and the value of the seasonal substance quotient itself are clearly important components of the overall reproductive capacity of the plant (Salisbury, 1942), while analysis of the seasonal

13.4

substance quotient could also be extended to take account of perennation (e.g., in the rootstock of *L. salicaria*) and vegetative reproduction (e.g., the rhizomes of *E. hirsutum*).

13.5   V. H. Blackman and the efficiency index

The second idea to be developed also has a long history. It is difficult to say who first put forward the idea of plants growing in a geometric progression, but it was well established by the time Chodat produced the second edition of his *Principes de Botanique* in 1911. On p. 136 he says, speaking of Monnier's work on the growth of *Avena sativa* (I translate), 'For us the active mass is the protoplasm in its entirety, for others it is the nucleus. At each mitosis the nucleus is doubled, but as far as is known, at least initially the periods of mitosis are equal, in consequence of which in successive equal time intervals the number of nuclei (or the active protoplasmic mass) increases in the progression 1, 2, 4, 8, 16, 32, 64, ...'.

In his work on sunflowers Gressler had had a similar idea. In obtaining an average substance quotient over a period, what he was in effect doing was to treat his results as a geometric progression, where $_nW$, the weight at the end of the nth week was related to $_0W$, the weight at the beginning, as $_nW = {_0W}\,(sQ)^n$, where sQ is the mean substance quotient. We need not concern ourselves with the propriety of this procedure (although the reader might like to consider what the consequences would be), beyond noticing that growth is being described by what is in effect a discontinuous function, the interest accruing week by week.

In 1919 V. H. Blackman published a paper on 'The compound interest law and plant growth', in which he said 'it is obvious, however, that during the daylight period the plant is adding new material continuously' and pointed out that what is needed is not a discontinuous function, but a smooth curve. He added, 'In many phenomena of nature we find processes in which the rate of change of some quantity is proportional to the quantity itself ... Lord Kelvin called the law which such processes follow "the compound interest law" ...'.

'The importance of this law for the proper appreciation of the growth of a plant was brought home to the writer in 1917 in connection with the results of some experiments on the growth of cucumbers carried out in association with Mr F. G. Gregory at the Cheshunt Experimental Station. ...'

13.5

'... If the rate of assimilation per unit area of leaf surface and the rate of respiration remain constant, and the size of the leaf system bears a constant relation to the dry weight of the whole plant, then the rate of production of new material, as measured by the dry weight, will be proportional to the size of the plant, i.e., the plant in its increase in dry weight will follow the compound interest law.'

Expressed in mathematical terms, if W is the dry weight, then

$$\frac{dW}{dT} = RW,$$
13.5.1

where R is a constant. Hence

$$\int_{1W}^{2W} \frac{dW}{W} = R \int_{1T}^{2T} dT,$$

and integrating

$$\log_e {}_2W - \log_e {}_1W = R({}_2T - {}_1T),$$
13.5.2

or

$$_2W = {}_1W e^{R({}_2T - {}_1T)}$$
13.5.3

where $_2W$, $_1W$ are the dry weights at times $_2T$ and $_1T$ respectively.

Blackman recalculated Gressler's figures using equation 13.5.2 and said, 'The rate of interest is clearly a very important physiological constant. It represents the efficiency of the plant as a producer of new material. It ... may be termed the efficiency index of dry weight production ...'.

As data accumulated and were examined it soon became apparent that the idea of R being a constant must be abandoned; and with this went the use of the term 'efficiency index', with its overtones of constancy, the name being replaced by the neutral term 'relative growth rate' used by West, Briggs and Kidd (1920).

However, Blackman's paper marked a notable advance in the arithmetic treatment of growth data, and added weight to the growing feeling of the importance of what is now called relative growth rate (RGR) as an overall measure, a summation or integration of all the processes bringing about increase in dry weight of the plant.

13.5

13.6 Relative growth rate

If we have abandoned the notion that relative growth rate (**R**) is a constant, we must also abandon our method of integration of the equation

$$\frac{dW}{dT} = \mathbf{R}W. \qquad 13.5.1$$

What we require is a true mean value over a period, $\bar{\mathbf{R}}$, such that

$$\bar{\mathbf{R}}(_2T - {_1T}) = \int_{_1T}^{_2T} \mathbf{R}\, dT = \int_{_1W}^{_2W} \frac{dW}{W}.$$

Then,

$$\bar{\mathbf{R}}(_2T - {_1T}) = \log_e {_2W} - \log_e {_1W}, \qquad 13.6.1$$

or

$$_2W = {_1W}\,e^{\bar{\mathbf{R}}(_2T - {_1T})}. \qquad 13.6.2$$

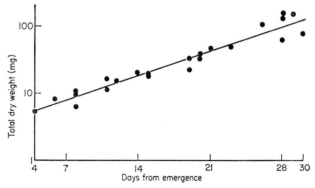

Fig. 13.6.1. Relationship between total dry weight, plotted on a logarithmic scale, and age from emergence of the cotyledons, for individual young plants of *Impatiens parviflora* cultured under constant conditions in growth cabinets. Four harvests at weekly intervals. From Lewis (1963).

Fig. 13.6.2. Relative growth rate in relation to age from germination for three annual crop plants. (a) Observations of Kreusler on *Zea mays* variety 'Badischer Früh' at Poppelsdorf near Bonn (51° N Lat.), beginning 11 May 1876; (b) observations of Briggs, Kidd and West on *Helianthus annuus* at Cambridge (52° N Lat.), beginning 27 May 1920; and (c) observations of Inamdar, Singh and Pande on *Gossypium arboreum* at Benares (25° N Lat.), beginning 15 July 1923.

13.6

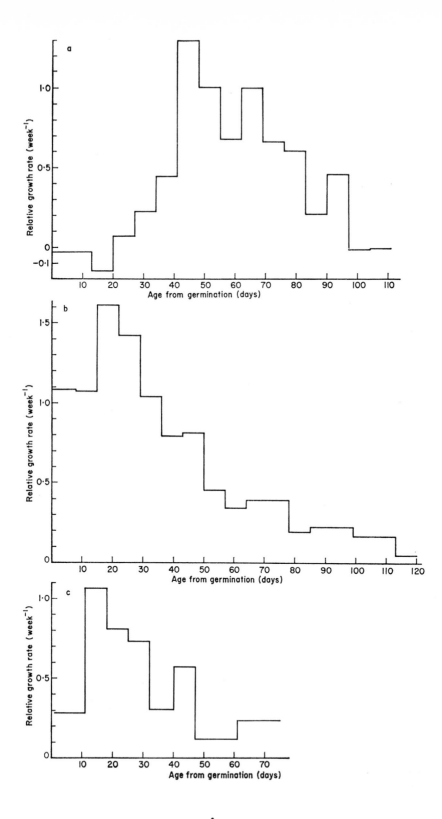

In fact the assumption of constancy is not necessary for this treatment. Either equation 13.6.1 or equation 13.6.2 will give us a true mean value of **R** for the period in question, irrespective of the variations of **R** during the period. Of course if **R** happened to be a constant, we should obtain the value of it, but it is not necessary that it should be so.

Equation 13.6.1 suggests a convenient rapid way of exploring the variations of **R** if we have a number of plants all of which belong to one population, but which have been growing over different periods, as in Fig. 13.6.1 (a set of data comparable to some that we have already seen in another connection in Fig. 5.8) where logarithms of total plant dry weight are plotted against the number of days which elapsed between germination and harvest. There is no sign of any systematic departure from a straight line, but a scatter around it. Then, from equation 13.6.1, the slope of this straight line is $\bar{\mathbf{R}}$, and in view of the number of points and the variety of periods, it does indeed appear to be constant in this instance. Cases of this kind are usually confined to the very early stages of seedling growth, and even then are by no means universal. Means of dealing with more complex, but regular, cases are considered in section 20.4.

Let us now consider briefly a few specific cases of studies of plants over the whole life cycle—a typical example of the observations of Kreusler and his co-workers on *Zea mays*, carried out at Poppelsdorf in 1876 (plotted from figures given by Kreusler *et al.*, 1877b); the observations of Briggs, Kidd and West on *Helianthus annuus* grown in Cambridge in 1920; and the observations of Inamdar, Singh and Pande (1925) on *Gossypium arboreum* grown at Benares in 1923. These are shown in Fig. 13.6.2, (a), (b) and (c) respectively. Here the relative growth rate is never constant for more than very short periods, if at all; and clearly in such cases this grand overall index is going to be a most difficult one to handle in relation to environmental changes. Where the ontogenetic drift is so marked, how is one to disentangle this from effects of changing environment? How to decide what the relative growth rate would have been in a given week, if the weather had been different?

13.7 Briggs, Kidd, West and Kreusler

The first steps towards answering these questions were taken almost at once. For 40 years there had been awaiting analysis a remarkable body

13.7

of data on the growth of maize, collected by Kreusler and a number of co-workers at the Agricultural Research Station at Poppelsdorf, near Bonn, and including some quantitative climatic observations (Kreusler *et al.*, 1877a, b, 1878, 1879; Fig. 2.7c, d). Ontogenetic drifts of relative growth rate similar to that shown in Fig. 13.6.2a are characteristic of these experiments, which were eventually analysed by Briggs, Kidd and West (1920a, b). One of their concerns was to distinguish the effects of weather on plant growth from the effects of these ontogenetic drifts. At an early stage of their analysis they saw that the ratio of leaf area to total dry weight L_A/W (Leaf Area Ratio), also showed an ontogenetic drift broadly parallel to that of relative growth rate. This parallelism can be seen in Fig. 13.7, where the corresponding figures for leaf area ratio are given for comparison with those for relative growth rate already shown in Fig. 13.6.2a. As they were searching for some aspect of plant growth relatively free from ontogenetic drift, in order to make progress with the study of the effects of weather, this led them to investigate the properties of the ratio of the two. Over an infinitesimally short period of time, dT,

$$\frac{\text{Relative Growth Rate}}{\text{Leaf Area Ratio}} = \frac{dW/dT \times 1/W}{L_A/W} = \frac{dW}{dT} \cdot \frac{1}{L_A} = E \quad 13.7.1$$

where W is total dry weight, dW is the increment of dry weight during the period dT, and L_A is leaf area. They gave the name 'unit leaf rate' to the quantity to which they had assigned the symbol E. The overall growth index, relative growth rate, was thus split into two

$$\text{Relative Growth Rate} = \text{Unit Leaf Rate} \times \text{Leaf Area Ratio}$$
$$13.7.2$$

and it can be seen that while leaf area ratio is a morphological index of plant form (leaf area per unit dry weight of the whole plant), unit leaf rate is a physiological index (rate of increase of dry weight of the whole plant per unit leaf area) closely connected with the photosynthetic activity of the leaves. Fig. 13.7 shows that each of these indexes has its own onto-genetic drift (together making up the drift in relative growth rate) and that the two are quite different in form. Leaf area ratio has a grand march, rising to a maximum between 40 and 60 days, and with relatively small fluctuations. Unit leaf rate, on the other hand, also shows an early rise, and then between days 42 and 90 shows marked fluctuations from week to

13.7

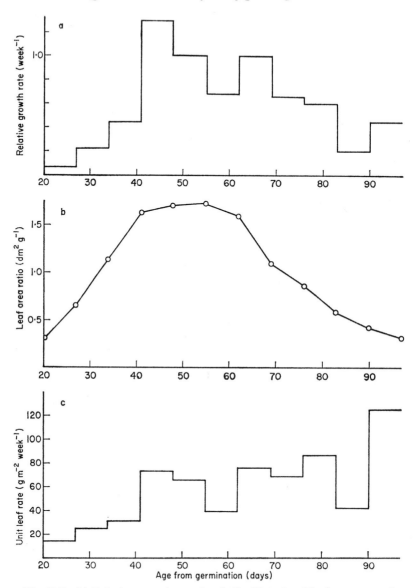

Fig. 13.7. (a) Relative growth rate (week^{-1}); (b) ratio of leaf area to total plant dry weight (dm^2 g^{-1}); (c) rate of dry weight increase per unit leaf area (unit leaf rate, g m^{-2} week^{-1}) in relation to age (between 20 and 97 days from germination) for plants of *Zea mays*, variety Badischer Früh, grown at Poppelsdorf in 1876. Evaluated from the data as given by Kreusler *et al.* (1877b), using for (c) the method discussed in 16.12 *et seq.* For the relative growth rate over the complete life of the same plants, see Fig. 13.6.2a.

13.7

week, but no marked upward of downward trend, during a period when leaf area ratio has dropped to a quarter of its value at day 42. (We need not now consider the causes of these fluctuations in **E**, which were only partly physiological, associated with the variable weather from week to week, and partly the consequences of experimental design based on the inadequate statistical apparatus available a century ago. This latter cause leads to progressively larger fluctuations as the plants become older, and the very large value of **E** between 90 and 97 days, followed by a fall to zero [Fig. 13.6.2a] is quite typical of this set of data as a whole (Briggs, Kidd and West, 1920b Figs. 3–6).

The splitting of relative growth rate into two components by bringing in the additional data on leaf area was also advantageous from a physiological viewpoint. It related dry weight increase, which in most plants has carbon assimilation as its most important component, to the area of those organs most concerned with carbon assimilation, the leaves. Ideas about this had been developing over an even longer period than the concepts so far discussed [13.4, 13.5].

13.8   Weber and Haberlandt: *Assimilationsenergie*

In 1878 Weber completed, and the following year published as an inaugural dissertation at Würzburg, a classical study in which he made estimates of what he called the '*Assimilationsenergie*' of a number of species of plants. This '*Assimilationsenergie*' was the mean rate of increase of dry weight per square metre of leaf area per 10 hours of daylight (*not* per day, the days being uniformly of 14 hours). Dry weight increase was determined over a single period, by the difference in weight between samples of the initial seed and that of plants at harvest. Daily measurements were made of areas of leaves (outlined on mica or glass), and in order to produce a succession of days all having an exact number of hours of illumination, the plants were covered at 19 h or 20 h each evening by shades which were removed at 05 h or 06 h the following morning.

This work attracted the attention of Haberlandt, who followed it up by anatomical studies of the species involved, and compared the *Assimilationsenergie* of Weber with the number of chloroplasts per unit area in the leaves of the same species. It is worth translating Haberlandt's account of this comparison, as it appeared in the first edition of *Physiologische Pflanzenanatomie*, published at Leipzig in 1884.

13.8

'The relations between *Assimilationsenergie* and the chlorophyll content of similarly constructed foliage leaves emerge from the comparison of observations which follow. The average amount of dry substance produced per unit leaf area per day of assimilation was ascertained by C. A. Weber, in order to express the efficiency or *Assimilationsenergie* of the leaf lamina. There proved to be a *"specifische Assimilationsenergie"* for each individual species of plant, which however was not further elucidated by Weber. In these circumstances the obvious next step was to determine the number of chloroplasts per unit area in the appropriate leaves, in order to see whether there was a fixed relationship between *Assimilationsenergie* and chlorophyll content. If now we set the values of these two quantities for *Tropaeolum majus* at 100, then the following values are produced for the remaining plants.

	Specifische *Assimilationsenergie*	Number of chloroplasts
Tropaeolum majus	100	100
Phaseolus multiflorus	72	64
Ricinus communis	118·5	120
Helianthus annuus	124·5	122

'The proportionality between chlorophyll content and *Assimilationsenergie* is therefore quite unmistakeable. The fact that this is not quite exact is explained by the incompleteness of the investigation, by the varying construction of the leaves and by the unequal size of the chloroplasts in the individual species.'

Haberlandt may at the time have thought the investigation incomplete, but there the matter rested for more than 30 years—*Physiologische Pflanzenanatomie* grew in size and complexity, edition followed edition, finally it was translated into English, but nothing was added to the above passage translated from the first edition. The work was before its time.

13.9 Gregory and apparent assimilation

The modern development of these studies began with Gregory's work at Cheshunt on cucumbers, published in 1918. He was at this stage concerned to estimate the apparent assimilation rate of the plants during the single period from germination to an age of 30 days. This was done by following the development of leaf area by daily measurements and taking account

13.9

of changing day length so as to be able to express the results as gain in dry weight per unit leaf area per hour of daylight. This brought the work into line with the substantial body of physiological measurements of apparent assimilation made by a variety of other techniques, and involving relatively short-term observations.

13.10   Unit leaf rate and net assimilation rate

When Briggs, Kidd and West's analysis of Kreusler's data appeared in 1920 it made an important advance in bringing together these different concepts [13.4, 13.5, 13.8] and applying them to the problem of onto-genetic drift. In the mean rate of increase of dry weight per unit leaf area they had found a quantity which over a substantial part of the life cycle of maize showed relatively little ontogenetic drift. They named this quantity 'unit leaf rate', and concentrated upon it in the latter part of their analysis. Before 1920 the only term in the literature for a similar quantity was Weber's '*Assimilationsenergie*' (five-sevenths of the value for a day with a constant light period of 14 hours, 13.8), which it would clearly have been unwise to attempt to translate. In his 1918 paper Gregory had not given the corresponding quantity on an hourly basis a special name, simply referring to it as 'average rate of assimilation', adding in a footnote '(neglecting the weight of material lost by respiration)'. In his work published from 1926 onwards Gregory preferred to use the term 'net assimilation rate' as synonymous with unit leaf rate. As a result of the great contributions which he and his co-workers made in the course of studying a wide range of agricultural and horticultural problems from a physiological viewpoint, and the great body of published work which emerged therefrom, 'net assimilation rate' has become widely used, particularly in agricultural and horticultural literature, so that questions inevitably arise in a practical form—should both terms continue in use? Should one be dropped? Or do we need rather a revision of terminology?

13.11   A revised terminology

Apart from any question of priority of origin, there are two objections to the continued use of the term 'net assimilation rate'. In the first place, to those not already familiar with its meaning it at once suggests comparability with the widely used terms 'apparent assimilation' and 'real

13.11

assimilation', which are solely concerned with carbon metabolism, and specifically with the photoreduction of carbon dioxide, net or with an allowance for respiration. But as has been already noted [4.3, 12.1], net dry weight change over a period involves two other important elements—uptake of mineral nutrients and overall metabolic balance. 'Unit leaf

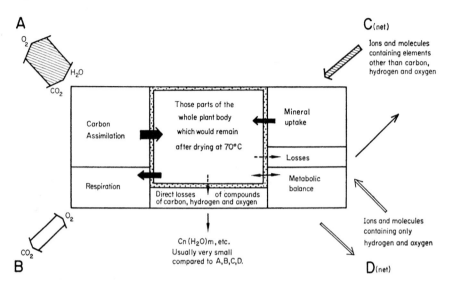

Fig. 13.11. Summary of exchanges between that part of the living plant body which would remain after drying, and its surroundings (including all the aqueous phases which would be driven off by drying at 70°C, 11.9), leading to changes in total plant dry weight. The summary excludes such chances as death, abscission or accidental loss of parts of the plant; interference by other organisms; and also any physical or chemical processes taking place during drying, and leading to changes in dry weight.

rate' has the merit of being meaningless without definition. But there is a second, more serious, objection. To a close approximation (Fig. 13.11)

Unit leaf rate = (A) Daily photosynthetic rate − (B) Respiration rate of the whole plant over 24 hours + (C) Daily mineral uptake + (D) Overall metabolic balance,

all expressed as dry weight change per unit leaf area [12.1]. However, when considering the carbon metabolism of the plant we need a term for (A) − (B) + (D), for which Carbon Dry-weight Rate (CDR) is suggested;

13.11

and when considering only the plant's stock of reduced-carbon units, we need a term for $(A) - (B)$, which is closely related to gain or loss of carbon atoms, which could be expressed as equivalent uptake or loss of carbon dioxide, and which is comparable with 'real' or 'apparent' assimilation. When considering relationships during the light phase, we also need terms for (A) daily photosynthetic rate $- (B_i)$ respiration rate of the whole plant during the photoperiod, which is a measure of the net change of reduced-carbon units in the plant as a whole; and (A) daily photosynthetic rate $- (B_{ii})$ respiration rate of the photosynthetic organs during the photoperiod; all these rates being again expressed per unit leaf area of the whole plant. These last two quantities are similar to, but not identical with, mean apparent assimilation rate, which concerns the gaseous exchanges of the particular photosynthetic organs being considered, and is usually expressed in terms of carbon dioxide or oxygen taken up or given out. But $(A) - (B_i)$ takes account of the respiration of the whole plant during the light phase, and both this and $(A) - (B_{ii})$ are expressed in terms of dry weight changes. However, these dry weight changes are notionally due to gain or loss of hexoses (dry weight changes due to other chemical transformations being included in (C) or (D)); consequently $(A) - (B_{ii})$ can be converted into net carbon dioxide intake (by weight) by multiplying by $1 \cdot 47$. It should be noted that all these quantities are net, and that in practice our analysis does not require at this point an estimate of any change in respiration rate of the photosynthetic organs in the light.

'Net assimilation rate' would be an excellent term for $(A) - (B)$, and it would conveniently express a most important biological quantity— the overall gain or loss of reduced-carbon units during a cycle of light and dark. But its use for this purpose has been blocked by using it as synonymous with unit leaf rate, and three new terms are therefore needed, for which are suggested Net Carbon Reduction Rate (NCRR) $= (A) - (B)$; Light-phase Whole-plant Carbon Balance (LWCB) $= (A) - (B_i)$ and Light-phase Leaf Carbon Balance (LLCB) $= (A) - (B_{ii})$.

Such distinctions may be of little importance if (A) is very large compared with $[(B) - (C)]$ or (D), which may well be the case under very favourable conditions of growth. The distinctions become increasingly important as conditions become less favourable, and (A) decreases without any necessary decrease in the other terms. Finally, as the net changes approach zero, we have five possible positions for the compensation point, respectively when ULR, CDR, NCRR, LWCB and LLCB $= 0$. The first three of these are likely to prove practically identical on many occasions, but there may well be substantial differences between the last three, where photosynthesis equals either the respiration of the whole plant during a complete cycle of light and dark (NCRR $= 0$); or the respiration of the whole plant during the light phase only (LWCB $= 0$); or the respiration of the photosynthetic organs alone during the light phase (LLCB $= 0$). Of these the first is likely to be of the greatest biological importance, corresponding to the point where there is no net uptake or loss of carbon dioxide during a single cycle of light and dark. Over the years neglect of these distinctions has led to a good deal of confusion in the literature dealing with compensation points.

Summing up the above definitions: if, in relation to the whole plant, we have the following rates expressed as changes in dry weight per unit leaf area; and if the dry weight changes involving photosynthesis and respiration be taken as gain or loss of hexoses; and

(A) = daily increment due to photosynthesis alone;
(B) = daily loss due to respiration over a complete cycle of light and dark;
(B_i) = loss due to respiration of the whole plant during the light-phase alone;
(B_{ii}) = loss due to respiration of the photosynthetic organs only during the light-phase;
(C) = daily increment due to uptake of elements other than carbon, hydrogen and oxygen;
(D) = daily increment due to metabolic changes within the plant other than (A), (B) and (C). This may be a positive or negative quantity;

Then

Unit Leaf Rate (ULR) $= (A) - (B) + (C) + (D)$	13.11.1
Carbon Dry-weight Rate (CDR) $= (A) - (B) + (D)$	13.11.2
Net Carbon Reduction Rate (NCRR) $= (A) - (B)$	13.11.3
Light-phase Whole-plant Carbon Balance (LWCB) $= (A) - (B_i)$	13.11.4
Light-phase Leaf Carbon Balance (LLCB) $= (A) - (B_{ii})$	13.11.5

It is worth noting that the net carbon reduction rate can be derived directly from the corresponding values of unit leaf rate, without making separate allowances for mineral uptake and metabolic balance, by determining the mean carbon dioxide equivalent of the dry weight [W_c,

13.11

29·4] at the beginning and end of the experimental period, using a combustion method. The procedure illustrates a practical advantage of the separation which has been notionally made between the changes in dry weight due to photosynthesis and respiration, A and B, working between carbon dioxide, hexose, and back to carbon dioxide (and hence involving constant dry weight factors of $1:0·68$ or $1:1·47$), and the changes in dry weight due to all other metabolic processes, C and D, which taken all together make up the measured value of the carbon dioxide equivalent of the dry weight W_C. As A and B are both expressed in terms of dry weight gained and lost in these processes, if W_C is found to be constant over the experimental period we at once obtain the value of the net carbon reduction rate by multiplying the value of unit leaf rate by $0·68\ W_C$. If W_C is found not to be constant we proceed as in a normal unit leaf rate calculation (Chapters 16 and 20), multiplying the values of total plant dry weight by the appropriate values of $0·68\ W_C$ throughout.

13.12   Problems of calculation

Measured over an infinitesimally short period of time, dT,

$$\text{Unit leaf rate, } \mathbf{E} = \frac{dW}{dT} \cdot \frac{1}{L_A} \qquad\qquad 13.7.1$$

where W is total dry weight and L_A total leaf area. The equation is useless for practical purposes as it stands, because the rate of change of dry weight with time cannot be determined directly over very short periods. To calculate $\mathbf{E}$ over longer periods requires that the expression be integrated. If $\mathbf{E}$, W, L_A and T are all independent variables, and nothing is known about the relationship between any two of them, no progress can be made with the integration. Depending on the nature and accuracy of our information there are many ways of tackling this problem, some requiring precise information on some specific relationship, others enabling us to make use of less precise information in order to set limits within which an estimate of $\mathbf{E}$ must lie. In order not to break the thread of the present discussion detailed consideration of these will be deferred to Chapters 16 and 20.

13.13   The unit of photosynthetic machinery

Before leaving unit leaf rate and passing on to other matters, it will be useful to consider shortly its relation to the photosynthetic mechanism.

The related questions of how far variations in unit leaf rate are to be attributed to changes in the environment, and how far to changes in the physiological mechanisms themselves, involve matters not yet discussed, and are therefore better postponed [20.6, 28.2–28.14, 30.2–30.9]. Of the four main branches of physiological activity which together determine total gain in dry weight over 24 hours [Fig. 13.11], carbon assimilation is normally the largest quantitatively. It is then the main component of unit leaf rate, and the question arises—is a unit of leaf area a suitable measure of a unit of photosynthetic machinery? Does the amount of machinery per unit area change with age? with position on the plant? with external conditions? It would be helpful to have answers to these questions when considering the relationships between unit leaf rate and external conditions. One would then know whether some specific change was connected with the amount of machinery present, or with the rate of working, or with both.

Haberlandt had this point in mind in counting chloroplasts per unit of leaf area, and comparing these numbers with '*specifische Assimilations-energie*' [13.8]. It is clear from the context that he was using chloroplasts as a means of assessing chlorophyll content, but nowadays we should consider the chloroplast itself to be a more useful unit. However, the matter can hardly rest there. For one thing, to make an estimate of the average number of chloroplasts per unit leaf area of a whole plant is a formidable task, involving much research into leaf structure and very many tedious measurements. Secondly, chloroplasts differ from each other in size and structure and probably also in physiological activity, and are thus inherently unsuitable as units of machinery for comparative purposes. This is not to say that more studies on the lines of those initiated by Haberlandt would not be very valuable: they would shed much light on the underlying mechanisms, and it is a pity that so little has been done along these lines in the eighty-odd years which have elapsed since the first appearance of *Physiologische Pflanzenanatomie*. It would be a great convenience to have some simple criterion which would make it possible to assess how much photosynthetic machinery is packed under unit leaf area.

13.14   Leaf protein–nitrogen

In 1946 Williams suggested that leaf protein–nitrogen was a suitable measure for this purpose, and used it to interpret data he had collected on the effects of mineral nutrient supply, and especially nitrogenous manuring, on the growth of *Phalaris tuberosa*. In practice this idea involves making an additional measurement on samples of the plant material, of

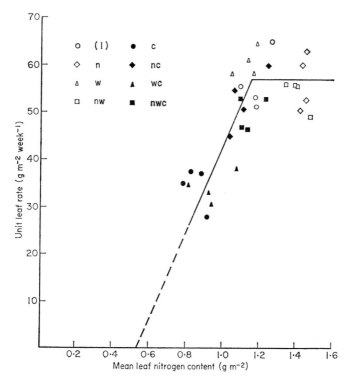

Fig. 13.14. Unit leaf rate of *Impatiens parviflora* compared with mean leaf nitrogen content during the second week of competition with *Agropyron repens*, using the experimental arrangement shown in Fig. 10.6. (1) control; n, added nitrogen; w, added water; c, competition. The fitted lines are constructed from the regression of leaf nitrogen content on unit leaf rate for all treatments except n and nw (ascending limb), and the mean unit leaf rate for the (1), n, w and nw treatments (horizontal limb). From Welbank (1962).

leaf protein–nitrogen per unit leaf area or per unit leaf dry weight, and then scaling up to give the total for the whole plant. The method therefore involves considerable additional analysis, and possibly for that reason it has been comparatively little used.

13.14

There is a further difficulty. If one could be sure that the division of leaf protein–nitrogen between photosynthetic machinery and other protoplasmic structures remained constant, the basis of the method would be established. Unfortunately what little evidence is available suggests that within a particular species or variety the proportion of the total leaf protein–nitrogen content contributed by the chloroplasts may vary with the age of the leaf, and also that it may be affected by nitrogenous manuring. It seems that as the supply of soil nitrogen increases, photosynthesis increases only up to a certain point (see, e.g., Fig. 13.14). When this point is passed, leaf nitrogen content continues to increase, and with it (though not necessarily proportionately) leaf protein–nitrogen, and also respiration. Thus nitrogenous manuring has different effects on the different components of overall dry weight gain.

The problem of finding a simple and convenient measure of the unit of photosynthetic machinery is still not solved.

13.15 Leaf area ratio

We have already compared the relatively small ontogenetic drift of unit leaf rate during the middle part of the life cycle of maize with the larger ontogenetic drift of relative growth rate [13·7]; and noted that in consequence it is here much easier to analyse the effects of environmental factors on the former than on the latter. The two are related by the leaf area ratio: instantaneously

$$\frac{dW}{dT}\cdot\frac{1}{W} \quad = \quad \frac{dW}{dT}\cdot\frac{1}{L_A} \quad \times \quad \frac{L_A}{W}.$$

Relative growth rate = Unit leaf rate × Leaf area ratio   13.7.2

RGR   ULR   LAR

The effect of taking in the extra datum, leaf area, is that it can be used to split relative growth rate into two derivates, which can then be studied separately. Each derivate has its own characteristic ontogenetic drift, and each is influenced in differing degrees by internal factors and by aspects of the environment. The forms of these drifts would be expected to vary considerably from one species to another, and also in seasonal climates with the time of year at which the plants are being observed. These points are illustrated by comparing the drifts of relative growth rate, unit leaf rate and leaf area ratio for Briggs, Kidd and West's experiments on *Helianthus annuus* in Fig. 13.15 with the corresponding figures

13.15

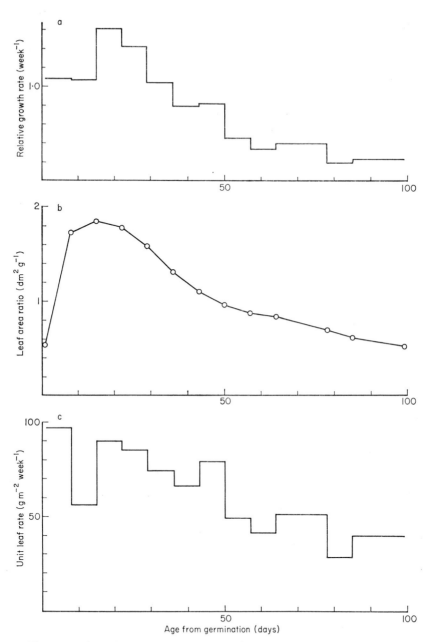

Fig. 13.15. (a) Relative growth rate (week^{-1}); (b) leaf area ratio (dm^2 g^{-1}); (c) unit leaf rate (g m^{-2} week^{-1}) in relation to age from germination of plants of *Helianthus annuus* grown by Briggs, Kidd and West at the Botanic Garden, Cambridge, between 27 May and 3 September 1920 (a repeats the data of Fig. 13.6.2b as far as 99 days).

13.15

for *Zea mays* given in Fig. 13.7 (for comparison of the relative growth rates alone, see Fig. 13.6.2). In maize the slow development of photosynthetic capacity by the young leaves led initially to a negative value of unit leaf rate, and hence of relative growth rate. During the fourth to the sixth week both unit leaf rate and leaf area ratio rose rapidly, bringing relative growth rate to a high value by the seventh week. Thereafter unit leaf rate fluctuated considerably but showed relatively little consistent drift until towards the end of the growing season; whereas by contrast leaf area ratio, remaining relatively constant from the seventh to the ninth week, fell to about a sixth of its former value during the following 5 weeks, bringing relative growth rate down again to a low value. By contrast Fig. 13.15 shows that in the sunflower there is no initial depression of unit leaf rate: its value fluctuates considerably from week to week, and after the summer solstice it falls gradually to a value in late August about half that in late June (cf. Fig. 13.19a, b). Leaf area ratio, on the other hand, starts low (accounting for the initial low value of relative growth rate), rises to a maximum in mid-June, and thereafter falls more or less steadily to reach in early September a value less than a third of the maximum. Although the form of the drifts is so different in the two species, yet during the later part of the life cycle of both, when the plants are large and the absolute growth rates high, leaf area ratio makes the major contribution to the fall of relative growth rate. When we think back to the type of organization envisaged in Fig. 2.8, it is not surprising that this should be so. Under favourable conditions for growth the rate of increase of dry weight is mainly determined by the carbon assimilation term; in most higher plants this is related directly to the activity of the leaves. For a plant growing in flushes we might reasonably expect that there would be a substantial period, between full maturity of the leaves and the onset of senescence, when there would be no marked drift in the functioning of the photosynthetic machinery under a particular set of environmental conditions. For a plant continually unfolding new leaves, there would similarly be a substantial period in the middle of the growing season when the proportions of young, middle-aged and old leaves would not be changing rapidly, and once again we might expect that there would be no marked drift in the rate of working of the photosynthetic machinery. At these times drifts of unit leaf rate would be expected to be associated with regular climatic changes (e.g., Fig. 13.19).

The case of leaf area ratio is much more complex. At any one moment

13.15

a little-understood internal correlation mechanism will determine what proportion of the new assimilates are translocated to the sites of development of new leaves, as opposed to all the other forms of consumption and storage. Once the assimilates have arrived in the young leaf, the area which this assumes in relation to the dry matter which it contains is determined by a different set of mechanisms, controlling leaf expansion. Finally, for an annual plant, the leaf area ratio at any time represents an integration of the effects of all these mechanisms, from germination up to that time. It is small wonder that it has an ontogenetic drift: or that it should be difficult to sort out the manifold effects of the environment on all these mechanisms—some additive, some interacting, some self-cancelling. It seems that a further analytical step is needed; it is desirable to distinguish the correlation mechanisms determining dry weight distribution between the various organs, from the mechanisms governing leaf expansion.

13.16   Leaf weight ratio and specific leaf area

This analytical step involves taking account of yet another datum, total leaf dry weight (L_W); but it is likely that this will already have been recorded [11.4, 11.6], and if so no extra work is involved here. We then have

$$\frac{L_A}{W} = \frac{L_A}{L_W} \times \frac{L_W}{W} \qquad 13.16.1$$

Leaf area ratio = Specific leaf area × Leaf weight ratio

LAR   SLA   LWR

Unlike unit leaf rate, where there may be problems in deciding how the derivative should be computed, these relationships can be determined at a specific time, that of harvest, when they are quite unambiguous, provided that leaf area can be readily defined [13.8, 18.1, 19.1]. We shall return to the question of how they have been changing between harvests in Chapters 18 and 19.

Putting together the ideas of the preceding paragraphs, the original overall measure of growth, relative growth in total dry weight, has been progressively split up by bringing in other quantities; first of all, leaf area, L_A;

$$\frac{dW}{dT} \cdot \frac{1}{W} = \frac{dW}{dT} \cdot \frac{1}{L_A} \times \frac{L_A}{W}.$$

Relative growth rate = Unit leaf rate × Leaf area ratio.   13.7.2

Each of these derivates has then been split into two again, by bringing in, first, some measure of the total amount of photosynthetic machinery, L_p (supposing that there were some way of measuring it); and second, the total dry weight of the leaves, L_W, which in any case would probably be part of our harvest data. In the present state of our knowledge it will not often happen that we wish to split unit leaf rate, but supposing that we did, we should then have split up relative growth rate into four complementary parts.

$$
\begin{array}{ccccc}
\text{RGR} & = & \text{ULR} & \times & \text{LAR} \\
\end{array}
$$

$$
\frac{dW}{dT} \cdot \frac{1}{W} = \overbrace{\frac{dW}{dT} \cdot \frac{1}{L_p} \times \frac{L_p}{L_A}} \times \overbrace{\frac{L_A}{L_W} \times \frac{L_W}{W}}, \qquad \text{13.16.2}
$$

where, instantaneously,

$\dfrac{dW}{dT} \cdot \dfrac{1}{L_p}$ represents the rate of increase of total dry weight per unit of photosynthetic machinery;

$\dfrac{L_p}{L_A}$ represents the photosynthetic machinery per unit leaf area: a physiological measure of leaf structure;

$\dfrac{L_A}{L_W}$ the specific leaf area, represents the average leaf expansion in area per unit dry weight;

$\dfrac{L_W}{W}$ the leaf weight ratio, represents the average fraction of the total dry weight in the form of leaves.

Each of these facets of growth can be studied, both separately, and in relation to each other.

13.17   Relative leaf growth rate

From the preceding paragraphs it will be apparent that relative leaf growth rate $dL_A/dT \times 1/L_A$, analogous to relative growth rate, $dW/dT \times 1/W$, is quite as difficult as the latter to handle from the point of view of ontogenetic drift. If anything, it is even more complex from the point of view of the effects of the external environment upon it. Consideration of dW/dT involves the four broad categories of physiological activity already mentioned [Fig. 13.11]—photosynthesis, respiration, mineral uptake and metabolic balance. dL_A/dT involves all this as far as the supply of new material is concerned; and it is further affected by internal correlation mechanisms [13.15], by meristematic activity and the mechanisms in-

13.17

volved in the origin of new organs [2.8], and by the mechanisms involved in leaf expansion (2.8, and see also Chapters 18 and 19). The extent to which one system or another predominates in determining dL_A/dT is partly a question of the species and its growth form, partly a matter of the phase of the growth cycle, partly a function of external conditions. In the case of many rapidly growing annuals the supply of new material predominates, and one not infrequently finds an early stage in the life cycle in which $dL_A/dT \times 1/L_A = dW/dT \times 1/W$, in other words the leaf area ratio, L_A/W, remains constant. During this phase the plant can be regarded as multiplying units of structure, all having the same ratio of leaf area to dry weight. Not unexpectedly this phase is always limited in length and is succeeded by a stage where $dW/dT \times 1/W > dL_A/dT \times 1/L_A$, that is, where a decreasing portion of new dry material is being converted to new leaf area, as the plant architecture becomes more complex and the correlation mechanisms become increasingly important [22.3, 22.6].

At the other end of the scale, in many deciduous tropical trees the first expanded flushes of new leaves are not green, but pink or red; they hang down limply when first expanded, and only gradually during the next week or so do they become green and assume their mature texture and posture. In those cases where photosynthesis is slow to develop (and in many species little is known about this), the whole of the material needed for the expansion of the flush must come from storage, so that growth of leaf area is almost wholly controlled by correlation and expansion mechanisms. The relations between relative growth rate and relative leaf growth rate, and their consequences, will be considered further in Chapter 22.

13.18 Leaf area index

So far in this general survey of developing methods of analysis we have been considering the growth of the individual plant, the increase in its stock of reduced-carbon units and the ways in which these are deployed in making up the plant body, as a system complete in itself, acted upon in a variety of ways by the external environment. The further step of including in consideration the area of ground occupied by the plant was taken by Watson, who introduced the concept of Leaf Area Index, and related it to the ideas already discussed.

The determination of leaf area index is not difficult for an agricultural crop with a known spacing, provided that leaf area can be readily defined

[13.18, 18.1, 19.1]. In addition to the harvest data already collected, we need the number of plants per square metre of ground area, N, and hence the area of ground per plant, N^{-1}. The leaf area divided by this area of ground gives the leaf area index (LAI), a pure number. If L_A be also in square metres, then $LAI = NL_A$. It is obviously the average number of complete layers of leaves produced by the plant; but in practice the leaf mosaic is never perfect, and even with an average of four layers of leaves over the whole area, looking upward from any particular position on the ground, there will still be gaps where there are no leaves, compensated for by areas where there are five or more layers. The point is well illustrated by the hemispherical photographs in Plate 13.18 of two crops of *Musanga cecropioides*, growing in pots, and spaced so that those in the upper photograph have a leaf area index of 1·0, and those in the lower photograph one of 4·0. As the position of these gaps changes with the point of view, the lower leaves will be subjected to alternating conditions of sun and shade as the sun moves over the sky, and the sunny spells will become less frequent, the higher the leaf area index, and the lower the position of the leaf in the canopy.

For a field crop, the leaf area index is at first low; in the cases of crops grown from seed, it may remain below 1·0 for many weeks. Eventually, the increasing absolute growth rate of the leaves carries the index rapidly through 1, and on up to a maximum. The form of the curve and the value of the maximum may vary considerably, as shown in Fig. 13.18.1, giving smoothed curves for the changes in leaf area index in a variety of crops, from Watson's classical paper based on work carried out at the Rothamsted Experimental Station (Watson, 1947). Here winter wheat shows a maximum value of around 3 in late May or early June, potatoes around 2 in late August, and sugar beet of nearly 3 in October. Of the instances given in Fig. 13.18.1, only for barley does the maximum value of the leaf area index occur close to the summer solstice.

In the case of a cereal crop, much photosynthesis may take place in the stems, the leaf sheaths, and parts of the inflorescences. Watson, Thorne and French (1958) considered this in the course of an investigation into barley varietal differences, and showed that if the area of these organs be added to those of the leaf laminae, the leaf area index in this sense may reach a value as high as 9 for a short time, in this particular case early in July, whereas LAI based on leaf laminae alone reached a maximum of 4·5 late in June. It is necessary to proceed with caution when making additions of this kind, on account of the large possible differences in rate of photosynthesis per unit area in the different organs concerned.

Broad-leaved trees in pure stand often reach a leaf area index of 4–6 (and cf. 31.13, 31.17 for an instance of a mixed tropical forest). In addition to varying with species and conditions of growth, the value of the

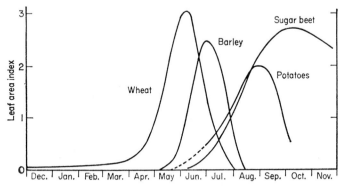

Fig. 13.18.1. Smoothed curves showing the changes with time of year of the leaf area index of four field crops grown at the Rothamsted Experimental Station (approximately in 51° 50′ N Lat, 0° 30′ W Long; from Watson, 1947). For the corresponding figures for unit leaf rate, see Fig. 13.19. Note that only for barley does the maximum value of leaf area index occur close to the summer solstice. Compare also Fig. 2.7 for maize, where the maximum L A I fell in mid-August.

index appears to be related to leaf attitude, which in turn affects the penetration of light into the stand. This point is most conveniently illustrated by comparing different forms within a single species, as in Fig. 13.18.2a and 13.18.2b for three stands of tea, *Camellia sinensis* [App. 3.1] (Hadfield, 1968 and personal communication). The form with the most upright leaves, (i), taxonomically near *C. sinensis* var. *sinensis*, had a leaf area index of 5–7; the form with the most nearly horizontal leaves, (iii), near *C. sinensis* var. *assamica*, a leaf area index of 3–4. Fig. 13.18.2b shows the most rapid falling off of irradiance in relation to L A I in the stand with the most nearly horizontal leaves, the differences being highly significant statistically (e.g., comparing any individual 20 cm point with any other, the probability of identity is very much less than 0·001).

These agricultural instances are the easy ones to measure, and the matter may be much more complex in a more natural situation, in which

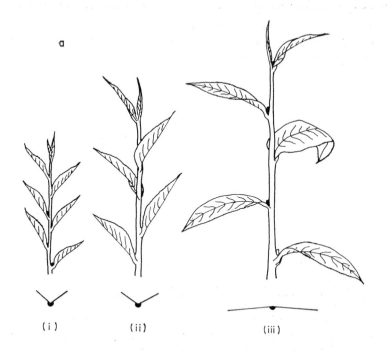

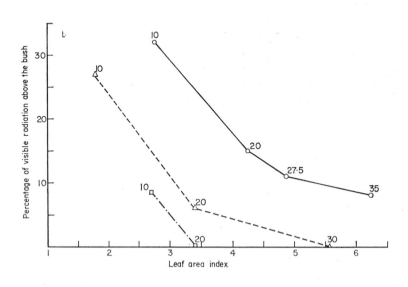

13.18

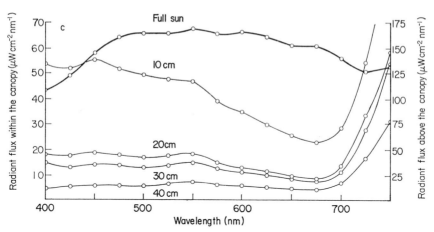

Fig. 13.18.2. Light inside the canopy of different types of tea bushes in Assam in relation to leaf area index above the plane of measurement. (a) Semi-diagrammatic drawings of shoots, all to the same scale, showing variations in leaf pose and foliar distribution (×0·2): type (i), near *Camellia sinensis* var. *sinensis*; type (ii), intermediate between types (i) and (iii); type (iii), near *C. sinensis* var. *assamica*: also showing in section the characteristic mid-leaf angle between the two halves of the lamina and the midrib, for each type. (b) Mean values of visible radiation measured on a horizontal surface within each type of canopy, in relation to the leaf area index above the plane of measurement (depth in cm below the top of the canopy shown beside each point), as percentages of the radiation above the bushes in full sun: ⊙——, type (i); ▵----, type (ii); ▢–·–·, type (iii). Fiducial limits, all type (i) means and type (ii) 10 cm, 0·13 times each individual mean value; type (ii), 20 cm, ±1·23 per cent, 30 cm ± 0·4 per cent; type (iii) 10 cm, ±2·2 per cent, 20 cm ± 0·2 per cent. (c) Spectral composition of radiation incident on a horizontal surface at different depths within the canopy of a type (i) bush, measured under sunny conditions: thick line and right-hand ordinates, immediately above the canopy of the bush; thin lines and left-hand ordinates, at depths of 10, 20, 30 and 40 cm below the top of the canopy (W. Hadfield, personal communication).

pure stands are the exception, rather than the rule. However, the importance of these notions in helping to elucidate a competitive situation can hardly be overestimated. It is possible to harvest and measure the vegetation species by species and layer by layer. Although this in its nature a destructive method, there are times when its use is unavoidable; e.g., in their study of tropical high forest in the Ivory Coast Müller and Nielsen (1965) estimated that the total leaf area index above a height of 15 m

13.18

was about 2·2 [13.13]. When dealing with vegetation on a smaller scale, and particularly when it is important to use non-destructive methods, vertical or inclined point quadrats give a means of estimating a quantity comparable to leaf area index, species by species (Greig-Smith, 1964, chapters 1 and 2).

In all these cases the position, in relation to the rest of the vegetation, of the foliage being considered is of prime importance. Fig. 13.18.2c shows, for the example previously considered, how both the intensity and the spectral composition of the radiation incident on a leaf can vary depending on its vertical position in the vegetational complex; just as Fig. 6.9 showed how the annual march of total daily short-wave radiation within a plant community can vary from place to place in a horizontal plane.

Natural vegetation poses further problems as compared with the single uniform crop. The crop can often be regarded as flat-topped, so that the maximum radiation available to each plant can easily be worked out from the appropriate area of the top of the crop. The same is not true of an irregular plant stand, and the difficulties reach their maximum in the case of an isolated tree. Here the amount of direct solar radiation (measured on a horizontal surface) intercepted by the foliage is given in effect by the shadow of the tree, which, for a given tree, varies in area with time of day and time of year. Unless the crown itself has a simple shape, approximating to a solid of revolution around a vertical axis, the estimation of radiation intercepted is exceedingly tedious.

Another illuminating use of the concept of leaf area index in relation to ecological situations was made by Watson and Witts (1959), who compared the growth of a cultivated sugar beet variety with strains of *Beta vulgaris* ssp. *maritima* collected from three different coastal localities. These strains differed considerably from each other, both in the development of leaf area index and in unit leaf rate; but when all were grown together under comparable conditions none of them produced such large plants as the cultivated sugar beet. One wild strain showed a development of leaf area index almost exactly the same as the cultivated beet, but had a lower weighted mean unit leaf rate [13.19]; one had as high a weighted mean unit leaf rate as the cultivated beet, but a much lower leaf area index at all stages; only the cultivated strain showed high values for both.

Thus while the concept of leaf area index had its origin in relation to

13.18

closed stands of vegetation of relatively simple structure, it has been possible by extension to apply it to a variety of ecological situations. Here also it has proved to be a most valuable means of relating the architecture, both of the vegetation complex and of its individual components, to the plant's opportunities for intercepting radiation.

13.19 Leaf area duration

Leaf area index is measured at a particular instant; but Watson (1947) pointed out that by integrating the area under an LAI/time curve (such as one of those in Fig. 13.18.1), the individual values could be summed over the whole growing season, giving a quantity which he called the leaf area duration. Leaf area index being a pure number, and time being measured in weeks, the area, and hence the leaf area duration, is also measured in weeks. It is as Watson put it 'a measure of the ability of the plant to produce and maintain leaf area, and hence of its whole opportunity for assimilation' during a growing season.

Watson illustrated the importance of this concept with reference to several field crops, as in Table 13.19.

These crops cover a wide range, both of taxonomic position and of habit of growth, yet it is clear that the final dry weight at harvest is related

Table 13.19. Final dry weight per unit area of ground, leaf area duration, and weighted mean unit leaf rate for the same four field crops as in Fig. 13.18.1, grown at Rothamsted Experimental Station. Based on data from Watson (1947).

	Mean dry matter at harvest		Mean leaf area duration (weeks)	Weighted mean unit leaf rate	
	cwt acre^{-1}	tonne ha^{-1}		cwt acre^{-1} week^{-1}	g m^{-2} week^{-1} = tonne km^{-2} week^{-1}
Barley	58	7·3	17	3·4	43
Potatoes	61	7·7	21	2·9	36
Wheat	76	9·5	25	3·0	38
Sugar Beet (at end October)	96	12·0	33	2·9	36

13.19

more closely to the leaf area duration than to the weighted mean unit leaf rate—in other words more closely to the plant's ability to maintain leafiness than to the weighted mean rate at which dry matter increases per unit leaf area. For example, comparing potatoes and sugar beet, which have the same weighted mean ULR, the sugar beet has 57 per cent more dry matter at harvest, and its leaf area duration is also higher by 57 per cent. Barley has the highest weighted mean ULR, 19 per cent higher than sugar beet, but sugar beet has a much larger LAD, 94 per cent higher, and a dry weight at harvest 66 per cent higher.

A word is necessary about the concept of 'weighted mean' unit leaf rate, which Watson obtained by dividing the crop yield per unit ground area by the leaf area duration. In a highly seasonal climate unit leaf rate has been found to vary within wide limits with time of year, and some of Watson's own smoothed curves for the same crops as those of Fig. 13.18.1 are shown in Fig. 13.19a (see also 20.6. We shall see in Chapter 16 that it is possible to produce the appearance of an ontogenetic drift of unit leaf rate where none exists by using faulty methods of calculation, although there is no question of this in the present case). Fig. 13.19b shows that there is a broad parallelism between these smoothed curves and changes in mean daily totals of short-wave radiation, and in particular that ULR is changing most rapidly around the equinoxes, when the daily radiation total is also changing rapidly. The changes of unit leaf rate with time of year were less noticeable in the examples previously examined [Figs. 13.7c, 13.15c]. These were centred on the summer solstice and did not extend as far as the equinoxes, although when examining Fig. 13.15c we concluded that for Briggs, Kidd and West's observations on sunflowers unit leaf rate had fallen by late August to about half its value in late June, which is in broad agreement with Fig. 13.19.

Therefore, when considering the development of a crop over a whole growing season, we must weight the values of unit leaf rate for individual periods to allow for the relative development of foliage during those

Fig. 13.19. (a) Smoothed curves of yearly march of unit leaf rate in four agricultural crop plants grown in the field at the Rothamsted Experimental Station, Hertfordshire, England (from Watson, 1947). (b) Yearly march of total short-wave radiation measured on a horizontal surface at the Meteorological Office Research Station at Cambridge, 36 miles NE of Rothamsted. Monthly means of daily totals for the years 1957–68 inclusive (data supplied by the Meteorological Office).

13.19

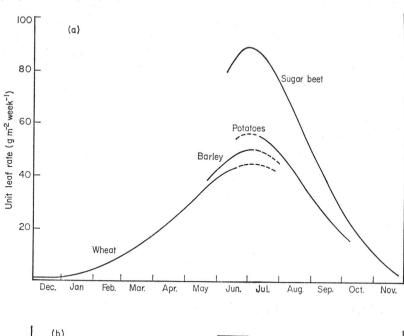

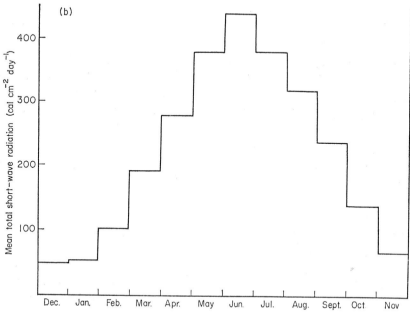

13.19

periods. The procedure underlying Watson's method of dividing the final dry weight by the leaf area duration would be as follows: if the mean ULR computed for the Xth period of growth is $_x\bar{E}$, then for computing the weighted mean ULR we multiply by $_x(\overline{LAI})/\overline{LAI}$, where $_x(\overline{LAI})$ is the mean leaf area index for the Xth period and $\overline{LAI}$ is the mean leaf area index for the whole period of growth, equalling LAD/M, where M is the number of weeks in the growing season. We have thus weighted the mean rate of increase of dry weight per unit leaf area per week to allow for the amount of leafage which the plant possessed in this particular week in relation to its available ground area. The weighted mean unit leaf rate is then given by the expression

$$\frac{1}{M}\left(_1\bar{E}\cdot\frac{_1(\overline{LAI})}{\overline{LAI}}+\dots {_x}\bar{E}\cdot\frac{_x(\overline{LAI})}{\overline{LAI}}+\dots {_M}\bar{E}\cdot\frac{_M(\overline{LAI})}{\overline{LAI}}\right)=$$

$$\frac{1}{LAD}\left(_1\bar{E}\cdot{_1}(\overline{LAI})+\dots {_x}\bar{E}\cdot{_x}(\overline{LAI})+\dots {_M}\bar{E}\cdot{_M}(\overline{LAI})\right) \quad 13.19.1$$

because $M(\overline{LAI})=$ LAD.

Let us suppose $_x\bar{E}$ to be calculated by the formula

$$_x\bar{E}\cdot\varDelta T=\frac{_xW-_{x-1}W}{_x(\bar{L}_A)} \quad 13.19.2 \text{ [cf. 16.1.3]}$$

where $_{x-1}W$ and $_xW$ are the mean dry weights of a plant at the beginning and end of the Xth period of growth, $_x(\bar{L}_A)$ is the mean leaf area for the same period, and $\varDelta T$ is the duration of the Xth period of growth. The validity of this widely used formula will be examined in Chapter 16; here it suffices to say that although there are theoretical objections to this method of estimating unit leaf rate [16.1], in practice the errors involved are large only if there are large proportional changes in leaf area during the period covered by the estimate [16.10, 16.12]. But as will be seen later in this section, the estimate of weighted mean unit leaf rate is dominated by those periods when the main absolute increase in dry weight of the plant is taking place, and for annual crops of the temperate zone this is usually a time when the proportional increase of leaf area during any one period between harvests is not large [e.g., Fig. 2.7a, d]. Therefore in an agricultural context the errors involved in the simple method of estimation of equation 13.19.2 will usually be found to be negligible (e.g., compare Fig. 2.7d with Table 16.18).

13.19

Let us further suppose that the durations of all the periods of growth are the same, and that they all equal one unit of time, so that $\Delta T = 1$ throughout. Now $_x(\overline{\text{LAI}}) = N \cdot {}_x(\overline{L}_A)$, where, as in 13.18, N is the number of plants per unit area. Therefore, from equation 13.19.2

$$_x\bar{\text{E}} \cdot {}_x(\overline{\text{LAI}}) = \frac{({}_xW - {}_{x-1}W)}{{}_x(\overline{L}_A)} \cdot N \cdot {}_x(\overline{L}_A) = N({}_xW - {}_{x-1}W),$$

and substituting in equation 13.19.1, the weighted mean ULR becomes

$$\frac{N}{\text{LAD}}[({}_1W - {}_0W) + ({}_2W - {}_1W) + ({}_3W - {}_2W) + \ldots + ({}_MW - {}_{M-1}W)] =$$

$$\frac{N({}_MW - {}_0W)}{\text{LAD}} \quad 13.19.3$$

$_0W$ is the mean weight of the seed or other propagule, which can be neglected in comparison with the final weight of the fully grown plant, $_MW$. There are N plants per unit area, and therefore the expression becomes

Weighted mean unit leaf rate =

$$\frac{\text{Final crop dry weight per unit area of ground}}{\text{Leaf area duration}} \quad 13.19.4$$

and it is not necessary to weight the individual terms of unit leaf rate for each period of growth.

Having watched this process of cancellation we now see another major restriction in the information yielded by the leaf area duration analysis—it is limited to the period of the plant's life cycle when the leaf area is comparatively large. The formula for weighted mean unit leaf rate, 13.19.4, takes no account either of the length of time which passed while the leaf area index was very small, or of the individual values of unit leaf rate during this period. These questions, which may be of great importance for the final development of the crop, must be tackled in another way.

Care is also necessary in interpreting analyses involving leaf area duration which include periods when the leaf area index was high. By integrating the area under the LAI/time curve we are assuming that all parts of the area are equivalent. While this may be true when self shading is minimal, as the value of LAI rises each further layer of leaves will contribute less and less to the overall photosynthesis of the plant, behaving like the artificially shaded plants of Fig. 3.4.1. Finally a point

13.19

may be reached where the lowest layer of leaves is near its own compensation point, where photosynthesis during the light phase equals respiration during the 24-hour cycle, and these leaves make no net contribution to the carbohydrate economy of the plant as a whole, although they increase the values of leaf area index and leaf area duration. Watson (1958) showed that this situation can easily arise in practice, producing evidence for the particular case of crops of kale that the maximum rate of dry matter production per unit area of ground occurred for values of leaf area index between 3 and 4, above which the rate fell, and concluding, 'For heavy kale crops leaf area index is already far in excess of the optimum, and it may be possible to increase the total dry matter yield of kale by repeated thinning or defoliation to hold LAI near the optimum'.

Thus, when allowances have been made for these various restrictions and cautions, the concepts of leaf area index and leaf area duration retain their interest and importance when considering the phase of substantial leaf area during which most of the final dry weight of an annual plant is built up.

13.20   Roots and mineral uptake

Up to this point our analytical discussions have been confined to the carbon part of the whole dry weight complex; we must now consider analogous means of analysing the process of mineral uptake.

When thinking about carbon assimilation, we concentrated on the leaves, as the main assimilatory organs: and considered various means of estimating and studying the actual amount of photosynthetic machinery present. We considered unit leaf rate, the mean rate of increase of dry weight per unit leaf area; leaf area ratio, the ratio of total leaf area to total dry weight, and leaf area index, the ratio of total leaf area to the ground area appropriate to that particular plant.

If it were possible, it would be desirable to set up an analogous series of concepts for uptake of mineral nutrients by roots, based on the actual area of root through which absorption of nutrients was going on. For this purpose we should disregard the non-absorbent parts of the roots, just as we disregard non-photosynthetic stems when discussing carbon assimilation. Our analogous concepts would then include unit root rate, the mean rate of increase of mineral dry weight per unit absorbent area of roots; root area ratio, the ratio of this absorbent area to total plant dry weight; specific root area, the ratio of the same absorbent area to total dry weight of absorbent roots, and root area index, the ratio of the same area to the ground area appropriate to this particular species of plant.

But valuable and interesting though these concepts would be, they are not at the present time practicable. The determination of leaf area is child's play compared to that of the absorbent area of roots, which, in the case of a natural, growing root system may be attended by very serious difficulties. Before measurements could be begun it would be necessary to carry out an elaborate experimental programme to discover the length of young growing root which is actually absorbing, and the stage of development at which absorption ceases (for an account of recent work along these lines see Scott Russell, 1970). Then it would be necessary to collect all the absorbing parts of the roots separately from the remainder, and these are often the most difficult part of the root system to recover from the soil. Measurement would be further complicated by the irregular growth of root hairs in most natural soils, and in some cases by ecto-trophic mycorrhizas (for accounts of these see Harley, 1969).

In practice the analysis has hitherto been confined to quantities which are simpler to handle, an interesting recent example being given by Hackett (1969). In 1948, in connection with studies of uptake of phos-phate, Williams suggested that average rates of absorption of mineral nutrients should be calculated per unit dry weight of the root system as a whole, and he gives some figures for these. In 1962 Welbank suggested that such rates should be called Specific Absorption Rates, giving them the symbol $\mathbf{A}$, with a suffix indicating the element concerned, thus $\mathbf{A_N}$ for the specific absorption rate for nitrogen. He also included in his analysis the root weight ratio, RWR, the ratio of total root dry weight to total plant dry weight. These measures have the merit of practicability, and their use can lead to interesting conclusions if we bear our earlier considerations in mind.

Welbank (1962) experimented on competition between *Impatiens parviflora* and *Agropyron repens* with and without added water and nitrogen (as ammonium sulphate, equivalent to 100 p.p.m. nitrogen), eight treatments in all. His figures for the effects of these on the root weight ratios and specific absorption rates of the plants of *I. parviflora* are shown in Fig. 13.20. The contrary effects of nitrogen and competition exemplify the value of this type of analysis. Added nitrogen caused a significant

13.20

increase in A_N, and either no effect on RWR (first week) or a decrease (second week), but a decrease much smaller than the increase in A_N, so that on balance nitrogen uptake was substantially higher in both weeks. This is all as one would expect.

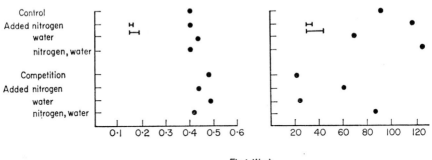

First Week

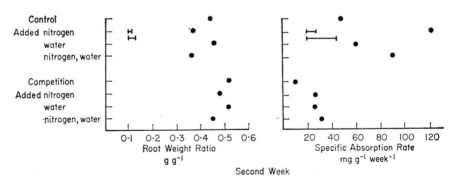

Second Week

Fig. 13.20. Root weight ratios and specific absorption rates (in mg nitrogen taken up per g dry weight of root per week) of *Impatiens parviflora*, growing alone and in competition with *Agropyron repens*, with and without added water and nitrogen (as ammonium sulphate), using the experimental arrangement shown in Fig. 10.6. Standard errors shown in each case, together with least differences significant at the 5 per cent level of probability. Redrawn from Welbank (1962).

Competition, on the other hand, caused a large decrease in A_N, to less than a quarter of that of the control, in both weeks; but this was accompanied by an increase in the growth of the root system of *I. parviflora* relative to that of the rest of the plant. Under competition alone, RWR was significantly higher than the control in both weeks.

13.20

We also see that in the first weeks, the addition of both extra nitrogen and extra water, under conditions of competition, restored both A_N and RWR to values close to those of the control. In the second week, this is also true of RWR, but not of A_N. The specific absorption rate with added nitrogen and water is more than three times as great as with competition alone, but it is now below the control, though not significantly so.

In the same paper is a discussion of several related questions—of depletion of soil nitrogen during the second week; of the effects of soil water on nitrogen availability; of the role of mineralization of organic soil nitrogen, and the effects of soil water on this; and it pointed to many problems still unsolved. With these we are not, however, at present concerned, our object in looking at these experiments having been to exemplify the utility of analysis of absorption data by this simple use of figures for total root dry weight.

13.21 Special problems of large and long-lived plants

The ideas and broad methods of analysis discussed in the foregoing sections have all been worked out in connection with the growth of annual plants, or of short-lived plants like sugar beet which are treated as annuals for agricultural purposes. For all these plants our biological unit of time is the growing season; we start from seed or some other perennating structure and follow the changes in size and in plant architecture during one such unit of time; in the next growing season a broadly similar cycle is repeated, and so on. But for large and long lived plants, of which trees form the most conspicuous category, there is a progression from one growing season to the next, and many of the concepts which we have been considering no longer apply in quite the same simple form.

A plant developing from seed starts its growth cycle with an initial partially differentiated structure, the embryo, and an initial store of reduced carbon compounds in the mobilizable food reserves. At least part of the dry weight of the seed, represented as a minimum by the testa, cannot be mobilized to form part of the growing plant, and this moiety will in due course be lost. The first photosynthetic structures are developed from the embryo and the food reserves; the plant then becomes autotrophic, and if it is an annual in an open situation, as were most of the examples we have so far considered, photosynthesis soon supplies the

13.21

bulk of the reduced-carbon units needed for continued increase in size. The plant then enters on its main phase of vegetative growth; and it is with this phase that our analysis has so far been mainly concerned. We have examined the concept of unit leaf rate, as relating the overall increase in dry weight to leaf area, considered as a simple measure of the photosynthetic machinery; we have also examined a variety of indexes connected with plant architecture, and particularly with the production, expansion and support of the developing leafage; and we have related these to the plant's dry weight, considered as a measure of its growing stock of reduced-carbon units. We have not so far been concerned with that part of the initial seed dry weight which was not available to the young plant, because throughout most of the vegetative phase it has been negligibly small.

When we come to consider a tree growing in a highly seasonal climate the concepts of unit leaf rate and specific leaf area can still be applied, but our other indexes need to be reconsidered. For example, are we to continue to express leaf area ratio as the ratio of leaf area to total plant dry weight? If so, it will change its meaning as the tree gets older. Even in the second season the total plant dry weight ceases to be almost wholly the production of the current season; year by year the non-mobilizable fraction of previous years' production increases. Furthermore, most trees of seasonal climates, and many of the wet tropics, grow in flushes, and these flushes are frequently of predetermined length. The leaf area of the current phase of the vegetative cycle is then related less to photosynthesis during the current phase than to stored materials remaining from previous vegetative phases; and both the size and structure of the flush (e.g., whether it produces flowers or not) may be determined by past events (for a survey of the many possible variants in growth habit of tropical trees see Koriba, 1958).

If total plant dry weight is not to be our basis, what can be used instead? In one sense the concept of leaf area ratio is as valid as ever—the plant functions as a whole; if it is to become a large tree massive supporting and conducting systems are essential, and these can only be developed over periods of years. When we consider a tropical tree such as *Musanga cecropioides* where growth is continuous from seed to death, producing a tree over 20 m in height in 12–14 years (or, as a more familiar example of the same growth form, the common house plant *Ficus elastica*), we see that the mode of growth is simply a continuation of that

13.21

which we have been considering in a plant such as the sunflower. There is no obvious reason why the whole growth cycle should be divided into arbitrary time intervals, annual or other. But here the growth of the plant is indeed one continuous process, and as with the annual sunflower, the structure *now*, the rate of production of new dry material *now*, and the rate of production of new leafage *now*, all depend on a subtle integration of the past; and they will all make their contribution to the corresponding integration in the future.

The plant growing in flushes, on the other hand, grows by a mechanism which itself naturally divides the whole growth period into a number of vegetative cycles, whether or not these correspond to climatic seasons. They thus afford us a simple opportunity for further analysis, and the problem is how this opportunity should be seized.

The simplest initial attack would seem to be by drawing up a balance sheet of the dry weight changes during a single vegetative cycle. Dry weight gain and loss will result from the processes previously considered [4.3, 12.1, 13.11]; if the tree is deciduous there may be a period when respiration of the trunk, branches, and roots is continuing in the absence of photosynthesis; whether the tree is deciduous or not there will be a loss of leaves, and probably also of other organs, branches, roots, flowers, etc. When all these have been allowed for a net gain will remain [Fig. 31.14]. The size of this net gain is likely to be an important factor contributing to the size of the next flush of growth. The main outlines of the size and structure of the flush are frequently laid down in the bud. In plants where photosynthesis is slow to develop in young leaves, the bulk of the initial material for leaf expansion will come from the net gain of the previous season. But there are other essential calls upon this net gain —the production of young roots (often linked to the flushing cycle of the leaves), and the addition of new material—xylem, phloem, bark—to the older parts of the plant. Finally, some of the net gain may remain in semi-permanent storage.

The collection of suitable quantitative data followed by an analysis along these lines should yield valuable additional information on the life of these long-lived plants in relation to their environment. In particular, it should be of value to study the ratios of leaf area and leaf weight during a particular vegetative period to the net gain from the previous vegetative period, two quantities analogous to leaf area ratio and leaf weight ratio during the main vegetative phase of an annual. These could be related to

13.21

the size of the plant, and to environmental conditions both during the previous phase of the vegetative cycle and during the period when the new flush of foliage is developing. It is not surprising, however, that as yet comparatively little has been done along these lines. This is not solely because of the long periods of time needed when working with long lived plants. Their size and structure pose a number of special technical difficulties of measurement which can often be avoided when working with annuals [e.g., 11.12, 12.12]. We shall consider a number of these difficulties in relation to specific examples in Chapter 31.

13.22   Reproductive structures

In the great majority of higher plants growth is at first vegetative, and then after a time comes a phase of formation of reproductive structures—spores in the Pteridophyta, seeds in the Gymnosperms and Angiosperms and in all groups a very great variety of organs of asexual reproduction. The relations between this phase and the earlier, purely vegetative one are extremely various. A number of plants will grow vegetatively for many years or even several decades, finally flowering, producing seed, and dying. They are said to be monocarpic (see also 23.3). Well-known examples include some Liliaceous plants such as species of *Aloë*, and many species of Bamboo. The latter are particularly spectacular as in places they form important components of the vegetation. Other plants start to flower very early in life, and thereafter flower continuously. Many long-lived plants flower every year; some plants growing in flushes flower on every flush. If different branches have different flushing cycles, this can lead in a non-seasonal climate to almost continuous flowering somewhere on the tree. Some plants flower only occasionally, and in many trees little is known about the causal mechanisms. At the same time, whether flowering is frequent and regular, or whether it is occasional, and possibly erratic, can be of great biological importance. On Poore's theory it supplies an essential part of the key to the old enigma of the mosaic of very many tree species in tropical rain forest (Poore, 1967, 1968).

The onset of flowering may be brought on by a mechanism wholly internal to the plant, such as the reaching of a particular morphological state—e.g., in *Impatiens parviflora*, the unfolding of the eighth leaf [27.9]; or it may be triggered by some wholly external factor such as daylength or temperature. It may be conditioned solely by conditions around the time of flowering; or the switch to the flowering state may have happened a few weeks earlier, as in chrysanthemums or lettuce; or it may have been already predetermined months or even (as in some bamboos, 23.3) years before. The flower primordia may have been laid down in the resting bud, but even when this is not so (as in the grape vine, *Vitis vinifera*), flowering or non-flowering may be wholly determined by conditions during the previous season. Sometimes the non-reproductive part of the plant in the flowering state appears indistinguishable, morphologically and anatomically, from a similar plant in the non-flowering state; often there are profound differences. These may be reversible, as in sugar beet, where the growing region can be switched from the rosette to the elongate flowering form and back to the rosette by applications of plant hormones; or they may be apparently irreversible, as in ivy, where cuttings of the flowering form will form upright bushes and grow on in this form without reversion for decades.

Asexual reproductive structures, which in different plants may take many forms, and appear in almost any position on the plant body, exhibit equally varied behaviour. Any attempt to assemble into a single conspectus all the more important facets of the formation of reproductive structures by plants would have to extend to many times the length of the above short survey. It is not surprising that no single unifying principle has been found to run through such diverse phenomena. In consequence, those working on quantitative aspects of the growth of reproductive structures have had to consider problems of their initial appearance, growth, and final size against the background of the relationships found in the particular plant concerned. The numbers, sizes and times of appearance of the reproductive structures themselves may be of the greatest importance, especially if these are related to some unusual combination of events in the environment, the plant, or both together. Alternatively, if the onset of the reproductive process is very regular, or related to some almost inevitable combination of events such as seasonal changes in the environment, the quantity of plant material involved may have greater biological importance, and this can be related to total plant dry weight by an index of plant architecture similar to the leaf, stem and root weight ratios [23.5–23.9]. For long-lived plants growing in flushes it could be related to the net gain in dry weight during the previous cycle of growth [13.21]. If we are concerned with only one kind of

13.22

reproductive structure, such as a seed, the number formed during a life cycle can be related to the *Substanzquotient* over the whole life cycle [13.4]. But under natural conditions many plants produce more than one kind of reproductive structure, and in general the most useful method of analysis can be chosen only when the main elements of the problem are known.

13.22

14

FIRST ANALYSIS OF
HARVEST DATA

14.1 State of plants at harvest

Our first analysis of the harvest data will be concerned with the state of
the plants as observed at the time of harvest, leaving consideration of
the rates at which the various processes of growth have been going on
between one harvest and the next to the following chapters.

14.2 A particular example

It will be convenient to analyse and comment on actual experimental
data, and two sets of results yielding interesting comparisons were
collected by the classes reading for Part II of the Natural Sciences Tripos
at Cambridge in the course of their normal field work in plant physiology
during the Long Vacations (so-called) of 1966 and 1967. These data have
already been referred to [7.6, 11.6, 11.21–11.23]. The work was done at
the University Botanic Garden, on plants which had grown in the open.
Three graded seeds were sown in each of a number of 25-cm pots of
coarse vermiculite, watered three times a week throughout the experiment
with a complete nutrient solution and at other times with distilled water
as necessary [6.6]. After a preliminary sorting and grading of the seedlings
to obtain plants all of which had germinated within 24 hours, the pots
were plunged 1·0 m apart in N-S rows 1·1 m apart on a site receiving
unobstructed sunlight from the southern half of the sky, but having
obstructions (distant trees, buildings) to the ENE and WNW, cutting out
low sun from the whole plot. There was thus no mutual shading [6.8]
and the positions of all plants were equivalent.

Two age groups were grown, sown about the middle and end of June
respectively, for harvesting in late July and early August. In both
years it was arranged that at the first harvest the younger plants were
25 days old and the older ones 39 days old from sowing. In each age

group 36 pots each containing one plant were set out in the field, and from these at the time of the first harvest were selected 21 plants, in 3 sets of 7, large (L), medium (M), and small (S), each set being as uniform as possible. Three of the seven plants were chosen by lot for immediate harvest, three for harvest 7 days later, the remaining one being a spare in case one of the chosen plants was damaged in any way during the week. The three comparable plants at each harvest carried the letters *ABD* for the older plants, *abd* for the younger ones, letters where there is no risk of confusion between upper and lower case. The students were divided into three groups, each of which harvested all the plants carrying one particular letter. In this way any systematic variations in harvesting technique between one group of students and another was evened out, and thrown into the general inter-plant variability.

14.3   Dry weights and leaf areas

In Table 14.3 are set out the harvest data for the 2 years, the two ages of plants, and the two harvests, designated as I and II.

In 1967 the 25-day-old plants are in all respects larger than the 46-day-old plants in 1966; in total dry weight they are nearly twice as

Table 14.3. Data from harvests of *Helianthus annuus*. Ages in days from date of sowing. All dry weights in grams per plant, leaf areas in cm² per plant. For details of experimental arrangement see text, 14.2.

| Sample | Age | Mean dry weights | | | | Mean leaf areas |
		Total	Leaf	Stem	Root	
1966						
abd I	25	1·568	0·789	0·324	0·455	256
abd II	32	4·508	2·366	0·844	1·298	717
ABD I	39	9·659	4·311	2·202	3·146	1432
ABD II	46	23·428	9·664	6·260	7·503	2760
1967						
abd I	25	45·420	25·540	9·677	11·203	6508
abd II	32	130·683	60·190	34·668	35·825	18739
ABD I	39	233·077	104·628	71·773	56·676	25713
ABD II	46	431·048	158·779	144·515	127·754	40918

14.3

large, and in leaf area and leaf dry weight more than twice as large. At the rate at which growth was going on in 1966, about another week would have elapsed before the oldest plants were approximately the same size as the youngest plants in 1967, making about 53 days from sowing as compared with 25. When we consider that 3 days of this period elapsed before the cotyledons emerged above the ground, the comparison is rather of 50 days' opportunity for carbon assimilation as compared with 22. Making the alternative comparison, at 46 days the plants grown in 1967 have more than 18 times the total dry weight, and roughly 15 times the leaf dry weight and leaf area, of those grown in 1966, thus affording a striking example of how natural variations in summer weather can affect plant growth.

Yet during the weeks which elapsed between the harvests in the two years the differences do not at first sight appear striking. In 1966 the *abd* plants increased in total dry weight by a factor of 2·9, in leaf area by 2·8. In 1967 the corresponding figures were 2·9 and 2·9. For the *ABD* plants in 1966 they were 2·4 and 1·9; for 1967, 1·85 and 1·6. In fact here the 1967 plants are not growing so fast relatively as those grown in 1966.

14.4   Comparison of dry weight distribution

In Table 14.4 are set out the distributions of dry matter between the different organs of the plant, not only in the plants as harvested, but also as the increments of dry matter added in the week between harvests. We see the interlocking effects of age, of size, and of previous environmental experience. The extreme differences in size for age are useful as indicating the range of distribution of material between the different types of organ over this considerable range of plant size. Leaf weight ratio rarely exceeds 55 per cent, or falls below 35 per cent; stem weight ratio rises from 18–20 per cent for the smallest plants to 30–34 per cent for the largest; root weight ratios are always much larger than stem weight ratios for the younger and smaller plants, but exhibit the smallest range of all, about 8 per cent, between extremes of 24·3 and 32·6. What is more, within any one year the range is only about half this, 3·8 per cent in 1966 and 5·3 per cent in 1967 (and see Fig. 26.4.2).

In one respect only do we find consistency between the two years, and that is when we compare the differences in distribution of the new dry matter produced during the week between the plants of the two age

Table 14.4. Data derived from the same harvests of *Helianthus annuus* as Table 14.3. Ages in days from date of sowing. Dry weights in grams per plant. Dry weight ratios in percentage of mean total dry weight.

Sample	Age (days)	Mean total dry weight (g)	Dry weight ratios			Mean leaf area ratio (dm² g⁻¹)
			Leaf (%)	Stem (%)	Root (%)	
1966						
abd I	25	1·568	50·31	20·66	29·01	1·63
abd II	32	4·508	52·48	18·72	28·79	1·59
abd II-I	—	2·940	53·6	17·7	28·7	—
ABD I	39	9·659	44·63	22·79	32·57	1·48
ABD II	46	23·43	41·24	26·72	32·02	1·18
ABD II-I	—	13·77	38·9	29·5	31·6	—
1967						
abd I	25	45·42	56·23	21·30	24·66	1·43
abd II	32	130·7	46·05	26·52	27·41	1·43
abd II-I	—	85·26	41·8	29·3	28·9	—
ABC I	39	233·1	44·88	30·79	24·31	1·10
ABD II	46	431·0	36·83	33·52	29·63	0·95
ABD II-I	—	198·0	27·4	36·7	35·9	—

groups (*ABD* II-I)–(*abd* II-I). For leaves the ratio fell by 14·7 and 14·4 per cent in 1966 and 1967 respectively, while for stems the ratio increased by 11·8 and 7·4 per cent, and for roots by 2·9 and 7·0 per cent. This consistency would be expected because here our comparison is mainly one of age and stage of development. For 3 weeks before the first harvest in each year the two age groups of plants have been exposed to an identical external environment, as they have also during the period between harvests. The differences arise because the older group have 14 days start. The particular change of growth pattern observed would also be expected, because of the changing plant architecture, which requires a more massive stem, and to a lesser degree also a more massive main root, as the plant becomes taller and the leaves larger.

Beyond this point the inconsistencies multiply. There may be a tendency for leaf weight ratio to fall with increasing size, but at both harvests the *ABD* plants of 1966 had smaller values than the *abd* plants of 1967, which were five times as big. Furthermore, of the dry weight

14.4

increase during the week between harvests, a larger proportion went into the new leaves of the larger, 1967 *abd* plants, than went into the new leaves of the smaller, 1966 *ABD* ones. Again, during the harvest week LWR always fell, except for the smallest plants of all, the *abd* batch of 1966, when it rose.

14.5 Leaf and stem weight ratios and plant size

All this suggests that we are dealing with two very different populations; populations which not only differ largely in dry weight, but which when considered on a basis of size, still do not fall into one single set of progressions. The point can be further investigated by plotting leaf weight ratio and stem weight ratio, which show the largest changes, against mean total plant dry weight, and this has been done in Fig. 14.5, using for convenience a logarithmic scale because of the great range of dry weight.

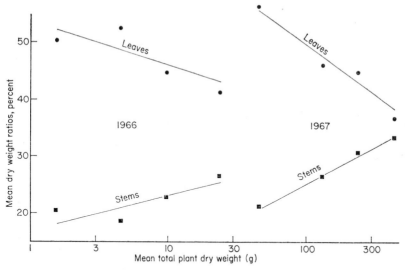

Fig. 14.5. Distribution of dry matter to leaves and stems plotted against the logarithm of mean total plant dry weight for the plants of *Helianthus annuus* discussed in 14.2–14.8. Means are of nine plants in each case, grown in the open at the Botanic Garden, Cambridge; 1966 compared with 1967. Ages in each case from date of sowing 25, 32, 39 and 46 days. ●, mean leaf weight ratio; ■ mean stem weight ratio, both in per cent. The straight lines are intended only to indicate trends, not to imply linear relationships.

14.5

There is a difference between this procedure and those which we have been using for checking purposes hitherto, which have involved comparisons between two sets of independent observations, say leaf area against leaf weight, or root weight against total top weight. In our present comparisons this is not so. For example, for leaf weight ratios, leaf dry weight enters both into the numerator and denominator, so that the effects of accidental variations are minimized; they would be larger if we were considering leaf weight divided by stem weight plus root weight. But also, by plotting the ratio against total dry weight we have created a built-in correlation—any accidental increase in leaf dry weight will increase both the LWR and also the total plant dry weight. It is necessary always to guard against such effects, created by the procedure being used to handle the data, and other examples will be mentioned [e.g., 18.13]. Examples of a more rigid alternative procedure, avoiding any possibility of the creation of automatic correlations, will be considered in 26.3–26.9.

However, we are on safe ground in our present enquiry, which is concerned with the point—is there evidence that the plants of the 2 years belong to totally distinct populations? Fig. 14.5 shows that they do. Within each year there is evidence of a regular progression, but the two progressions are quite distinct. Incidentally, in this case as the correlation of leaf weight ratio with total dry weight is negative, the only effect of the built-in positive correlation which we have just noticed is to increase the scatter.

Finally, it should be mentioned that the straight lines in Fig. 14.5 are there merely to emphasize the general trends, and not to suggest that linear relationships exist.

14.6 Comparisons involving leaf area

So far we have been dealing entirely with the dry weights of the various parts. The next step is to examine relationships involving leaf area—the leaf area ratio (ratio of leaf area to total plant dry weight) and specific leaf area (ratio of leaf area to leaf dry weight). The two are connected by the leaf weight ratio [13.16.1],

Leaf area ratio = Specific leaf area × Leaf weight ratio.

The three derivates are shown in Fig. 14.6, once again plotted against the logarithm of the total plant dry weight. The differences in leaf area ratio and in specific leaf area between the two groups are much less marked than those of leaf and stem weight ratios shown in Fig. 14.5. Furthermore, there is a downward trend within each group and in the

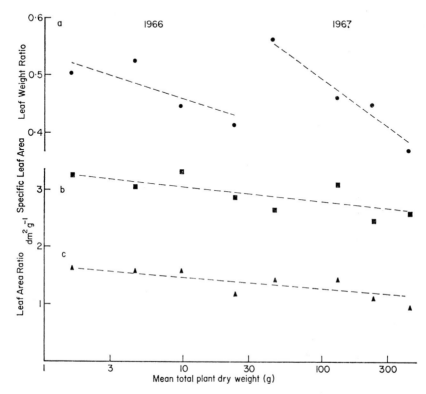

Fig. 14.6. Mean values for (a) leaf weight ratio (●), (b) specific leaf area (■) and (c) leaf area ratio (▲) for the same plants of *Helianthus annuus* as in Fig. 14.5, also plotted against the logarithm of mean total plant dry weight. Means are of nine plants in each case, values observed in 1966 are all below total plant dry weights of 30 g, those observed in 1967 are above. The straight lines are intended only to indicate trends, not to imply linear relationships.

heavier, 1967, plants this roughly continues that for the lighter, 1966, plants, so that it is possible to link the ratios for all the sets of plants by two straight lines, without the appearance of marked divergences from one single set of relationships.

In a very broad general way such a relationship would be expected.

14.6

As the sunflower plants become bigger and heavier their leaves become larger, and a necessary part of the architecture of these larger leaves is a more massive structure of midrib and main veins, whose presence naturally lowers the leaf area per unit leaf dry weight. Accordingly, a falling specific leaf area is commonly observed as seedlings grow up to maturity. Also, individual leaves, once they have ceased to expand, tend to become gradually heavier, and this reinforces the fall in SLA.

However, we ought not solely on these general grounds to conclude that our figures have revealed a single set of relationships running through the observations of both years; the apparent agreement is more likely to be fortuitous, and later the matter will be pursued in more detail, making use of the figures for other years.

14.7 A further analysis

Considering simply the data for 1966 and 1967, we remember that the points in Fig. 14.6 are not single observations, but each is a mean of observations on nine plants; and the experimental design links these sets of observations firstly into four pairs of means made on paired samples from the same population, and then into two pairs of paired plants which had been growing side by side for two or three weeks, subjected to the same external conditions. A convenient method of proceeding to a further analysis here is first to compare the individual members of each pair using 'Student's' t test. If we are comparing two mean estimates a and b each based on the same number of observations, with estimated standard errors s_a and s_b, then the difference between them, $a-b$, has a standard error $s_{a-b} = \sqrt{(s_a^2 + s_b^2)}$. It is necessary in the first instance to decide whether $a-b$ is significantly different from zero, calculating the probability that a difference as large as that observed might arise at random. We calculated $t = (a-b)/s_{a-b}$. These particular comparisons are based on two groups each of nine plants, 18 plants in all. Having calculated 2 means, 16 degrees of freedom are left, and there is a 1 in 20 chance of $t \geqslant 2 \cdot 2$, and a 1 in 100 chance of $t \geqslant 2 \cdot 9$. (If there is reason to suppose that the variance is homogeneous, in which case s_a will not differ significantly from s_b, we can use a somewhat more convenient procedure of which an example is given in 15.2–15.5.)

Table 14.7 shows that in 1966 there is a highly significant fall in specific leaf area for both the young and the old plants, there being less

Table 14.7. Comparison of specific leaf areas of plants of *Helianthus annuus* of different ages grown at the Botanic Garden, Cambridge, in two consecutive years. Means of observations on nine plants in each case, with standard errors. Same experiments as Tables 7.6, 14.3, 14.4, and Figs. 11.6, 11.21–11.23, 14.5, 14.6.

Sample	Age (days)	SLA ($dm^2\,g^{-1}$)	1966 Difference	t	SLA ($dm^2\,g^{-1}$)	1967 Difference	t
abd I	25	$3\cdot24 \pm 0\cdot0464$			$2\cdot66 \pm 0\cdot0571$		
			$-0\cdot20$	$3\cdot3$		$+0\cdot45$	$6\cdot5$
abd II	32	$3\cdot04 \pm 0\cdot0393$			$3\cdot11 \pm 0\cdot0315$		
ABD I	39	$3\cdot33 \pm 0\cdot0646$			$2\cdot46 \pm 0\cdot0410$		
			$-0\cdot47$	$6\cdot6$		$+0\cdot12$	$2\cdot0$
ABD II	46	$2\cdot86 \pm 0\cdot0301$			$2\cdot58 \pm 0\cdot0457$		

than 1 chance in 100 of such a fall occurring at random for the *abd* plants, and much less than 1 chance in 1000 for the *ABD* plants. There is accordingly less than 1 chance in 10^5 of the simultaneous occurrence of the two falls, by random variation alone. In 1967 there is an equally significant rise in specific leaf area, and though here for the *ABD* plants such a rise might have been observed at random in 1 case in 10, nevertheless, taking the two rises together, the combined chances of such being observed at random are less than 1 in 10^4.

When dealing with two simultaneous probabilities as small as these, there is no doubt that the plants were behaving differently during the week between harvests in the 2 years—so differently that in 1966 there was a very highly significant fall in specific leaf area, and in 1967 a very highly significant rise. The specific leaf area figures thus bear out the previous conclusions based on dry weight distributions.

14.8   A useful moral

These figures point a useful moral, because the very highly significant differences of behaviour just discussed were concealed within what appeared, at first sight of Fig. 14.6, to be a single overall trend.

Detailed examination has thus confirmed the impression of the interlocking effects of age, of size, and of previous environmental experience which was gained by inspection of Tables 14.3 and 14.4. It has at the same time revealed other highly significant differences between the 2 years, as well as the obvious one of size.

15

RELATIVE GROWTH RATE

15.1 Computation and units

By arranging the observations made on plants at the time of harvest in sequences of time and size, it has been possible to draw certain broad inferences about the course of development. The next question is a more detailed enquiry into the processes which have been going on between the two harvests.

We will take first the rate of relative growth in dry weight, usually called simply relative growth rate (RGR), and given the symbol **R**. (The corresponding quantity for leaf area is usually called relative leaf growth rate (RLGR), and given the symbol $\mathbf{R_L}$; see Chapter 22. When necessary, rates of relative growth of other plant attributes are similarly named, with suitable qualifications.) We have already discussed methods of computing relative growth rates [13.6], and have concluded that the true mean value of relative growth rate ($\bar{\mathbf{R}}$) over a period from time $_1T$ to time $_2T$ is given by

$$\bar{\mathbf{R}} = \frac{\log_e {_2W} - \log_e {_1W}}{_2T - {_1T}},\qquad \text{15.1 [cf 13.6.1]}$$

where $_2W$ and $_1W$ are the total plant dry weights at the beginning and end of the period. The usual units are g g^{-1} week^{-1}, or more simply week^{-1}, although occasionally other units of time are used; the value of RGR is unaffected by the unit of mass. In the case of an experiment employing weekly harvests, $_2T - {_1T} = 1$ week, and figures for $\bar{\mathbf{R}}$ in week^{-1} are obtained simply by subtracting the natural logarithms of the appropriate figures for dry weights. We also see from equation 15.1 that the computed value of $\bar{\mathbf{R}}$ has the useful property of being directly proportional to the duration of the time unit used; it is easy to convert from one unit to another if necessary.

The theory is thus quite straightforward; the practical difficulties arise when one begins to set values on $_2W$ and $_1W$, the total plant dry weights at the beginning and end of the period. In the nature of the case it is hardly ever possible to use the same plant to determine both $_2W$ and $_1W$ [4.3, 4.4]. But if different plants are used, it cannot be proved with certainty that the plant used at time $_2T$ for the determination of $_2W$ was, at time $_1T$, identical with the plant used for the determination of $_1W$, because such a proof would involve the destruction of the $_2W$ plant at time $_1T$. One or both of the two quantities $_1W$ and $_2W$ must therefore be an estimate, whose accuracy can only be assessed by considering the statistical properties of the population from which the two plants were derived. Consequently estimates of the value of mean relative growth rate are hardly ever based on two plants alone, because such an estimate would be of little value in the absence of some information on its probable accuracy.

The problems of computation of relative growth rate are therefore in practice statistical ones, which tend to be bound up with the experimental design. The object is to provide the maximum accuracy (which demands, amongst other things, large samples on which the estimates of probability can be based) with the minimum effort (which demands small samples).

15.2 Estimation of fiducial limits

As the individual plants in a population grow in size, the variability among them, measured in absolute units, tends to increase; but fortunately it is often found to bear a roughly constant proportion to plant size. If so, the variance of the logarithms of measured quantities such as plant dry weight may be found to be roughly constant over a period of time. In the most favourable cases, we can then use the whole of a body of data spread over many harvests and a considerable period of time to produce a single estimate of variability. In section 20.4 is given an example where tests showed that homogeneity of variance continued for a period of 39 days, during which there were 12 harvests. However, in this very favourable case all these harvests were taken from one single population growing under constant, controlled, environmental conditions.

More usually in the field the fluctuating conditions of growth are found to affect to a greater or lesser degree the variability of the population, as well as the various rates of growth. Homogeneity of variance may then persist only for short periods, and it is necessary in each case to investigate whether this is so. These arguments apply with even greater force to similar, related but not identical, populations, such as the two age groups of young plants of *Helianthus annuus* which we discussed in the last chapter [14.2–14.8, and also 7.5, 7.6, 7.13 and 8.8]. Here there were two elements of similarity, a common seed source and a common environment during the 3 weeks before harvest. There were also three elements of disparity: one group of plants was older; early in life it had experienced an environment which did not necessarily bear any relationship to the environment shared with the younger plants; and the shared environment was acting on plant structures whose form and functioning had been affected by this earlier, unshared environment. The elements of similarity would have been expected to produce homogeneity of variance of the logarithms of plant size; the elements of disparity might or might not have produced differences. Once again, tests are necessary, but we must regard it as fortunate if tests show that variance is homogeneous among populations of this kind, and it would be most unwise to base experimental planning on the expectation that it will prove to be so. The data given in 14.2–14.8 provide a convenient illustration of these last considerations.

15.3 An example

The mean relative growth rates for the young and old plants, and the two years 1966 and 1967, are set out in Table 15.3, where figures are given for the large, medium, and small samples, and also for the mean of the three representing the population as a whole. Considering first the figures for the populations as a whole, the *abd* plants of both years show closely similar relative growth rates, the difference (1967 5 per cent higher) obviously not being significant. The rates for the *ABD* plants are somewhat lower in 1966, and considerably lower in 1967. We should have expected this from our inspection of Table 14.3, where the dry weights increased 2·9 times for the *abd* plants in both years, and 2·4 and 1·85 times for the *ABD* plants in 1966 and 1967 respectively. These ratios are $_2W/_1W$ and our figures for relative growth rate are $\log_e(_2W/_1W)$, so that for single pairs of plants the two are immediately interconvertible. Comparisons of means of two populations are less exact, because there is no reason to expect that the difference of the means of the natural logarithms of individual plant dry weights would necessarily equal the natural logarithm of the ratio of the means. However, $\log_e 2·9 = 1·06$, $\log_e 2·4 = 0·88$ and $\log_e 1·85 = 0·62$, so that in this instance the rough approximation

15.3

Table 15.3. Mean relative growth rates during the week between harvests, for plants of *Helianthus annuus* grown at the Botanic Garden, Cambridge, and used to give the data set out in Table 14.3. Figures are differences between the mean logarithms of dry weight for three large, three medium and three small plants of two age groups, grown during June and July in the years 1966 and 1967. Ages in days from date of sowing.

Samples	Age range	Mean relative growth rate		
		1966	1967	Difference, 1966–1967
abd	25–32			
Large		1·09	0·94	
Medium		0·95	1·11	
Small		1·01	1·15	
Mean of all		1·014	1·065	−0·051
ABD	39–46			
Large		0·96	0·63	
Medium		1·00	0·55	
Small		0·74	0·66	
Mean of all		0·900	0·611	+0·289
Difference, *abd-ABD*		+0·114	+0·454	
Variance of mean of all				
abd		0·001130	0·0005825	
ABD		0·002170	0·0007392	
F		1·92	1·27	
Least significant difference				
Within years		0·117	0·074	
Between years		0·098		

differs from the more exact method by less than one standard deviation in three cases out of four.

15.4 Variances and variance ratios

The relative growth rates set out in Table 15.3 were derived from eight sets, each made up of nine plants (three large, three medium and three small), the eight sets comprising two harvests of two age groups in two years. Before embarking on comparisons we must ask to what extent the variances are homogeneous, and to investigate this we use the so-called

15.4

variance ratio or F test. F is the ratio of the larger (s_1^2) to the smaller (s_2^2) of the two variances being compared, so that it is never less than unity. We might compare the variances for the large, medium and small plants within a set, but unless some very large discrepancy is noted this is hardly worth while. A large variability would be expected among estimates of variance based on such small samples. Comparing two samples each with four degrees of freedom, both taken from the same large normally distributed population, a value of F as large as 6·4 would be expected on one occasion in 20, and 16·0 on one occasion in 100, due to sampling variation alone.

We will therefore consider only the whole groups of even-aged plants. Our estimate of the variance of the mean relative growth rates for the groups of 9 is based on our information about variability of the natural logarithms of total plant dry weight within the sub-groups of comparable large, medium and small plants in the usual way [adding the squares of the 18 individual deviates from the appropriate sub-group mean and dividing by 12 (the number of degrees of freedom; 18 observations minus 6 sub-group means) times 9/2 (for a mean value for the difference between two sets of 9)]. The resulting variances are set out in Table 15.3. Comparing the younger and older plants within years, when all had experienced the same weather conditions from 3 weeks before the first harvest, the values of F are 1·92 for 1966 and 1·27 for 1967. Neither difference reaches significance, the 5 per cent point for comparison of two variances each with 12 degrees of freedom coming at 2·69. We can therefore make a common estimate of the variance for all the plants in each year, with 24 degrees of freedom, and then proceed to compare 1966 with 1967, when F comes out at 2·5, the 5 per cent point for a comparison where each variance has 24 degrees of freedom being 1·96, and the 1 per cent point 2·67. It would therefore be unwise in this case to combine the 2 years. The fiducial limits for the mean relative growth rates then work out at ±0·084 for 1966 and ±0·053 for 1967, while the least significant differences between two means are 0·117 for 1966, 0·074 for 1967, and 0·098 for comparisons between the 2 years.

15.5   An alternative treatment

The above procedure relies on comparability between all six plants of each initial sample of large, medium and small plants. Should it turn

15.5

out in the course of the experiment that some systematic difference has arisen—due, for example, to differences in technique between different students—this basic assumption is destroyed. In a bad case, where the differences in technique are so large that the plants can no longer be regarded, for statistical purposes, as part of a single population, it may be best to discard the corrupt part of the data altogether, and proceed as before with fewer basic observations, and hence with fewer degrees of freedom. In such a case the gain in accuracy from pooling data may be greater than in the instance just examined.

On the other hand, with less marked, but nevertheless systematic differences we may be able to proceed by treating the data not as comparable sixes, but as comparable pairs. The relative growth rates worked out pair by pair then form the basis of our statistical procedure. If the data for 1966 and 1967 are treated in this way we obtain values of F of 1·32 for 1966 and 1·18 for 1967, neither being near significance. Comparing the combined estimates of variance for 1966 with 1967, F is 1·38, also nowhere near significance, the 5 per cent point for two variances each with 16 degrees of freedom being 2·34. We can therefore safely combine all the data to give a common fiducial limit for each mean relative growth rate of 0·09, and a least significant difference between any pair of means of 0·13. It will be noticed that we have lost some accuracy. When working with small samples this is not an unexpected result of a procedure which involves losing a number of degrees of freedom. The broad result, however, is unchanged: there is no significant difference between the relative growth rates of the younger, *abd*, plants in the 2 years. In 1967 the relative growth rate of the *ABD* plants was very significantly lower ($t > 6$, $p \ll 0.001$) than for any other group, younger or older, while in 1966 the older, *ABD*, plants had a somewhat lower relative growth rate than the younger ones, near significance at the 5 per cent level. Taking the two years together, there is no doubt even from this limited collection of data that relative growth rate was falling with age, a result which agrees with much other information [e.g., Fig. 13.6.2, Table 26.6].

15.6   Implications for experimental design

Where homogeneity of variance can be demonstrated we gain in two ways, both in an increase in the number of degrees of freedom (and hence in the reliability of the estimates), and in a decrease in the number of separate fiducial limits which must be calculated in comparing one set of data with another. On the other hand, in the worst possible case, where all the variances turn out to be different, the results of the experiment are not vitiated. It is still possible to make all the necessary comparisons, combining the variances in the usual way for each comparison, but the number of degrees of freedom will be smaller.

The experimental design thus permits alternative treatments, depending on the properties of the data collected, which cannot necessarily be foreseen when the experiment was planned. As a general proposition such a feature is obviously desirable. There are many other possible experimental designs which have similar properties, and which enable us to take full advantage of homogeneity of variance of logarithms of such quantities as dry weight and leaf area, as and when the data exhibit this property [e.g., Welbank (1961), which includes a method, devised by Prof. Daniels, for combining the data from three successive harvests]. The particular design which we have just considered also provides a simple instance of another desirable property—if by some accident part of the data is lost, or turns out to be corrupt, both types of treatment are still available. For example, if by some muddle student group A lost all three of their older plants at one harvest, the set would be reduced from 9 to 6, the degrees of freedom of the estimate of variance from 12 to 9. But we can still make the variance ratio test and in the most favourable case, where the variances of both age groups prove to be homogeneous, we still have a common estimate of variance with 21 degrees of freedom instead of 24, which hardly affects the reliability of the estimate.

15.7 Accidents and experimental design

The particular instance just considered illustrates an important general principle of experimental design applying particularly to experiments extending over prolonged periods of time or involving a great deal of effort. Any experiment can be upset by accidental factors outside the scope of the experiment itself. If the investment of time, effort or scarce material is small, the experiment is abandoned and another one started. Otherwise it is desirable to be able to salve as much useful information as possible from the wreck of the original scheme. By preparations such as those outlined in Part II we endeavoured to provide against as many interfering factors as possible, within the general scope of the intentions underlying the investigation. But however careful we are, accidents of many kinds are inseparable from field experimentation, and there is always the possibility of human error. It is therefore desirable, firstly to provide insurances against accident, where this can be done without much additional effort; and secondly, wherever possible to choose experimental designs which provide for alternative treatments in the event of the loss of part of the basic data. In the particular experiment just discussed we met the first point in our scheme of grading plants [8.8 (ii)], by selecting seven comparable ones, allotting at random three plants to each harvest, and having one fully comparable spare which, in case of accident, could be harvested in place of a corresponding second harvest plant. In practice this has proved to be worth the small extra labour. We have now seen [15.4, 15.5, 15.6] how alternative statistical treatments of the data are possible, and how if needs be whole sections of corrupt data can be removed without destroying the value of what remains. This is only one simple example, adapted to a class of students without previous training in this particular type of work. The principle is, however, of general application and worth bearing in mind when selecting an experimental design, be it simple or complex.

15.8   Analysis of relative growth rate

Relative growth rate provides a valuable over-all index of plant growth [13.5], but in the nature of the case its make-up is complex, and almost inevitably involves a complex ontogenetic drift, with all the consequential problems of untangling the operation of the various determining factors [3.6, 7.1, 17.1]. The time progressions of relative growth rate given in Fig. 13.6.2 showed how large can be the changes from week to week, and how difficult it would be to decide what the relative growth rate in a particular week might have been, had the weather been different in some specific way.

However, when a plant is growing in a specific, closely controlled, environment its relative growth rate may go through a reproducible time progression, which can be characterized and compared with the progressions under other specific conditions. The method of 15.4 can then be extended to a series of harvests. If it can be shown that the variance of the natural logarithms of the dry weights is homogeneous over the experimental period, we can treat the whole of the data together, calculate a single polynomial regression for the logarithms of dry weight on time, and hence by differentiation obtain the time progression of relative growth rate. This method is examined in section 20.4.

Usually it is not profitable to embark on a direct study of the relationships of relative growth rate with various aspects of the interlocking complex of genetic make-up, developmental sequence and environmental influence. More progress is made towards this goal by breaking up the relative growth rate (considered as an overall index) into components which can more readily be related to particular aspects of the structure and functioning of the plant. These further steps in analysis will be considered in Chapters 16, 18, 19 and 20, while their relations with each other, the synthesis of the results achieved and applications to particular problems form the subject of Part IV.

15.9 Identity of relative growth rates

The labyrinthine intricacy of the plant/environment complex [3.6, 7.1] lends particular interest to those cases where two sets of plants having the same relative growth rate also have in common some aspects of the complex, while others are markedly different. We shall examine several such examples as we proceed: plants of different species, growing side by side in an identical environment, and having the same relative growth rate in spite of large morphological and physiological differences [Table 26.5]; plants of identical genetic constitution growing in different environments and having identical relative growth rates, the environmental differences compensated by morphological and physiological changes [Figs. 17.6, 20.4.1]; cases where different forms of ontogenetic drift bring together for longer or shorter periods relative growth rates which at one time were very different [Figs. 20.4.1, 26.6, 26.12]; and so on [Fig. 28.6.2]. All these cases are of biological importance, as instances of identical results being produced by different mechanisms, involving quantitative compensations. In turn these quantitative compensations themselves provide clues to the operations of the genetic—development—environment complex.

15.9

THE COMPUTATION OF
UNIT LEAF RATE

We have already seen that instantaneously the unit leaf rate,

$$\mathbf{E} = \frac{\mathrm{d}W}{\mathrm{d}T} \cdot \frac{1}{L_A},$$ 13.7.1

where W is total dry weight and L_A leaf area (in the sense mentioned in 2.6 and further discussed in 18.1, 19.1). One way of evaluating this is to estimate a value for $\mathrm{d}W/\mathrm{d}T$ at a time when L_A is also known. We shall consider this method later [16.19; 20.4]. However, mean values must still be estimated over periods of time [13.12], and if $\mathbf{E}$, W and L_A are all independent variables, nothing being known about the relationships of any of them with each other or with time, then no progress can be made with the integration. We will begin by considering some simple limiting cases.

16.1 A   E assumed constant

This is an assumption which is most unlikely to be realized under natural conditions, although we shall find instances where it fluctuates around a constant value, depending on external conditions. It is much more likely to be constant in experiments carried out under constant artificial conditions, although even here the assumption should not be made without justification. We shall consider later [16.11] methods of checking whether the assumption is justified. We will, however, consider first the cases based on the assumption that $\mathbf{E}$ is constant, because it underlay the pioneer work of Weber (1879) [13.8] and, whether consciously or not, has been used a good deal since that time.

If $\mathbf{E}$ is constant, then from equation 13.7.1 we have, within the limits of time $_1T$ and $_2T$, and of dry weight $_1W$ and $_2W$,

$$\mathbf{E} \int_{_1T}^{^2T} L_A \, \mathrm{d}T = \int_{_1W}^{^2W} \mathrm{d}W = {_2W} - {_1W}.$$ 16.1.1

To evaluate this expression, we require to evaluate $\int_{1T}^{2T} L_A \, dT$. If we can obtain a mean value for leaf area during the period, $\bar{L}_A$, such that

$$\bar{L}_A(_2T - {}_1T) = \int_{1T}^{2T} L_A \, dT, \qquad 16.1.2$$

then

$$E = \frac{{}_2W - {}_1W}{\bar{L}_A(_2T - {}_1T)}. \qquad 16.1.3$$

16.2 A(i) Measurement and numerical integration

For plants having leaves of a regular shape, L_A can be measured at intervals, and a value for $\bar{L}_A$ derived by numerical integration, regardless of the actual form of the relationship between L_A and T. If L_A is repeatedly measured midway through periods of time ${}_1\varDelta T$, ${}_2\varDelta T$, etc., such that $({}_1\varDelta T + {}_2\varDelta T + \dots {}_n\varDelta T) = {}_2T - {}_1T$, then

$$\bar{L}_A(_2T - {}_1T) = \int_{1T}^{2T} L_A \, dT \simeq ({}_1L_A \, {}_1\varDelta T + {}_2L_A \, {}_2\varDelta T + \dots {}_nL_A \, {}_n\varDelta T).$$

$$16.2.1$$

And in the special case where all the values of $\varDelta T$ are equal and represent unit time interval (as for leaves measured every day throughout the period)

$$\bar{L}_A(_2T - {}_1T) \simeq ({}_1L_A + {}_2L_A + {}_3L_A + \dots {}_nL_A),$$

and

$$E = \frac{{}_2W - {}_1W}{{}_1L_A + {}_2L_A \dots {}_nL_A}. \qquad 16.2.2$$

It will be noticed that in this method n, the number of leaf measurements, is equal to the number of days in the period ${}_2T - {}_1T$, and E is expressed as increment per unit area per day. The method also has a formal similarity to the method used by Watson to calculate 'weighted mean unit leaf rate' over the growing season of an annual crop plant [13.19].

This was in effect the method used by Weber. Gregory (1918) improved on it by taking account of the changing length of day, and working out the result per hour. A number of other workers since then have used the same technique of daily, or at least frequent, leaf measurements. This is essential if the relationship between L_A and T is to be determined

16.2

directly, whether for this particular purpose or some other (e.g., 16.11, where the information is used to infer the relationship between **E** and time when unit leaf rate is not assumed to be constant). However, it is not always practicable to make frequent leaf measurements, and in the case of leaves having a difficult shape it may be impossible. Here, if we are to make progress with this line of attack we must make some assumption about the relationship between L_A and time.

16.3   A(ii)   Linear increase of leaf area

Ths simplest assumption is that leaf area is increasing linearly with time, an approximation to which is shown by the data given in Fig. 16.3, derived from Kreusler's observations on maize, variety 'Badischer

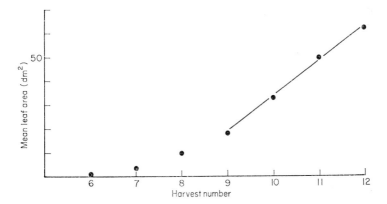

Fig. 16.3. Mean leaf area of maize plants as a function of time (interval between harvests constant at 7 days). Variety 'Badischer Früh' grown at Poppelsdorf near Bonn in 1876. Plotted from the data given by Kreusler *et al.* (1877b). Note that this is a portion of a curve similar to that given in Fig. 2.7d.

Früh', grown at Poppelsdorf in 1876, from harvests 9 to 12. There would be no difficulty in multiplying instances from the middle life of many plants. Like a number of other plant attributes, leaf area frequently shows a sigmoid relationship with time, so that there is almost inevitably a period around the point of inflexion, when the increase approximates to a linear one.

16.3

If $L_A = \mathbf{u} + \mathbf{v}T$, where $\mathbf{u}$ and $\mathbf{v}$ are both constant, then $\bar{L}_A = \frac{1}{2}(_1L_A + _2L_A)$ and hence from 16.1.3

$$E = \frac{_2W - _1W}{\left(\dfrac{_1L_A + _2L_A}{2}\right)(_2T - _1T)}. \qquad 16.3$$

16.4 A(iii)   Exponential increase of leaf area

Another possible assumption is that leaf area is increasing exponentially, in which case relative leaf growth rate would be constant, and if we plot the logarithms of the leaf areas against time, we should get a straight line, as in Fig. 16.4. Here we are once again using Kreusler's data for the

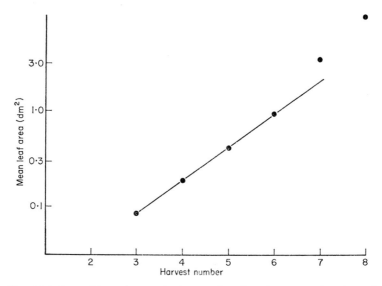

Fig. 16.4. Logarithm of mean leaf area of maize plants as a function of time (intervals between harvests constant at 7 days). Same plants as in Fig. 16.3; variety 'Badischer Früh' grown at Poppelsdorf in 1876. Based on the data given by Kreusler *et al.* (1877b).

same maize variety in the same year, but this time between harvests 3 and 6. A constant relative leaf growth rate is not infrequently found early in the growth of the seedlings of a number of species, and these observations would also be easy to parallel.

16.4

If $L_A = pe^{qT}$, where p is constant, and q (the relative leaf growth rate) is also constant, then

$$_1L_A = p\,e^{q_1T}$$

and

$$_2L_A = p\,e^{q_2T}$$

and hence

$$\frac{_2L_A}{_1L_A} = \frac{p\,e^{q_2T}}{p\,e^{q_1T}} = e^{q(_2T - _1T)}.$$

Therefore

$$q(_2T - _1T) = \log_e {_2L_A} - \log_e {_1L_A}$$

and

$$q = \frac{\log_e {_2L_A} - \log_e {_1L_A}}{_2T - _1T}. \qquad \text{16.4.1 compare 13.6.1}$$

Now

$$\int_{_1T}^{_2T} L_A\,dT = p \int_{_1T}^{_2T} e^{qT}\,dT$$

$$= \frac{p}{q}(e^{q_2T} - e^{q_1T}),$$

and substituting for q and pe^{qT}, this

$$= \frac{(_2L_A - _1L_A)(_2T - _1T)}{\log_e {_2L_A} - \log_e {_1L_A}}.$$

Hence

$$E = \frac{_2W - _1W}{_2T - _1T} \cdot \frac{\log_e {_2L_A} - \log_e {_1L_A}}{_2L_A - _1L_A}. \qquad \text{16.4.2}$$

16.5   A(iv)   Other possibilities

It would be possible to make similar calculations on other assumptions, but it is not worth while doing so because of the limited utility of the method. The examples we have covered include the classical assumptions and formulae which have been widely used in the literature. We shall also find them of considerable value in another context, below [16.11].

16.5

16.6 B E not assumed constant

If we do not make the assumption that $\mathbf{E}$ is constant during the period between harvests, our first requirement is a method of calculating a true mean value of $\mathbf{E}$ over the period, $\bar{\mathbf{E}}$, such that

$$\bar{\mathbf{E}}(_2T - {}_1T) = \int_{_1T}^{_2T} \mathbf{E}\,\mathrm{d}T,$$

and, from equation 13.7, this

$$= \int_{_1W}^{_2W} \frac{\mathrm{d}W}{L_\mathrm{A}}. \qquad 16.6$$

16.7 B(i)   Leaf area constant

This can be integrated at once if L_A is constant, as frequently happens during specific periods for plants growing in flushes. Then

$$\bar{\mathbf{E}} = \frac{1}{L_\mathrm{A}(_2T - {}_1T)} \cdot \int_{_1W}^{_2W} \mathrm{d}W = \frac{_2W - {}_1W}{L_\mathrm{A}(_2T - {}_1T)}. \qquad 16.7$$

If, however, L_A is not constant, in order to integrate expression 16.6 we need to know the relationship between W and L_A. This relationship can usually be determined only by a series of harvests; and if, as in the experiments considered in Chapter 14, there is one pair of harvests only, it is necessary to make some assumption about the relationship.

16.8 B(ii)   Dry weight linearly related to leaf area

Let us assume first that $W = \mathbf{c} + \mathbf{d}L_\mathrm{A}$, where $\mathbf{c}$ and $\mathbf{d}$ are both constant. Data on the relationship of dry weight with leaf area over a large fraction of the life cycle are available for a few species only; in these it seems that a linear relationship is usually confined to a short period early in the life of the seedling. We have already noticed in Fig. 4.5 one such example for harvests 1 to 4 of Briggs, Kidd and West's observations on *Helianthus annuus*, where the approximation to linearity was very close. In the data of Kreusler for Badischer Früh maize in 1876, there cannot be said to be a marked linear phase at all (Fig. 16.8, and see 16.13). Another marked

16.8

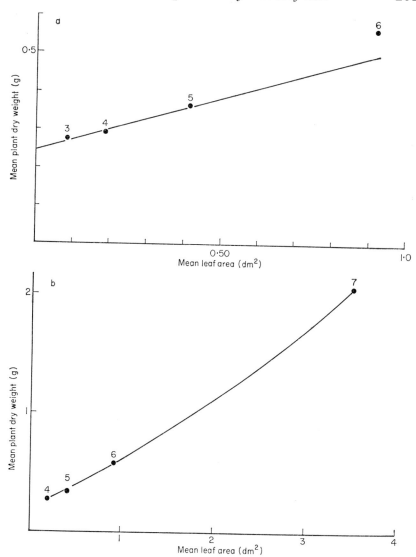

Fig. 16.8. Mean dry weight of maize plants as a function of mean leaf area (numbers against points are those of the harvests of Figs. 16.3, 16.4). Variety 'Badischer Früh', grown at Poppelsdorf in 1876. Plotted from the data as given by Kreusler *et al.* (1877b). Note that there is no substantial period of linear relationship, and that the straight line of best fit to the points for harvests 3, 4 and 5 has a substantial intercept on the axis of ordinates (see text).

16.8

difference between the two species is in the value of the parameter **c**, the intercept of the straight line on the dry weight axis. For *Helianthus annuus* **c** is negligibly small; over the period from harvests 1 to 4 dry weight and leaf area are proportional to each other. For maize, on the other hand, if one put a best straight line through the points for harvests 3 to 5, **c** would have the considerable value of 0·24 g, not much less than the total dry weight at harvest 3. Some such relationship is almost inevitable in the early stages of the growth of plants where the bulk of the food reserves of the seed are not stored in the embryo (and where the interesting question can arise of how much of the dry weight of the germinating seed should actually be regarded as part of the young plant, as germination proceeds). It is also not uncommon in later stages of the life cycle, whether or not the early stages approximate to linearity [Fig. 16.9.1].

Turning to the computation of a mean value of unit leaf rate on the assumption of linearity, if

$$W = c + dL_A,\qquad\qquad 16.8.1$$

where **c** and **d** are both constant, then

$$dW = d\, dL_A$$

and

$$\bar{E}(_2T - {_1T}) = d \int_{_1L_A}^{^2L_A} \frac{dL_A}{L_A}$$

$$= d(\log_e {_2L_A} - \log_e {_1L_A});$$

but

$$d = \frac{_2W - {_1W}}{_2L_A - {_1L_A}},$$

therefore

$$\bar{E} = \frac{_2W - {_1W}}{_2T - {_1T}} \cdot \frac{\log_e {_2L_A} - \log_e {_1L_A}}{_2L_A - {_1L_A}}.\qquad\qquad 16.8.2$$

16.9 B(iii)   Quadratic relation of dry weight with leaf area

The plots of dry weight against leaf area of Figs. 4.5 and 16.8 show that after the linear phase, if any, the curve tends to turn up—dry weight is increasing even more rapidly than leaf area. Is it now proportional to the

16.9

square of the leaf area ? It seems that in some plants there is a considerable phase during which dry weight increase approximates closely to a linear relationship with the square of the leaf area. For Briggs, Kidd and West's sunflowers (Fig. 16.9.1) it holds from harvests 9 to 13, rather beyond halfway through the life cycle. For Kreusler's data on maize, the relationship is similar. There is a good approximation to a straight line between harvests 8 and 11 (Fig. 16.9.2), after which the relationship

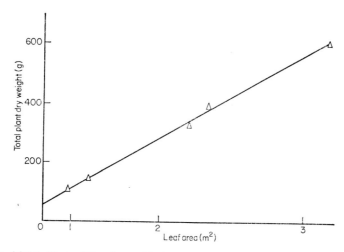

Fig. 16.9.1. Part of the data of Briggs, Kidd and West for the growth of *Helianthus annuus* at the Cambridge University Botanic Garden in 1920, showing mean total plant dry weight plotted against mean total leaf area squared, for harvests 9 to 13 (data for the remaining harvests in Figs. 4.5 and 16.13). From Evans and Hughes (1962).

again becomes curved, rising above the straight line extrapolated, and indicating a power greater than 2. If we assume that

$$W = r + sL_A^2, \qquad 16.9.1$$

where r and s are both constant, we notice that in both the specific cases we have looked at, r, the intercept of the straight line on the dry weight axis, is considerable, reaching about 50 g for *Helianthus annuus*, considerably more than half the total plant dry weight at harvest 9, when this phase begins. Thus here dry weight was not proportional to the square of the leaf area. Turning to the computation of $\bar{E}$ in these

16.9

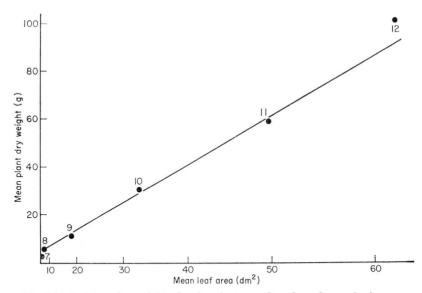

Fig. 16.9.2. Mean dry weight of maize plants as a function of mean leaf area squared (numbers against points are those of the harvests of Figs. 16.3, 16.4, 16.8). Variety 'Badischer Früh', grown at Poppelsdorf in 1876. Based on the data given by Kreusler *et al.* (1877b).

cases, from equation 16.9.1 $dW = 2sL_A \cdot dL_A$, and, from equation 16.6,

$$\bar{E}(_2T - _1T) = 2s \int_{_1L_A}^{^2L_A} \frac{L_A\, dL_A}{L_A} = 2s \int_{_1L_A}^{^2L_A} dL_A$$

$$= 2s(_2L_A - _1L_A);$$

but, substituting the particular values $_1W$, $_2W$, $_1L_A$ and $_2L_A$ in equation 16.9.1,

$$s = \frac{_2W - _1W}{_2L_A^2 - _1L_A^2};$$

therefore

$$\bar{E} = \frac{2(_2W - _1W)}{(_2T - _1T)(_2L_A + _1L_A)}. \qquad 16.9.2$$

16.10 B(iv) Interrelations

Equations 16.8.2 and 16.9.2 are the generally accepted formulae which have been used for most of the computation of mean values of unit leaf

16.10

rate to be found in the literature. But, as we have seen, adequate data are not always available to enable one to decide whether the one or the other expression is appropriate. The question then arises as to how large are the errors involved in the wrong choice. Coombe (1960) pointed out that the ratio of the values obtained using the two expressions depends only on the ratio of $_2L_A/_1L_A$. Let us call $\bar{E}$ calculated from equation 16.8.2, on the basis of a linear relation between W and L_A, $\bar{E}_1$; and $\bar{E}$ calculated from equation 16.9.2 $\bar{E}_2$. Then, from equations 16.8.2 and 16.9.2

$$
\begin{aligned}
\frac{\bar{E}_1}{\bar{E}_2} &= \frac{\dfrac{_2W - _1W}{_2T - _1T} \cdot \dfrac{\log_e {_2L_A} - \log_e {_1L_A}}{_2L_A - _1L_A}}{\dfrac{_2W - _1W}{_2T - _1T} \cdot \dfrac{2}{_1L_A + _1L_A}} \\[2ex]
&= \frac{\log_e {_2L_A} - \log_e {_1L_A}}{2} \cdot \frac{_2L_A + _1L_A}{_2L_A - _1L_A} \\[2ex]
&= \frac{\log_e P}{2} \cdot \frac{P+1}{P-1},
\end{aligned}
\qquad 16.10
$$

if

$$
P = \frac{_2L_A}{_1L_A}.
$$

It is therefore possible to tabulate either the ratio itself, or the percentage difference between the two values,

$$
100 \left(\frac{\bar{E}_1 - \bar{E}_2}{\bar{E}_2} \right),
$$

simply as a function of P. This difference will always be positive, because for any positive value of P,

$$
\frac{\log_e P}{2} \cdot \frac{P+1}{P-1} > 1.
$$

Coombe made the latter tabulation, as in Table 16.10. It will be seen that so long as the leaves do not change in area during the period between harvests by more than a factor of 2, the error through a completely wrong choice of formula for calculating $\bar{E}$ does not exceed 4 per cent, provided that we can be sure that the relationship between total dry weight and leaf area lies within the range from linear (as in equation 16.8.1) to quadratic (as in equation 16.9.1). We shall consider later [16.12] methods

16.10

Table **16.10.** The percentage differences between mean unit leaf rate calculated from equations 16.8.2, $\bar{E}_1$, and 16.9.2, $\bar{E}_2$, as a function of the ratio $_2L_A/_1L_A$, of leaf area at the second and first harvest times. Adapted from Coombe (1960).

$\dfrac{_2L_A}{_1L_A}$	1·0	1·5	2·0	2·5	3·0	4·0	5·0	7·5	10·0
$100\left(\dfrac{\bar{E}_1 - \bar{E}_2}{\bar{E}_2}\right)$	0·0	1·37	3·97	6·90	9·86	15·5	20·7	31·7	40·7

of proceeding if there is reason to suppose that the relationship does not lie within this range.

In much field experimentation an error of 4 per cent is acceptable, and here this particular uncertainty as to the correct method of working out the true mean unit leaf rate is not of great consequence. However, this error of 4 per cent depends on the ratio $_2L_A/_1L_A$ not exceeding 2·0, and we now see the reason, in our earlier experimental planning [7.10], for so arranging harvest frequencies as to limit leaf area increases between harvests to roughly two-fold. Table 16.10 shows that if it increased ten-fold, the difference between the two estimates increases to 40 per cent, a very serious matter. Such a case necessarily also implies considerable uncertainty as to the relationship between total dry weight and leaf area, thus making it difficult to estimate the actual magnitude of the error involved, and leaving the whole question of the computation of a true mean value of unit leaf rate in a very uncertain and unsatisfactory state. In the intermediate cases, where leaf area has increased by say three- or four-fold, we may have sufficient data to be able to estimate the relationship between dry weight and leaf area. We shall consider below [16.12–16.16] how corrections can be estimated to bring the uncertainty here also to within tolerable limits.

16.11 B(v) Relationships between E and time

It will have been noticed that although they rest upon different conditions, equation 16.4.2, for the determination of unit leaf rate if it is constant, has the same form as equation 16.8.2, for the determination of a true mean value of a changing ULR. The first is applicable if leaf area is increasing exponentially with time; the second if the dry weight is linearly

16.11

related to leaf area. The implication is that if both these conditions are met, then ULR has indeed been constant during the inter-harvest period, so that the true mean is equal to the constant value. It can easily be shown that this is so.

If

$$L_A = p\,e^{qT},$$

and

$$W = c + dL_A,$$
$$= c + dp\,e^{qT},$$

where **c**, **d**, **p** and **q** are all constant, then

$$\frac{dW}{dT} = dpq\,e^{qT},$$

and

$$E = \frac{1}{L_A}\cdot\frac{dW}{dT} = \frac{dpq\,e^{qT}}{p\,e^{qT}} = dq. \qquad 16.11.1$$

E is therefore indeed constant.

Similarly, if $L_A = u + vT$ and $W = r + sL_A^2$, where **u**, **v**, **r** and **s** are all constant, it can be shown that

$$E = 2vs, \qquad 16.11.2$$

again constant.

We have already come across instances of the double conditions being satisfied. For example, in Kreusler's maize data for 1876, the second pair of conditions are both satisfied between harvests 9 and 11 [16.3, 16.9]. In the case of these plants, therefore, we have reason to believe that ULR was indeed effectively constant at this part of the life cycle, whatever may have happened at other times. During these two weeks there presumably were day-to-day fluctuations in the actual value of **E**, but these would be fluctuations about a mean value, not a systematic rise or fall.

On the other hand, during the week between harvests 4 and 5 in the data for maize, we noticed that leaf area was still increasing exponentially with time, whereas as we shall see [Fig. 16.13.2] dry weight had already begun a linear relationship with leaf area raised to the power 1·25. For this week, then, we have simultaneously

$$L_A = p\,e^{qT}$$

and

$$W = \mathbf{h} + \mathbf{k}L_A^{1.25} = \mathbf{h} + \mathbf{kp}\,e^{1.25\mathbf{q}T},$$

where $\mathbf{p}$ and $\mathbf{q}$, $\mathbf{h}$ and $\mathbf{k}$ are all constant.
Therefore

$$\frac{dW}{dT} = 1.25\,\mathbf{qkp}\,e^{1.25\mathbf{q}T},$$

and

$$E = \frac{1}{L_A}\cdot\frac{dW}{dT} = \frac{1.25\,\mathbf{qkp}\,e^{1.25\mathbf{q}T}}{\mathbf{p}\,e^{\mathbf{q}T}} = 1.25\,\mathbf{qk}\,e^{0.25\mathbf{q}T}. \qquad 16.11.3$$

During this particular week, therefore, unit leaf rate was increasing, but much less rapidly than leaf area.

These three particular instances illustrate the general proposition that if we have information on the relationships both of leaf area with time and of dry weight with leaf area, the relationship between ULR and time can be inferred (Evans and Hughes, 1961).

16.12　B(vi)　A more general case

Up to 1962 the two equations 16.8.2 and 16.9.2 remained the only guides to a complex situation. In favourable cases it might be possible to show that the relationship between W and L_A was of such a form that these two equations set limits within which the true mean value of ULR must lie (see, e.g., Evans and Hughes, 1961, p. 159); but these limits might be uncomfortably wide if leaf area had increased during the period between harvests by a factor of much more than 2. However, the examples we have already examined make clear that for substantial periods the total plant dry weight is not simply related either to the first or second power of leaf area. If it could be shown that the relationship took the relatively simple form of $W = \mathbf{k}L_A^n$, where $\mathbf{k}$ and $\mathbf{n}$ are constant, the problem could be readily resolved, as $\mathbf{n}$ can be determined from the ratio of relative growth rate to relative leaf growth rate [22.3]. In favourable cases such a relationship may indeed hold approximately for a few weeks after germination, but it does not necessarily hold even then, as was shown by our examination of Kreusler's data for maize in Fig. 16.8. In any case these are the simplest instances posing the least difficulties. On the other hand, later in life, as our examples have shown, extrapolation of the relation between W and L_A is likely to produce a substantial intercept on the W-axis (e.g., Fig. 16.9.1). Evans and Hughes (1962) therefore investigated the computation of $\bar{E}$ for the series of cases given by

$$W = h + kL_A^n,$$ 16.12.1

where h, k and n are all constant, for any positive value of n. Their objects in doing so were two-fold. Firstly, to attempt to provide a guide to the selection of most probable values of unit leaf rate, a problem to which, as we have seen, they had already given attention; and secondly, to assist in estimating the probable magnitude of the errors due to uncertainty as to the correct method of computation in a particular case.

If

$$W = h + kL_A^n,$$ 16.12.1

where h, k and n are all constant, then

$$dW = nkL_A^{n-1}\,dL_A,$$

and

$$\bar{E}(_2T - {}_1T) = \int_{_1W}^{_2W} \frac{dW}{L_A}$$ 16.6

$$= nk \int_{_1L_A}^{_2L_A} \frac{L_A^{n-1}}{L_A}\,dL_A$$

$$= nk \int_{_1L_A}^{_2L_A} L_A^{n-2}\,dL_A$$

$$= \frac{nk}{n-1}(_2L_A^{n-1} - {}_1L_A^{n-1}).$$

But substituting the particular values $_2W$, $_1W$, $_2L_A$, $_1L_A$, in equation 16.12.1,

$$k = \frac{_2W - {}_1W}{_2L_A^n - {}_1L_A^n}.$$

Therefore

$$\bar{E} = \frac{_2W - {}_1W}{_2T - {}_1T} \cdot \frac{n}{n-1} \cdot \frac{_2L_A^{n-1} - {}_1L_A^{n-1}}{_2L_A - {}_1L_A},$$

$$= \frac{_2W - {}_1W}{_2T - {}_1T} \cdot \frac{n}{n-1} \cdot \frac{P^{n-1} - 1}{P^n - 1} \cdot \frac{1}{_1L_A},$$ 16.12.2

16.12

if

$$P = \frac{{}_2L_A}{{}_1L_A}.$$

In this treatment it is immaterial whether the values of **h** and **k** are positive or negative; and although strictly ${}_1L_A$ should be the smaller and ${}_2L_A$ the larger of the two leaf areas, because of the form of equations 16.12.2 we can preserve our normal notation, making ${}_1L_A$ the leaf area at the first of the two harvests, irrespective of whether leaf area is increasing or decreasing with time. These particular approximations can therefore be fitted to a wide variety of specific relationships.

Let us call the value of $\bar{E}$, computed according to equation 16.12.2 for a particular value of **n**, $\bar{E}_n$; and let us compare $\bar{E}_n$ with the value of $\bar{E}$, $\bar{E}_2$, computed from the same values of W and L_A on the assumption that

$$W = r + sL_A^2,\qquad\qquad 16.9.1$$

in which case

$$\bar{E}_2 = \frac{{}_2W - {}_1W}{{}_2T - {}_1T} \cdot \frac{2}{{}_2L_A + {}_1L_A}.\qquad\qquad 16.9.2$$

Then

$$\frac{\bar{E}_n}{\bar{E}_2} = \frac{\dfrac{{}_2W - {}_1W}{{}_2T - {}_1T} \cdot \dfrac{n}{n-1} \cdot \dfrac{P^{n-1}-1}{P^n-1} \cdot \dfrac{1}{{}_1L_A}}{\dfrac{{}_2W - {}_1W}{{}_2T - {}_1T} \cdot \dfrac{2}{{}_2L_A + {}_1L_A}}$$

$$= \frac{n}{n-1} \cdot \frac{P+1}{2} \cdot \frac{P^{n-1}-1}{P^n-1},\qquad\qquad 16.12.3$$

that is to say, the ratio depends only upon **n** and **P**, and if they are known, it can readily be computed in a particular case. The situation is thus exactly analogous to the particular case of **n** = 1 worked out by Coombe (1960) [16.10]; incidentally, inspection will show that this particular case cannot be worked out from equation 16.12.3 (because it involves **P°**), so that it is necessary to compute it separately, by the method already given [16.10]. We can now extend the table to cover any desired value of **n**. A convenient property of equation 16.12.3 is that

$$\frac{\bar{E}_n(P)}{\bar{E}_2(P)} = \frac{\bar{E}_n(P^{-1})}{\bar{E}_2(P^{-1})},$$

i.e. the value of $\bar{E}_n/\bar{E}_2$ is not changed by substituting P^{-1} for **P** throughout. This means that we need tabulate only for values of **P** greater than 1, and if ${}_1L_A$ is greater than ${}_2L_A$ we use the column corresponding to P^{-1}

16.12

$(=_1L_A/_2L_A)$. We notice also that for values of **n** greater than 1·0 the expression $[(P + 1)(P^{n-1} - 1)]/[P^n - 1]$ tends to a value of 1·0 as **P** becomes very large, in which case the whole expression of equation 16.12.3 tends to a value of $n/[2(n - 1)]$, and this limit can be inserted. It is most convenient to follow Coombe's mode of presentation and to tabulate

$$100\left(\frac{\bar{E}_n - \bar{E}_2}{\bar{E}_2}\right).$$

This is presented in Table 16.12, and a fuller tabulation is given in Appendix 2.

It will have been noticed that

$$\bar{E}_2 = \frac{_2W - _1W}{_2T - _1T} \cdot \frac{2}{_2L_A + _1L_A}$$

(equation 16.9.2) is much the simplest to evaluate of all the various expressions for $\bar{E}$ which we have hitherto derived. Our most convenient

Table 16.12. Percentage differences between mean values of unit leaf rate calculated on the assumption that $W = h + kL_A^n$ and the corresponding value for the assumption that

$$W = r + sL_A^2,$$
$$\left(100\left(\frac{\bar{E}_n - \bar{E}_2}{\bar{E}_2}\right)\right),$$

as a function of n and of P, the ratio of total leaf area at the second harvest to that at the first. Note that for any particular value of P the tabulated differences are identical with those for P^{-1}, so that for values of $P < 1$ the column corresponding to P^{-1} is used, i.e., always the ratio of the larger to the smaller leaf area, whether leaf area is increasing with time or not. From Evans and Hughes (1962). For a fuller tabulation see Appendix 2.

n	P 1·0	1·5	2·0	2·5	3·0	4·0	5·0	7·5	10·0	∞
0·5	0	+2·06	+6·07	+10·68	+15·47	+25·0	+34·2	+55·2	+73·9	—
1·0	0	+1·37	+3·97	+6·90	+9·86	+15·53	+20·7	+31·7	+40·7	—
1·5	0	+0·68	+1·94	+3·32	+4·68	+7·14	+9·28	+13·45	+16·51	+50·0
2·0	0	0	0	0	0	0	0	0	0	0
2·5	0	−0·66	−1·84	−3·04	−4·12	−5·91	−7·29	−9·57	−10·95	−16·6
3·0	0	−1·32	−3·57	−5·77	−7·69	−10·71	−12·90	−16·31	−18·24	−25·0
3·5	0	−1·95	−5·18	−8·20	−10·75	−14·57	−17·21	−21·1	−23·2	−30·0
4·0	0	−2·56	−6·67	−10·34	−13·33	−17·65	−20·5	−24·6	−26·7	−33·3

means of calculating $\bar{E}$ is therefore to use equation 16.9.2, and apply to the result a correction derived from Table 16.12 or Appendix 2 for the particular values of **P** and **n** appropriate to the period of growth in question, interpolating if needs be.

Table 16.12 has another use. If we are in doubt as to the exact value of **n**, but can define limits within which it must lie (as not infrequently happens in practice), then from the table we can convert these limits into limits of uncertainty attached to the value of $\bar{E}$. We can then apply a correction midway between these limits, and thus obtain both a best value and probable limits of error.

16.13 Graphical estimation of n

If data from a long series of harvests are available, in suitable cases it is a simple matter to obtain a value of **n** with sufficient accuracy, by an extension of the graphical method which we have already used to estimate the periods during which the total plant dry weight is related to leaf area

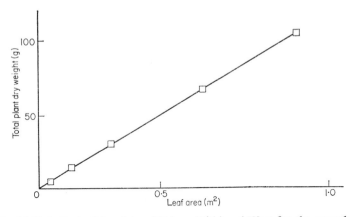

Fig. 16.13.1. Part of the data of Briggs, Kidd and West for the growth of *Helianthus annuus* at the Cambridge University Botanic Garden in 1920, showing mean total plant dry weight plotted against mean total leaf area raised to the power 1·25, for harvests 5 to 9, at weekly intervals (data for the remaining harvests in Figs. 4.5 and 16.9.1). From Evans and Hughes (1962).

either by a linear or quadratic expression [16.8, 16.9, Figs. 4.5, 16.9.1 and 16.9.2]. It appears that from germination to attainment of the maximum leaf area in an annual plant, **n** in the expression $W = \mathbf{h} + \mathbf{k}L_A^n$

16.13

shows a tendency to increase. Thus, taking Briggs, Kidd and West's data for *Helianthus annuus* as an example, we have already seen from Fig. 4.5 that between harvests 1 and 4 **n** was very close to 1·0. If we go on and plot the point for harvest 5, we find it above the best straight line through the points for harvest 1 to 4. Harvests 6 onward show larger and larger positive deviations, showing that **n** has increased. Similarly, in

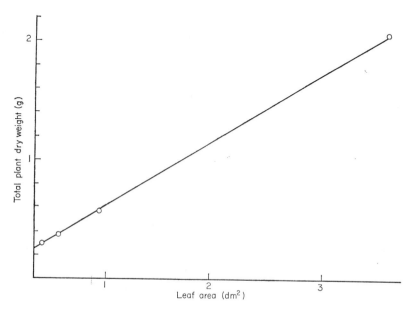

Fig. 16.13.2. Part of the data of Kreusler, Prehn and Becker for the growth of maize (variety Badischer Früh) at Poppelsdorf in 1876, showing mean total plant dry weight plotted against mean total leaf area raised to the power 1·25, for harvests 4 to 7, at weekly intervals. For the same data plotted on a linear scale of leaf area, see Fig. 16.8; for the data of later harvests, see Fig. 16.9.2. Calculated from Kreusler *et al.* (1877b).

Fig. 16.9.1 we saw that from harvests 9 to 13 there was a linear relationship between W and L_A^2. If we plot on this graph the points for harvests 5 to 8 we find them below the line; at that time **n** was less than 2. Therefore during the period between harvests 4 and 9 **n** must have lain between 1 and 2. We now plot W for this period against $L_A^{1·5}$. We find a curve of decreasing slope; our estimate of **n** is too high. We replot W against $L_A^{1·25}$, and obtain the relationship of Fig. 16.13.1. Clearly during the period

16.13

between harvests 5 and 9 (between total plant dry weights of 4·82 to 105 g) **n** was very close to 1·25. The point for harvest 4 lies below this line. Therefore during the period between harvests 4 and 5 **n** must have lain between 1 and 1·25. It is not worth proceeding further, because although between harvests 4 and 5 L_A increased from 209 to 760 cm², giving a value of **P** of 3·64, if one adopts a value for **n** of 1·125, the remaining uncertainty is less than 1·8 per cent. We find in Fig. 16.13.2 a similar relationship, with **n** close to 1·25, between harvests 4 and 7 of Kreusler's observations on Badischer Früh maize in 1876 (cf. Fig. 16.9.2 for the data of harvests 8 to 11); but there is an important difference from Fig. 16.13.1, in that the straight line of best fit has a substantial intercept on the dry weight axis, of 0·24 g, more than 80 per cent of the mean total dry weight at harvest 4. Fig. 16.13.2 should also be compared with Fig. 16.8, where the same data are plotted on a linear scale of leaf area.

An interesting case of the parameter **h** being negative was observed for *Filipendula ulmaria* by Faheemuddin (1969). Young plants were grown in 16-hour days of partly natural and partly artificial illumination. There was a linear relationship between W and L_A (**n** = 1) over a period of 60 days during which there were 5 harvests, W increasing from 0·13 to 6·43 g and L_A from 0·18 to 5·50 dm². **h** had a value of −0·09 g, corresponding to a positive intercept on the leaf area axis of 0·08 dm².

16.14 Computation of n

(i) If the number of harvests is small and the value of **n** is seen to be greater than 1, it may be better to proceed by calculation, although it must be remembered that variation in the figures for an individual harvest arising from whatever cause may produce a substantial change in the calculated value of **n**. However, seeing that three harvests are needed for the calculation, if there are more we can bracket the values, calculating one for harvests 1 to 3 and one for harvests 2 to 4. We then have two values for the second period between harvests, and discrepancies can be detected (for an example see 16.16).

It is convenient to proceed as follows: if the dry weights at the three harvests are $_1W$, $_2W$, $_3W$, and the corresponding leaf areas are $_1L_A$, $_2L_A$, $_3L_A$, we compute values of **P**, **Q**, and **Z**, where

$$P = \frac{_2L_A}{_1L_A}, \qquad Q = \frac{_3L_A}{_1L_A}, \qquad Z = \frac{_3W - _2W}{_2W - _1W}.$$

Then, as

$$W = h + kL_A^n,$$ 16.12.1

$$_1W - _1W = k(_2L_A^n - _1L_A^n),$$

$$_2W - _2W = k(_3L_A^n - _2L_A^n),$$

and

$$Z = \frac{_3W - _2W}{_2W - _1W} = \frac{_3L_A^n - _2L_A^n}{_2L_A^n - _1L_A^n}$$

$$= \frac{Q^n - P^n}{P^n - 1}.$$

Then

$$Q^n - P^n = Z(P^n - 1)$$

and

$$Q^n - (1 + Z)P^n + Z = 0.$$ 16.14.1

Values of this expression for several values of n are worked out and plotted against n. The point where the curve passes through zero gives the value of n.

(ii) In favourable cases, when it is known that there is no large change in the slope of the relationship between dry weight and leaf area between one harvest and the next, we can obtain an approximate value of n less tediously by assuming that the slope of the tangent at the midpoint between harvests is approximately equal to that of the chord joining one harvest with the next, i.e., that when

$$L_A = \frac{_1L_A + _2L_A}{2},$$

$$\frac{dW}{dL_A} = \frac{_2W - _1W}{_2L_A - _1L_A}$$

approximately. But, if

$$W = h + kL_A^n,$$

$$\frac{dW}{dL_A} = knL_A^{n-1};$$

therefore

$$\frac{_2W - _1W}{_2L_A - _1L_A} = kn\left(\frac{_2L_A + _1L_A}{2}\right)^{n-1},$$

16.14

approximately. Therefore

$$\log\left(\frac{_2W - {}_1W}{_2L_A - {}_1L_A}\right) = \log \mathbf{kn} + (\mathbf{n} - 1)\log\left(\frac{_2L_A + {}_1L_A}{2}\right),$$

and

$$\log\left(\frac{_3W - {}_2W}{_3L_A - {}_2L_A}\right) = \log \mathbf{kn} + (\mathbf{n} - 1)\log\left(\frac{_3L_A + {}_2L_A}{2}\right).$$

Therefore

$$\mathbf{n} - 1 = \frac{\log\left[\left(\dfrac{_3W - {}_2W}{_3L_A - {}_2L_A}\right)\left(\dfrac{_2L_A - {}_1L_A}{_2W - {}_1W}\right)\right]}{\log\left(\dfrac{_3L_A + {}_2L_A}{_2L_A + {}_1L_A}\right)} \qquad 16.14.2$$

approximately. In cases of doubt, this simplified procedure can be checked by substitution in equation 16.14.1 (for an example, see 20.3).

These two computational procedures requiring data from three successive harvests are most suited to work on uniform plants growing under relatively constant conditions. For less uniform populations, or for plants grown under fluctuating conditions, these methods are likely to show very substantial fluctuations in the computed value of the constants, and sometimes prove to be quite useless.

There is also no guarantee, even under the most uniform conditions, and after every precaution has been taken, that an equation of the type $W = \mathbf{h} + \mathbf{k}L_A^n$ will be a good approximation to the relationship between dry weight and leaf area. We have examined a few instances in which the approximation is as good as could be desired, and indeed, in Fig. 16.13.1, Briggs, Kidd and West's data for *Helianthus annuus* growing in open ground show a closeness of approach to a straight line rarely found in biological experimentation, with a correlation coefficient of 0·999969. On the other hand, we shall examine in 20.3 an example of growth under constant, controlled conditions where it can be shown that the approximation was not good, and in 16.16 we shall consider what can be done, in suitable cases, to estimate the magnitude of the uncertainty.

16.15 The meaning of n

The procedure adopted in section 16.13, using Figs. 4.5, 16.9.1, 16.13.1 and 16.13.2 clearly simplified the problem of selecting suitable bases for the computation of unit leaf rate in these particular instances. The close fit to the relationships plotted there prompts questions on the meaning of the parameters, the sharpness of the transitions, and on whether the particular values of the parameters encountered in the different phases, and the transitions between them, are likely to be reproduced in other years, under other conditions of growth. Fig. 4.5 represents the particularly simple case of $W = \mathbf{d}L_A$ (equation 16.8.1, when $\mathbf{c}$ is negligibly small) where $\mathbf{d}$ is obviously the reciprocal of the leaf area ratio. As $\mathbf{n}$ rises, the points for the later harvests rise more and more above this initial straight line, and the leaf area ratios at these harvests are correspondingly lower. Fig. 4.5 showed that the point for harvest 4 would lie above a straight line joining the origin to the point for harvest 3, indicating a fall in leaf area ratio between dry weights of 0·2 and 1·2 g, and suggesting that $\mathbf{n}$ is perhaps already beginning to rise in this region. This suggestion is confirmed by the more extended observations made on plants of *Helianthus annuus* 2 to 4 weeks old grown in the field at Cambridge each summer between 1952 and 1960, already seen in Table 7.5. During the third or fourth week of growth these plants showed a fall in leaf area ratio in all 9 years, averaging 19 per cent, and to this the average contribution [13.16] of leaf weight ratio was 10 per cent and of specific leaf area 9 per cent; but the variability between years was large, ranging from a fall of 1 per cent to one of 37 per cent. Sampling in individual years accounts for only part of this large variability, and the explanation of the remainder no doubt lies in the differing effects of environmental change and ontogenetic drift on leaf weight ratio, and particularly on specific leaf area. Thus, in the extreme cases mentioned, in 1953 specific leaf area fell by 41 per cent while leaf weight ratio rose by 4 per cent; while in 1954 specific leaf area rose by 10 per cent and leaf weight ratio fell by 9 per cent. This illustrates the difficulty of interpreting the changing values of $\mathbf{n}$ encountered in field studies, involved as they are both in changes in the proportion of new assimilates being converted into new leaves, and in changes in expansion of the leaf material. For *H. annuus*, then, we conclude that while $\mathbf{n}$ shows an early tendency to rise, the particular values encountered in 1920, and the transitions between them, were probably peculiar to this particular year.

16.16   Limitation of uncertainty

Another difficult example was illustrated in Fig. 16.8, from Kreusler's data on the early stages in the growth of maize in 1876. Here it seems that

n is rising continuously. This is confirmed by the corresponding data for 1877, harvests 4 to 9, which we shall consider in Table 16.18, where harvests 4, 5 and 6 yield a value of $\mathbf{n} = 1{\cdot}00$; 5, 6 and 7 one of $1{\cdot}23$; 6, 7 and 8, $1{\cdot}13$; and 7, 8 and 9, $1{\cdot}42$. This means, in effect, that our equation $W = \mathbf{h} + \mathbf{k}L_{\mathrm{A}}^{\mathbf{n}}$ is an unsuitable approximation to the growth of these particular plants at this particular stage. The equation contains three arbitrary constants; it is therefore bound to fit any three points (giving a value for **n**), and no other significance can be attached to the fact that it does so, unless there is supporting evidence of some kind. In the case of the Briggs, Kidd and West data, this support comes from the coincidence of consecutive values of all the constants. But in the absence of such support, continuously changing values of the constants would suggest that growth is not approximating at all closely to our assumptions, so that we can no longer rely on this method of computation to give us a close approximation to the true mean value of **E**, even though in suitable cases it may still be valuable as setting limits to the uncertainty.

Kreusler's data for the growth of maize in 1877, harvests 4 to 9, just mentioned (and Table 16.18) provide a convenient example. During this period all the values of **n** calculated from three consecutive harvests lies between 1 and 2: they start near 1, the first three are all below $1{\cdot}25$, only the fourth reaches $1{\cdot}42$. We should therefore conclude that the first two values of $\bar{\mathbf{E}}$ are likely to be close to $\bar{\mathbf{E}}_1$, the third is likely to be closer to $\bar{\mathbf{E}}_1$ than to $\bar{\mathbf{E}}_2$, and the last two will also lie somewhere between $\bar{\mathbf{E}}_1$ and $\bar{\mathbf{E}}_2$. The values of **P** for the five consecutive intervals are $3{\cdot}7$, $2{\cdot}8$, $2{\cdot}2$, $1{\cdot}6$ and $1{\cdot}9$, the corresponding differences between $\bar{\mathbf{E}}_1$ and $\bar{\mathbf{E}}_2$, as in Tables 16.10 and 16.12, being $13{\cdot}9$, $8{\cdot}7$, $5{\cdot}1$, $1{\cdot}8$ and $3{\cdot}4$ per cent. We can thus make good estimates of the likely values of $\bar{\mathbf{E}}$, and the uncertainty will be limited to a few per cent—perhaps 5 per cent for the first interval and less for the others. For many purposes such an estimate would suffice. In the next two sections we shall consider a way of checking it.

This setting of limits to the uncertainty involved in a specific method of computation is also useful when dealing with plants whose annual growth cycle involves at first a phase of increasing leaf area, then a slowing down and the attainment of a maximum leaf area, followed finally by a phase of decrease, while total plant dry weight is still increasing. This is often encountered in cereal crops (e.g., Fig. 2.7) but can apply equally to a number of other growth forms (e.g., Fig. 13.18.1). The phase of dry weight increase and leaf area decrease may be prolonged; for example, it ex-

16.16

tended over the whole 10 weeks of the observations of Vernon and Allison (1963) on the post-flowering phase in maize.

For such a plant a plot of dry weight against leaf area will pass through the stages we have already noticed, of increasing values of $\mathbf{n}$; then, as the maximum leaf area is approached, $\mathbf{n}$ will increase more and more rapidly until the maximum leaf area is reached, and the curve is parallel to the y-axis. Then, as W continues to increase and L_A begins to decrease, the curve bends back towards the y-axis. However, in the nature of the case, when $\mathbf{n}$ is very large $\mathbf{P}$ is very small, and reference to the table in Appendix 2 shows that under these conditions the deviations from $\bar{\mathbf{E}}_2$ are also very small, more or less irrespective of the value of $\mathbf{n}$. (For small values of $\mathbf{P}$, say less than 1·2, and values of $\mathbf{n}$ which are not very large, the correction approximates to $[(2 - \mathbf{n})(\mathbf{P} - 1)^2]/12$, so that for $\mathbf{P} = 1·2$ the correction does not exceed 1 per cent until $\mathbf{n}$ exceeds 5; for $\mathbf{P} = 1·1$ until $\mathbf{n}$ exceeds 20, and so on.)

We can therefore separate the whole curve into three parts, the first of which would be dealt with in the way already discussed [16.12–16.15]. The second would begin when $\mathbf{P}$ was small enough for uncertainties about the exact value of $\mathbf{n}$ to have a negligible effect upon the estimate of $\bar{\mathbf{E}}$ (a judgment which would depend on the acceptable limits of uncertainty in the particular experiment), and would continue as long as this condition was met. In the third part leaf area is decreasing more rapidly, and we now enter a phase of approximation where $\mathbf{k}$ in the expression $\mathbf{h} + \mathbf{k}L_A^n$ is negative, and in the table of corrections (16.12 and Appendix 2) we use the column corresponding to $\mathbf{P}^{-1}$ ($=_1L_A/_2L_A$).

16.17 B(vii) Numerical integration

In Chapter 20 we shall explore some other methods of approach, but meantime we should notice a method, published by Williams in 1946, of proceeding by a method of numerical integration, which does not involve any assumptions about the actual form of the relationship between total plant dry weight and leaf area.

Williams's method involved (i) plotting mean total dry weight per plant at harvest against time, and drawing a smooth curve by eye through the points; (ii) by interpolation on this curve, obtaining values for W at 2-day intervals; (iii) plotting mean leaf area per plant at harvest against mean total dry weight, and drawing a smooth curve by eye through the points; (iv) by interpolation on this curve, obtaining the values of L_A corresponding to the 2-daily values of W already obtained in (ii); (v) for each 2-day interval, working out a value of $\bar{E}$, from equation 16.8.2, i.e., on the assumption that for each interval W is a linear function of L_A (we notice that this is not necessarily justified, but procedure (iii) provides a check, and the short time interval makes for a small proportional increase in leaf area, giving a small value of **P**. As we have seen (16.16) if **P** is small enough, the value of **n** is immaterial); (vi) by using the appropriate values of $\bar{E}_1$, obtaining means for each interval between harvests.

Clearly the main assumption on which this method rests is that the plants have indeed been growing during the intervals between harvests in the way indicated by the smooth curves drawn in processes (i) and (iii). This is most likely to be true for constant, controlled environmental conditions; it is likely to be true for stable climates, where the weather on one day is practically identical with the weather the day before. It is unlikely to be true in a fluctuating climate, where a harvest interval may contain periods of warm and sunny weather, and of cold and dull weather. The nature and disposition of these fluctuations in time will determine the extent, direction and timing of deviations from a smooth growth curve passing through the harvest points, and hence the reliance which can be placed on figures worked out by this means. Granted the assumption that the smooth curves are an accurate presentation of the actual growth, the uncertainties involved in the method of working out $\bar{E}$ in process (v) can be kept within any desired limits by ensuring that when **n** does not equal 1 **P** is small enough, which in effect means using a short enough time interval in processes (ii) and (iv). In a particular case a series of values of **P** can be worked out, or the overall effect of lengthening the time interval can be tested by using shorter intervals and comparing the results. Williams made such a comparison over a 56-day period for one of his experiments. This particular experiment was concerned with unit leaf rate on a basis of leaf protein nitrogen (see 13.14, above), but this does not affect the principle we are considering. He made daily interpolations in processes (ii) and (iv), thus obtaining 57 values covering 56 daily intervals. He then worked out a set of 56 values of $\bar{E}_1$ from the interpolated daily values of W and L_A, and took a mean of all; worked out a set of 28 values of $\bar{E}_1$ taking the values of W and L_A interpolated at intervals of 2 days, and took a mean; and repeated the process for a series of longer periods. His comparison of these mean values is set out in Table 16.17. He pointed

16.17

Table 16.17. Comparison of mean values of unit leaf rate of *Phalaris tuberosa* computed over a series of different time intervals. Adapted from Williams (1946).

ΔT, days	1	2	4	7	14	28	56
$\bar{E}$, g m^{-2} week^{-1}	34·60	34·60	34·68	34·79	35·39	38·37	55·29
Difference (%)	—	0	0·23	0·55	2·28	10·9	59·8

out that the closeness of agreement between 1- and 2-day intervals was fortuitous, but even so there is obviously nothing to be gained in this case by using intervals shorter than 2 days. However, this is not necessarily so, and if this method of numerical integration is to be used, it is wise to check that the interval to be used is of a suitable length—not so long as to introduce errors, not so short as to make unnecessary work.

16.18   Another look at Kreusler's data for maize

Williams also used his method to rework some of Kreusler's data, and gives a comparative table of the results obtained by numerical integration as compared with those given by equation 16.8.2, ($\bar{E}_1$, assuming a linear relation between total dry weight and leaf area), and equation 16.9.2 ($\bar{E}_2$, assuming a quadratic one). In view of the conclusions which we had already reached [16.16] when examining some of Kreusler's data on the growth of maize in connection with the assumption that $W = h + kL_A^n$ (equation 16.12.1), it is interesting to see that $\bar{E}$ computed by Williams's method lies consistently between the estimates of equations 16.8.2 and 16.9.2. Furthermore, the estimate is at first closer to that of equation 16.8.2, but by July 10 it has become very close to that of equation 16.9.2. In 1877 the maize was sown on May 17, and the first harvest was on May 29. The seventh harvest was therefore on July 10. We had concluded that it was at about this stage that the growth pattern began to approximate closely to $W = r + sL_A^2$, corresponding to the use of equation 16.9.2, while earlier it seemed that the power of L_A was somewhat above 1, but considerably below 2 [16.16]. This fits in very well with Williams's computation. It will be noticed that in the fourth interval the difference between the estimates of equations 16.8.2 and 16.9.2 is very small; for the third interval it was 5·0 per cent, for the fifth 3·5 per cent, but for the fourth only 1·6 per cent. This corresponds to the small increase in leaf

16.18

area during this interval between harvests—P, $(_2L_A/_1L_A)$, being only 1·58 as compared with 2·22 for the third interval and 1·93 for the fifth. The differences are thus what we should expect from Coombe's relation between the two estimates, as set out in Table 16.10 (and see the line for $n = 1$ in Appendix 2).

Table 16.18. A comparison of unit leaf rates for 'Badischer Früh' maize growth at Poppelsdorf in 1877 using three different methods of calculation in g m^{-2} week^{-1}. Adapted from Williams (1946). Differences from the result of the numerical integration given in percent.

| Interval between harvests | Numerical $\bar{E}$ | Method of Calculation | | | |
| | | Equation 16.8.2 | | Equation 16.9.2 | |
		$\bar{E}_1$	Difference	$\bar{E}_2$	Difference
12 June–19 June	67·7	69·2	+2·2	60·8	−10·2
19 June–26 June	51·6	52·6	+1·9	48·3	−6·4
26 June–3 July	52·5	52·9	+0·8	50·3	−4·2
3 July–10 July	31·7	31·8	+0·3	31·3	−1·3
10 July–17 July	56·7	58·5	+3·2	56·5	−0·3

Thus in suitable cases—particularly where there is a long series of harvests, and where n in the equation $W = h + kL_A^n$ does not remain for long approximately constant, Williams's method can be of great assistance.

16.19   Instantaneous values of unit leaf rate

Related to Williams's method, but much more restricted in scope and in the occasions on which it can be profitably used, is a method described by Vernon and Allison (1963). They plotted mean total plant dry weight and leaf area against time, but instead of putting curves by eye through the points they worked out quadratic regression equations. By differentiating the equation relating dry weight and time they obtained dW/dT, and then by dividing by the appropriate value of L_A interpolated on the regression line, they obtained instantaneous values of E. These served the purposes of their investigations, which were to compare the effects of

16.19

two treatments on maize plants during the 10 weeks following flowering, but for any purpose involving growth during a period it would still be necessary to integrate, and unlike Williams's method this procedure gives no information on the relationship between total plant dry weight and leaf area.

In a sense they were fortunate in the data with which they had to deal, but these same features of the data are rare enough to be among the causes which greatly restrict the utility of their method. In the first place the range of dry weights was very small—from about 270 to 610 g per plant, an increase by a factor of only around 2·2 in 10 weeks. Secondly, although the variances of their six samples, harvested at fortnightly intervals, were not homogeneous, tending to increase as the plants became larger, the small size range made possible the calculation of regressions of dry weights against time: more usually one would have to calculate regressions of $\log_e W$ against time, as in the method of Hughes and Freeman (1967) which we shall consider later [20.4]. Finally, the change of leaf area (a decrease from 1·5 to just over 1 m² per plant over 8 weeks) was even smaller than the change of dry weight, and as we have seen [16.12, 16.16], when the leaf area changes so little the worst difficulties of computation disappear. Consequently, if it is desired to calculate instantaneous values of unit leaf rate, the method of Hughes and Freeman (1967) [20.4], which incidentally has a sounder statistical basis, is generally to be preferred, or more simply, if the criteria discussed in 16.17 are met, they can be derived by Williams's method, taking tangents to the smooth curve of W against T and omitting the integration.

16.20   Conclusion: further developments

The examples we have discussed illustrate that by one or another of these means we can usually make an estimate of $\bar{E}$ within the desired limits of accuracy for most field experimentation; and we can at least in most cases set limits to the probable inaccuracies attached to a particular method of computation.

There remain in field work some difficult cases which do not fit closely into any of the frameworks which we have considered. There are also experiments made under controlled conditions in the laboratory, where the aim is a high degree of accuracy, and which also do not fit. We shall turn in Chapter 20 to special methods of dealing with these.

Yet other developments [20.4] flow from modern computing facilities, which make possible experimental designs previously impracticable because of the labour involved in working out the results. Such is the work of Hughes and Freeman (1967), making use of more frequent, smaller harvests, involving no more plants in total, but producing closer approximations to continuous functions than does the usual procedure of harvests more widely spaced in time.

16.20

A FIRST LOOK AT THE EFFECTS OF SPECIFIC ENVIRONMENTAL CHANGES

17.1 Seven facets of the plant-environment complex

Throughout our discussions we have seen how closely linked are the processes of plant growth with the external environment in which the plant is growing. Yet many practical problems involving the growth of plants cannot be solved unless the factors influencing some particular facet of growth can be identified, and their relative importance estimated in quantitative terms. These are among the important aims of interpretative analysis; and the process of analysis is likely itself to suggest modifications both in the characterization of the processes of growth and in the procedures used for analysis. Before pursuing our discussion of analytical procedures any further it will be convenient to pause and to ask how far we can get at this stage in disentangling the interlinking of the onto-genetic processes of the plant with the multifarious effects of the environment.

In section 7.1 we distinguished seven facets of the plant-environment complex;

(a) genetic factors may influence the growth of the plant in a specific environment;

(b) the response of a plant to a change in a particular aspect of the environment is usually not linear;

(c) frequently the responses of the plant to changes in two different aspects of the environment interact, so that the overall effect is not a simple addition of individual responses;

(d) parts of the plant itself may react on its surroundings, producing environmental changes which may immediately affect other parts of the same plant, or produce effects remoter in space or time;

(e) under natural conditions, changes in different aspects of the environment are frequently correlated;

(f) the plant is subject to inherent ontogenetic drift—its responses to specific changes in the environment vary throughout the life of the plant;

(g) when grown in a fluctuating environment, those parts of the plant which became mature at various stages in the past embody the past reactions of the plant, (b) and (c) under the influence of (f), to past states of the environment.

How are these interlocking influences to be disentangled? Clearly it will not be easy, and it may be impossible, to state the magnitude of one particular influence in isolation from the others; statements, to have any meaning, will probably have to include the full context of other influences. It will then be necessary to establish, not some unique value applicable to a particular species or variety, but a set of relationships. Other sets of relationships, which may be more or less closely related to each other, will apply to other species and varieties, and comparisons between them, and the detailed working out of the consequences in a particular situation, will be important analytical tools.

Meantime we should obviously start with the simplest possible situation, and our previous discussions suggest how simplifications may be achieved. Selection of a genetically uniform strain of a particular species can be used to take care of (a), and we have seen in Chapter 7 that much can be done with (b), (c) and (e) by working under controlled environmental conditions. Such conditions also simplify the effects of (g); the plant is still a kind of integration of its past experience, but that experience is much less complex, and readily reproducible. Controlled environmental conditions may also on occasion simplify the effects of (d), the reactions of the plant on its environment, and the consequent problems may also be simplified by the selection of suitable species, forms or parts of the life cycle [17.3]. There remain the effects of (f), ontogenetic drift.

17.2 Comparisons at a fixed total plant dry weight

The problems associated with ontogenetic drift will be deferred to the further consideration of analytical procedures in Chapters 18–24, and we will now consider a possible method of temporarily setting aside the complications arising from ontogenetic changes by making comparisons at a fixed total plant dry weight. This involves looking over the range of dry weights in a number of experiments and choosing one common to all; and then estimating by interpolation the values of the various measures of growth at this particular dry weight. The detailed conclusions

17.2

reached from such comparisons will not necessarily be valid at other stages in the life of the same species of plant; but at least if the plants have been grown in carefully controlled conditions we can in this way specify all the pre-conditions of the particular observations, so that if needs be they can readily be linked with others.

17.3   An example

Results obtained by studying a particular species growing in a particular range of conditions cannot be extended to other species and other conditions unless there is reason to suppose that similarities of behaviour exist, but for our present purpose of considering a single simple instance, and to illustrate the use of the method, an example from a single species will suffice. The particular observations to be considered (Hughes, 1959a) were made on *Impatiens parviflora*, chosen as being the only native or naturalized British woodland annual which also forms natural communities in full sun. It is thus a very plastic plant, producing under natural conditions a wide range of plant form; and the woodland forms can readily be generated under completely artificial conditions in the laboratory. It has the added advantages that in the young plant, roughly up to the onset of flowering, there is little or no branching, an effective leaf mosaic [Fig. 18.5], short upper internodes, and (when lit from above) horizontal leaves, so that when cultured well spaced-out in growth cabinets the whole foliage can be considered as subjected to the same, readily characterized, light regime (thus reducing the complications of 17.1(d)). It has therefore proved a most useful subject for investigating in detail the relations of a plant and its environment.

17.4   Plants grown under natural and artificial conditions compared

We have already seen [3.6] how important it is that plants grown under controlled environmental conditions should correspond in structure and functioning with plants grown under natural conditions (or, at the least, semi-natural conditions) if valid inferences are to be drawn about the relationships of growth and environment. Hughes therefore first made several such comparisons between plants grown under a variety of conditions in his cabinets (described in Evans, 1959; Hughes, 1959a), and plants grown in the open under a variety of artificial screens. The first was concerned with the distribution of dry matter between the leaves and the rest of the plant (Fig. 17.4.1), and it will be seen that there was a general agreement between the responses of the plants to the two sets of conditions.

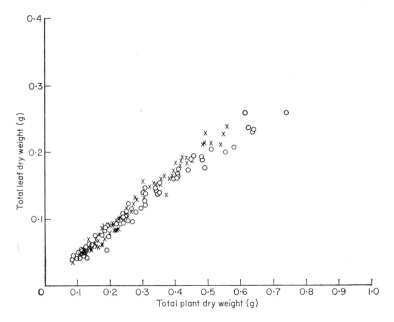

Fig. 17.4.1. Relationship of total leaf dry weight to total plant dry weight for individual plants of *Impatiens parviflora* grown out of doors near midsummer at Cambridge under a variety of neutral shades (×), and for plants grown in controlled environment cabinets, with a 16-hour day and a variety of conditions of temperature and intensity of visible radiation (○). From Hughes (1959a).

The second comparison was concerned with the expansion in space of the dry matter of the leaves, as expressed by the specific leaf area (leaf area per unit of leaf dry weight). Specific leaf area is a quantity having a very marked ontogenetic drift, while it is also extremely sensitive to the total daily light. The comparison in Table 17.4 was therefore made at a total plant dry weight of 0·15 g, and expressed as a function of total daily visible radiation. Other observations had shown that over this particular

17.4

range the relationship of SLA with total daily light was roughly linear. The particular light values appropriate to the cabinet-grown plants were therefore paralleled by linear interpolation between mean totals for the field-grown plants. Hughes notes that in view of the differences in

Table 17.4. Specific leaf areas of plants of *Impatiens parviflora* grown under screens in the field compared with those of plants of the same total dry weight (0·15 g) grown under artificial conditions in the laboratory, as a function of total daily visible radiation. Adapted from Hughes (1959a).

| Mean total daily light (cal cm^{-2}) | Specific leaf area dm^2 g^{-1} | | Cabinets |
| | Field experiment | | |
	Observed	Linear interpolation	Observed
13	12·7	—	—
15	—	11·6	10·9
29	—	7·3	7·4
36	—	6·2	6·2
48	4·9	—	—

spectral composition and in spatial and temporal distribution of the radiation in the two cases, it would be unwise to rely on the comparison of light conditions to an accuracy greater than 15 per cent. In view of this, the agreement between the two sets of conditions is better than would be expected, and is no doubt partly coincidental.

Finally, as a combined physiological and anatomical comparison, the relationships between unit leaf rate and specific leaf area for the two sets of conditions are given in Fig. 17.4.2. Specific leaf area is influenced in a complex manner by many factors, including ontogenetic drift [Figs. 19.4.2, 19.6, 19.8], and we shall return to this particular comparison later [28.6]; but we are now asking the relatively simple question—is there evidence of systematic differences between the plants grown under artificial conditions, and those grown in the field? It seems clear from all these three comparisons that the plants grown in the cabinets fall within the naturally occurring range of plant form, structure, and, at least as far as unit leaf rate is concerned, physiological performance.

17.4

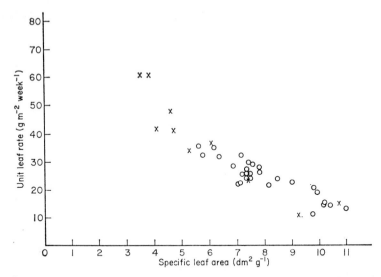

Fig. 17.4.2. Relationship of mean unit leaf rate during a period of a week to mean specific leaf area at the end of the same period, for samples (usually of eight) plants of *Impatiens parviflora* grown out-of-doors near midsummer at Cambridge under a variety of neutral shades ($\times$), and for similar samples grown in controlled environment cabinets, with a 16-hour day and a variety of conditions of temperature and intensity of visible radiation ($\circ$). From Hughes (1959a).

17.5 Effects of specific environmental changes

The way was therefore clear for a pilot study comparing the effects on the plant of specific changes in environmental conditions. For this purpose the plants were grown from seed onwards in pots in a controlled environment, at fixed temperatures and humidities, under radiation from fluorescent lamps fixed in daylength, intensity and spectral composition. Comparisons were made of the structure and performance of young plants weighing 0·1 g, i.e., varying in age from about 25 days at the highest total daily light to about 45 days at the lowest (under which the plants had not reached 0·15 g at the final harvest, necessitating the choice of 0·1 g as the common dry weight for this series of comparisons), the standard plant having been grown in a 16-hour day. Hughes compared the effects of five specific changes in the aerial environment.

(i)   A 24-hour with a 16-hour day at the same irradiance, i.e., a 50 per cent increase in total daily light, but the same temperature and humidity.

17.5

(ii) An 8-hour with a 16-hour day at the same total daily light, i.e., double the irradiance, and the same temperature and humidity.

(iii) An increase in total daily light from 15 to 36 cal cm^{-2} day^{-1} at a daylength of 16 hours in both cases, and the same temperature and humidity.

(iv) A continuous temperature of 20°C with one of 17·5°C, with in both cases the same lighting conditions and saturation deficit (chosen as being the aspect of humidity most closely related to rates of transpiration).

(v) A saturation deficit of 4·8 mm Hg with one of 1·5 mm, with the same temperature and lighting conditions in both cases.

Each comparison thus deals in the classical way with one facet of the aerial environment at a time, leaving the others fixed.

The comparisons were set out in the form of percentage increase or decrease of the particular quantity being considered, from the value for means of eight plants grown under a 16-hour day, or for the lower level of the last three variable conditions. Significant differences were in the neighbourhood of 10 per cent. Five quantities were compared:

RGR   relative growth rate;
ULR   unit leaf rate;
LAR   leaf area ratio;
SLA   specific leaf area;
T   the time interval between germination and 0·1 g total plant dry weight.

The first four quantities represent a progressive analysis of functions of growth. It will be recalled that

$$\text{relative growth rate} = \text{unit leaf rate} \times \text{leaf area ratio,} \qquad 13.7.2$$

and

$$\text{leaf area ratio} = \text{specific leaf area} \times \text{leaf weight ratio.} \qquad 13.16.1$$

As Fig. 17.4.1 showed, and as we shall see later in more detail [18.3–18.15], for *Impatiens parviflora* leaf weight ratio is a relatively invariant quantity under the conditions of these experiments, so that most of the change in leaf area ratio was brought about by changes in specific leaf area, i.e., in leaf expansion.

17.6 (i) Increase in irradiance

The five comparisons are set out in Fig. 17.6. Considering first the simple increase in irradiance at a fixed daylength (iii), we see that an increase of

17.6

140 per cent in total daily light has caused an increase of 117 per cent in unit leaf rate, rather less than a proportional increase. Fig. 3.2.1 showed that for this plant unit leaf rate is very nearly proportional to total daily light up to about 20 cal cm^{-2} day^{-1}, but that it ceases to be proportional above this value. However, this increase of ULR (to rather more than double) was accompanied by a decrease in LAR (to rather more than

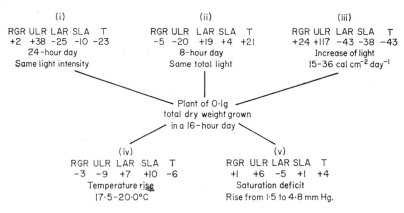

Fig. 17.6. Comparisons of the effects of specific differences in the aerial environment on various aspects of the growth of *Impatiens parviflora* cultured from seed under constant conditions in controlled environment cabinets. Differences between values for different treatments expressed as a percentage of the value for a 16-hour day or for the lower level of the variable condition. RGR, relative growth rate; ULR, unit leaf rate; LAR, leaf area ratio; SLA, specific leaf area; T, time to reach 0·1 g total dry weight. Adapted from Hughes (1959a).

half). The net effect on relative growth rate was therefore an increase of only 24 per cent. We also see that the bulk of the 43 per cent decrease in LAR can be attributed to a fall of 38 per cent in SLA; the leaves have become heavier for a given leaf area. Later we shall examine the anatomical basis of this change [27.2–27.6]. Finally we notice that although at 100 mg RGR is only 24 per cent higher, nevertheless the time taken for the plant to reach 0·1 g from seed was 43 per cent less, little over half the time taken at the lower irradiance. The plants grown in the higher irradiance must therefore have had a relatively higher RGR earlier in life, for

$$\text{Mean RGR} = \frac{\log_e {}_2W - \log_e {}_1W}{{}_2T - {}_1T},$$
 13.6.1

17.6

therefore

$$\text{Mean}\ \underset{\text{seed}\,\to\,100\ \text{mg}}{\text{RGR}} = \frac{\mathbf{m}}{T}, \qquad\qquad 17.6$$

where T is the time taken to grow from seed to 0·1 g total dry weight, and $\mathbf{m}$ is a constant, because $_2W = 0\!\cdot\!1$ g and $_1W = $ the graded seed weight, also constant for any given experiment. The mean RGR for the period of growth from seed to 0·1 g total dry weight is thus inversely proportional to the time taken. If this was reduced by 43 per cent, the mean RGR over the period must have been increased by 75 per cent. But at 0·1 g, the difference was only plus 24 per cent. Therefore, earlier in life the difference must have been considerably more than 75 per cent, and at some stage under the higher light the instantaneous value of the relative growth rate may well have been double that in the lower light.

17.7 (ii) Daylength changes

Considering now the daylength differences, (i) and (ii), we see in consequence of the increase of daylength from 16 to 24 hours a series of changes essentially similar to those just examined, but less marked, as might be expected for an increase of total daily light of 50 per cent as compared with 140 per cent. ULR increases by 38 per cent, once again less than proportional, but now LAR decreases by 25 per cent, almost exactly compensating, and producing the very small (and not statistically significant) difference of only 2 per cent in RGR. We therefore have the very interesting case of two sets of plants, grown from the same seed but under different conditions, having the same total dry weight but substantially different structure (witness the difference in LAR) and yet both growing at the same rate. Clearly there is a very interesting compensating mechanism here. Equally clearly it takes time to operate, because once again we see the difference in time taken to reach 0·1 g. When *they have once reached* 0·1 *g*, the two sets of plants are growing at the same rate, but *on the way there*, the ones growing in the 24-hour day had a mean relative growth rate 30 per cent higher. We also notice a significant difference between the change in LAR (−25 per cent) and in SLA (− 10 per cent), indicating that here there must have been a much larger change in leaf weight ratio than we observed in the case of the much larger change in intensity alone.

The change from the 16- to the 8-hour day is complementary, and interesting in that here total daily light was substantially unchanged.

The doubling of intensity and halving of daylength produces a reduction (of 20 per cent) in ULR, as might be expected; but the almost compensating increase of 19 per cent in LAR reduces the change in RGR to 5 per cent. The change in specific leaf area is a mere 4 per cent, consonant with our earlier remark that it appears to be related to total daily light, so that there must have been an increase in leaf weight ratio. The time taken to reach 0·1 g has increased by 21 per cent, indicating a decrease of mean relative growth rate of 17 per cent, once again much larger than the difference between the relative growth rates of the two sets of plants at 0·1 g.

Taking together these three sets of comparisons, all involving changes in the light climate of different sorts, we have noticed differences in mean RGR from seed up to 100 mg of +75 per cent, +30 per cent and −17 per cent, reduced by the time the plants reached 100 mg to +24 per cent, +2 per cent and −5 per cent. The compensating mechanism must thus be a powerful one, and it is no doubt connected with the plasticity of the plant. We shall return to this subject later [17.6, 17.10; 26.14; 28.2 et seq.].

17.8 (iii)   Increase in temperature

Considering now the temperature rise of 2·5°C, from 17·5 to 20°C, we see that the changes brought about are much smaller, and that in this single experiment they only just approach significance. However, they are highly significant when considered in conjunction with a number of other similar experiments in the range 15–22·5°C, all of which gave results of similar magnitude. We see a fall of unit leaf rate of nearly 10 per cent, presumably due to respiration having a higher temperature coefficient than photosynthesis under these conditions of low carbon dioxide concentration. This is practically completely compensated by an increase in leaf area ratio (due to a corresponding increase in specific leaf area), and the net change in relative growth rate is very small.

17.9 (iv)   Increase in saturation deficit

The comparison involving an increase of more than three-fold in saturation deficit is interesting, as the corresponding changes in the pattern of growth nowhere reach significance. Considered in terms of relative

17.9

humidity at 15°C, the temperature at which the plants were grown, it involves a change from a relative humidity of 88 per cent to one of 62 per cent, and thus covers a large part of the natural range encountered by plants growing in a wood, in all but the wettest and driest weather. This was an interesting result at the time, as there was then little information of this kind available. The implication was that very fine control of saturation deficit in the cabinets was unnecessary, and that the system of dew-point control, adopted as a basic feature of the cabinet design (Evans, 1959), was adequate in itself. This system involved a slow cycle over a range of about 2–3 per cent in relative humidity, but for all the experiments we have just been considering, it had been backed up by an additional humidifier with a short cycle, which in effect filled up the troughs of the long cycle and maintained humidity within 1 per cent.

17.10 (v)   Evidence of a compensation mechanism

Taking all these comparisons together, we have seen that every change in unit leaf rate was accompanied by a change in leaf area ratio in the opposite sense, and of much the same magnitude, so that the net effect on relative growth rate was always minimized. We have also seen that where there were significant changes, specific leaf area always changed in the same sense as leaf area ratio, making the largest contribution wherever no change of daylength was involved. It therefore seems likely that the compensating mechanism we discussed earlier operates on leafiness, tending to increase leaf area when increase in dry weight per unit leaf area falls, and vice versa. Such a change would, however, take time to come into operation. If specific leaf area, the expansion in space of a given weight of leaf tissue, is involved, we have the time scale of leaf development; if leaf weight ratio, the distribution of dry material between leaves and the rest of the plant, we have in addition the time taken to translocate the extra material to the leaves.

We have already seen evidence of a reciprocal relationship between unit leaf rate and specific leaf area at a particular total plant dry weight in Fig. 17.4.2, and such a relationship is no doubt another reflection of the compensating mechanism. It might at first appear that such a relationship implied that unit leaf rate on a leaf weight basis was constant;

$$\frac{dW}{dT} \cdot \frac{1}{L_A} \times \frac{L_A}{L_W} = \text{a constant} = \frac{dW}{dT} \cdot \frac{1}{L_W} \, ;$$

17.10

but it must be remembered that specific leaf area is subject to a very marked ontogenetic drift, and that indeed this was the main reason for making the comparison with unit leaf rate at a fixed total plant dry weight. The above relationship, observed at a particular plant dry weight, therefore does not imply that unit leaf rate on a leaf weight basis would remain constant as total plant dry weight changed.

17.11 Back to ontogenetic drift

At this stage it is not profitable to consider compensating mechanisms any further, because inevitably they involve problems of ontogenetic drift, to which as yet we have given relatively little attention. In this chapter we have been deliberately attempting to avoid such problems by working at a fixed total plant dry weight. But our discussions of the data obtained in this way have shown that under different conditions the plant deploys its resources in different ways. The mere existence of these compensating mechanisms reveals how various and far-reaching are the plant's reactions, both in structure and function, to environmental change. Our attempts to avoid the consideration of ontogenetic drift have thus brought us back to it in both its aspects—that of changes in the instantaneous responses of the plant to its environment as it moves through its life cycle, and that of the accumulation of past experience in the plant body itself, which in turn affects the form of the instantaneous response. These are the problems which we must now consider in more detail, and we can then return to the questions connected with compensating mechanisms.

18

LEAF WEIGHT RATIO—
PRODUCTIVE INVESTMENT

18.1 Photosynthetic organs

The study of how the plant's total stock of organic material is divided
between photosynthetic systems and the rest must be an important
element in any extended analysis of growth. In principle there appears
a simple distinction between tissues and organs capable of photosynthesis,
and those not capable; and in some plants the distinction is equally simple
in practice. It must, however, be borne in mind that in many plants no
such simple distinction is practicable. The most obvious photosynthetic
organs, the leaves, may provide the major part of the production of new
material, yet there may still be a substantial contribution by other organs,
stems, parts of flowers and fruits, and so on [13.18]. How large a contribution these latter organs make to the carbohydrate economy of the
plant can usually be decided only by physiological experiments specially
designed to investigate the systems in question. Such investigations can be
exceedingly difficult, and for many experimental plants there is little
information.

In considering the consequent problems it must be borne in mind that
the wide variations in the structure and functioning of non-foliar photosynthetic organs can lead to wide variations in the rate of photosynthesis
per unit dry weight—the problems involved are similar to, but usually
more difficult to tackle than, those encountered in leaves, where there
may also be substantial differences in rate due to leaf structure [27.4–27.6;
28.6–28.14], age [29.4–29.9], or position on the plant [13.18, 13.19]. In
particular cases it may be possible to show that the total area of all other
available photosynthetic organs is small in relation to that of the leaves,
and a comparison of structures may lead to the conclusion that the rate
per unit area in these organs is unlikely to be any higher than the average
rate for the leaves, so that analysis can safely be concentrated on the leaves
alone. It is, however, always wise to check these points, as organs other

than leaf laminas may have unexpectedly large areas [13.18], or they may assume relatively great importance at particular parts of the life cycle, such as when the main foliage is dying back. On the other hand, we must also remember that, as seen in Chapter 16, there are likely to be substantial errors in the estimation of the overall rate of accumulation of new dry weight in any field experiment, so that in a particular investigation very great refinement of analysis may well be unprofitable.

Finally, there may be doubts about the figure to be used to represent the plant's total stock of dry material, available for distribution between photosynthetic organs and the rest of the plant. Some consequent problems have already been considered [13.21] in relation to large and long-lived plants, where they are especially acute, and we shall return to them in Chapter 31.

Let us now assume that we know the answers to these questions, and that in the particular plants we are studying the great bulk of photo-synthesis is going on in the leaves. We have seen [13.16] that the leafiness of a plant—the area of leaves which the plant produces at a given total plant dry weight—can be regarded as the product of the operation of two systems. The first controls the distribution between the leaves and the rest of the plant of the plant's stock of reduced carbon compounds, of which for the time being we are taking dry weight as a measure. It depends on what may be termed a 'correlation mechanism' of the plant as a whole, about which very little is known. The second controls the extension in space of the material forming the leaves, and at any particular stage in the plant's development it is expressed in the leaf anatomy. This is inherently a simpler problem than that of correlation in the plant as a whole, and light can be shed on it by suitable anatomical studies of leaf development. *A priori* we should expect these two systems to show different and possibly independent ontogenetic drifts, and to be related to the external environment in different ways. We will therefore consider them separately, make comparisons, and hope in this way to break down the complex problems posed by the ontogenetic drift of leafiness into more manageable parts.

18.2 Leaf weight and total plant weight: Sunflower

It is convenient first to consider the increase of leaf weight in relation to increase of total dry weight for two species of annual plants having different growth forms. Fig. 18.2 shows Briggs, Kidd and West's data for *Helianthus annuus* up to total plant dry weights of 600 g. We see that for the first 3 weeks, up to a dry weight of 1·2 g, the relationship is only

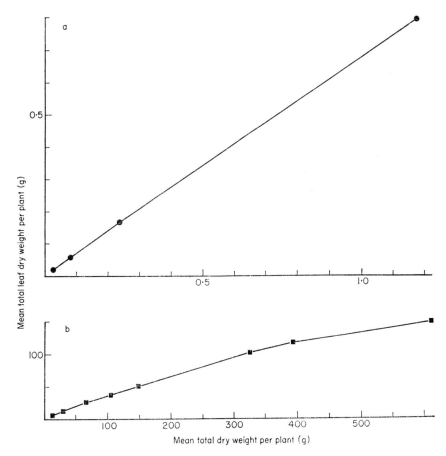

Fig. 18.2. Mean total leaf dry weight per plant in relation to mean total dry weight per plant for *Helianthus annuus* grown by Briggs, Kidd and West at the Botanic Garden, Cambridge, in 1920. (a) Harvests 1–4, 28 May to 18 June. (b) (to 500 times the scale on both axes) Harvests 6–13, 2 July to 3 September.

very slightly curved, leaf dry weight being almost proportional to total plant dry weight; the straight line of best fit would pass very close to the origin. However, as the dry weight increases the relationship curves over;

the proportion of dry matter going to new leaves steadily declines throughout the life of the plant, producing what appears to be a regularly curved
relationship. This was a case of a plant with little branching, which
produces larger and larger leaves supported by a very massive stem as it
gets older.

18.3   Cotton

Cotton, on the other hand, is richly branched and has a much thinner
stem and more numerous, smaller leaves. In 1923 Inamdar, Singh and
Pande worked at Benares on the growth of the well-established annual

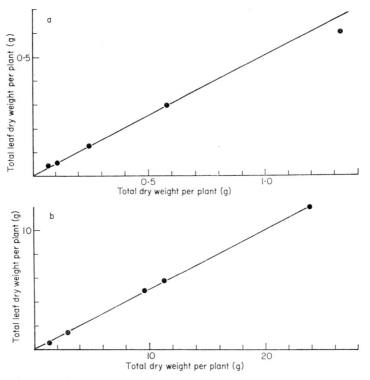

Fig. 18.3. Leaf dry weight as a function of total plant dry weight for plants
of cotton (*Gossypium arboreum*) grown at Benares in 1923. (a) Up to an
age of 42 days from germination; (b) 42 to 83 days (when flowering was
beginning), plotted to 20 times the scale on both axes. Straight lines represent
a constant leaf weight ratio of 0·5. Plotted from data given by Inamdar,
Singh and Pande (1925). See also Fig. 23.5.2.

18.3

variety of what is now classified as *Gossypium arboreum*, known at that time as *roseum*. Fig. 18.3 gives some of their data for two ranges of plant size, (a) up to a total plant dry weight of 1·3 g, (b) from 1·3 up to 24 g. The straight line representing a constant leaf weight ratio of 0·5 is given for both ranges. We see that in this case the phase of a close approximation to a linear relationship between leaf weight and total plant dry weight extends over a substantial fraction of the life cycle. There was no marked departure up to a total plant dry weight of about 24 g, nearly three quarters of the final dry weight (33·6 g) of the mature plants at the end of the experiment, at an age of 17 weeks.

18.4   Ontogeny and environment

The regularity of these relationships is most interesting. It must be remembered that each point represents a mean for a harvest of plants grown in the field; not only does it relate to a different set of plants from all other points, but as the experiment progresses we should expect an accumulation of past environmental experience, which might well produce irregularities in the relationships. Yet such irregularities seem to be minimal, which is encouraging. It leads us to think that here, although we are dealing with a complex correlation phenomenon about which little is known, yet perhaps it is one relatively little affected by the normal range of natural environmental fluctuations. If so, it may present a very favourable opportunity for disentangling the interrelations of plant responses to the environment from ontogenetic drift and its consequences.

   These ontogenetic changes may well be of multiple origin. No doubt to a large extent they reflect the changing architecture of the plant as the size and arrangement of the leaves changes; but we cannot exclude the possibility that they also reflect the changing environment of the leaves themselves. As the plant grows in nature, its new leaves occupy new positions in the complex plant canopy; whereas for cultivated plants grown singly with ample spacing, the lateral branches need not behave in the same way as the main stem, and also self-shading may become considerable as the plant grows towards maturity. For these reasons it seems that in disentangling the effects of the environment on the leafiness of the growing plant it would be best to study first the early stages, while the form of the plant is simple, and the effects of self-shading are minimal (thus avoiding as far as possible this aspect of plant reaction, (d) of 7.1 and 17.1).

18.5   *Impatiens parviflora*

Impatiens parviflora happens to be a very suitable plant in this respect. The hypocotyl is elongated (excessively so in very low light, in the absence of blue light, or at high temperatures, so that the plant falls over unless supported). On the other hand, in the early stages of growth the stem internodes are relatively short, and the leaves are arranged with a minimum of overlap in a rosette-like form, as shown in Fig. 18.5. Furthermore, under favourable conditions *I. parviflora* comes into flower early, within 5–6 weeks of germination, and at a total plant dry weight of less than 250 mg [27.8]. Under less favourable conditions flowering may be delayed until about 9 weeks from germination, but the total plant dry weight is considerably less, and lateral branching is much restricted. Accordingly, although under very favourable conditions the plant may later form a considerable bush, and go on flowering and fruiting as long as favourable growing conditions continue, nevertheless the early stages form a considerable fraction of the minimum possible life cycle. From the standpoint of the biology of the plant they therefore occupy a much more prominent place than in a plant such as the sunflower which flowers only much later, after great morphological and anatomical changes in the plant, and at a dry weight many hundred times greater. For such a plant, large changes due both to mechanical causes and to self-shading are inevitable before the plant flowers.

Our present analysis will therefore be restricted to plants up to about 6 weeks old, which in practice even in favourable environments means a dry weight of little over half a gram. It is to be hoped that the insight thus gained into the effects of external environmental conditions on the development and form of the leaves may in due course make possible an attack on the much greater problems posed by the later stages of growth; but, as always in scientific investigations, it is wise to take the simple cases first.

18.6   Comparison of growth in the field—

For the above reasons, and also because it has been extensively studied under controlled environmental conditions, we will therefore turn once

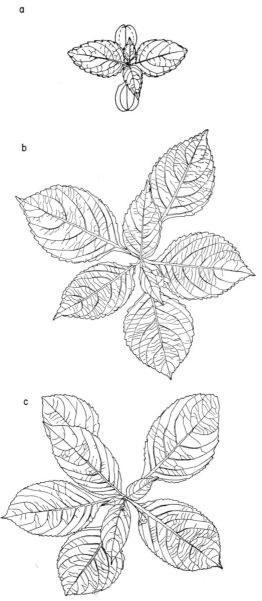

Fig. 18.5. Leaf mosaics in the early stages of growth of *Impatiens parviflora*. (a) A typical young plant viewed from above, showing cotyledons, the first pair of opposite leaves, and two alternate leaves just beginning to expand (from Coombe, 1956). (b) Stage when the seventh leaf is beginning to expand: the plant is coming into flower. (c) The ninth leaf is expanding, and self-shading is beginning. a, $\times\frac{4}{9}$; b and c, $\times\frac{1}{3}$.

again to *Impatiens parviflora*. Fig. 17.4.1 has already shown a broadly linear relationship between leaf weight and total plant dry weight, but it included a very miscellaneous collection of experiments, both in the field and in the laboratory; its aim was to demonstrate the broad correspondence of the two sets of figures, and to show that they covered similar

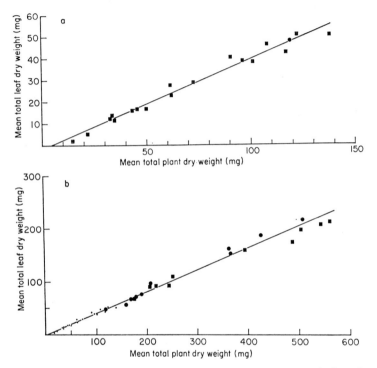

Fig. 18.6. Mean total leaf dry weight in relation to mean total plant dry weight for samples of *Impatiens parviflora* grown with a 16-hour day under a variety of artificial conditions in the laboratory (■) and in the open or under neutral screens in the field (●): (a) for mean total plant dry weights up to 150 mg; (b) up to 600 mg, the laboratory points from (a) being shown as dots.

ranges. For a closer comparison, and to decide whether here also we are dealing with an essentially linear relationship in the earlier stages of growth, we will examine these observations made in the field and under artificial conditions in the laboratory on occasions when the daylength was close to 16 hours. In Fig. 18.6 are shown data for the relationship

18.6

between leaf dry weight and total plant dry weight for means of 6–8 plants, aged up to 6 weeks. The circles indicate plants from an initially uniform population grown in the field between 17 May and 3 June 1957, with a daylength from sunrise to sunset ranging from 15 h 41 min at the first harvest to 16 h 22 min at the last. The plants were harvested on 27 May and 3 June, the mean daylength (sunrise to sunset) for the two periods being 15 h 55 min and 16 h 16 min. Some were grown in the open, fully exposed except for an arrangement of baffle boards 15 cm high placed north, south, east and west at a distance of 60 cm from the nearest plants. The remainder were grown under artificial shades of a fine enough mesh to avoid the formation of distinct sunflecks when the sun was shining [9.15], and transmitting approximately 67 per cent, 42 per cent, 25 per cent and 7 per cent of full daylight (for details see Evans and Hughes, 1961). From the lightest to the darkest the light intensity thus varied by a factor of about 14. Although at the end of 18 days' exposure to these conditions the plants showed large differences in size (ranging from a dry weight of 208 up to 510 mg), yet within their limits of growth all follow much the same course in the proportion of dry matter forming the leaves. It was found that the heaviest plants grew under the 67 per cent shade, the mean dry weight of these being 190 mg on 27 May and 510 mg on 3 June.

18.7   —and under controlled conditions

The squares in Fig. 18.6 indicate mean values for 6 or 8 plants grown from seed under constant, artificial conditions in the laboratory, as set out in Table 18.9a. The range of these conditions covers temperatures between 15°C and 20°C, saturation deficits from 1·5 to 4·8 mm (corresponding respectively to 88 per cent and 62 per cent relative humidity), irradiances from 1·2 to 2·7 cal dm^{-2} min^{-1} of visible radiation, and two different combinations of fluorescent tubes. The combination of artificial and semi-natural conditions thus covers a wide range of combinations of conditions of the aerial environment; yet within the size limits of about 30 up to 500–600 mg one single linear relationship appears to be an equally good fit to all. Fig. 18.6a shows it up to 150 mg; in this range most of the sets of observations were made under artificial conditions. Fig. 18.6b shows the continuation of the same straight line up to 500–600 mg.

It thus appears possible that for the early stages of growth in this plant, the distribution of dry matter between the leaves and the rest of the plant is unaffected, not only by the extensive range of artificial conditions just mentioned, but also by the change from the laboratory to the field. This involves a number of very substantial environmental differences [18.15, (ix)–(xii)]: (a) from the constant conditions of temperature and humidity in the laboratory to the fluctuating conditions of the field, involving weather changes, daily cycles, and short-period variations; (b) from the constant, particular, spectral composition of the fluorescent tubes used for illumination in the laboratory to the varying spectral composition of natural daylight; (c) from the simple, rectangular, on–off daily lighting cycle in the laboratory to the complex daily march and short period fluctuations of irradiance in the field [7.3]. However, the fact that field conditions differ from conditions in the laboratory in so many different ways imposes caution. In so complex a situation it would be well to bear in mind the possibility that opposing tendencies may cancel each other out, and to plan observations to bridge the gap. Meantime, the conventions of scientific method require us to accept the simplest explanation which will fit all the available facts. Here, the simplest explanation is certainly that the transfer from the laboratory to the field, with all its attendant consequences, has produced no change in the distribution of dry material between the leaves and the rest of the plant, within the limits which we have been considering. These cover approximately a twenty-fold increase in the total dry weight of the plant, up to a size at which it is nearly always in flower. However, considered against the scale of the life of the plant from seed through to seed, this almost linear region is only a part of a more complex relationship. This is at first curved immediately after germination, and later settles to the steady progression we have just seen, finally curving again. The central conducting and supporting regions of the stem and root system absorb an increasing proportion of the total dry weight as the plant continues to grow, differentiate, branch and increase in complexity. However, from the standpoint of analysing the combined effects of individual environmental factors and ontogenetic drift in the distribution of dry weight, it is fortunate that the almost linear region should cover so wide a range. We have here one of the naturally occurring simple cases which are always worth examination. The grand progressions of ontogenetic drift have not ceased to operate, but there has been a pause, as it were, lasting long enough for us to be able to take account of perhaps four harvests with their intervening growth periods, and to view the

18.7

weight ratios were higher in September than in June and July, and consideration of all the possible sources of difference suggested that day-length was probably involved. A laboratory experiment was therefore carried out at a temperature of 15°C and a saturation deficit of 2·0 mmHg. Conditions were otherwise generally similar to those of Table 18.9a and two sets of plants were grown from seed, one in a 16-hour day with an

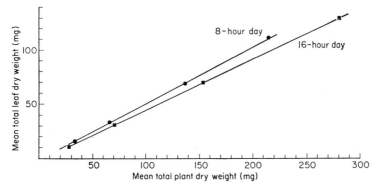

Fig. 18.12. Mean total leaf dry weight in relation to mean total plant dry weight for samples of *Impatiens parviflora* grown under artificial conditions in the laboratory, ● with an 8-hour day and ■ with a 16-hour day, and harvested at weekly intervals. For other conditions, see text.

intensity of visible radiation of 1·44 cal dm^{-2} min^{-1}, the other with an 8-hour day and 2·82 cal dm^{-2} min^{-1}, giving closely similar total daily light. The mean leaf weights for four harvests in each case are plotted against mean total plant dry weight in Fig. 18.12. The correlation coefficients are 0·99997 for the 16-hour day and 0·99974 for the 8-hour day, a good example of how extraneous variability can be reduced by careful attention to the principles of experimental planning and preparation and handling of material which we have discussed in earlier chapters.

18.13   Regression equations

This is a case where the method of analysis by the calculation of regression equations is justified, although we have noticed earlier [14.5] an objection to a procedure of this sort: that it is unwise to plot, as we are doing, leaf dry weight against [leaf dry weight + stem dry weight + root dry weight], because any random or systematic fluctuation in leaf dry weight will affect both variables simultaneously. Later, when considering distribution of dry material again, in Chapter 26, we shall use a different procedure avoiding any possible confusions of this kind. Here, however, is an instance where the procedure of relating leaf dry weight to total plant dry weight is immediately justified by its results. At any particular fixed total of [stem dry weight + root dry weight], the effect of a fluctuation in leaf dry weight, from whatever cause, will be to move the point up or down a line with a slope of +1. When, as in the previous instance which we noticed, the relationship which we are investigating has a negative slope, such a movement can only add to the scatter. The same is true to a lesser degree for positive slopes other than 1. In the particular instance which we are now considering the slope is about 0·5, and the scatter is so small that our conclusions have no need to depend on a nice balancing of statistical probabilities. We are therefore free to work on a basis of total plant dry weight, whose significance we have already discussed in detail in Chapter 4. We then have, for the 16-hour day,

$$L_W = 0{\cdot}463W - 2{\cdot}14 \qquad\qquad 18.13.1$$

and, for the 8-hour day,

$$L_W = 0{\cdot}519W - 1.76 \qquad\qquad 18.13.2$$

where L_W is leaf dry weight and W is total plant dry weight in mg.

The coefficients of W represent the fraction of new dry weight appearing in the form of leaves. The two lines cut the x-axis at about 4 mg (4·6 and 3·4 respectively), about twice the dry weight of the embryo. This is interesting, because as we have already noted, the whole relationship between leaf (including cotyledon) weight and total plant dry weight has a curved portion below total plant dry weights of around 30 mg, where these two straight lines begin.

The standard deviations of the two regression coefficients are 0·002 (for the 16-hour day), 0·007 (for the 8-hour day), and hence also 0·007 for the difference between the two. This difference, 0·056, is eight times as large as its standard deviation, and the probability of such a difference arising by chance is negligibly small. We have thus established that for this plant, up to the time of flowering, a reduction of daylength from 16 to 8 hours increases the proportion of new assimilates remaining in or passing to the leaves, and that the amount of this increase is about $5\frac{1}{2}$ per cent of the total increment of dry weight.

18.13

18.14 Leaf weight ratio and plant size

The accuracy with which these regression equations represent this aspect of growth over the range of observation makes it worth while to look back on leaf weight ratio as a function of plant size, in view of the fact that the regression line does not pass through the origin. Dividing through equation 18.13.1 by W we have

$$\text{Leaf weight ratio} = \frac{L_W}{W} = 0.463 - \frac{2.14}{W}. \qquad 18.14$$

We see that at 30 mg total dry weight the leaf weight ratio thus works out at 0.39, rising to 0.42 at 50 mg, 0.44 at 100 mg and 0.45 at 200 mg. It would not reach 0.46 until a total dry weight of about 300 mg, just outside the range of this particular experiment. It is quite likely therefore that the first signs of the curvature of the leaf weight/total plant dry weight relationship begin to be apparent between 400 and 500 mg, because we see from Fig. 18.6 that the four points for cabinet experiments around 500 mg are all below the line. By rejecting these four points and correcting for the intercept on the x-axis we could produce a more consistent version of Table 18.9, but our main conclusions would not be altered thereby. If all the experiments showed a consistency equal to the two daylength ones it would no doubt be worth making the corrections, but at the general level of variability of the whole series of experiments it is hardly worth while. The main conclusions from Table 18.9 can be accepted without further refinement as an example of what can be achieved by an analysis of leaf weight ratio itself.

18.15 Summary of environmental effects

We may sum up by (a) listing the factors which have been proved to influence the distribution of dry matter between the leaves and the rest of the plant, up to the flowering stage; (b) giving the magnitude of the observed effect of a given change in each of these, expressed as a percentage increase of total plant dry weight added to the leaves; and (c) listing those factors where a change appears not to influence the distribution.

Aspects of the environment influencing the distribution of dry matter between the leaves of Impatiens parviflora *and the rest of the plant*

18.15

(i) *Daylength.* At 15°C and for plants growing in loam and sand, a reduction in daylength from 16 to 8 hours causes an increase of $5\frac{1}{2}$ per cent.

(ii) *Temperature.* Although changes in temperature between 15°C and 20°C appear to have a negligible effect, for plants growing in vermiculite lowering the temperature from 15°C to 10°C causes a decrease of about 4 per cent.

(iii) *Soil factors.* A change from loam and sand with added mineral nutrients to vermiculite with a culture solution caused an increase of nearly $12\frac{1}{2}$ per cent. Here it is not clear whether one or several factors are operating; increased availability of nutrients and improved aeration are obvious possibilities.

Aspects not influencing the distribution

(iv) *Constant temperature.* Changes between 15°C and 20°C.

(v) *Saturation deficit* of the air. Changes between 1·5 and 4·8 mm. *Relative humidity.* Changes between 88 and 62 per cent at 15°C.

(vi) *Daylength.* Changes between 16 and 24 hours.

(vii) *Intensity of visible radiation.* Under artificial conditions, changes between 1·2 and 2·6 cal dm^{-2} min^{-1}.

(viii) *Spectral composition of visible radiation.* Fluorescent tubes with widely different spectral patterns of emission.

Aspects probably not influencing the distribution

The simplest hypothesis would be that none of the following aspects of the environment had an influence on the distribution of dry matter between the leaves and the rest of the plant. But together they constitute a very complex series of changes, and in the nature of the case all of them have happened together. The possibility therefore remains that the coincidence between the plant grown in the open field and that grown under artificial conditions may in part be due to a cancelling of two or more effects working in opposite directions.

(ix) *Temperature and humidity.* Changes from constant conditions in the laboratory to fluctuating conditions, within a similar range, in the field.

(x) *Spectral composition of radiation.* Change from the various artificial conditions in the laboratory to natural daylight. This involves not only a change in the visible part of the spectrum, but a much larger change in the infra-red. Probably around 5 per cent of the total short-wave radiation

18.15

(wavelengths less than 3μ) from the standard combinations of fluorescent tubes is in the infra-red. For natural daylight the proportion is around 55–60 per cent (but also see 28.9).

(xi) *Irradiance.* Under artificial conditions, the daily totals of visible radiation mentioned in Table 18.9 (up to 26 cal cm^{-2}) produced a plant having a leaf weight ratio within $2\frac{1}{2}$ per cent of one grown in full daylight (at maximum about 210 cal cm^{-2}), conditions (i), (ii) and (iii) also being satisfied. Balancing the probabilities, it is more likely that the difference was due to some difference in soil conditions than to differences in irradiance, large though these were.

(xii) *Daily march of irradiance.* Change from the rectangular on–off day of the controlled cabinets to the regular march combined with the short period fluctuations of natural daylight.

18.16 Conclusion

We have already noticed that the correlation mechanisms controlling dry weight distribution in actively growing higher plants are complex, and that little is known about them. They are, however, of central importance to an understanding of the growth processes of the plant as a whole, and we shall return to the matter in Chapter 21 (stems and roots), Chapter 23 (reproductive structures) and in Chapter 26, when considering the relationships of these problems to studies of plant morphology.

In this chapter we have been able to take advantage of a fortunate instance. By choosing a closely restricted case, we have been able to establish a linear relationship between leaf dry weight and total plant dry weight, in other words, a simple ontogenetic relationship. One strand of the Gordian Knot of genetics, development and environment [3.5, 7.1 and 17.1] once loosened, we have been able to make progress in disentangling others. In our particular case we have been able to show that the effects of changes of environment on the form of the linear relationship are few and circumscribed, thus providing some clues to the nature of the underlying mechanisms. It is clear that much more work will be needed before these mechanisms can be elucidated, but we have here one promising approach.

SPECIFIC LEAF AREA

19.1 Ontogenetic changes

Briggs, Kidd and West (1920a, b), in the course of their analysis of the complex ontogenetic drift of relative growth rate in maize, noticed that leaf area ratio showed a similar drift. They therefore split RGR into two derivates, of which one was leaf area ratio, while the other, unit leaf rate, showed relatively little drift [13.7]. In section 13.16 we discussed the splitting of leaf area ratio into two, leaf weight ratio and specific leaf area. In Chapter 18 we have now considered some favourable instances in which leaf weight ratio showed relatively little drift, and we have been able to use these to draw inferences on environmental effects on the distribution of dry matter between photosynthetic organs (specifically, leaves, 18.1) and the rest of the plant.

We have now to give similar consideration to specific leaf area, which measures the mean over the whole plant of the extension of this leaf dry matter in space. We have to ask how far the effects of the various aspects of the environment on the value of specific leaf area can be distinguished from inherent changes due to ontogenetic drift, and the available means by which, in favourable cases, they can be disentangled. In the resolution of practical problems these questions share the importance already assigned to leaf weight ratio [18.1], and the cautions, problems and complications there mentioned apply equally here.

19.2 Measures of 'physiological age'

Experience shows that ontogenetic drifts of specific leaf area are frequently complex, and this immediately raises the problem of what is to be the basis of measurement of such a drift. How are we to measure the 'physiological age' of a plant? Indeed, is there any evidence that such a concept has any utility for a plant as a whole? May it not be that different organs reach a particular physiological stage at different times from other organs? May it not be that the relationships between these times are not solely determined internally by the ontogeny of the plant, but are also subject to environmental influences? Common knowledge of flowering behaviour would suggest that this is so. If the developing relations between the different plant organs, and hence the form of the overall ontogenetic drift, are partly under environmental control, the search for a simple measure of ontogenetic drift will be difficult, and we may find it necessary to use as our baseline some attribute of the organ itself, rather than of the plant as a whole. This is no doubt a consequence of some of the properties of plant growth mentioned in Chapter 2; the absence of a set adult form in most higher plants [2.1], in distinction to the pattern of growth of the higher animals, which usually leads by definite steps to the establishment of an adult form which is successful only within narrow limits of size and shape.

The questions raised are difficult ones, to which as yet no precise answer can be given; and it seems likely that even if one had a full knowledge of the ontogeny/environment relations of one specific plant, the investigation of another species would disclose significant differences. Accordingly we must consider a variety of different ways of making the analysis, and in this chapter we shall consider three possible bases—age in days [19.3]; total plant dry weight [19.4–19.8]; and the dry weight of the particular organs themselves (here the leaves; 19.9–19.11). The relation of specific leaf area to the physiological measure of unit leaf rate has already been mentioned [17.4], and we shall consider this in more detail later [28.6–28.14]. Many other bases could be suggested, although as yet they have been little applied to the problems relating to specific leaf area with which we are now concerned. Later we shall use some of them in other connections, for example the morphological indexes of leaf plastochron [27.8] or stage of flower development [23.8]; leaf water content, which could be regarded either as an anatomical or a physiological measure [24.17, 24.19]; or even a pathological index of the progress of a disease [30.16]. However, the three bases considered in this chapter provide scope for a variety of treatment, and similar methods employing other bases could be used in appropriate cases.

19.3 Age in days

The problems raised by drifting values of specific leaf area often pose themselves in the first instance against a chronological age scale measured in days. Fig. 19.3a shows the changes in specific leaf area and leaf weight

19.3

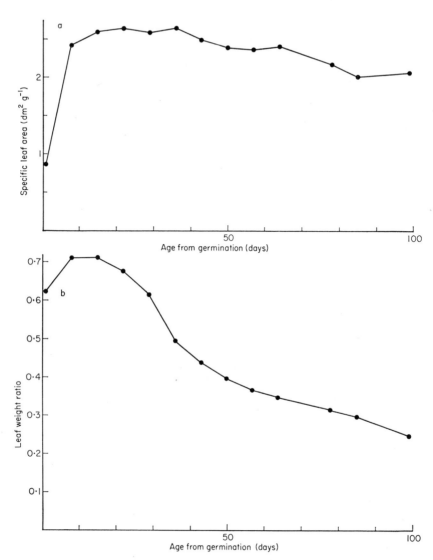

Fig. 19.3. (a) Specific leaf area, and (b) leaf weight ratio in relation to age of plant for *Helianthus annuus* grown by Briggs, Kidd and West at the Botanic Garden, Cambridge, in 1920 (compare (a) with Fig. 19.4.1). For the comparable data on relative growth rate, leaf area ratio and unit leaf rate see Fig. 13.15. For further data on dry weight distribution see Fig. 23.6.

19.3

ratio with time for Briggs, Kidd and West's sunflower plants, and the progress of the ontogenetic drift can be compared with Fig. 13.15 for relative growth rate, leaf area ratio and unit leaf rate. Taken together these five graphs give an example of how useful a simple time plot can be in helping to break down the overall problem of the ontogenetic drift of relative growth rate, and in assessing the relative importance of the different systems contributing to it. Fig. 13.15 showed how for this particular sunflower crop relative growth rate rose from an initial value of rather over $1 \cdot 0$ week^{-1} to a maximum of around $1 \cdot 6$ during the third week, and then fell to a value of around $0 \cdot 2$ between 80 and 100 days. Unit leaf rate was high in the first week, low in the second, and then fluctuated about a falling line ranging from 90 g m^{-2} week^{-1} during the third week to 40 between 85 and 100 days. Thus during this long falling phase relative growth rate fell to one-eighth of its maximum value, unit leaf rate to just under a half, while leaf area ratio [Fig. 13.15b] fell from a maximum of around $1 \cdot 8$ dm^2 g^{-1} to around $0 \cdot 5$ at 99 days, a fall to between a third and a quarter. Here changes in leafiness are clearly of great importance, and we can now see that the predominant element in these particular changes is leaf weight ratio, which fell from a maximum of $0 \cdot 71$ at 15 days to $0 \cdot 25$ at 99, while specific leaf area fell in the same period only from around $2 \cdot 6$ dm^2 g^{-1} to around $2 \cdot 05$. On the other hand, the very low value of relative growth rate during the first week of growth is seen to be associated mainly with a very low value of specific leaf area. Having broken the problem down so far we can suggest hypotheses to account for the various phenomena observed, and each of these must then be tested by some appropriate means. For example, it seems likely that the early very low value of specific leaf area was due to a higher fraction of leaf material than usual in an early, unexpanded, stage of development. This can be tested by anatomical investigation; and so on. Plots against time are thus useful as an analytical tool, but they usually provide rather little direct evidence on the underlying causal mechanisms, because the baseline is not related either directly or indirectly either to changes in the environment, or to the accumulating effects of previous experience on the form and functioning of the plant.

19.4   Total plant dry weight

To use total plant dry weight as a basis for the assessment of ontogenetic change is more hopeful than simply using the lapse of time. Much of our thinking about plant development has taken total dry weight as a basis, but at the same time, we have exposed many inadequacies in the use of total dry weight as an index of development. It has the merit of simplicity, and also the merit of presenting progressions against a baseline which is

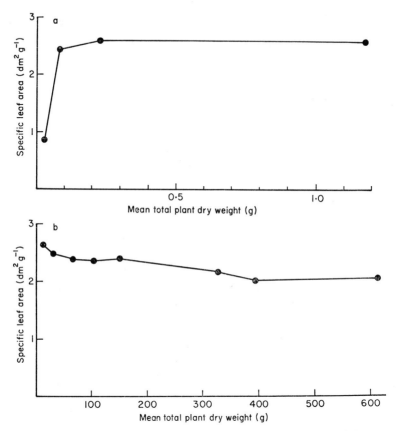

Fig. 19.4.1. Specific leaf area in relation to mean total dry weight per plant for *Helianthus annuus* grown by Briggs, Kidd and West at the Botanic Garden, Cambridge, in 1920. (a) Harvests 1–4, 28 May to 18 June; (b) (to 500 times the scale of dry weight), harvests 6–13, 2 July to 3 September. See also Fig. 19.3a.

related in some way to the plant's previous experience. But to take only one of many possible objections; we already know that for a specific plant under different conditions a given total dry weight can be reached by a number of different routes, producing at this particular dry weight plants

19.4

differing in anatomy, morphology, and physiological performance. This may well vitiate an individual comparison, although used over a period as a measure of the physiological progress of the plant it may have more utility.

Fig. 19.4.1a and 19.4.1b show the corresponding changes to Fig. 19.3a, but this time expressed on a total dry weight basis. At once one of the major disadvantages of the total dry weight basis becomes apparent— the extent of the scale, and the increasing intervals between one value and the next. These could be overcome by using a logarithmic scale for dry weight: but a suggested relationship involving the logarithm of plant dry weight is likely to be very difficult to express in simple physico-chemical terms. Generally, such plots are most useful when summarizing information, or when, as in Fig. 14.5, the aim is merely to demonstrate that two relationships are different, without wishing to express either of them in precise terms.

Fig. 19.4.1 does, however, emphasize how early in its ontogeny this plant reaches a comparatively stable value of specific leaf area. The dry weights of the embryos in a sample of the seed used lay between 20·8 and 31·6 mg. By the time this initial dry weight had roughly trebled, at a mean plant dry weight of 80·7 mg, specific leaf area had already reached a value of 2·42 dm^2 g^{-1}. Thereafter, over a 200-fold increase in dry weight, from roughly 0·15 up to 30 g, its value changes very little, lying between 2·49 and 2·64 dm^2 g^{-1}, a range of less than 6 per cent. During the next twenty-fold increase, up to 600 g, large changes are taking place in the size of individual leaves, and hence in the massiveness of the midrib and petiole. Yet even here the value of specific leaf area covers a range from 2·49 to 2·02 dm^2 g^{-1}, changing by only about 20 per cent. We should conclude that over the whole 4000-fold range, covering the bulk of the ontogeny, the structure of the leaf laminas must be remarkably stable: once again, this can readily be checked by anatomical investigation.

Not all plants, however, show such stability, and in some species leaf structure goes through ontogenetic progressions which are at the same time very sensitive to environmental change, leading overall to large changes in specific leaf area. *Impatiens parviflora* provides such an instance, and as for leaf weight, it is no doubt best to begin a detailed discussion of the complexities involved by considering data from plants grown under controlled environmental conditions. Fig. 19.4.2 presents some of Evans and Hughes's data for *Impatiens parviflora*, three of the

19.4

experiments being ones we have already seen in Table 18.9 (with light intensities of 1·17, 2·15 and 2·73 cal dm^{-2} min^{-1}, corresponding to 11·2, 20·6 and 26·2 cal cm^{-2} day^{-1}). All plants were grown from seed under the specified conditions, of a standard rooting medium; 15°C temperature; 1·5 or 2·0 mmHg saturation deficit; with De Luxe Warm White and Blue fluorescent tubes in a ratio of 3:1 (in other words a fixed spectral composition); and a 16-hour day. The differences were in intensity of visible radiation, and therefore, the daylength being constant, of total daily light.

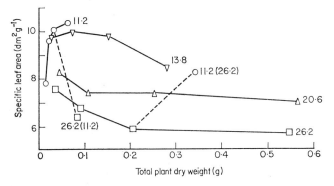

Fig. 19.4.2. Relationships between specific leaf area and total plant dry weight for means of eight plants (except for 3 points, where the number was 6) in four experiments on *Impatiens parviflora* growing under constant conditions in growth cabinets under different daily totals of visible radiation, ranging from 11·2 to 26·2 cal cm^{-2} day^{-1}. Harvests at weekly intervals, ages of plants between 17 and 44 days. Points marked 26·2 (11·2) and 11·2 (26·2) give the mean values for batches of plants transferred from the latter to the former condition for the third week of growth. From Hughes and Evans (1962).

19.5 Complex relationships

Fig. 19.4.2 shows the complex nature of the relationships involving both ontogenetic drift and this one aspect of the aerial environment. There is first a phase when specific leaf area is rising; the rise is more rapid the lower the irradiance. This corresponds to the stage where the first formed leaves are beginning to expand. Secondly, the rate of rise decreases until a maximum specific leaf area is reached. This maximum is higher, the lower the irradiance, but it is also attained later. Beyond about 30 and

19.5

40 mg respectively, the specific leaf areas of the plants growing in 26·2 and 20·6 cal cm^{-2} day^{-1} are already declining. At 13·8 cal cm^{-2} day^{-1} the maximum is reached at about 80 mg; at 11·2 the maximum had not been reached, and specific leaf area is still rising markedly, at about 70 mg, the maximum total dry weight reached after 40 days in this low intensity. Thirdly, there is a falling phase, with an increasing gradient until a point of inflexion is reached. This point of inflexion appears to be reached at lower total dry weights, the higher the total daily light. In a fourth phase the gradient decreases again, until finally it is declining only very slowly with increasing dry weight. This phase is reached only by the plants having the two highest daily light totals. When it is once established, the plants in 20·6 cal cm^{-2} day^{-1} have a markedly higher specific leaf area (about 7·0 dm^2 g^{-1}) as compared with those in 26·2 (about 5·9 dm^2 g^{-1}). In this fourth phase, specific leaf area appears to be nearly inversely proportional to the total daily light, over this very limited range.

Fig. 19.4.2 also shows what happens if plants are transferred from one light regime to another. These transfers were all made at the time of the third harvest, so that the transferred plants spent the third growth period of 7 days under the new conditions, and they were harvested at the same time as the fourth harvest of the control plants. For the transfer from the lowest to the highest total daily light, we see that full compensation had taken place, and the specific leaf area is if anything slightly below that of the control plants grown continuously in the high intensity. For the transfer down, we cannot be sure how far the plant has adapted itself within the week, although it will be noticed that the specific leaf area has risen to a value near, or if anything above, an extrapolation of the line of the plants grown in 13·8 cal cm^{-2} day^{-1}.

19.6 Interpretation of a field experiment

With this picture in mind, we can now turn to a series of much higher values of total daily visible radiation in two field experiments with artificial screens. We have already examined the leaf weight figures for one of these in Fig. 18.6. Approximate values of total daily visible radiation, for comparison with the cabinet figures, are given in Table 19.6. They are derived by taking 42 per cent of the mean daily totals of short wave radiation for the periods concerned (30.6–30.8).

Table 19.6. Approximate mean daily totals of visible radiation, in cal cm^{-2} day^{-1}, under artificial screens of various percentage transmissions during three experimental periods. Calculated from data given by Evans and Hughes (1961).

Artificial screens transmitting (%)	7	25	42	67	100
Experiment I, Period (ii)	8	27	46	73	109
Experiment II, Period (i)	12	43	73	116	173
Period (ii)	14	50	84	134	200

It will be seen that the highest daily totals are some eight times higher than for the cabinet experiments, while the lowest totals, under the screens with 7 per cent transmission only, are comparable with the lowest totals under artificial conditions, and in one case, below them.

In both experiments the young plants were grown in a greenhouse, in Experiment I for 9 days until the cotyledons had expanded; in Experiment II for 23 days until the first pair of leaves had expanded. In each case the experimental plants were then selected and graded from a large population of seedlings, and put out in the field under the screens. In the first experiment the second and third harvests were at 29 and 36 days; in the second at 33 and 40 days. Thus we have here examples of the second of the two techniques discussed in sections 7.6 and 7.7. In the previous, cabinet, experiments the plants were always adapted to the conditions in which they were growing (except for the deliberate transfers), but each batch always differed from all the others both in structure and in range of size. Here we have started off with uniform young plants of two ages, and allowed, in one case 10 days before the first harvest in the field, in the other 20. From what we have already seen, the plants having had 10 days exposure to field conditions are likely to have assumed the structure characteristic of the conditions under which they are growing; after 20 days exposure we should expect adaptation to be complete in all cases.

Turning to Fig. 19.6a, in which the specific leaf areas for Experiment I are plotted against total dry weight, we see that at 29 days, and a total dry weight of about 90 mg, the SLA of the plants under the 7 per cent screen had reached 8·0 dm^2 g^{-1}, and it is still rising rapidly, reaching 11·0 dm^2 g^{-1} at 36 days. The plants growing in the cabinets under 11·2 cal cm^{-2} day^{-1} never reached so high a specific leaf area, but this would be expected from our previous analysis, because the total daily light under the 7 per cent screen was considerably lower. The plants under the 25 per cent screen, on the other hand, had a mean total daily light very close to the highest value for the cabinet plants, yet at a total dry weight of about 160 mg the SLA had reached 7·0 dm² g⁻¹, and it was still rising, being about 8·0 at 390 mg. This does not agree with the cabinet plants grown under a total daily light of 26·2 cal cm⁻² day⁻¹ where the SLA was about 6·0 dm² g⁻¹ and slowly falling, between 200 and 400 mg [see 28.10].

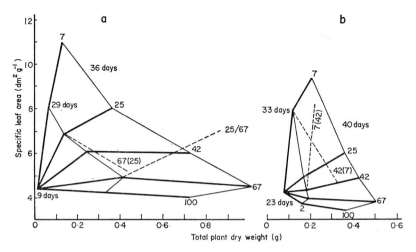

Fig. 19.6. Relationships between specific leaf area and total plant dry weight for means of six plants in two experiments on *Impatiens parviflora* growing under neutral screens in the field, with mean daily totals of visible radiation as in Table 19.6. Ages of plants in days, transmission of screens in percent. (a) Experiment I and (b) Experiment II. From Evans and Hughes (1961).

Otherwise there is a broad correspondence of form with the picture we built up in studying the cabinet plants—under the 42 per cent screen the peak SLA of at least 6·0 dm² g⁻¹ is probably reached at a total plant dry weight of at least 500 mg; under the 67 per cent screen a maximum of 5·0 dm² g⁻¹ around 400 mg, while the full daylight plants alone show continuously falling values of specific leaf areas from about 4·4 dm² g⁻¹ at the very low initial dry weight of about 20 mg. Thus we see that under the screens in the field, although the broad relationships correspond in form, the actual values do not agree. For a given total daily light the peak values of specific leaf area are higher, and are attained at much larger total plant dry weights. These conclusions are confirmed by Experiment II, where, however, we might reasonably expect more discrepancies. In this experiment the plants were grown in relatively high light for 23 days, until they reached a total dry weight of around 80 mg, and only then did they begin to undergo the structural changes associated with growth in low light.

19.7   Transfers

The transfers, shown by dotted lines in Fig. 19.6, indicate that for plants transferred to higher light the appropriate SLA was assumed within a week (as we noticed for the plants grown in the cabinets). For the downward transfer, where the changes needed were much larger, the appropriate SLA had not been reached after 7 days under the 7 per cent screen, although perhaps three quarters of the change was accomplished. For the downward transfer to the 25 per cent screen it also seems that the bulk of the change was over in 7 days. Accordingly it seems very probable that after 20 days under the screens, the plants in Experiment I were indeed fully adapted to the light regimes in which they were growing. The subsequent large differences in their behaviour, as compared with the cabinet plants, can therefore not be attributed to a continuing process of adaptation, but rather to the accumulated effects of all the various differences between conditions in the cabinets and those in the field which we have already considered in section 18.15 (ix)–(xii). Our preliminary discussion [13.16] had suggested that specific leaf area is sensitive to many environmental influences, and the above examples show how difficult it can be to disentangle these influences from each other without the assistance of experiments under closely controlled conditions.

19.8   Rooting conditions

The difficulties of using a method of this kind to obtain more precise information is well shown in Fig. 19.8, reproduced from Hughes (1965c). Hitherto we have been looking at means for all the plants of a particular batch at harvest. Fig. 19.8 presents specific leaf areas for individual plants, plotted against the individual plant dry weight. Three of these sets, +, ○, and ×, are from experiments we have already seen in Fig. 19.4, and relate to plants grown in soil; the other two, △ and ●, are for

19.8

plants grown in a mixture of vermiculite, sand and gravel, watered with culture solution. All had the same daylength, spectral composition of

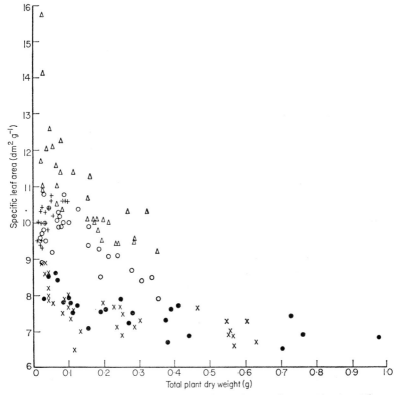

Fig. 19.8. Specific leaf areas of individual plants of *Impatiens parviflora* grown under a variety of constant, controlled conditions, in relation to total plant dry weight. All were grown at 15°C and a water vapour pressure deficit of 2·0 mmHg, and all had 16 hours per day of illumination of the same spectral composition. Daily totals of visible radiation in cal cm^{-2} were, △ 7·8, + 11·2, ○ 13·8, ● 16·0, and × 20·6. The substrate for △ and ● was a mixture of vermiculite, sand and gravel, with added culture solution; for +, ○ and × it was a mixture of loam and sand, with added artificial fertilizers. From Hughes (1965c).

radiation, temperature, and humidity. We see that the plants distinguished by triangles, grown at 7·8 cal cm^{-2} day^{-1} of visible radiation, agree with the 36-day harvest of the plants grown under the 7 per cent screen, shown in Fig. 19.6a. These field grown plants had reached a specific leaf area of about $11 \cdot 0$ dm^2 g^{-1} at a total plant dry weight of about 120 mg, a point which falls near the centre of the comparable distribution of the cabinet plants. But on the other hand it is not easy, from a plot of this kind, to reach any quantitative conclusions about the effects of culture in vermiculite. The solid points represent plants grown at $16 \cdot 0$ cal cm^{-2} day^{-1}, closer to the total daily light of the circles ($13 \cdot 8$ cal cm^{-2} day^{-1}) than the crosses [20.6]. Judging by eye, the actual distribution of the solid points is much closer to the crosses than the circles, and we should therefore conclude that specific leaf area is reduced, along the whole of this part of the ontogenetic drift, by culture in vermiculite. However, it is not worth pursuing this point statistically, because the variances in the x and y axes are interlinked by a common variable, leaf dry weight. We have been able to reach useful conclusions in similar situations before because the random errors were very small. Here they are considerable, and apart from using such plots to reach broad conclusions, it is a sounder practice to eliminate the dependence of the variables before beginning a statistical treatment. We shall consider another approach to the matter in section 28.8.

19.9 Individual plants and ontogenetic drift

However, another interesting suggestion emerges from Fig. 19.8. Each set of points presents the appearance of a regular progression through which the plants are all passing, rather than a series of sets of random scatter about the individual harvest points. If this is so, and if, as seems quite possible under these constant cultural conditions, the plants grown in one particular set of conditions are all progressing along one grand line of development, we should expect that variation in dimensions would be only partly at random, and mainly associated with the particular part of the progression which the individual plant had reached. A further corollary is that in a situation of this kind, where the progression may not be linear, we may be making unnecessary difficulties for ourselves by confining our attention to means of all the various plant attributes at harvest.

19.10 The daylength experiment

This point can be illustrated by reference to the pair of experiments on daylength considered in the last chapter. They are likely to furnish a

19.10

useful test, as we have already seen that, at least as concerns leaf weight ratio, the variability of this particular set of data was remarkably low. The leaf areas of individual plants plotted against the corresponding values of leaf dry weight are shown in Fig. 19.10a relating to plants grown in an 8-hour and Fig. 19.10b in a 16-hour day respectively. Measuring L_A in square decimetres and L_W in grams, the appropriate regression lines were for the 8-hour day,

$$L_A = 11 \cdot 9 L_W - 21 L_W^2, \qquad\qquad 19.10.1$$

and for the 16-hour day,

$$L_A = 10 \cdot 7 L_W - 16 L_W^2. \qquad\qquad 19.10.2$$

There appears to be no doubt that the suggestion was justified, and that the bulk of the variability associated with differences in plant size is due to movement of the plant up a regular progression, rather than to random scatter.

These relations imply that specific leaf area is declining linearly with increase of leaf dry weight, for, dividing through by L_W, the two equations become; for the 8-hour day,

$$\text{SLA} = \frac{L_A}{L_W} = 11 \cdot 9 - 21 L_W, \qquad\qquad 19.10.3$$

and for the 16-hour day,

$$\text{SLA} = 10 \cdot 7 - 16 L_W, \qquad\qquad 19.10.4$$

SLA being measured in $dm^2\ g^{-1}$, and L_W in g.

At the same time the relation between the two is not constant, the specific leaf area in the 8-hour day being 11 per cent higher at a leaf dry weight of 10 mg, declining to 8 per cent at 110 mg. If one extrapolated, the difference would decrease still further, becoming zero at 240 mg leaf dry weight, when both would have a specific leaf area of $6 \cdot 9\ dm^2\ g^{-1}$. But such an extrapolation is clearly not justified. From what we have already seen of the relationship between specific leaf area and total plant dry weight, there is no doubt that what we are dealing with here is an almost linear part of a much more complex relationship. In this particular experiment it is easy to make the transition from leaf dry weight to total plant dry weight as a baseline, because of the extraordinarily close

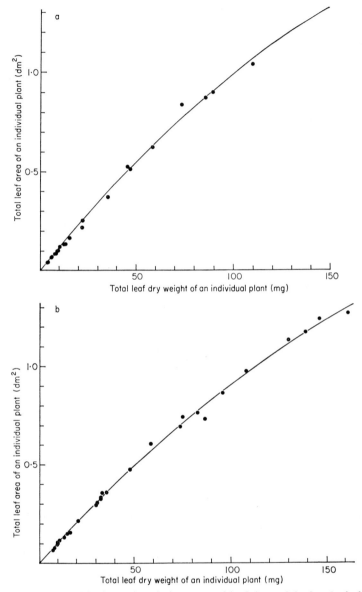

Fig. 19.10. Total leaf area in relation to total leaf dry weight for the individual plants of *Impatiens parviflora* comprising the samples of the day-length experiment considered in 18.12, grown under artificial conditions in the laboratory (a) with an 8-hour day, and (b) with a 16-hour day. For equations of the regression lines shown, see text.

19.10

approximation to a linear relationship between the two which we have already seen [18.13].

This is a good example of an interaction between a specific change in a factor of the aerial environment and the ontogenetic drift of plant response, the difference between the two treatments declining by nearly a third over the range of size which we have considered.

19.11 Comparison with changes in daily light total

However, this daylength change is large—larger, indeed, than a plant of *Impatiens parviflora* growing in England would encounter during the whole of the possible growing season. And the change in specific leaf area, although significant, is small when compared with the ontogenetic changes themselves, and even more so when compared with the effects of total daily light. There, at comparable total plant dry weights around 150 mg (corresponding to a leaf dry weight of about 60 mg, near the middle of our range), we have seen an increase in SLA from 6.3 dm^2 g^{-1} under 26.2 cal cm^{-2} day^{-1} to about 9.8 under 13.8 cal cm^{-2} day^{-1}, an increase of 55 per cent for a decrease in light of less than half, in a range commonly encountered under woodland conditions. This difference is so large that it is worth asking whether the small difference in total daily light between the 8- and 16-hour day treatments is likely to have contributed significantly to the observed difference in specific leaf area between them. We have already seen that the relationship between specific leaf area and total daily light is roughly linear in this region [Fig. 3.2.2]; a linear interpolation will give us at least an indication of the order of magnitude of the answer. The 8-hour day had about 0.25 cal cm^{-2} day^{-1} less visible radiation than the 16-hour day: this would correspond to an increase in specific leaf area of about 0.07 dm^2 g^{-1}, rather less than 1 per cent out of the 9 per cent difference observed. Thus our conclusions are not disturbed by this calculation, but we can see how essential close control of the environment is when dealing with an index as sensitive as specific leaf area.

19.12 A very sensitive index

For leaf weight ratio we could draw up a list of environmental factors, changes in which hardly affected the value of LWR. It would be profitless to attempt to do the same for specific leaf area, as there has been at least some influence of the environment in every instance which we have examined. Generally, it seems likely that such complex interrelationships as are observed in the field could not be interpreted in the absence of carefully arranged observations under controlled conditions in the laboratory.

19.12

UNIT LEAF RATE—
OTHER APPROACHES

20.1 Implications of the analysis of leafiness

The relationships established in equations 18.13.1, 18.13.2 and 19.10.1, 19.10.2 make possible a new approach to the computation of unit leaf rate. This is based neither on the requirement that the pattern of growth should follow one of a variety of simple models [16.1–16.11], nor on assessment of the magnitude of deviations from a specific model [16.12], but rests on the relations between leaf area, leaf weight and total dry weight which have been uncovered during the discussion of leaf weight ratio and specific leaf area. These make possible the establishment of direct relationships between leaf area and total dry weight, during a period when the plants were not growing in exact accordance with any of the model systems which we examined in Chapter 16.

It will be seen that this change of approach involves a change in the nature of our assumptions as to what was going on during the period between harvests. In Chapter 16 we examined the consequences of formulating a number of mathematical models, and investigated ways of assessing how closely the growth of the experimental plants approximated to one or another of these, using data on leaf area and total plant dry weight at harvest, in the form of means of harvested samples, each of several plants. In Chapter 19, on the other hand, we have been making use also of the variability observed at harvest in the plants constituting the sample. Under the carefully controlled conditions of the experiments, and for the species examined, we were led to conclude that the individual values of leaf area and leaf dry weight were connected by a regular relationship. There was some random scatter about the line of best fit, but this represented only a small part of the total variability at harvest, most of which was due to scatter up and down the line itself. Furthermore, the scatter up and down the line was large enough to leave hardly any gap between one harvest and the next, so that the data for four successive harvests presented the appearance of one single relationship. Here it is reasonable to assume that in the period between harvests all the plants were moving along this single progression, and we can then use the relationship revealed in this way as a basis for evaluating unit leaf rate.

In the case examined in Chapter 19 the relationship between leaf area and leaf dry weight was quadratic, and at present this particular form can be applied only to the particular experimental plant, *Impatiens parviflora*, and then only over the limited range during which the basic expressions are valid. It can, however, be used as an example of a method of procedure, readily applicable not only to similar cases in other species, but also to other relationships found to represent the patterns of growth of other plants under other conditions.

Two cautions must be borne in mind. The mere fact that regular relationships have been found between leaf area, leaf dry weight, and total plant dry weight is no reason for assuming that there may not be large variations in unit leaf rate. It is possible that all the plants are travelling along the same track, but at different rates. Such differences of behaviour can be detected and studied using the method of Goodall [8.10], and this was done, for example, in compiling Table 20.3. Secondly, it is probable that for many species the methods based on the assumption of a single basic progression can be applied only to plants grown under very closely controlled conditions, when steps have been taken to reduce as many as possible of the sources of variability [17.1] to a minimum, by methods such as those we have discussed. Otherwise our basic assumption itself will not be valid. Genetic variability alone could provide a whole family of tracks, and if these are caused to waver by environmental fluctuations, there would be no sort of approximation to the kind of close relationships which we have examined in Chapters 18 and 19. The very foundations of the method are missing, with the consequences which we have seen in the field experiments discussed in Chapter 14, in particular in relation to Figs. 14.5 and 14.6.

20.2   An example from Chapters 18 and 19

In equations 18.13.1 and 18.13.2 we showed that for these particular experiments the relationships between leaf dry weight and total plant dry weight took the form

$$L_W = \mathbf{a}W - \mathbf{b}, \qquad\qquad 20.2.1$$

20.2

where **a** and **b** are constant. We also showed, in equations 19.10.1 and 19.10.2, that

$$L_A = L_W(f - gL_W),$$ 20.2.2

where **f** and **g** are also constant.
Now

$$\bar{E}(_2T - {_1}T) = \int_{_1W}^{_2W} \frac{dW}{L_A};$$ 16.6

and substituting,

$$= \int_{_1W}^{_2W} \frac{dW}{(aW - b)\,[f - g(aW - b)]}.$$

But

$$dW = \frac{d(aW - b)}{a}.$$

Therefore

$$\bar{E}(_2T - {_1}T) = \frac{1}{a} \int_{_1W}^{_2W} \frac{d(aW - b)}{(aW - b)\,[f - g(aW - b)]},$$

$$= \frac{1}{a} \left[-\frac{1}{f}\log_e \left(\frac{f - g(aW - b)}{aW - b} \right) \right]_{_1W}^{_2W},$$

$$= \frac{1}{af} \left[\log_e \left(\frac{(aW - b)^2}{(aW - b)\,(f - g(aW - b))} \right) \right]_{_1W}^{_2W},$$

$$= \frac{1}{af} \left(\log_e \left(\frac{(_2L_W)^2}{_2L_A} \right) - \log_e \left(\frac{(_1L_W)^2}{_1L_A} \right) \right),$$

$$= \frac{1}{af} \log_e \left[\left(\frac{_2L_W}{_1L_W} \right)^2 \cdot \frac{_1L_A}{_2L_A} \right].$$ 20.2.3

This is an unfamiliar form, not depending, as have all the other formulae for unit leaf rate we have examined, on a straightforward difference of dry weights between the second and first harvests. Because of the known interrelations between total plant dry weight, leaf dry weight, and leaf area, the result is expressed, as it happens, only in terms of the last two. Equally the relations between the value of $\bar{E}$ obtained from equation 20.2.3 and those obtained from equations 16.8.2 or 16.9.2, are not so simple as that which we noted between the values from equations 16.8.2 and 16.9.2 themselves, depending solely on P, the ratio of $_2L_A/_1L_A$.

20.3 The daylength experiment further considered

This can best be shown by comparing the results of the three methods of computation using the data from the daylength experiment we have been discussing, working out separate values of $\bar{E}$ for each plant in a sample, and taking means for each group. The result is set out in Table 20.3,

Table 20.3. Weekly mean unit leaf rates, with fiducial limits, for an experiment on the growth of *Impatiens parviflora* under controlled conditions in the laboratory (the 16-hour day experiment of Figs. 18.12 and 19.10b), worked out using several different methods.

| | Mean unit leaf rate, g m^{-2} week^{-1} | | |
| | Intervals between harvests, weeks | | |
Method of computation	First	Second	Third
1. From equation 20.2.3	$20{\cdot}4 \pm 1{\cdot}7$	$19{\cdot}3 \pm 2{\cdot}1$	$17{\cdot}7 \pm 2{\cdot}1$
2a. On the assumption that total dry weight is linearly related to leaf area ($\bar{E}_1$, 16.8.2)	$22{\cdot}0 \pm 1{\cdot}4$	$18{\cdot}6 \pm 1{\cdot}7$	$17{\cdot}8 \pm 1{\cdot}7$
2b. Assuming that total dry weight is linearly related to the square of leaf area ($\bar{E}_2$, 16.9.2)	$20{\cdot}1 \pm 1{\cdot}2$	$17{\cdot}6 \pm 1{\cdot}4$	$17{\cdot}2 \pm 1{\cdot}4$
3a. Correcting for departures from assumption 2b by the method of 16.12.3 and 16.14.1 ($\bar{E}_n$). Value of n derived from the data of harvests 1, 2 and 3	$21{\cdot}8 \pm 1{\cdot}4$	$18{\cdot}5 \pm 1{\cdot}6$	—
3b. As 3a, but using the value of n derived from the data of harvests 2, 3 and 4	—	$18{\cdot}4 \pm 1{\cdot}6$	$17{\cdot}7 \pm 1{\cdot}6$
4a. As 3a, but computing n by the approximate method of 16.14.2	$21{\cdot}8 \pm 1{\cdot}4$	$18{\cdot}5 \pm 1{\cdot}6$	
4b. As 3b, but computing n by the approximate method of 16.14.2		$18{\cdot}1 \pm 1{\cdot}5$	$17{\cdot}5 \pm 1{\cdot}6$

which shows that the value of mean unit leaf rate calculated using equation 20.2.3 is close to $\bar{E}_2$ for the first interval, that it rises to a value

20.3

somewhat above $\bar{E}_1$ in the second, falling finally to a value close to $\bar{E}_1$. Remembering that the treatment of the equation $W = h + kL_A^n$ in section 16.12 depended upon n being constant, at least during the period between one harvest and the next, the values of Table 20.3 would imply a value of n near 2 in the first week, less than 1 in the second, and about 1·2 in the third. Thus in this case the basic assumption of the constancy of n within each week, used in 16.12–16.16, is almost certainly untrue.

There is no doubt that in this particular experiment the best available estimate of mean unit leaf rate is given by equation 20.2.3, and the figures therefore provide a good opportunity to investigate the performance of the approximate method of computation of 16.12 in an unfavourable case, where we know that one of the basic assumptions is untrue. (It will be recalled that the evaluation of n depends on the data of three successive harvests, and the additional assumption that n is constant for two successive inter-harvest periods, which is not the case in Table 20.3.)

Lines 3a and 3b of Table 20.3 set out the values of mean unit leaf rate calculated by the method of 16.12.3 and 16.14.1, using respectively the data for the first three and the last three harvests. The values of 3a are high by 7·2 per cent in the first week and low by 3·9 per cent in the second; the values of 3b are low by 4·4 per cent in the second week and are almost exactly correct in the third. For some purposes such deviations would be acceptable, but for critical work an investigation along the lines of that leading up to equation 20.2.3 would be worth while.

Lines 4a and 4b show the results of using the further computational approximation of 16.14.2, giving a value for n without requiring the graphical solution of equation 16.14.1. Working to three significant figures there is no difference between lines 4a and 3a (the systematic difference being about 0·2 per cent) while line 4b is 1·8 per cent below line 3b for the second interval, and 1 per cent below for the third. The approximation is thus a worthwhile one for normal use, while the more rigorous procedure of 16.14.1 can be used as a check in very critical work or at times when there is reason to suppose that the apparent value of n is changing rapidly.

Table 20.3 thus illustrates how easy it is, using a carefully collected but unfavourable set of original data, to produce fallacious figures for unit leaf rate by choosing the wrong assumptions as the basis for calculation. This has a bearing on another interesting aspect of the Table, the falling value of mean unit leaf rate, which is large enough to appear consistently

20.3

throughout, irrespective of the method of computation. However, lines 2a, 2b and 3a, 3b all suggest a larger and more significant fall in ULR between the first and second periods (p around 0.01), the differences between the second and third periods being small and insignificant. There is, however, no such difference in line 1, but a much steadier fall, the only significant difference (at the level of $p = 0.05$) being that between the values for the first and third periods. Thus the overall fall is there, and proves to be significant whatever the method of computation: but the apparently more rapid fall in the early stages of the experiment is shown to be an artefact of calculation.

In these particular experiments growth conditions were constant throughout, and there is no reason to suppose that even at the end of the experiment the falling values were due either to self- or mutual shading. We have already had occasion to analyse unit leaf rate into broad groupings of physiological components [4.3, 12.1, 13.11], and later we shall consider quantitative examples [28.2–28.14]. Here it need only be said that the most likely explanation of the fall in unit leaf rate with time in Table 20.3 is some ontogenetic change in the rate of functioning of the photosynthetic apparatus per unit leaf area. We shall pursue this matter later also [20.6, 29.6, 29.10].

20.4 Continuous harvesting

The methods discussed in the last three sections lead also to the possibility of a different approach to the collection of data. The methods discussed earlier involve harvests so spaced as to allow time for substantial changes to take place between one harvest and the next; and samples large enough to make it possible to demonstrate that these changes are statistically significant. But we could arrange to harvest the same total number of plants in smaller batches at more frequent intervals, say every two or three days. Under favourable circumstances this might have two substantial advantages. Firstly, it would overcome the difficulties we have encountered about the determination of the course of growth when a considerable period has elapsed between one harvest and the next; and secondly it might fit in better with laboratory organization when much material has to be handled with limited facilities.

However, the change to continuous harvesting does raise statistical problems, which are discussed by Hughes and Freeman (1967). If suffici-

20.4

ent computing facilities are available, it is possible to make a series of instantaneous determinations of both relative growth rate and unit leaf rate, together with the appropriate estimates of error. A convenient example is provided by Hughes's experiments in growth cabinets on the effect of varying atmospheric carbon dioxide concentration on the growth of *Callistephus chinensis* cultivar Johannistag, which we shall discuss later [23.7, 24.12, 24.13, 24.17]. Three plants from each of three treatments were taken at each harvest, and the total dry weight (W) and leaf area (L_A) of each plant were determined in the usual way. In order to make the variability approximately homogeneous over the whole experimental period, all the statistical work was done on the logarithms of the dry weights and leaf areas, natural logarithms being used. The polynomial of sufficient fit to the logarithms of the weights and areas on time was determined by the least squares method. A cubic was found adequate in both cases, giving

$$\log_e W = z_1 + z_2 T + z_3 T^2 + z_4 T^3, \qquad\qquad 20.4.1$$

and

$$\log_e L_A = z_5 + z_6 T + z_7 T^2 + z_8 T^3, \qquad\qquad 20.4.2$$

where T is time and z_1, z_2, z_3, z_4, z_5, z_6, z_7, z_8 are all constant. In this particular instance it turned out that z_3 and z_7 were negative in all cases, all the other regression coefficients being positive.

The fitted curves for the logarithms of the dry weight against time are shown in Fig. 20.4.1a for three treatments involving atmospheric carbon dioxide concentrations of 325, 600 and 900 parts per million, with fiducial limits for the lowest concentration. The mean dry weights for each harvest are also shown for the highest and lowest concentrations.

The progressions of relative growth rate against time shown in Fig. 20.4.1b were then derived by differentiation from equation 20.4.1, for

$$\frac{d(\log_e W)}{dT} = \frac{1}{W} \cdot \frac{dW}{dT}. \qquad\qquad 20.4.3$$

The progression for relative leaf growth rate was derived similarly from equation 20.4.2.

The progressions of leaf area ratio emerged as

$$\frac{L_A}{W} = \text{antilog}_e (\log_e L_A - \log_e W). \qquad\qquad 20.4.4$$

20.4

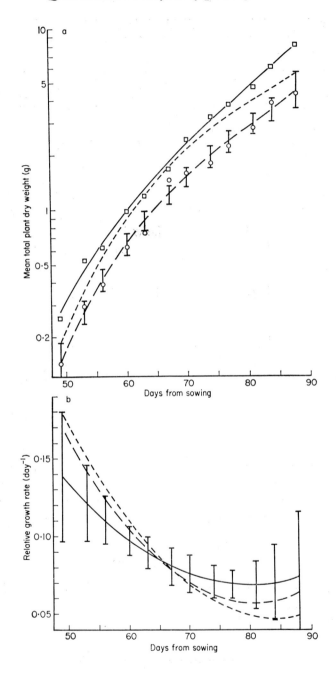

They are shown in Fig. 20.4.2a, together with the fiducial limits for the lowest concentration. As in Fig. 20.4.1a, we can compare the fit with the mean values of leaf area ratio for the sample at each harvest. It was found that the variance of leaf area ratio varied in a complex manner with time, so that nothing would have been gained by attempting to fit a regression directly.

Finally, the progress curves for unit leaf rates were derived by dividing the expression of equation 20.4.3 by that of 20.4.4, for

$$\frac{1}{L_A}\cdot\frac{dW}{dT} = \frac{1}{W}\cdot\frac{dW}{dT} \times \frac{W}{L_A} = \frac{d(\log_e W)}{dT} \times \frac{1}{\text{antilog}_e (\log_e L_A - \log_e W)}.$$
20.4.5

These are shown in Fig. 20.4.2b, together with the fiducial limits for the lowest concentration, as before. It will be noticed that these are instantaneous values of E; if one wished to calculate $\bar{E}$, the true mean over a period, integration would be necessary, as in the related method of Vernon and Allison (1963) [16.19].

Hughes and Freeman's example also illustrates neatly a problem inherent in curve fitting using polynomials and the least squares method, which is ill-adapted to giving a good fit to a curve most of which exhibits a regular progression, but which has a sudden change of curvature near one end. An inspection of the plotted dry weights of the plants grown in 900 p.p.m. carbon dioxide in Fig. 20.4.1a shows no sign of a point of inflection before 82 days: but the exigencies of fitting a cubic to the data create a point of inflection in the fitted curve around 80 days producing a minimum in the relative growth rate curve at the same time (Fig. 20.4.1b). Furthermore, an inspection of the leaf area ratios for the same plants in Fig. 20.4.2 shows that the latter part might well be linear: but again the fitted curve has a point of inflection around 70 days. The combined effect, in Fig. 20.4.2b, is to produce an apparent sharp rise in unit leaf rate

---

Fig. **20.4.1.** Relationships with time of (a) mean total plant dry weight plotted on a logarithmic scale and (b) relative growth rate per day for plants of *Callistephus chinensis* cultured in growth cabinets with three different concentrations of atmospheric carbon dioxide: □—— 900 parts per million; ------ 600 ppm; ○—— 325 ppm, with fiducial limits for the 0·05 probability level, in (a) for the 325 ppm regression line, in (b) for the 900 ppm one. Points are means for harvests of three plants, twice weekly. From Hughes and Freeman (1967).

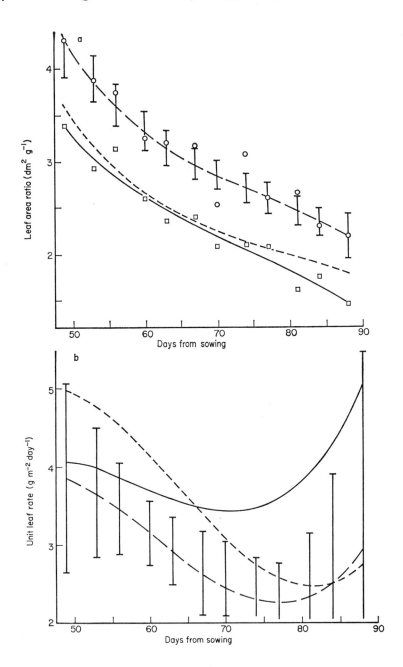

towards the end of the experimental period, when, however, the fiducial limits are also increasing rapidly, placing the duration and extent of the final rise in doubt. Before embarking on the fitting of polynomials it is accordingly always advisable to consider whether the fitting is likely to generate an adequate generalized description of the original data, or whether these contain features which are best dealt with in some other way.

However, if these cautions are borne in mind, a procedure of this kind can be very simple and straightforward for those with ample computer facilities available, involving only the punching of a relatively small amount of harvest data. For those less well endowed the work involved can easily become impossibly tedious, as a good fit to a long sequence of harvest data may easily involve polynomials of more terms than those which we have considered. At the other extreme we may know enough about the plants we are working with to be sure beforehand that both relative growth rate and relative leaf growth rate will be effectively constant, as happens early in the life of many seedlings. Equations 20.4.1 and 20.4.2 then reduce to linear expressions and the statistical work becomes simple enough to work out if necessary by hand without undue effort. In such cases the application of these statistical methods to a scheme of continuous harvesting is well worth consideration in the stage of experimental planning, even by those without extensive computing facilities.

20.5   Relation to analytical methods previously discussed

In the methods of computation discussed earlier we have contrived to avoid very heavy numerical work, and these methods can also be adapted to cope with continuous harvesting. Our estimates of error may be less reliable from a straightforward statistical point of view, but at times this is less of a handicap than may at first sight appear. Situations can easily arise where curves may be very difficult to fit by a polynomial expression,

---

Fig. 20.4.2. Relationships with time of (a) leaf area ratio and (b) unit leaf rate on a daily basis for the same plants of *Callistephus chinensis* used for Fig. 20.4.1, cultured in growth cabinets with three different concentrations of atmospheric carbon dioxide: □ and ——, 900 parts per million; ------ 600 ppm; ○ and — — 325 ppm, with fiducial limits for the 0·05 probability level. Points are mean ratios for harvests of three plants, twice weekly. From Hughes and Freeman (1967).

or where the very basis of the statistical procedure itself is in doubt, as we noticed for the relationship of leaf area ratio and time in Fig. 20.4.2a. We have also to bear in mind that the establishment of empirical parameters on a most probable statistical basis is itself only a means to an end; a convenience in the formulation of generalizations on a sound basis. We are still a long way from the stage in the development of scientific explanation where both the general forms of the equations, and the values to be assigned to the principal parameters, can be predicted on the basis of scientific knowledge of the system and its components.

The simplest scheme is to adapt the methods devised by Williams (discussed in section 16.17), and proceed by drawing freehand curves through the points followed by numerical integration.

However, data from continuous harvesting are obviously adapted to the kind of ontogenetic studies which we discussed in Chapters 18, 19, and 20.1–20.3. They are also suitable for the method which we discussed in section 16.13, for the graphical determination of a suitable value of n for the equation $W = h + kL_A^n$. If we use the methods of estimation of leaf areas and dry weights from measurements taken, say, a week before harvest (discussed in section 8.10 and used in the computations for Table 20.3), we are then in a position to work out the mean relative growth rate and unit leaf rate for each plant during the week before harvest, and these can then themselves be used to determine the form of the ontogenetic drifts.

There is one circumstance in which this last mode of procedure may even be preferable to the straightforward, if heavy, computational procedure which we mentioned in section 20.4. This is the case where a number of plants are all passing through a similar morphological progression, as in Fig. 19.10, but at different rates—where the machinery for distributing new dry matter among the various organs, and the machinery controlling development and expansion of the organs themselves, are all functioning in a regular manner, but the new dry matter itself is being supplied at somewhat varying rates. If we start with a large uniform population, sort the plants and grade them, and then harvest at intervals, such variations will appear as scatter of total plant dry weight at harvest, and will become merged in the general experimental error. By measuring the plants a week before harvest we have additional data which enable us to allow for plant to plant variations of this kind, and to study them further if need be.

20.5

20.6 The constancy or inconstancy of unit leaf rate

The question of the relative constancy of unit leaf rate [13.13] has been the subject of much discussion. In their analysis of the growth of maize Briggs, Kidd and West (1920b) pointed to the slow development of photosynthetic activity of the young leaves as a reason for the initial very low (at first negative) values of unit leaf rate and hence of relative growth rate [Fig. 13.6.2], although later in life the drift of leaf area ratio in this plant was greater than that of unit leaf rate [Fig. 13.7]. Heath and Gregory (1938) took the view that in a range of plants (predominantly crop plants) unit leaf rate was relatively constant, and that differences in leafiness between species were much more important in determining the differences in relative growth rate. This is at first sight born out by Watson's figures for weighted mean unit leaf rates of a variety of crop plants in Table 13.19, where the differences are much smaller than those for leaf area duration, which can be considered as a measure of the extension of leafiness in time. But Watson (1947) was concerned to point out how greatly the value of unit leaf rate is affected by time of year in a climate as seasonal as that in southern England [Fig. 13.19]. Heath and Gregory's findings are readily reconciled with this when it is remembered that crop plants are normally grown during the height of the favourable season; and that in England, at this time of year, a wide range of weather is possible in any particular year (see, e.g., the records for total radiation on 19 June 1956 compared with 19 June 1957, in Fig. 7.3). To show the underlying steady march of any particular aspect of so variable a climate, one has to look at weekly or monthly means of a long sequence of years, as in Fig. 13.19b. These same causes no doubt account for the relatively small range of values shown by Watson's weighted mean unit leaf rates in Table 13.19. At the same time Heath and Gregory's species list comprised only a very limited range of ecological types (in particular, they included no woody plants), and when in 1960 Coombe surveyed the information available at that time he concluded that in woody plants as a class the evidence indicated that unit leaf rate was much lower than in the herbaceous plants previously investigated. The evidence which has accumulated since 1960 has confirmed this view.

Unit leaf rate can thus show a wide range of values, and be affected by ontogeny and by ecological type as well as by environment. In the nature of their work ecologists have to deal with a wide range of growth forms and with a wide range of seasons, and in this context it is just as important to consider possible differences in unit leaf rate and the reasons for them as it is to consider the various factors affecting the extent and maintenance of a plant's leafiness. These matters will be discussed in more detail in Part IV.

20.6

STEM AND ROOT WEIGHT RATIOS—
PLANT ARCHITECTURE

21.1 Interrelationships

The methods which we have already discussed at length for the study of leaf weight ratio apply equally to the study of stem and root weight ratios. Plastic though it usually is, the plant as an organism functions as a whole, and events leading to a higher or lower proportion of new material finding their way to one particular type of organ necessarily imply a lower or higher proportion finding its way elsewhere. This is obvious enough, and the difficulty may lie in detecting the decisive order of events, and which way the causal connection lies. Has an increase in size of root reduced the material available for developing new leaves? Or does the causal relationship operate the other way round, delay in leaf development from some other cause making more material available for the roots? In some cases such questions are relatively easy to answer, and we have already noticed an example in the important effect of a change in rooting medium on leaf weight ratio (18.10, and see Fig. 26.13).

21.2 Effects of light climate on young plants

Such reciprocal relationships are well illustrated by the effect of varying radiation climates on the proportions of stem and root. Fig. 21.2 shows stem weight and root weight plotted against total plant dry weight for the field experiments on *Impatiens parviflora* which we have already examined in section 18.6. It will be recalled that when the experiment began one uniform population of plants was distributed under five different conditions—full exposure to daylight, and screens transmitting 67 per cent, 42 per cent, 25 per cent and 7 per cent. In Experiment I (Fig. 21.2a, c) the next harvest was 20 days later, and there is no doubt that all the plants were fully adapted; in Experiment 2 (b and d), the initial exposure under the screens was 10 days, and the plants were probably all fully adapted,

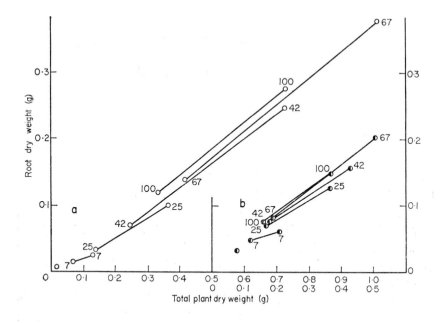

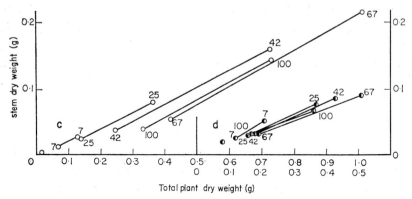

Fig. 21.2. Relationships between root (a, b) and stem (c, d) dry weights and total plant dry weight for means of six plants in two experiments on *Impatiens parviflora* growing under neutral screens in the field, with mean daily totals of visible radiation as in Table 19.6. (a, c) Experiment I, harvests at 9, 29 and 36 days; (b, d) Experiment II, harvests at 23, 33 and 40 days. Common initial sample, then lines join the points for harvests 2 and 3 at each level of shading (screen transmission in per cent). From Evans and Hughes (1961).

21.2

the only likely exception being the 7 per cent screen. Lines join the points for the second and third harvests, under the various screens.

Looking first at the relations of stem weight, we see in Fig. 21.2c a number of more or less parallel lines, following the order of growth rate (this is the same as the order of total dry weight at the second harvest, because all started at the same dry weight), and the same relationship is also seen in Fig. 21.2d. Clearly we have here a different type of relation from that of leaf dry weight, where differences in light intensity affected the rate of growth and total plant size, but not the basic progression up which all the plants were moving. Here we seem to have a number of parallel progressions, determined by the effects of the various screens, and arranged, not in order of total daily light, an environmental measure, but in the order of plant size at second harvest, a physiological measure.

The progressions of root dry weight necessarily follow the opposite pattern, the plants grown under the 7 per cent screens below, and the others in order, with only one difference; here the plants grown in the open, without screens, have the largest proportion of root, in three cases out of four.

21.3 Natural and artificial light climates

In the fluctuating climate of the field there is always the possibility that relationships of this kind are fortuitous, brought about by changes of weather during the course of the experiment. The agreement between only two experiments does not rule out coincidence. We therefore examine the series of cabinet experiments under constant conditions of all aspects of the environment except for different intensities of visible radiation. The relevant figures are shown in Fig. 21.3a for stems, and Fig. 21.3b for roots. Over the corresponding range of dry weights, above 0·1 g, we see the same series of progressions, arranged in the same way, with the lowest intensity above for stems, and below for roots. It is clear, therefore, that unlike the leaf weight, which is affected only at the highest intensities, the proportions of stem and root are influenced by the intensity of visible radiation throughout the range used in these experiments. It is also clear that, whatever may be the individual effects of the various environmental changes which we listed in section 18.15 (ix to xii) as accompanying the change from the cabinets to the field, when operating together they do not alter the pattern of response occasioned by changes in intensity of visible radiation alone.

21.3

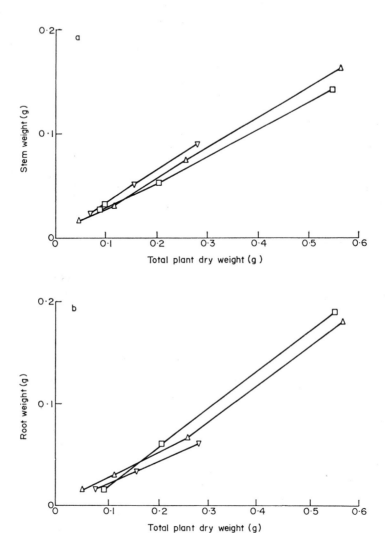

Fig. 21.3. Relationships between stem (a) and root (b) dry weights and total plant dry weight for means of eight plants (for three points, six plants) in three experiments on *Impatiens parviflora* growing under constant conditions in growth cabinets under different daily totals of visible radiation, $\triangledown$, $\triangle$, $\square$ representing 13·8, 20·6 and 26·2 cal cm^{-2} day^{-1} respectively. Harvests at weekly intervals, ages of plants between 17 and 44 days. From Hughes and Evans (1962).

21.3

21.4 Progressions up to maturity in *Helianthus annuus*

These observations covered only the early stages in the life of an annual, in which no large changes in the proportion of root, stem and leaf have been taking place. In Fig. 21.4 are set out the whole progression of stem and root weight ratios for Briggs, Kidd and West's observations on the growth of sunflowers. Taken in conjunction with Fig. 19.3b we see how, in a plant with this kind of architecture and mode of growth, the

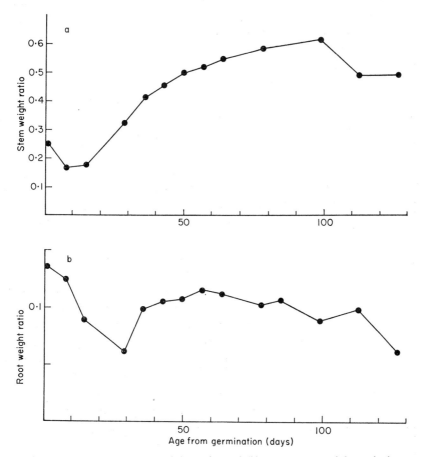

Fig. 21.4. (a) Mean stem weight ratio and (b) mean root weight ratio in relation to age from germination for plants of *Helianthus annuus* grown by Briggs, Kidd and West at the Botanic Garden, Cambridge, in 1920. For the corresponding figures for mean leaf weight ratio, see Fig 19.3b, and for the reproductive organs, see Figs. 23.5.1 and 23.6.

pattern of dry weight distribution passes through a number of phases. At first, the proportion of dry matter in the leaves is increasing at the expense of both stem and root [Fig. 19.3b]. It reaches a maximum of over 70 per cent around 15 days from germination, when the plant is still small, with a total dry weight of less than 0·25 g. Thereafter there is a progressive shift in dry matter distribution in favour of the stem, which around 40 days begins to exceed the leaves in total dry weight, and which finally exceeds 60 per cent of the total plant dry weight about 100 days. In the last phase, as the massive capitulum develops [Fig. 23.6], the stem weight ratio falls to a lower value of just below 50 per cent.

Root weight ratio undergoes a comparable series of changes—an initial fall, followed by a rise up to about 60 days, and then a progressive fall throughout the remainder of the life of the plant. We notice that this progressive fall is much less regular that the corresponding progressions for leaf [Fig. 19.3b] and stem weight ratios. This may partly be due to inherent variability—an interaction of the plant/soil system with a fluctuating environment; and partly due to the difficulties involved in extracting the root systems from the soil, particularly for large plants with an extensive root run. But this irregularity is not disturbing. At its maximum in the middle of the life cycle the root weight ratio in this particular case never exceeds 12 per cent, and as we shall see in Fig. 23.5.1 the fluctuations in root weight ratio towards the end of the plant's life are very minor as compared with the major changes going on elsewhere in the plant, following the onset of the flowering phase.

Nevertheless, the main progressions of both stem and root weight ratios are themselves complex, each involving several main phases. The investigation of the effects of environment on relationships of this kind poses problems similar to those which we have discussed for leaves, and requires similar methods of analysis. The examples from young plants which we have seen, together with the evidence which we looked at when discussing leaves, makes it very likely that the basic progressions will be much influenced by environment, as well as by plant size. We can illustrate this from the series of experiments on the growth of a uniform strain of sunflowers made at the end of July in the years 1961 to 1967.

21.5 Variations from year to year

Table 21.5 shows this data, arranged in order of increasing amount of growth in dry weight during the experimental week, for convenience of

21.5

Table 21.5. Growth of stems and roots of *Helianthus annuus* at the Botanic Garden, Cambridge during one week at the end of July, 1961–67. Two age groups in each year (age ranges in days from sowing), arranged in order of increasing mean increment in total dry weight per plant. The fifth and sixth columns show the mean percentage of this total increment which was found at the end of the week to have been added to the stems and roots, respectively. In the last column, the ratio of stem increment to leaf increment. All values are means from 9 pairs of plants. Least significant differences between two mean values in the last three columns: stems, 5·4 per cent; roots, 8·7 per cent; stem/leaf increment ratio, for younger plants 0·04, for older plants 0·35, between age groups 0·26.

Year	Age range (days)	Initial dry wt. (g)	Mean increment in dry wt. (g)	Percentage of increment in Stems	Roots	Stem Increment / Leaf Increment
1966	25–32	1·568	2·940	18·6	27·6	0·35
1965	25–32	2·017	3·802	20·4	26·9	0·39
1962	22–29	3·324	7·784	20·9	33·1	0·45
1966	39–46	9·659	13·769	29·9	29·4	0·75
1965	39–46	16·811	14.541	29·1	18·5	0·63
1961	22–29	4·837	15·444	20·5	23·4	0·37
1964	22–29	4·939	15·512	20·3	36·0	0·48
1963	22–29	5·391	20·484	23·1	29·2	0·50
1964	36–43	12·056	20·653	26·2	37·2	0·72
1962	36–43	47·559	76·989	30·5	35·5	0·95
1967	25–32	45·420	85·263	29·6	28·6	0·73
1963	36–43	61·690	120·612	27·5	43·3	0·92
1967	39–46	233·077	197·971	36·6	35·0	1·36
1961	40–47	240·129	313·060	48·5	28·6	2·29

comparison. The table also gives the year, the age range in days, and the mean plant dry weight at the beginning of the experimental period.

We notice, firstly, the very large range in amount of increment of dry weight among plants of comparable ages, a thirty-fold range in the case of the younger plants (25 to 32 days, 2·9 g in 1966, 85·3 g in 1967), and more than a twenty-fold range in the case of the older ones (39–46 days, 13·8 g in 1966; 40–47 days, 313 g in 1961). This results in considerable overlapping of the age groups, the first four entries in the table including

21.5

one of the older group, the last four entries one of the younger group. From this point of view we can regard the middle of the table as coming between entries 9 and 10, with roughly a seven-fold increase between entries 1 and 9, a four-fold increase between entries 10 and 14, and as it happens a large gap between entries 9 and 10.

In the first section of the table, the mean percentage of the increment which accrued during the experimental week going to stems was 23·2, and to roots 29·0. In the second section, the mean percentage of the increment going to stems was 34·5, and to roots 34·2. The expected increase in both percentages is there, and, also as expected, it is much larger in stems than in roots. But both increases are overlaid by very large differences from year to year, even between plants comparable both in age and in mean increment in total plant dry weight. Thus, in 1964, plants aged 22–29 days and with an increment of 15·5 g had 36 per cent of this going to roots. In 1961, on the other hand, plants of the same age and with an increment of 15·4 g had 23 per cent of this going to roots (and compare Fig. 26.4.2).

It is clear that the variability both within years and from year to year is much larger in roots, and that the magnitude of the percentage increment changes less with plant size than does that for stems. It is accordingly worth looking at another index of developing plant form, the ratio of stem increment to leaf increment, which is, of course, independent arithmetically (though probably not physiologically) of the variations in increment of the roots. It is set out in the last column of Table 21.5.

21.6 Relations of stem and leaf increments

Once again, we see the general tendency for this ratio to rise as the plant gets larger, and once again the very large variations from year to year among plants having comparable total increments. There appears to be a decided tendency for the value of the ratio to be affected not merely by the size of the overall increment, but also by the age of the plant; we notice that the three highest values of the ratio in the upper part of the table are shown by plants of the older age group, and the one set of young plants in the lower part of the table has the lowest value there. The mean value for the younger age group in the upper part of the table is 0·42; and the mean for the three ratios of the older age group is 0·70. The difference between the two means is highly significant statistically, the probability

21.6

of so large a difference occurring by chance being very much less than 0·001.

The problem can be approached in another way. The standard deviation of a single mean ratio for the older plants in the upper part of Table 21.5 is 0·0764. The three deviates of the values for the older plants from the line of best fit to the values for the younger plants are +0·310, +0·184 and +0·245, respectively 4·06, 2·42 and 3·21 standard deviations away from the line, all in the same sense. The chances of finding such a combination of deviations at random (with an estimate of the standard deviation based on 24 degrees of freedom) are considerably less than one in 10^6, so that we are safe in concluding that the three sets of values for the older plants are genuinely different from those for the younger ones with which we have grouped them on the basis of amount of increment during the week. It will be seen that this conclusion is not affected by the relationship between total increment for the week and the ratio of stem increment to leaf increment [14.5, 18.12, 18.13, 26.2].

Clearly, then, we have here a good example of the interaction of the effects of environmental conditions during a particular week, with the previous experience, which influenced the form of the plant at the beginning of the week, and with the plant's progress along its inherent ontogenetic drift during the week. This is reinforced if we consider the final dry weight at the end of the week (the sum of columns 3 and 4). As examples, the young plants in 1963, with a final dry weight of 25·9 g, had a stem to leaf increment ratio of 0·50; the older plants in 1966, with a final weight of 23·4 g, had a ratio of 0·75: and the younger plants in 1967, with a larger final weight than the older plants in 1962, had a ratio of 0·73, as compared with 0·95. Mere size alone, in fact, is only part of the key to these differences.

21.7   Combination of field and laboratory experimentation

Taken together, the three different types of experiment which we have looked at in this chapter form a good example of the value of combining field and laboratory experimentation. The kind of understanding of the effects of changing light climate yielded by the laboratory experiments which we considered [21.3] could not have been achieved by field experiments. This is particularly true of the last example [21.5, 21.6], where the experiment is planned in advance, the plants are put out in the field,

21.7

and have to take their chance with the weather. Here relationships are bound to be obscured by the great variety of combinations between weather during a particular experimental period, and the plant size, structure and physiological capacity present at the beginning of the experimental period, which are in themselves in part integrations of environmental conditions earlier on.

Experiments under partially controlled conditions in the field [21.2] extended this understanding based on laboratory studies. They showed that the relationships which had been established also held good in the field, and that they were not easily disturbed by the variations in climate between one experimental period and another.

On the other hand, the field experiments listed in Table 21.5, carried on over a number of years, to a common plan and using the same variety of plant, provide information which could not be obtained by the other methods. They provide the plants with the opportunity of acting as natural integrators of the wide range of natural climatic variation from year to year. This is reflected in the very wide range of size and structure observed in even-aged plants at the beginning of the experimental period, representing a range of plant dry weights as large as thirty-fold for 25-day-old plants, and over twenty-fold for 39-day-old ones. It is also reflected in the wide range of plant response during the experimental period—in one year plants weighing initially 4·8 g added 15·4 g to their dry weight during a week; in another year plants weighing initially 16·8 g added only 14·5 g; and so on.

These wide ranges of structure and response then provide us with the varied background needed to establish generalizations about the development of plant structure and response on a firm basis. We saw, for example, that in spite of a thirty-fold range of initial size of the young group of plants, and a similar range in the amount of increment during the week, nevertheless, the ratio of stem increment to leaf increment for all these plants stood at 0·41, with a standard deviation for a single observation of only 0·047

The three methods of investigation are thus not merely complementary; but also each is in its own way indispensable.

21.7

RELATIVE LEAF GROWTH RATE

22.1 Analytical difficulties

One of the concerns of Gregory's observations on the cucumber, made at Cheshunt in 1917, was the development of leafiness, and he followed up and elaborated this in his 1926 paper on barley, extending the analysis to relative growth rate in leaf area,

$$\frac{1}{L_A} \cdot \frac{dL_A}{dT},$$

a quantity analogous to relative growth rate in dry weight. As time went on, and information about the comparatively small variations in unit leaf rate of crop plants accumulated [20.6], the importance of leafiness became increasingly apparent. It also became apparent that relative leaf growth rate is a difficult concept to handle analytically—every bit as difficult as relative growth rate itself, to which we have given so much attention. This will be clear if we refer again to Fig. 2.8. There we saw, at the top of the figure, the increment in new dry matter, related to leaf area by unit leaf rate, with all the consequent relations to ontogeny and environment that that implies. As we pass down the figure this new dry matter is divided among the various plant organs, the proportion retained in the old leaves, or going to new leaves, being controlled in the manner which we discussed in Chapter 18 on leaf weight ratio. The expansion of the increment in leaf dry matter into new leaf area was discussed in Chapter 19 on specific leaf area; and it will be recalled how sensitive this quantity is to a great variety of environmental changes, and how complex is its ontogenetic drift. In short, relative leaf growth rate involves the whole complex of plant-environment relationships extending in time.

As we have already seen, direct attack by field studies on so complex a situation is rarely profitable; it is best tackled by breaking the system down into more readily handleable components in the way which we have already been doing. This is one reason for postponing the discussion of relative leaf growth rate to this point, when we have these other, complementary, discussions behind us.

22.2 Relative leaf growth rate and time

On the other hand relative leaf growth rate is a most useful analytical tool, finding a place in most investigations of this kind. In the first place, as for relative growth rate in dry weight, if relative leaf growth rate (R_L) is constant, then the logarithms of the leaf areas will yield a straight line when plotted against time; for, if

$$\frac{1}{L_A} \cdot \frac{dL_A}{dT} = R_L,$$

and if R_L is a constant, then

$$\log_e {}_2L_A - \log_e {}_1L_A = R_L({}_2T - {}_1T). \qquad \text{22.2, cf. 13.5.2}$$

Therefore, if we have a sequence of three or more determinations of L_A, one of the first things that we do is to plot their logarithms against time in order to determine whether this is so.

It may well be that we have much more frequent observations of leaf area than we have of dry weight, because, as we have seen, leaf area can readily be determined with sufficient accuracy without a harvest in the case of plants with regular leaves—Weber, for example, measured the leaves of his plants every day during his experiments in 1878, and so did Gregory in 1917.

If, on the other hand, we have only two observations separated by a period of time, then, as for relative growth rate [equation 13.6], equation 22.2 gives us a true mean value of the relative leaf growth rate ($\bar{R}_L$) during the period.

22.3 Relations with relative growth in dry weight

If we find that R_L is effectively constant over a specific period (and this frequently happens during the early life of plants grown from seed, especially under controlled conditions), we then examine the corresponding behaviour of relative growth rate in dry weight, R, during the same period, as shown by a plot of the logarithms of total plant dry weight against time. If R_L is constant, it is commonly found that R is constant too; indeed, when there are linear phases for both sets of logarithms, leaf area is usually the first to show a departure.

If both relative growth rate in dry weight and relative leaf growth rate are constant, then the ratio of the two gives us **n** in the equation

$$W = \mathbf{h} + \mathbf{k}L_A^n:$$ 16.12.1

for if

$$\frac{1}{L_A} \cdot \frac{dL_A}{dT} = \mathbf{R}_L,$$

then

$$_T L_A = {_0}L_A \, e^{\mathbf{R}_L T}:$$

and if

$$\frac{1}{W} \cdot \frac{dW}{dT} = \mathbf{R},$$

then

$$_T W = {_0}W e^{\mathbf{R} T},$$

where $_0L_A$, $_0W$, are the values of L_A and W at time 0; $_TL_A$, $_TW$, at time T; and $\mathbf{R}_L$ and $\mathbf{R}$ are constant; then

$$e^T = \left(\frac{_TL_A}{_0L_A}\right)^{1/\mathbf{R}_L};$$

and

$$_T W = {_0}W \left(\frac{_TL_A}{_0L_A}\right)^{\mathbf{R}/\mathbf{R}_L} = \frac{_0W}{(_0L_A)^{\mathbf{R}/\mathbf{R}_L}} \cdot (_TL_A)^{\mathbf{R}/\mathbf{R}_L}.$$ 22.3

Therefore, in equation 16.12.1, $\mathbf{h} = 0$,

$$\mathbf{k} = \frac{_0W}{(_0L_A)^{\mathbf{R}/\mathbf{R}_L}},$$

and $\mathbf{n} = \mathbf{R}/\mathbf{R}_L$.

In practice it is found that these conditions are usually met when $\mathbf{R}_L = \mathbf{R}$ and $\mathbf{n} = 1$, which, as we have already seen [16.12 and 16.15], is a state of affairs commonly encountered in the early stages of seedling growth, and here again **h** commonly equals zero. If we have enough values of both L_A and W to be sure that $\mathbf{R}_L$ and $\mathbf{R}$ are indeed constant, even if they are not equal, we can conveniently use this method to give us a value of **n**, and we also know that $\mathbf{h} = 0$. If, however, the points are either few or scattered, so that an approximate fit to a straight line might conceal some other relationship, it is always wise to plot W against L_A^n, using the value of **n** just determined, in order to check that the result is a straight line passing through the origin. If it is not, then we must proceed by one or other of the methods given in sections 16.13 or 16.14.

As the seedlings grow larger, in the instances which we have examined in Chapter 16, not only does **n** increase, but **h** commonly increases to substantial values [16.9 and 16.12]. Here the ratio of the two relative growth rates no longer gives us a true value of **n**. Equally, growth no longer meets our first test for the use of the method, because here it is not possible that both relative growth rate in dry weight and relative leaf growth rate should both be constant.

22.4 Field experimentation: variability from year to year

Substantial variations in both **R** and $\mathbf{R}_L$ are liable to be encountered in field experimentation even using a single variety and a limited age range, and the relationship between them can also vary considerably. Such

Table **22.4.** Mean relative growth rates in dry weight and leaf area of *Helianthus annuus* growing at the Cambridge University Botanic Garden during a week at the end of July, 1961–67. Simultaneous harvests of two age groups, arranged in order of increasing mean increment in total dry weight per plant, as in Table 21.5. Age ranges in days from sowing.

Year	Age range (days)	Initial dry weight (g)	Initial leaf area (cm²)	Mean relative growth rates, week⁻¹		
				Dry weight ($\mathbf{\bar{R}}$)	Leaf area ($\mathbf{\bar{R}}_L$)	$\dfrac{\mathbf{\bar{R}}}{\mathbf{\bar{R}}_L}$
1966	25–32	1·568	256	1·056	1·030	1·025
1965	25–32	2·017	367	1·060	1·004	1·056
1962	22–29	3·324	639	1·207	1·015	1·189
1966	39–46	9·659	1,432	0·886	0·656	1·351
1965	39–46	16·811	2,395	0·623	0·624	1·00
1961	22–29	4·837	842	1·433	1·322	1·084
1964	22–29	4·939	778	1·421	1·349	1·053
1963	22–29	5·391	845	1·569	1·267	1·238
1964	36–43	12·056	1,585	0·998	0·903	1·105
1962	36–43	47·559	6,204	0·963	0·781	1·232
1967	25–32	45·420	6,508	1·057	1·058	1·00
1963	36–43	61·690	6,820	1·083	0·638	1·697
1967	39–46	233·08	25,710	0·615	0·464	1·325
1961	40–47	240·13	22,450	0·834	0·481	1·734

variations are seen in Table 22.4, which is based on the same experiments on the growth of sunflowers in the Cambridge University Botanic Garden, in the years 1961–67, which we saw in Table 21.5. The order of entries is the same as in Table 21.5, in increasing increment of dry weight during the experimental week. The range of both relative growth rates is seen to be considerable; for dry weight from 0·62 to 1·43, for leaf area from 0·46 to 1·35 week^{-1}. The tendency of both relative growth rates to fall with age and increasing plant size is at first sight obscured by the year to year variation, but a comparison of the two age groups in each year shows the older one to be lower in every one of the fourteen comparisons. The ratio of the two relative growth rates shows two cases of equality, corresponding to the simple case of $n = 1$ which we have just discussed. Otherwise relative growth rate in dry weight is always higher than in leaf area, although in five of the twelve cases the difference does not exceed 10 per cent. For computing values of $\bar{E}_n$ we can take n as effectively equal to 1 in half the cases, as the errors involved in the deviation of less than 10 per cent would be inconsiderable (the worst case is the 22–29 day plants in 1961, where the error involved in assuming $n = 1$ would be 1·1 per cent. For both sets of plants in 1964 it would be 0·7 per cent. The other departures are smaller).

22.5 Comparison with Briggs, Kidd and West's experiment

It is interesting to notice that there is no general tendency for the establishment of a phase where $n \simeq 1{\cdot}25$, as happened for the Briggs, Kidd and West observations on sunflowers in 1920 [16.13, and Fig. 16.13.1]. It will be recalled that we had concluded [16.15] that this particular growth pattern was a product of a particular year, by comparing the figures for 1920 with corresponding data for the 9 years 1952–60 (Table 7.5). The figures of Table 22.4 for the years 1961–67 confirm this. There are five instances within 10 per cent of 1·25; 1·19, 1·23, 1·24, 1·32 and 1·35, the first two in 1962 and the others in 1963, 1967 and 1966; in the last 3 years the other age group gave quite a different ratio, 1·70, 1·00 and 1·02. Thus it seems clear that the values of 1920, when for four successive weeks n was close to 1·25, were the product of the particular weather of that year acting on a system in which the ratio of relative growth rates shows a general tendency to rise. On the other hand, of our 14 ratios the highest is 1·73. As we have seen, although when the ratio rises it becomes an increasingly untrustworthy guide to the value of **n**, nevertheless it is clear that **n** has not yet reached 2 in the largest and oldest plants. This is in accord with the Briggs, Kidd and West data, where **n** did not reach 2 before harvest 9, 2 weeks beyond the age of our oldest plants.

22.6   Growth rates and plant form

In view of the substantial range of both relative growth rates it is interesting that we should not encounter a value for the ratio of the two of less than 1·00; that a value of 1·00 should be relatively uncommon; and that there should then be a complete gradation of values, 1·02, 1·05, 1·06, 1·08, 1·10, 1·19, 1·23, and so on. The simplest explanation for a value of 1·00 would be the simple case which we considered in Chapter 2, when the plant grows by reproducing identical units of plant form. But as we then saw, the continuance of this process in a plant with a vertical stem, like *Helianthus annuus*, inevitably implies in time the creation of a scaffolding holding the units apart; and when the creation of this begins the relative leaf growth rate will tend to fall below the relative growth rate in dry weight—a larger and larger proportion of the new increments in dry weight will be diverted to the creation of the scaffolding [Figs. 18.2.1, 19.3b]. It seems that this is the process exemplified in Table 22.4; in half the cases it has not gone very far, and the ratio of growth rates has not risen above 1·10, but the process is already beginning in a small way, and this may well account for there being no values below 1·00. *A priori* we might have expected a number of values scattered about 1·00, including values below 1 representing weeks particularly favourable to leaf expansion, when specific leaf area was high; because we are dealing with relative growth rate in leaf *area*, and the value of R_L will be just as much a function of leaf expansion as of the distribution of new dry matter to the leaves. If, however, the 7 years in Table 22.4 are representative, such conditions must be unusual in this climate, and furthermore Fig. 19.3a showed that for plants of *H. annuus*, in the age range of Table 22.4, week-to-week variations in specific leaf area were relatively small.

22.7   Useful in three ways

Thus we see firstly, that relative leaf growth rate has an intrinsic interest as an index of growth in leafiness. Secondly, when compared with relative growth in dry weight, it can be a useful aid in the computation of unit leaf rate. Thirdly, the same comparisons offer a useful guide to the onset of changes in plant form. However, like relative growth rate, relative leaf growth rate is related to the environment in very complex ways, and the effects of the natural environment upon it can only be elucidated by methods similar to those which we have used for relative growth rate in dry weight.

22.8   The 'photosynthetic entity'

At this point we must consider the interesting discussion by Whitehead and Myerscough (1962) of the question of the growth of plants in terms of the reproduction of identical units of form, which they called 'photosynthetic entities'. They introduced the notion of 'surplus' production of dry weight, over and above that 'used up in maintaining the proportions of the plants as a photosynthetic entity', and pointed out that 'this "surplus" dry weight is what is required to produce flowers, fruits, organs of vegetative reproduction and perennation, etc.'

If the mean total plant dry weights at successive harvests are $_1W$, $_2W$, $_3W$, ... $_xW$; and the corresponding leaf areas are $_1L_A$, $_2L_A$, $_3L_A$, ... $_xL_A$, they then computed α (equivalent to our $\mathbf{n}$), at the time of the xth harvest as

$$_x\mathbf{n} = \frac{_x\bar{R}}{_x\bar{R}_L} \qquad \text{(in our notation, compare 22.3)}$$

$$= \frac{\log_e {_xW} - \log_e {_1W}}{\log_e {_xL_A} - \log_e {_1L_A}}. \qquad 22.8.1$$

Their equation for the 'surplus' weight, S, was

$$S = (\mathbf{n} - 1)\,W. \qquad 22.8.2$$

At the xth harvest we should thus have

$$_xS = {_xW}(_x\mathbf{n} - 1),$$

and substituting for $_x\mathbf{n}$ (22.8.1),

$$\frac{_xS}{_xW} = \frac{\log_e\left(\dfrac{_xW}{_1W}\right)}{\log_e\left(\dfrac{_xL_A}{_1L_A}\right)} - 1. \qquad 22.8.3$$

22.8

Thus at harvest x, $_xS/_xW$, the 'surplus' fraction of the total dry weight, would be a function of the ratio of the logarithms of the proportional increases in dry weight and leaf area over those at harvest 1.

But let us assume that at harvest 1 we have a photosynthetic entity represented by $_1W$, $_1L_A$. Then, at the xth harvest, assuming (as White-head and Myerscough do) that this photosynthetic entity is to be re-produced both as regards leaf area and dry weight, the leaf area $_xL_A$ will correspond to a number $_xL_A/_1L_A$ of 'photosynthetic entities', and these will have a dry weight of $_1W \cdot _xL_A/_1L_A$. The 'surplus' over and above that 'required to maintain the photosynthetic entity' would then be given by

$$_xS = _xW - _1W \cdot \frac{_xL_A}{_1L_A},$$

$$= _xW\left(1 - \frac{_1W}{_xW} \cdot \frac{_xL_A}{_1L_A}\right);$$

and the surplus fraction,

$$\frac{_xS}{_xW} = 1 - \frac{\dfrac{_xL_A}{_1L_A}}{\dfrac{_xW}{_1W}}. \tag{22.8.4}$$

It would thus be a function of the ratio of the proportional increases in leaf area and dry weight, rather than of the ratio of their logarithms.

At the same time there are several difficulties in the way of using the developing leaf area of a plant as a guide to the development of 'photosynthetic entities' as the plant grows older. First is the possibility that there may be complex changes in specific leaf area associated with changing environment and developing ontogeny [Fig. 19.4.2, 19.5–19.8]. Secondly, even when, as in *Helianthus annuus* [Fig. 19.3a], there is an extended phase when SLA changes relatively little, nevertheless how much of the 'surplus' is available to produce flowers, fruits, organs of vegetative reproduction and perennation, and so on, may depend on how much is needed to sustain the developing photosynthetic part of the plant, in other words on what we have been calling plant architecture [22.6 and Figs. 18.2, 19.3b]. Thirdly, as we shall see in Chapter 23, it is often not practicable to distinguish the source, in the vegetative plant body, from which the material used in the developing reproductive structures is derived. The plant is functioning as a whole, and when this

22.8

whole enters the reproductive phase the vegetative progressions, themselves complex, may undergo a variety of changes.

The idea of the 'photosynthetic entity', and the trains of reasoning connected with it, thus raise important questions, whose answers may well have to wait on an accumulation of favourable instances, as has been the case for so many of the questions we have already examined. One possible favourable instance would be offered by a plant with a relatively stable specific leaf area (like the sunflower over most of its life cycle, Fig. 19.3a), combined with a relatively stable leaf weight ratio (like cotton during its vegetative phase, Fig. 18.3), and other favourable combinations of characteristics could be suggested. It would also be possible to have a series of values of W and L_A which would produce a simple and regular relationship between the expressions of equations 22.8.3 and 22.8.4, and something of the kind may have happened in the figures for *Epilobium roseum* examined by Whitehead and Myerscough, where it was possible to demonstrate a linear relationship between 'surplus' dry weight and the number of capsules produced, which was highly significant statistically. Yet despite this tantalizing glimpse, these important ideas remain difficult to handle in terms of attributes of the plant which can readily be measured, and it seems that at the present state of our knowledge we must be content in most cases to study the distribution of dry matter to the reproductive structures directly.

22.8

REPRODUCTIVE STRUCTURES

23.1 Asexual reproduction

For reasons which we have discussed, no individual higher plant can
continue vegetative growth indefinitely; there must be reproductive
structures, by which the individual passes on its basic organization to
other individuals. If reproduction be by a sexual or pseudosexual process,
it is a relatively simple matter to distinguish the offspring from the parent;
with asexual reproduction, it may be much more difficult to decide
the precise moment at which one individual becomes two, especially
when organic connection is maintained for a long time, as happens for
example in bracken, *Pteridium aquilinum*. At times, with asexual reproduction, there may be the minimum of specialized organs, or diversion of
material from the normal pattern of vegetative growth to the production
of specialized 'reproductive structures'. Bracken is a good case in point.
It has an underground, branching stem, bearing adventitious roots along
its whole length; only the leaves are above ground. Formally, growth is
similar to that of the vast majority of higher plants, consisting in extension
of the stem, which branches and produces leaves as it grows. However,
the fact that the stem is underground, and is growing parallel to the soil
surface, creates a situation in which mechanically every leaf is identically
supported; there is no necessity for an elaborate stem structure bridging
the gap between the individual photosynthetic leaf and the ground, with
all that this implies in mechanical and physiological relations. In consequence there is no formal barrier to the indefinite onward growth of an
individual stem, and Watt has measured them up to 21 m in length.
After about 50 years the stem dies, and this gradual movement of the
living part of the plant through the ground inevitably produces many
individuals out of the branches of one, without there ever having been
any specifically 'reproductive structures' involved. In this way, what
started as a single individual, from a spore, can soon cover large areas
and ultimately become many individuals. A slightly more complex case

is provided by the blackberry, *Rubus* spp, where the arching branches become stolons and root when they reach the ground. Here there is a definite change of structure, thickening, loss of chlorophyll, production of adventitious roots, but the amount of plant material involved is very small compared with that comprised in the normal vegetative structure of the particular branch involved.

In other cases the organs of vegetative reproduction may absorb a large fraction of the total production of dry matter during a growing season. They are also often organs of storage and perennation. Some of these—potatoes in temperate regions, yams and cassava in the tropics— are important food crops, and the potato in particular has been extensively studied. There exist a wide range of intermediate forms.

23.2   Flowering and production of seed

A similar range is found in the case of seeds, although here obviously the seed must be a separate entity and cannot be an integral part of the vegetative plant body. Some plants continue vegetative growth for many years, and attain a great size, before producing seed. The seed may then be only a small fraction of the total annual production of dry matter, or it may be substantial. When it has begun to flower and produce seed, the plant may do so several times a year, as in the relatively non-seasonal wet tropics. Here flowering may occupy a regular place in the flushing cycle of plants growing in flushes. In those plants where the flushing cycles of different branches become out of phase, the different branches also flower and fruit at different times, in harmony with their individual cycles. *Peltophorum pterocarpum* affords a striking example in some parts (but not at all its sites) in Malaysia (T. C. Whitmore, personal communication), another, more widespread but less spectacular, is the mango, *Mangifera indica*; although trees of the latter species growing in a relatively non-seasonal climate usually flush the whole crown of leaves at the same time, it is not uncommon to observe individual limbs flowering and fruiting independently of each other.

In more seasonal climates flowering is usually reduced to once a year, but many long-lived plants vary very much in the amount of flowering and fruiting in individual years. Some have a 2-year cycle, as in many cultivated apples (the so-called 'biennial croppers', such as Blenheim Orange), and the apples are also a good example, in that the inherent cycle is readily modified by the weather of an individual year, which tends to bring all the biennial croppers in an individual orchard into phase, irrespective of variety and age of tree. Other long-lived plants flower much less frequently, and in some climates some plants will grow well vegetatively, but flower very rarely, or not at all, a good example in England being the Jerusalem artichoke, *Helianthus tuberosus*.

The problems involved in the apparently simple question of why a particular plant comes into flower are many and various, and an enormous literature has grown up around the subject. It is conveniently entered through reviews by Lang (1965), Searle (1965), Chailakhyan (1968), Hillman (1969), Mohr (1969) and Siegelman (1969), the last three collected together in Wilkins (1969). The impossibility of a short treatment is sufficiently indicated by the fact that the six references just cited would themselves make a book of 300 pages. Yet much of the work done so far has been concentrated on a limited number of relatively simple cases—inevitably so, since progress would never have been made with problems of such complexity by any other means. Consequently, as with seed germination [5.7–5.9], the investigator beginning work on a hitherto unstudied species must be prepared for problems connected with flowering behaviour which may be complex and difficult to tackle, and which will not necessarily fall into any of the well-recognized categories. For this reason, in addition to dealing with some results from species where the mechanisms involved have been much studied and are relatively well known (e.g. *Chrysanthemum morifolium*, 23.8–23.9) we shall notice more unusual instances (e.g. *Impatiens parviflora* 27.8–27.9) and cases where behaviour is complex and the underlying mechanisms far from clear (e.g. *Phaseolus vulgaris*, 23.10). For many species the sheer difficulties of experimentation impede the progress of investigation, and important among these are many trees of wet tropical forest, which appear to be very erratic in flowering and producing seed. For the great majority of these virtually nothing is known about the factors which bring an individual tree into flower. The phenomenon is, however, so common as to play an important part in Poore's theory of tropical forest mosaic (Poore, 1967, 1968).

23.3 Monocarpy: annuals and biennials

Many long-lived plants, having begun to flower and produce seed, continue to do so for an indefinite number of years. Others are monocarpic

[13.22]. They flower only once, set seed, and die. A number of species of bamboo afford examples which are spectacular, because of the additional habit of simultaneous flowering. This can have profound effects on the whole ecosystem, and, combined with their economic utility, also have important repercussions for man. In places where life is very dependent on the bamboo much hardship may be caused by the simultaneous disappearance of groves over a wide area. On the other hand, famine may be alleviated by the massive seed production; yet this latter may induce serious plagues of rats. Indeed, changes in the populations of seed-eating animals may be an important factor in keeping the members of such a simultaneous-flowering species in phase. For all these reasons an extensive literature has grown up on the flowering of bamboos (conveniently summarized by Arber, 1934, who after ten pages concludes 'This brief summary ... shows that the problem is a most intricate one'). The behaviour of the climbing bamboo *Chusquea abietifolia* affords an interesting and well-documented example. 'Apparently every plant in the island' of Jamaica flowered, set seed and died during a period of rather over a year, starting in the autumn of 1884 (Morris, 1886). Plants transferred from Jamaica to Kew in a Wardian case behaved in exactly the same way, at the same time (Hooker, 1885). In 1886 Jamaica was thus refurnished with an even-aged population from seed. Nearly all of these in turn flowered, set seed, and died between 1916 and 1918 (Seifriz, 1920), indicating a vegetative cycle of just over 30 years, not exceptionally long for bamboos. The mechanism underlying these phenomena is still largely obscure, but it is clear that in many bamboos a prolonged, but limited period of vegetative growth is necessary before the flowering phase can begin. *Raphia monbuttorum* behaves in a similar way in Kenya and Uganda: a fine palm, up to 10 m or more in height, with leaves 7–8 m long, it is monoecious and produces a large pendulous terminal inflorescence and then dies, the majority of the adult trees in a district flowering and dying in the same year (Eggeling and Dale, 1951; Dale and Greenway, 1961). Malaysian palms of the genus *Arenga* afford a complementary example. *A. retroflexens* and *A. westerhoutii* have the habit common to so many trees of starting to flower when young and thereafter continuing to flower and fruit at intervals throughout life. On the other hand *A. pinnata* is monocarpic. It grows for 7–10 years and attains a height of several metres before beginning to flower, nearly always at the apex. For more than a year thereafter, flowering and fruiting spread basipetally

23.3

down the stem to the ground, when the plant dies, having burnt down like a candle (Burkill, 1935; J. Dransfield and T. C. Whitmore, personal communication).

In the British Isles one is chiefly familiar with the monocarpic habit in the less spectacular cases of annuals and biennials, but even here the situation is complex. The majority of annuals are truly monocarpic, and die after setting seed; but some, including many garden plants are only functionally monocarpic. They die because of the onset of winter, not because of a pattern of growth inherent in the plant, and in a suitable climate can grow as short-lived perennials, examples being *Tropaeolum majus* and *Cobaea scandens*. The same is true of many biennials; among garden plants some short-lived perennials are normally grown as biennials; but most biennials require the stimulus of the annual climatic cycle before they will flower (although in some, but not all cases, they can be brought into flower by the application of hormones). The responses involved are many and various—some require a long day, some a short; some require to be frosted, and so on.

The biennial habit is also commonly found in the individual branches of long-lived plants. The obvious examples are all those garden plants which are said to flower 'on last year's wood'. In some of these cases what is involved is a year's delay before the establishment of perennial flowering, often involving short-shoot systems (see section 2.3), as in some varieties of apple. In other cases the branches are monocarpic, and die after setting seed, or there may be partial die back, as in some species of *Rubus*.

If we consider annuals only, we are still faced with a great variety of behaviour. Some annuals are always small and short lived, as, for example, *Cicendia filiformis* or *Ionopsidium acaule*. Others, as *Euphorbia peplus*, can be small and short-lived, producing a flower and seed on two to four leaves, if conditions are unfavourable; or, under favourable conditions, they can grow a hundred times larger, flowering and seeding all the time. Others, as *Helianthus annuus*, require to make a plant of much larger minimum size before they will flower at all, although even here a large size range is possible (Clements *et al.*, 1929, Pls 23, 24, Tab. 50). Some flower as long as growing conditions are favourable; others flower with great regularity at a particular time of year, and there may be a variety of reasons for this. In short, a very great range of behaviour is possible (for further examples see Salisbury, 1942), and of the very numerous types,

23.3

only a few have as yet been subjected to quantitative study in any detail.

23.4 Mobilization efficiency

For annual plants, an idea similar to that underlying Noll's Substanz-quotient, but extended from germination over the whole or part of the life cycle, can provide useful information. It gives the number of multiples of the original seed weight which the plant has produced during the specified period, which obviously sets an upper limit to seed production. When the actual seed production during the same period is compared with this, we have a measure of the efficiency with which the plant has mobilized its available resources in subserving the function of reproduction, provided that there has been no systematic change in seed size from one generation to the next. The ratio can be called the 'mean mobilization efficiency' over the life cycle up to the point considered: and by extension the perennating organs of perennials could be dealt with in a similar way. It is also possible to derive an instantaneous or short period value of the mobilization efficiency by considering, at a particular time, the rate of increase of dry weight of seeds or other reproductive bodies of a plant as a fraction of the rate of increase of total dry matter. This instantaneous value can easily be greater than 1, although the mean mobilization efficiency from germination onwards must be less than 1. These instantaneous or short period values are useful when dealing with plants where seed production is going on for most or nearly all of the vegetative life, and in such cases use can conveniently be made of the techniques of frequent small harvests [23.7]. These concepts have obvious affinities with the 'migration coefficient' of Beaven (1947) (which, however, neglected the roots, and was calculated on a basis of total top dry weight) and the 'harvest index' of Donald (1962), which relates economic yield of a crop to total dry matter at harvest. It will have been noticed that in general we have endeavoured to avoid the introduction of new terms, but for ecological purposes it seems that 'mobilization efficiency' will be more useful here. It is a more neutral term than 'migration coefficient' (in some plants substantial contributions to dry matter production are made by the reproductive structures themselves, so that 'migration' might here be thought to be a misnomer); it can be used equally for reproductive bodies of all kinds; and although at the end of the life cycle of monocarpic annuals its value may be identical with Donald's 'harvest index', it is more readily adapted to a variety of ecological situations, such as daily seed production rates by plants with a long fruiting season, and so on. It has the added advantage of facilitating a distinction between the seeds and other reproductive bodies themselves, and the reproductive organs and associated structures of the parent plant. However essential these latter may be in the formation and development of the seed, etc., they often play no part in dispersal and make no direct contribution to the nutrition or establishment of the next generation. Here individual cases must be considered on their merits: for example, when considering indehiscent single-seeded fruits dispersed as a whole it might be pedantic to insist on a distinction between the fruit and the seed coat, and so on. Considerations of this kind make it possible to link the concept of mobilization efficiency with studies of partition of dry weight which we shall consider in the next section. These concepts will also link up with Whitehead and Myerscough's idea of 'surplus' dry weight, when it becomes possible to overcome the difficulties in the way of estimating it in terms of the material available over and above that needed to maintain the 'photosynthetic entities' [22.8].

For plants with small seeds the substance quotient over a whole growing season can reach large figures: for example, Shamsi (1970) gives an example of plants of *Epilobium hirsutum* in their first year of growth which had multiplied the original seed dry weight by a factor of around 500,000 by August, at which time they had each already produced on average 78,000 seeds (average dry weight 0·18 mg). At this stage the mean mobilization efficiency on account of seed was about 15 per cent, while the formation of vegetative perennating organs had already begun, and would continue. *Lythrum salicaria* gave similar figures, with an even higher SQ and production of a larger number of smaller seeds (average dry weight 0·06 mg). These plants were standing well spaced out, but density of planting is no doubt among the numerous factors which may affect the mobilization efficiency. Under certain circumstances, however, it may be relatively constant over a wide range of plant size, as illustrated in Table 23.4, giving some results from a spacing experiment on barley. Here although the range of mean shoot dry weight per plant extended from 1·22 to 17·8, and the mean number of seeds per plant from 16 to 189, the ratio of seed weight to shoot weight ranged only from 0·47 to 0·51. We also notice that for the three densest plantings both the mean shoot

23.4

dry weight and the mean dry weight of seed per square metre of ground surface are practically constant.

Table 23.4. Effect of spacing of barley plants (cv. Maris Concord), grown in the field during 1967, at the Plant Breeding Institute, Trumpington, Cambridge, on grain yield and dry weight of shoot, excluding root (data of E. J. M. Kirby, personal communication).

(a) Number of plants per square metre	800	400	200	100	50
(b) Mean number of seeds per plant	16	32	62	112	189
(c) Mean dry weight of seed per plant, g	0·59	1·25	2·55	4·78	8·44
(d) Mean shoot dry weight per plant, g	1·22	2·51	5·01	9·95	17·8
(e) c/d	0·48	0·50	0·51	0·48	0·47

23.5 Partition of dry weight

In plants where seed production starts early, is continuous and is linked with continuing vegetative growth, the plant may achieve a more or less steady state of partition of material between reproductive structures and the rest of the plant. In other cases, however, there may be a great change in the pattern of growth, and the bulk of new assimilates may be switched to reproductive structures from some other organ, which may be root, stem, or leaf, any of which may accordingly suffer a marked check to growth.

The magnitude of this switch can be illustrated from Briggs, Kidd and West's data on *Helianthus annuus*. Fig. 23.5.1 shows the distribution between root, stem, leaf, and reproductive structures in relation to the age of the plant. We see that the fraction going to the terminal capitula starts as a small amount around 57 days (↑), and that from 90 days onwards the fraction increases rapidly. At the same time the changing pattern of distribution among the other plant organs [Figs. 18.2, 21.4], established before the reproductive structures appear, makes it very difficult to say that the bulk of the new material going to the capitula had been diverted from any specific organ.

In other instances we have clearer indications; in section 18.3 we saw that in the case of cotton the data of Inamdar *et al.* (1924) showed a comparatively steady leaf weight ratio up to an age of 12 weeks, when the dry weight of flowers begins to become significant. Fig. 23.5.2 gives

23.5

the relative distribution of plant material between the various types of
organ from 7 to 17 weeks, as a function of total dry weight. We see that
more than two-thirds of the final dry weight has been formed before flowering begins, most of it during a period when stems plus roots form an
almost constant fraction of the total, slowly increasing from 47 to 50 per

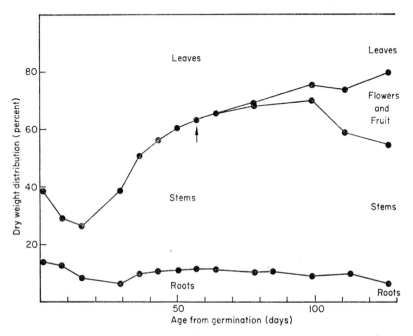

Fig. 23.5.1. Relative distribution (in per cent) of mean total plant dry
weight between the various categories of organs, in relation to age, for
plants of *Helianthus annuus* grown by Briggs, Kidd and West at the Botanic
Garden, Cambridge, in 1920. ↑, First appearance of terminal capitula.
For discussion of changes in leaf weight ratio, see 18.4, 19.3; for stem and
root weight ratios, see 21.4.

cent. Thereafter the proportion of total dry weight in stems and roots
continues its slow increase, finally reaching 55 per cent at 17 weeks.
Here, during the period of formation of flowers and fruit, the reproductive
organs together with the leaves form an almost constant proportion. We
cannot conclude that there has been a simple switch of material between
the two categories of organ, but the analysis forms a useful basis for
further investigations into the mechanisms involved.

23.5

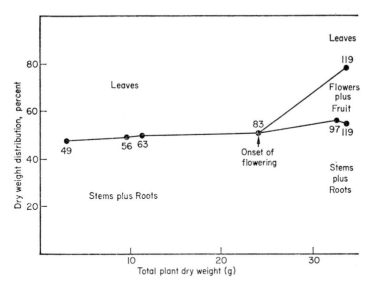

Fig. 23.5.2. Relative distribution of dry matter between leaves, reproductive structures and the rest of the plant for the latter half of the life cycle of cotton (*Gossypium arboreum*) grown at Benares. Figures beside points give ages in days from germination on 14 May 1923. Points for days 83 and 97 'represent both the reproductive organs and the leaves, which were dried together by mistake' [cf. 11.6]. Plotted from data given by Inamdar, Singh and Pande (1925). See also Fig. 18.2.2.

23.6   Absolute and relative increments

Such clear indications are, however, not common, and when investigating a new species we may well be confronted by a more complex situation, as in the sunflower data of Fig. 23.5.1. It is then worth examining two other sets of relationships. We first consider the absolute amounts of dry weight in flowers and fruit, in comparison with the other organs of the plant [Fig. 23.6a]. We saw in Fig. 23.5.1 that the relative amounts of total dry weight in the roots was declining from about day 60, and in the stems in a much more marked degree from about day 100: yet Fig. 23.6a shows that the roots were still increasing in total dry weight up to day 113, and the stems continue to increase right up to the end.

Further information is provided by plotting the distribution of the dry weight increment between one harvest and the next, as in Fig. 23.6b. We then see that between days 57 and 99 a roughly constant fraction of

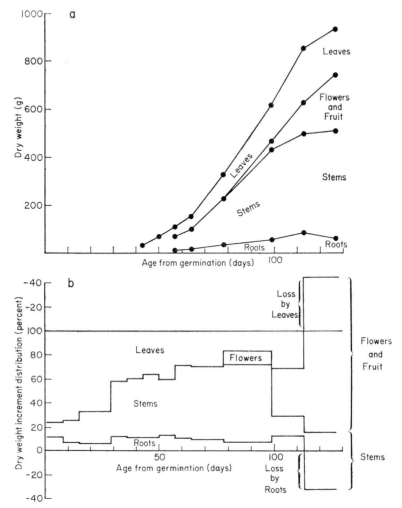

Fig. 23.6. (a) Distribution of mean dry weight at harvest between the various categories of organs, in relation to age, for plants of *Helianthus annuus* grown by Briggs, Kidd and West, at the Botanic Garden, Cambridge in 1920 (compare Figs. 19.3a and 23.5.1); (b) for the same plants, the relative distribution (in per cent) between organs of the increment in dry weight accruing between one harvest and the next, also in relation to age. Note that the two parallel lines for 0 and 100 mark the limits of positive dry weight increments by the plant as a whole. Losses by roots are shown below 0; losses by leaves above 100, on a reversed scale. In this way the very large gain by flowers and fruit, of more than 100 per cent of the total increment *by the plant*, is shown directly.

23.6

about 70 per cent of the increment is going to stems plus roots, the remainder at first to leaves, and then to leaves plus flowers. Only around day 100 is there a sudden, marked change in the pattern of distribution of new assimilates, the fraction going to stems falling to less than a third of what it had been, and remaining low but positive thereafter, while in the last fortnight there are substantial losses in weight by both leaves and roots. When considering such changes we must remember that a loss of weight by a non-assimilatory organ such as the root may be due mainly to respiration, and need not necessarily imply translocation away to other organs. We must also remember that in a case such as the sunflower, where a large quantity of fatty seeds are being formed, there may be a substantial loss of dry weight due to changes in the chemical composition of the plant [App. 1.12]. Here the metabolic balance term (D, of 13.11) becomes of importance, and allowance must be made for it in assessing the amounts of movement of carbon compounds from one part of the plant to another. Thus to elucidate the mechanisms underlying the partition of dry weight may require a complex investigation. A powerful tool for this purpose is provided by supplying some or all of the assimilatory organs for a short period with $^{14}CO_2$, and tracking the destinations of the labelled assimilates (see e.g., Thaine *et al.*, 1959; Stoy, 1963; Webb and Gorham, 1964; Lupton, 1966; Ryle, 1970a). Such investigations are, however, exceedingly time-consuming (Ryle, 1970b), and overall growth studies of the types examined in the last two sections help to focus attention on those organs and stages of development most worth detailed investigation (for further discussion of the utility and limitations of these methods see 32.8).

23.7   Information from frequent small harvests

It should be noted that if use has been made of the methods of continuous harvesting previously discussed [20.4, 20.5], histograms such as those of Fig. 23.6b can be replaced by smooth curves. In particular, the curve fitting programme of Hughes and Freeman (1967) is very readily modified to produce curves of distribution of dry weight increment, as pointed out by Hughes and Cockshull (1969). Fig. 23.7 shows an example of such a treatment, from some experiments on *Callistephus chinensis* which we shall discuss later [24.12, 24.13]. The fractions of the total daily increment of the plant as a whole going to the different categories of organ is shown

23.7

in relation to total plant dry weight plotted on a logarithmic scale, for three experiments carried out with different concentrations of atmospheric carbon dioxide. The main features of these curves are broadly similar to those of Fig. 23.6b, allowing for the different mode of presentation, and it is not surprising that this should be so in another member

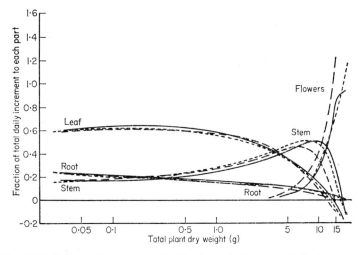

Fig. 23.7. Information on the distribution of the total daily increment in dry weight by plants of *Callistephus chinensis* in relation to the logarithm of total plant dry weight, derived by fitting curves to the data from frequent small harvests. Plants cultured in growth cabinets at three levels of atmospheric carbon dioxide concentration, 325 (— —), 600 (------) and 900 (———) parts per million. From Hughes and Cockshull (1969).

of the same family of Compositae, also having large terminal capitula. The differences in root behaviour between the two species may well be accounted for partly by cultural conditions, which as we have seen can cause wide variations in the proportion of root in *Helianthus annuus* [21.5], and partly by the fact that the observations on *H. annuus* were carried through to ripening of the seed, while those on *C. chinensis* were terminated when the plants were in full flower. The most marked difference between the two species appears to be that in *C. chinensis* there was in the last 10 days of the observations a marked loss in weight by the stems, whereas the stems of *H. annuus* showed a gain in every period.

23.7

23.8 Plant size and plant organization

As seen in our earlier short survey [23.2, 23.3], whether a plant flowers at all, and the extent to which it flowers, may be affected by the age of the plant, and its past and present environments, in a great variety of ways. It is obvious, however, that if the plant is to flower and set seed, the maximum possible performance, within which all the other factors must act, is set by the size of the plant and its basic organization. Thus in *Helianthus annuus*, where the main stem normally terminates in a single capitulum, the maximum possible size for this is set by the size of the plant. Granted that most of the dry matter of which the capitulum is composed was produced by the plant as assimilates during the period when the capitulum was growing, nevertheless the possible maximum production of new assimilates during this period was set by the size of the plant at the beginning; and hence was very much affected by growth conditions from the seed up to this point.

The point has been clearly made by Cockshull and Hughes (1967, 1968) in experiments on the growth and flowering of *Chrysanthemum morifolium* (the two commercial varieties Golden Princess Anne and Bright Golden Anne). Fig. 23.8 shows the results of two experiments carried out in growth cabinets, one a 3×3 factorial design with atmospheric carbon dioxide concentrations of 300, 600 and 900 parts per million and 8-hour days with total visible radiation of 30, 60 and 90 cal cm^{-2} day^{-1}. The second experiment repeated this, and added a further carbon dioxide concentration of 1500 ppm at all three light levels. The plants were grown from uniform batches of cuttings, all of which had been rooted in long days. The short day treatment, bringing about flower initiation, began on their transfer to the stipulated conditions in the cabinets.

In Fig. 23.8 the flower weight ratios for the different treatments are plotted against the stage of flower development (for a detailed description of these stages see Cockshull and Hughes, 1971), and it will be seen that at a given stage of development all treatments show much the same flower weight ratio. This happened in spite of considerable differences both in the length of time taken for the plants to come into flower, and in their dry weights when they did so.

In all cases we see that at stage 8 of the flower development, when the flower is open, the flower weight ratio is very close to 0·4. We have

23.8

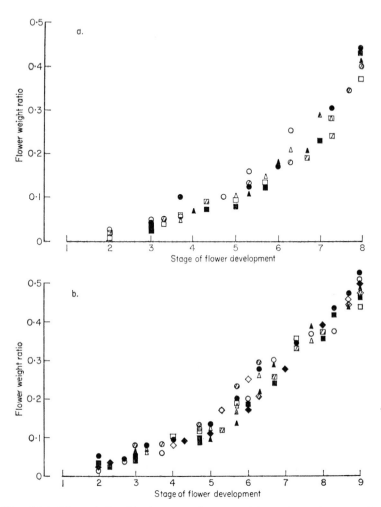

Fig. 23.8. The relationship between flower weight ratio and stage of flower development in *Chrysanthemum morifolium* (commercial variety 'Golden Princess Anne'), in two sets of experiments in growth cabinets, (a) begun on 18 January 1966, (b) begun on 22 September 1966. Each point represents the mean for three plants. Circles, 300 parts per million atmospheric carbon dioxide; triangles, 600; squares, 900; diamonds, 1500. Open symbols, 30 cal cm^{-2} day^{-1}; hatched symbols, 60 cal cm^{-2} day^{-1}; solid symbols 90 cal cm^{-2} day^{-1}. From Cockshull and Hughes (1967).

23.8

here, then, not just a case where the size of the capitulum varies generally with the size of the plant, but a more precise mechanism ensuring that when the flowers open their combined dry weight shall be close to 40 per cent of that of the plant, irrespective of substantial differences in conditions of growth, and consequently of plant size.

23.9 Further observations on *Chrysanthemum*

Cockshull and Hughes made a number of other observations and experiments designed to shed light on this interesting quantitative mechanism. Fig. 23.9.1 shows the distribution of dry matter between roots, stems, leaves and flowers between 5 and 10 weeks from potting up, by which time the plants were in full flower. At 5 weeks the flower

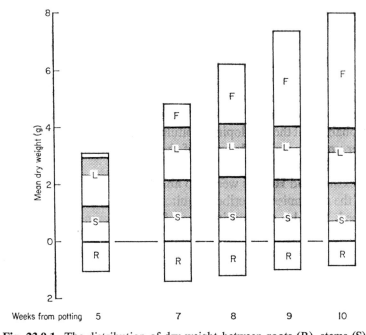

Fig. 23.9.1. The distribution of dry weight between roots (R), stems (S), leaves (L) and flowers (F) as a function of time in *Chrysanthemum mori-folium* (commercial variety 'Bright Golden Anne') cultured in growth cabinets. Final harvest, 26 July 1967. Means of five plants on each occasion. Shaded areas, branch; white areas, main axis. From Cockshull and Hughes (1968).

weight ratio is small, and the vegetative parts are still growing. At 7 weeks the plants are between stages 3 and 4 of flower development, and the flower weight ratio has risen to about 10 per cent. Thereafter changes in the dry weights of the vegetative parts are very small (that of the roots actually falls) and virtually the whole of the net production of dry matter goes to the developing capitula. It seems that downward translocation is barely keeping pace with respiratory loss.

In other experiments they showed that continued accumulation of dry matter did not depend on floral sinks (the development of alternative sinks, such as meristems producing side branches, had been prevented in the flowering plants by removal of buds, following normal commercial practice). If the flowers were removed after $5\frac{1}{2}$ or 8 weeks the plants developed as shown in Fig. 23.9.2. More than half of the assimilates which would have gone to flowers were found in the roots, most of the remainder in the leaves, the stem weights being little affected. What is particularly interesting, however, is that by $5\frac{1}{2}$ weeks the leaves were all fully mature, so that the leaves of the non-flowering plants almost doubled in dry weight with hardly any increase in area. The plants whose flowers were removed at $5\frac{1}{2}$ weeks produced the highest dry weight total, adding the weights of flowers removed to the corresponding final plant weights. This does not necessarily imply any increase in total carbon assimilation. It is probable that the developing capitula had a relatively high respiration rate per unit dry weight, because of their relatively high proportion of meristematic tissue.

Cockshull and Hughes were also able to show that successful operation of the mechanism described requires not only the conditions already described, but also the continuance of an environment favourable to growth. By giving periods of low light (equivalent to dull midwinter days) at different periods during the 10 weeks or so that the plants were coming into flower they found that synchronization of flowering in all the branches of a plant was determined in the first week, and floret number (appropriate to plant size at the time) in the second, low light in the first week disturbing synchronization, and in the second causing a low floret number. Thereafter, up to the 8th week, 2 weeks of low light simply delayed development without any other obvious effect. With low light during the 8th and 9th week, on the other hand, opening of the outer florets proceeded normally, while the central florets of the capitulum died.

23.9

This is a good example of the complexity of the mechanisms controlling flowering, several subsidiary ones coming into play at various stages of development, in addition to the main daylength mechanism bringing

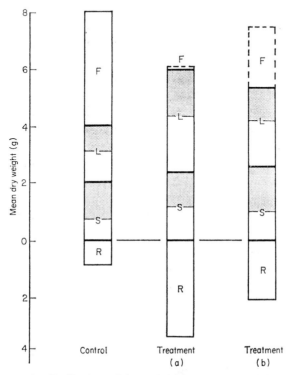

Fig. 23.9.2. The distribution of dry weight between roots (R), stems (S), leaves (L) and flowers (F) in *Chrysanthemum morifolium* (means of five plants in each case). Control represents the final harvest in Fig. 23.9.1, on 26 July 1967. Treatment (a): flowers removed 24 June 1967, plants harvested 27 July 1967. Treatment (b): flowers removed 14 July 1967, plants harvested 27 July 1967. The dotted lines enclose the weights of flower removed. From Cockshull and Hughes (1968).

about flowering, and the quantitative distribution mechanism determining the size of the capitula. Together they produce the kind of interlocking effects of past and present environment with the inherent ontogenetic drifts of the plant with which we have become familiar in our studies of vegetative growth.

23.9

23.10 Other examples: available reproductive sites

If the overall plant size, establishing the maximum material available for reproduction, is one side of the coin, then the availability of sites where reproduction can take place is the other side, and the two are inseparable in considering quantitative reproductive problems in many plants. Some instances are relatively simple—the data examined in 23.8 established quantitative relations between the dry weights of the fully opened flowers and the rest of the plant in *Chrysanthemum morifolium* (although even here we noticed in 23.9 complications concerning floret number); but if the growing plants had been kept in long days they would never have entered the flowering phase at all, and whatever the plant size became there would have been no sites available where flowering and seed production could take place. The references mentioned above [23.2] deal with many such instances.

Mechanisms of this kind can have important effects on plant distribution—for example, *Epilobium hirsutum*, which has a very wide temperate distribution including upland sites, may be excluded from otherwise suitable areas on tropical mountains by requiring a daylength for flowering longer than the tropical solar day provides (Shamsi, 1970); similarly plants which require a period of frosting at some point in the life cycle may be excluded from frost free areas, and so on; and such restrictions may affect only certain varieties within a species [23.11].

Such differences in behaviour in different environments are often qualitative and follow an all-or-none pattern. There are, however, important cases of quantitative differences, and these are frequently more complex and obscure. Barley affords an example of a growth habit common among annual grasses, where the number of inflorescences produced depends on the survival of 'tillers', shoots produced from the base of the stem. For isolated plants growing in very favourable environments these may be very numerous: under normal field conditions most of the young tillers produced die before they reach maturity. Under certain conditions of spring drought very many young tillers may die, and the final weight of seed produced after more favourable weather later in the season may be closely related to the number of surviving tillers, and hence the number of possible inflorescences (for a similar phenomenon in wheat, see Bingham, 1966). On the other hand, in other years under superficially similar

23.10

conditions there may be a large measure of compensation, the surviving ears containing both more and heavier seeds (O. Rackham, personal communication). The causes of these differences of behaviour are exceedingly complex and still obscure.

23.11 *Phaseolus vulgaris*

Complementary examples are provided by *Phaseolus vulgaris*, where recent work by Zehni (1969, and Zehni *et al.*, 1970) has shown how large quantitative differences between varieties of a single species may be, and how complex a situation investigation of these differences may uncover.

Flowering in many varieties of *Ph. vulgaris* is unaffected by day length, but it has long been known that there are sensitive varieties in which flowering can be delayed or suppressed by long days, and more recent work on the collection of bean varieties maintained by the School of Agriculture at Cambridge had already shown that in particular varieties from Colombia and Peru the initiation of flower buds is unaffected by day length, floral differentiation proceeding normally up to a certain stage, after which the flower buds begin to respond differently to long and short days (Ojehomon *et al.*, 1968). By dissection of developing flower buds Zehni showed that the first detectable sensitive stage is not always the same; it may cover a wide developmental range between a very early stage in which bracteole initials are distinguishable but not calyx or corolla, and a much later stage where the ten anthers are separately visible. In the subsequent development many other quantitative differences in response to daylength changes were observed; on the rate of development of individual flowers; on the rate and on the pattern of flower formation by the plant as a whole; on the behaviour of terminal and lateral inflorescences; and so on. Individual quantitative effects might range from a mere delay of a day or two in the opening of a flower, to cessation of development at an immature stage and abscission of the aborted bud. The degree of sensitivity of the developing flower buds to particular changes of conditions was found to vary considerably from one part of the plant to another, the first-formed flower buds being most sensitive, followed by all the buds of the terminal inflorescence, while the later-formed buds on lateral inflorescences were least sensitive. On occasion perfect flowers would develop from these last, when the

23.11

plant was growing in continuously controlled conditions which had already caused the abortion and abscission of the earliest formed buds.

Comparative studies on the effects of night breaks on different varieties showed effects broadly similar to those already well established for flower-bud initiation in other species (short days with a night break simulating the effects of long days), but the effects varied quantitatively with the duration of the break and its position in the night. Experiments on the interaction of long and short days given in different sequences showed that there is here no simple triggering action, such as has been observed in some other species, but a variety of quantitative differences. The role of the leaves in mediating the daylength effects was also investigated, and once again quantitative differences were found, a single leaflet of one of the first trifoliate leaves being more potent than the large primary pair of simple leaves, particularly in suppressing flowering. Finally, a series of experiments on the effects of applying growth substances showed that both auxins (IAA and NAA) and an anti-auxin (TIBA) all produced long-day effects, while a gibberellin (GA3) produced simultaneously a long-day effect on vegetative growth (longer internodes) and a short-day effect on flowering (earlier flowering and more flowers). However, none of these, even in very high concentrations, produced effects of the same order of magnitude as the long- and short-day treatments themselves.

When considering vegetative growth we have already become familiar with varying quantitative responses to the interlocking elements of genetics and ontogeny with past and present environments [7.1, 17.1]. These varieties of *Phaseolus vulgaris* provide an example of similar complexity in flowering behaviour.

23.12   Importance of the whole ontogeny

Putting these various examples together we conclude that, when viewing flowering and fruiting against the background of the plant of a particular species or variety and its environment, it is essential to consider whether it should not be related, both qualitatively and quantitatively, not merely to the size of the plant and to the environment just before and during flower and fruit formation, but also to much remoter influences far earlier in the plant's ontogeny, and to the accumulated effects of earlier environmental conditions upon it. We have considered some examples among annual herbs, but such remoter influences are particularly obvious in plants where the vegetative and flowering phases are separated by a gap, as in deciduous woody species which come into flower on otherwise bare wood, or in bulbous or tuberous plants which flower and produce leafage at different times of year. Even if there is no gap between vegetative and reproductive phases, in long-lived plants there are common and well-known examples of the effects of a previous season's growth in such phenomena as biennial cropping in apples; and some plants must be able to store such remote influences for longer periods, as in the example described earlier, when plants of *Chusquea abietifolia* originally from the same population flowered simultaneously in Jamaica and at Kew Gardens [23.3]. Clearly reproductive behaviour must be considered as part of the whole complex of the plant's life form (Raunkiaer, 1937), its ontogeny in relation to past and present environments, its longevity and mode of perennation. Even then we may be far from being able to consider questions of adaptive significance, which may involve many aspects of the community in which the plant grows.

Thus we conclude this chapter as we began, by stressing the very great diversity of reproductive behaviour among plants. Qualitatively much is already known about the morphological, anatomical and physiological mechanisms involved, although far more remains to be found out. But as yet quantitative studies have been relatively few, and it has been possible here to do little more than give a few samples of different modes of attack, which must be adjusted in relation to the particular species being studied.

23.12

DRY WEIGHT/FRESH WEIGHT RATIOS:
WATER CONTENT

24.1 A valuable anatomical index

The vacuolate nature of most plant cells makes the ratio of dry weight to fresh weight, at full turgor, a valuable anatomical index. The fresh weight at full turgor represents the whole system of the living plant organ; cell walls, protoplasts and vacuoles of living cells, walls and contents of dead cells, all in the normal, fully functional, state of the particular organ at that time. We have earlier discussed [11.19] the difficulties in the way of accurate determination of dry weights, and means of estimating dry weights in awkward cases. When these precautions have been taken, the dry weight gives us, within the limitations which we have also discussed [4.3], a measure of the quantity of reduced-carbon units which have gone to building up this total structure. Variations in the amounts of reduced-carbon units for a given fresh weight, in comparable plant organs, can yield interesting information on the conditions of growth.

The difficulty is that all plant cell walls are more or less extensible, leading to short-term variations in the amount of water accompanying a given amount of dry matter in the active plant. Some are very inelastic—Krasnoselsky-Maximov (1925) found that *Impatiens parviflora* may show a decrease of fresh weight of only 1–3 per cent between full turgor and wilting (these observations must have been made on a shade form), and in many observations on *Eupatorium adenophorum* Knight (1922) found decreases of less than 2 per cent. For such plants a standard state of turgor is easy to obtain; all one has to do is to avoid visible wilting. But these plants are exceptional. Mesophytes usually have much more elastic cell walls; Maximov (1929) recorded changes in fresh weight of mesophytes ranging up to about 30 per cent, in different species, between full turgor and wilting. His records of xerophytes (in suitable cases, when wilting could be observed) were even higher, between 30 and 40 per cent. An extreme instance of very low water contents in physiologically active plants was provided by the studies of Thoday (1921) on *Passerina filiformis* and *P.* cf. *falcifolia* growing in southern Africa, which showed that the turgor mechanism controlling inrolling of the leaves began to operate only when the water content of the leaves had fallen to between 37 and 35 per cent of the fresh weight, i.e. when the dry weight was roughly double the water content. This represented a decrease to around 50 per cent of the fully turgid water content. These differences between plants clearly have important influences both on the magnitude of the water deficit likely to be encountered in a particular case, and also on the physiological effects of a specific value of the deficit. Later work has added largely to the number of cases investigated, without changing this general picture (for entry to the extensive literature, see Walter (1960) II B and Slatyer (1967) chapter 9. These works also provide a general background. For a shorter account see Dainty (1969)).

24.2   Effects of water status

When dealing with any of these plants with extensible cell walls, the fresh weight at harvest will depend not just on the conditions of growth, but also on the water status at that moment. In a fluctuating climate this can vary widely from day to day and from hour to hour. Consequently for a long time many workers on plants growing in the field have doubted the values of fresh weight determinations; thus Hackenberg (1909), whose work on *Cannabis sativa* has already been discussed in other connections [11.19, 13.4] (I translate), 'In the determination of weight the fresh weight could not be used as a criterion because the water content of plants in different stages of development varied a great deal, and was dependent upon the daily weather conditions....' However, in many programmes of research, study of the water status of the plants is important. Then, if plants are to be grown in the field and harvested from there, and if determinations of both dry weight and fresh weight are important parts of the programme of observations, it may be necessary to grow spare plants, comparable and side-by-side with those to be harvested, and to use these for determinations of relative turgidity at the time of harvest by one of the standard methods (see, e.g. Slatyer 1967, 5.3A). Such observations do, however, add considerably to the burden of work at the time of harvest, and in the absence of some special reason it is often not worth while making them.

24.2

On the other hand, in the absence of such information, data on the fresh weights of field-grown plants may be difficult, if not impossible, to understand in any detail.

24.3   An example from field experimentation

Table 24.3 sets out mean ratios of dry weight to fresh weight (in per cent) for the stems and leaves of the plants of *Helianthus annuus* and *H. debilis* discussed in sections 7.5, 7.6, 21.5, grown at the Botanic Garden in Cambridge in the years 1961–68; two age groups in each year, two harvests for each age group, 128 cases in all.

It will be recalled that the main reason for collecting this fresh weight data at the time was as part of the checking procedure used to control the conduct of the harvest [11.6]; and that each year three new groups of students made their first harvests without any preliminary practice. It is accordingly interesting that in any one year, for any particular group of plants, the standard error of the mean for *H. annuus* should in all sixty-two cases lie within the range 0·05 to 0·38, with its own mean for any particular column of Table 24.3 within the limits 0·1 to 0·2. For *H. debilis*, where the plants obviously belonged to a much more variable strain, the corresponding figure is around 0·2 to 0·3. Generally, for any particular set of groups of plants (the columns of Table 24.3), the variance appears to be reasonably homogeneous. In two cases only was the variance for a single harvest very significantly higher ($p < 0.01$) than that for the other seven instances in the particular column. These have been excluded from the table, as pointing to variability in harvesting techniques rather than to any real differences between the plants themselves.

The first thing to spring to the eye from the table is the variability from one year to another. This seems to be greater for the leaves than for the stems, a not unexpected result. Secondly, without any elaborate analysis, it is clear that for any particular group of plants, the leaves have a higher dry weight/fresh weight ratio, in all 62 cases.

24.4   Comparisons of plants of different ages

For other comparisons, it will be convenient to work by differences, in order to eliminate the variations from year to year. Our first comparison, in Table 24.4.1, will be by age of plant, minimizing the effects of

24.4

Table 24.3. Dry weight as a percentage of fresh weight of stems and leaves of two species of *Helianthus* compared, in a number of different years. The intervals between the ages of plants in adjacent columns was in all cases 7 days. For the ages in days (which differed somewhat from year to year), and for other details of these plants see Tables 7.5, 7.6, 21.5. Figures are means of determinations on nine plants.

	Stems							
Species:	*H. annuus*				*H. debilis*			
Age (weeks):	3	4	5	6	3	4	5	6
Harvest:	I	II	I	II	I	II	I	II
1961	6·6	6·7	6·9	—	9·8	8·3	6·8	7·4
1962	6·1	5·7	6·1	5·6	8·3	7·8	7·4	6·1
1963	7·1	7·5	6·2	7·4	8·8	7·5	7·0	6·4
1964	7·0	6·7	7·2	7·5	10·5	7·5	8·8	8·4
1965	7·0	6·9	6·7	6·3	8·9	8·8	8·8	9·4
1966	—	7·4	7·0	7·2	8·4	8·3	9·2	10·0
1967	6·3	6·2	6·9	7·7	8·0	7·4	8·6	10·2
1968	6·9	6·0	6·0	6·0	7·9	7·8	8·1	8·4

	Leaves							
Species:	*H. annuus*				*H. debilis*			
Age (weeks):	3	4	5	6	3	4	5	6
Harvest:	I	II	I	II	I	II	I	II
1961	8·9	10·4	13·4	15·0	10·6	10·0	12·9	12·5
1962	8·5	9·2	11·3	11·0	9·8	8·6	11·3	10·6
1963	10·0	11·0	10·9	13·8	10·8	10·8	9·4	10·8
1964	9·0	10·3	11·3	11·6	10·4	9·2	10·8	10·4
1965	8·9	10·0	11·0	11·7	9·5	10·1	11·4	11·4
1966	8·4	10·0	10·2	12·1	10·8	11·6	11·5	13·4
1967	12·1	11·4	13·0	12·9	12·3	10·7	12·4	12·3
1968	9·6	10·0	10·7	11·1	10·7	10·4	10·9	10·8

24.4

environmental differences in individual years by subtracting the dry weight percentages of the younger plants at a particular harvest from those of the older plants. In this way we are comparing plants which have grown for several weeks side by side in the same fluctuating, natural environment, and which have been harvested simultaneously. We see at

Table 24.4.1. Change in dry weight as a percentage of fresh weight with age of plant. Stems and leaves of two species of *Helianthus* compared. Figures are differences in mean percentages for nine plants (older minus younger) harvested simultaneously, from Table 24.3. Age difference 14 days.

	Stems				Leaves			
	H. annuus		*H. debilis*		*H. annuus*		*H. debilis*	
Harvest:	I	II	I	II	I	II	I	II
1961	+0·3	—	−3·0	−0·9	+4·5	+4·6	+2·3	+2·5
1962	0	−0·1	−0·9	−1·7	+2·8	+1·8	+1·5	+2·0
1963	−0·9	−0·1	−1·8	−1·1	+0·9	+2·8	−1·4	0
1964	+0·2	+0·8	−1·7	+0·9	+2·3	+1·3	+0·4	+1·2
1965	−0·3	−0·6	−0·1	+0·6	+2·1	+1·7	+1·9	+1·3
1966	—	−0·2	+0·8	+1·7	+1·8	+2·1	+0·7	+1·8
1967	+0·6	+1·5	+0·6	+2·8	+0·9	+1·5	+0·1	+1·6
1968	−0·9	0	+0·2	+0·6	+1·1	+1·1	+0·2	+0·4

once that as concerns stems, there is little overall difference with age in either species, but for leaves the picture is quite different. For *Helianthus annuus* the differences in dry weight percentage are invariably positive, averaging 2·1 per cent. For *H. debilis* 14 out of 16 cases are positive, but the difference is much smaller, averaging 1·0 per cent. In all cases the differences in behaviour from year to year are striking, and they are often seen in both age groups and both species.

These changes no doubt reflect changing leaf structure. As the plants become older, the leaves of *H. annuus* increase considerably in size, and with this goes a more massive venation. In *H. debilis* there is also a size increase, but to a much less marked degree, and the plants never produce the individual very large leaves characteristic of *H. annuus*.

It is of interest to contrast this clear result for leaves with that of the other possible way of expressing the effect of age, by taking the difference

24.4

between the first and second harvests from the same population of plants. Here the older plants, a week before harvest, should on average have been identical with the younger ones harvested at that time, because both were similar graded samples from the same population, and chance decided which should be harvested first. The two sets of plants being compared shared a common early life, which was not the case for the earlier comparison, where plants of different ages shared a common external environment. On the other hand, the external environment during the week before harvest would be quite different.

Table 24.4.2. Change in dry weight as a percentage of fresh weight with age of plant. Stems and leaves of two species of *Helianthus* compared. Figures are differences (older minus young) in mean percentages for nine plants at two successive harvests, separated by a week, from Table 24.3.

| | Stems | | | | Leaves | | | |
| | *H. annuus* | | *H. debilis* | | *H. annuus* | | *H. debilis* | |
Age (weeks):	4–3	6–5	4–3	6–5	4–3	6–5	4–3	6–5
1961	+0·1	—	−1·5	+0·6	+1·5	+1·6	−0·6	−0·4
1962	−0·4	−0·5	−0·5	−1·3	+0·7	−0·3	−1·2	−0·7
1963	+0·4	+1·2	−1·3	−0·6	+1·0	+2·9	0	+1·4
1964	−0·3	+0·3	−3·0	−0·4	+1·3	+0·3	−1·2	−0·4
1965	−0·1	−0·4	−0·1	+0·6	+1·1	+0·7	+0·6	0
1966	—	+0·2	−0·1	+0·8	+1·6	+1·9	+0·8	+1·9
1967	−0·1	+0·8	−0·6	+1·6	−0·7	−0·1	−1·6	−0·1
1968	−0·9	0	−0·1	+0·3	+0·4	+0·4	−0·3	−0·1

The result is given in Table 24.4.2. The increase in dry weight percentage of the leaves of *H. annuus* with age is reduced to less than half; it now averages 0·9 per cent for a 7-day age difference, as compared with 2·1 per cent for a 14-day difference in Table 24.4.1. For *H. debilis* the effect is equally striking—the significant increase of dry weight percentage with age has disappeared. As it happens, in 10 of the 16 cases there is a decrease, and only 4 are positive, the mean being −0·1 per cent. The importance of the particular external environmental conditions during the week before harvest is plain, and we can see the advantage of proceeding as in the first comparison, and exposing plants of different ages to the

24.4

same environment for as long as possible before the harvest—in this case for the whole period after the germination of the younger plants.

24.5 Comparison of related species

Finally, we can compare the two species, as in Table 24.5. The stems of *Helianthus debilis* can be seen to have a higher dry weight percentage than those of *H. annuus*. There is also some slight evidence

Table 24.5. Difference in dry weight as a percentage of fresh weight between two species of *Helianthus*. Comparison of figures for *H. debilis* minus *H. annuus* for stems and leaves in plants of different ages, grown side by side and harvested simultaneously, from Table 24.3.

	Stems				Leaves			
Age (weeks):	3	4	5	6	3	4	5	6
Harvest:	I	II	I	II	I	II	I	II
1961	+3·2	+1·6	−0·1	—	+1·7	−0·4	−0·5	−2·5
1962	+2·2	+2·3	+1·3	+0·5	+1·3	−0·6	0	−0·4
1963	+1·7	0	+0·8	−1·0	+0·8	−0·2	−1·5	−3·0
1964	+3·5	+0·8	+1·6	+0·9	+1·4	−1·1	−0·5	−1·2
1965	+1·9	+1·9	+2·1	+3·1	+0·6	+0·1	+0·4	−0·3
1966	—	+0·9	+2·2	+1·7	+2·4	+1·6	+1·3	+1·3
1967	+1·7	+1·2	+1·7	+2·8	+0·2	−0·7	−0·6	−0·6
1968	+1·0	+1·8	+2·1	+0·6	+1·1	+0·4	+0·2	−0·3

of a tendency for the difference to decline with age, the mean falling from 2·17 per cent for the youngest plants, through 1·31 and 1·46 for the 32- and 39-day-old plants to 1·23 for the oldest plants. A similar but even more marked effect is observable in the leaves, a consistently positive difference in the leaves of the youngest plants (averaging +1·19 per cent) declining to a negative difference in the leaves of the oldest ones (averaging −0·88 per cent).

24.6 Conclusions from the field experiments

In spite of the large variations from year to year we are thus able to demonstrate an interesting combination of ontogenetic drift with genotypic differences between closely related species. Systematic differences

24.6

between the ontogenetic drifts of the two species are superimposed on the difference in the value of dry weight percentage at a particular age. These differences in the ontogenetic drift of both stems and leaves are no doubt connected with the differences in the development of plant form, the increases in size of leaves and massiveness of stem in *H. annuus* being accompanied by increasing amounts of woody tissue, with corresponding increases in dry weight percentages.

On the other hand, it is clear that for plants with extensible cell walls, if we wish to reach useful conclusions by comparisons based on field experimentation, the sets of plants must be grown side by side for as long as possible before harvesting, which should be done simultaneously. The importance of repetition of experiments of standard design under a variety of external environmental conditions is also clear. This enables us to distinguish inherent plant responses from the chance effects associated with the environment during any particular field experiment; and also gives us some idea of the range of response likely to be encountered over a particular range of climate. If for some reason experiments cannot be repeated, it is highly desirable to collect information on relative turgidity when working with plants where there is a large change of water content between full turgor and wilting [24.2].

24.7   Another field experiment on plants with less extensible cell walls

In contrast, for plants such as *Impatiens parviflora* with relatively inextensible cell walls, the turgid fresh weights determined at harvest from field experiments can yield much interesting information, without any attempt being made to correct for variations in relative turgidity. Fig. 24.7 shows such a series of observations, made in connection with an experiment (see 18.6 and 21.2 for details) on the effects of growing plants in the field under artificial shades. The ratios for two harvests are shown, at 33 and 40 days from sowing. A uniform population had been grown under light shade in a glasshouse, sorted and graded, and comparable plants had been placed at 23 days in pots plunged in the open (marked 100), and under shades transmitting 67, 42, 25 and 7 per cent of full daylight. The ratios are given for leaves, stems and roots, the uniform plants at 23 days being shown in each case as a point.

We see at once the very systematic consequences of these treatments on the dry weight/fresh weight ratios. For stems and leaves the ratios

24.7

rise in full daylight, and decline under all other conditions, the decline being steeper, the lower the transmission of the screen. The whole picture is so systematic that no line crosses any other. The changes brought about

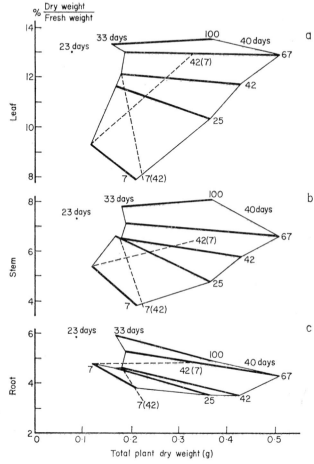

Fig. 24.7. Ratios of dry weight to fresh weight in percent for leaves, stems and roots of *Impatiens parviflora* growing under neutral screens in the field, with mean daily totals of visible radiation as in Table 19.6, Experiment II. Common initial sample at 23 days, then lines join the points for harvest 2 (33 days) and 3 (40 days) at each level of shading (screen transmission in per cent). At 33 days mutual transfers of samples between the 7 and 42 per cent screens; points at 40 days marked 42 (7) and 7 (42) connected by broken lines to the points for the corresponding samples at 33 days. From Evans and Hughes (1961).

24.7

by transferring plants from the 7 per cent screen to the 42 per cent one, and vice versa, are shown by dotted lines. It will be seen that the downward transfer is close to the 7 per cent line at the end of the week for stems and leaves, and has overshot it for roots; the upward transfer, from 7 per cent to 42 per cent, overshoots in all cases, and for leaves and roots is very close to the 67 per cent line.

If the plant had extensible cell walls we might wonder whether this whole picture was a consequence of the weather on the 2 days of harvest —on a sunny day the shaded plants would be expected to be relatively more turgid, and hence to have a lower dry weight/fresh weight ratio. But for this species, the observed differences are an order of magnitude larger than any change which could have been brought about by variations in turgor between full turgor and wilting. They must reflect changes in cellular anatomy.

24.8   Comparable laboratory experiments

We can check this by reference to plants grown under controlled conditions in the laboratory, when it is a simple matter to harvest the plants in a standard state of turgor. Also the plants can be grown from seed under constant conditions, and their behaviour can be compared with that of plants transferred from one condition to another, so that allowance can be made for the effect of adaptation in interpreting the results of the field experiments. The cabinet obervations are set out in Fig. 24.8. The lowest total daily light, of $11 \cdot 2$ cal cm^{-2} day^{-1}, is somewhat below the mean daily total under the 7 per cent screen in the field of $13 \cdot 7$ cal cm^{-2} day^{-1} for the 2 weeks taken together, although the unit leaf rates were closely similar. The unit leaf rate under $26 \cdot 2$ cal cm^{-2} day^{-1} in the cabinets was also closely similar to that under the 24 per cent screen in the field, where, however, the mean total daily light was almost double at 47 cal cm^{-2} day^{-1}. In general, therefore, it may be possible to interpret the observations under the 7 per cent and 24 per cent screens by means of the cabinet studies, although field conditions must have created many differences in water regime as compared with the cabinets.

Comparing first the field observations on roots with the cabinet-grown plants of more than 120 mg total dry weight, we see that both have a similar range of values, but the roots give systematically lower figures in the field ($4 \cdot 8$ to $3 \cdot 2$, as compared with $6 \cdot 3$ to $4 \cdot 3$ for the cabinets).

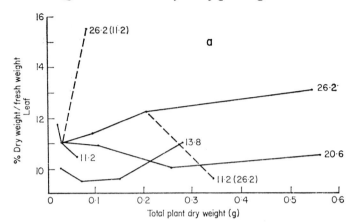

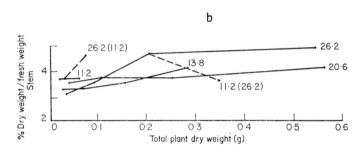

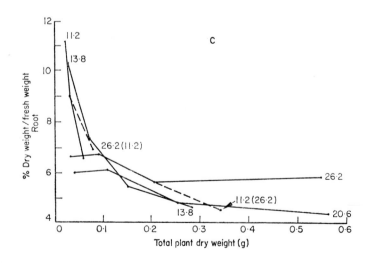

24.8

This difference may possibly be connected with an effect of adaptation to a lower light regime, as we have already noticed that the downward transfer from the 42 per cent to the 7 per cent screen caused overshooting; or it may be connected with a difference in water regime generally, because we also notice that the roots of the glasshouse-grown plants at 90 mg give a dry weight/fresh weight ratio of 5·8 per cent, whereas at 90 mg the highest light total in the cabinets gave one of 6·7 per cent, again about 1 per cent higher. Secondly, in both sets of conditions the dry weight percentage tends to fall with increasing dry weight at all light totals, except for the highest one in the cabinets, where the last two harvests gave almost identical values. Thirdly, in both the percentage tends to be lower, the lower the light.

24.9 Relationships of field- and laboratory-grown plants: stems and roots

We can now see that the field observations represent only a part of a much more elaborate and complex relationship which can be studied much more readily in the cabinets. There, at the lowest daily light totals, the roots of the small plants have very high dry weight percentages, which fall very rapidly; this does not seem to happen under the highest daily light totals. The rapidity of the changes, and the paucity of the data make it impossible to decide what is the relationship between dry weight percentage and total daily light for these very small plants, but there is a strong suggestion that here the order observed among the larger plants is reversed, the highest dry weight percentages corresponding to the lowest light totals. It seems almost certain that these high values in small plants in low light are due to a relatively high proportion of immature cells, not yet fully expanded.

For the stems of the field-grown plants, the process of adaptation obviously involved very large changes, which may well have taken more

Fig. 24.8. Ratios of dry weight to fresh weight in per cent for leaves, stems and roots of *Impatiens parviflora* (means, usually of eight plants) growing under constant conditions in cabinets under different daily totals of visible radiation, ranging from 11·2 to 26·2 cal cm^{-2} day^{-1}, in relation to total plant dry weight. Harvests at weekly intervals, ages of plants between 17 and 44 days. Points marked 26·2 (11·2) and 11·2 (26·2) give the mean values for batches of plants transferred from the latter to the former condition for the third week of growth. From Hughes and Evans (1962).

than 10 days to complete. We notice in the cabinets that the larger plants give higher dry weight percentages with higher light totals: the difference between 20·6 and 26·2 cal cm^{-2} day^{-1} is nearly 1 per cent. The glasshouse-grown plants at 90 mg had a dry weight ratio of more than 7 per cent, much higher than any observed in the cabinets (not exceeding 5 per cent). After 17 days' exposure under the 25 per cent screen, the field-grown plants with a dry weight of nearly 400 mg had a dry weight percentage almost identical with that of plants of comparable size grown under the highest daily light total in the cabinets; and these are the points which we might expect to be most closely comparable.

The effects of transferring plants from one light regime to another are also in agreement. In the field, the transfers show a substantial increase of dry weight percentage for plants transferred to a higher intensity, and a comparable fall for plants transferred to a lower one. This is confirmed by the changes consequent on the initial setting up of the experiment. The transfers in the cabinets show similar relationships.

The longer run of adapted plants in the cabinets shows a further interesting difference between stems and roots—for young stems there is no sign of the phase of very high values of dry weight percentages so obvious for young roots: in fact, the stems of the smallest plants all give values lying between 3 and 4 per cent, whereas for roots the corresponding values go up to over 13 per cent, and there are no values below 6 per cent. These large initial differences have, however, become reduced to 1 or 2 per cent by the time the plants have reached a total dry weight of 300 mg.

24.10   Relationships of field- and laboratory-grown plants: leaves

For leaves the picture is more complex still, and once again we have the impression, as for roots, that the field observations show only part of a larger and more complex relationship revealed by the cabinet observations. Again, in both cases the observations on the effects of transferring plants agree. Also, in both cases the dry weight percentages rise in the highest intensities and fall in the lowest. But from the cabinet observations we can see that in the medium intensities the percentage falls to a minimum and then rises again. It appears that this minimum is lower, the lower the intensity. These observations fit in well with the general pattern of leaf development; at low total dry weights a relatively high proportion of the total leaf tissue is meristematic; as the cells mature and expand, water is taken up, and the dry weight percentage falls, until the plant has reached a steady state of growth where the proportion of meristematic to expand tissue is relatively constant. Meantime, a further phase is becoming established, in which cell wall thickening and other changes bring about an increase in the dry weight of individual mature cells, without any change in volume at full turgor. This will tend to make the dry weight percentage rise again.

The field observations also fit in well with this general pattern. The leaves of the 90 mg glasshouse-grown plants had a dry weight percentage of 13, as compared with 11·3 for the highest intensity in the cabinets; but there is no discrepancy here, as we see that at 90 mg increasing total light brings about a higher dry weight percentage, and the glasshouse plants were grown under much higher daily light totals than any in the cabinets. For the larger, adapted, plants in the field the observations may well represent part of this relationship: with only two points we cannot, of course, detect a minimum. We notice that the lowest per centages attained in the field, around 8 per cent, were for plants of over 200 mg grown under the screen of lowest transmission. In the lowest intensity in the cabinets, although the percentage was falling rapidly, it never became as low as 10 per cent; but on the other hand, the plants were much smaller, not attaining 100 mg. Once again, there is no necessary discrepancy between the two sets of observations.

24.11 Relations between plants transferred from one condition to another

One final agreement concerning the transferred plants is of interest, even if it is at present difficult to interpret. For stems and leaves, in all cases both in the field and in the cabinets, the downward transfer, although it involved a larger plant, produced a percentage which was not necessarily discrepant from the relationship for plants grown at this lower intensity. The corresponding upward transfer, on the other hand, produced in all cases a substantial overshoot (larger in the cabinets), to a percentage well above that observed for plants of a comparable size. This obviously concerns the balance between the supply of fresh dry matter and cell expansion, which we have already touched on, and suggests a comparison with the effect of the same transfers on specific leaf area. If we examine these transfers [19.5, and Fig. 19.4], we find agreement to the extent that the large overshoot of dry weight percentage of leaves in the cabinets, of about 4 per cent, corresponds to an overshoot of specific leaf area (in the direction of a heavier leaf) of about 10 per cent; whereas the smaller overshoot in the field, of about 1 per cent, corresponds to an almost exact agreement in specific leaf area between the transferred plants and those grown throughout the experiment under the conditions concerned. This is interesting, because although there is, of course, no *a priori* reason why there should be an obvious correspondence between the two, as such a correspondence would depend on the type of relationship between wall expansion and cell contents, as modified by any differences in water regime, nevertheless where correspondences of this kind are noted, they may be expected ultimately to assist in unravelling these relationships in growing organs (and see 24.21).

24.12 *Callistephus chinensis*: vegetative organs

The changes in dry weight as a percentage of fresh weight later in the life of an annual plant with a massive flower head are well shown in some data of Hughes and Cockshull (1969, and unpublished data from the same experiment). In Fig. 24.12 we see the changes in dry weight percentage of the vegetative organs as a function of total plant dry weight, for two experiments on plants of *Callistephus chinensis* grown in controlled environment cabinets with three levels of atmospheric carbon dioxide—325, 600 and 900 parts per million. The right-hand axis of ordinates gives the corresponding scale for grams of water per gram of dry matter. The flower buds had grown large enough to be removed from the stem and dried and weighed separately about 12 weeks from sowing, when the plants had a total dry weight varying from 3g (for the lowest carbon dioxide concentration) to 4·5 g (for the highest). From a total plant dry weight of about 0·2 g, over a twenty-fold increase in dry weight, up to this age of about 12 weeks the dry weight percentage of the leaves had been slowly rising, by between 1 and 2 per cent, a development similar to that which we noticed in sunflowers, though on a smaller scale. At the same time there were only small changes in the dry weight percentage in the stems and roots.

From 12 to 18 weeks the flower head developed steadily, the flower dry weight ratio increasing to over 0·4 at the end of this period, making it

24.12

the dominant organ (see Fig. 24.13a). The dry weight of the leaves continued to increase up to about 16 weeks, when the total plant dry weight
was around 15 g, after which it declined. Fig. 24.12 shows that the progressive slow rise in dry weight percentage of the leaves continued during
this period, with no obvious break. In contrast there was a great change in

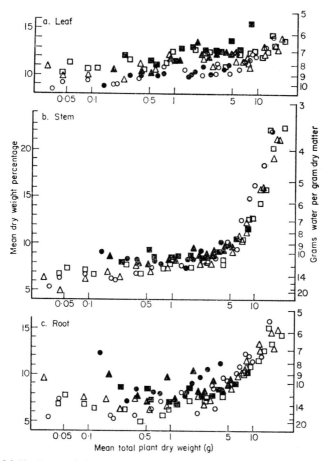

Fig. 24.12. Dry weight percentages, and grams water per gram dry matter,
of (a) leaves, (b) stems, and (c) roots in relation to total plant dry weight,
for two experiments on plants of *Callistephus chinensis* grown in controlled environment cabinets with three levels of atmospheric carbon
dioxide, 325 (circles), 600 (triangles) and 900 parts per million (squares).
Solid symbols, Experiment 1; open symbols, Experiment 2. (Graphical
presentation of data summarized in Hughes and Cockshull, 1969, para (xi).)

24.12

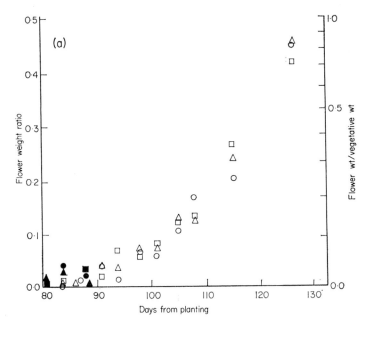

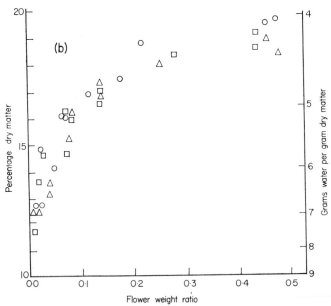

24.12

the dry weight percentage of both stem and root, which rose rapidly to 23 and 15 per cent respectively, reflecting the increase in mechanical tissue of low water content, associated with support of the massive capitulum.

24.13   *Callistephus chinensis*: capitula

Similar changes took place in the capitula themselves, as shown in Fig. 24.13b. When the flower heads were first separated from the stems the latter had a dry weight percentage of about 8. For the flower heads this increased rapidly to more than 16 while the flowers were still quite immature, the flower weight ratio itself not yet having reached 0·1. Thereafter, as the flower heads developed and the flower weight ratio rose to between 0·4 and 0·5, there was a much slower rise in dry weight percentage, to nearly 20 when the plants were in full flower and the experiment ended.

24.14   Dry weight/fresh weight ratios: conclusions

Here the study of dry weight of the various parts as a percentage of fresh weight has served to emphasise the great changes in form and structure, in the proportions of the different types of cell and in their biochemical makeup, associated with the onset of flowering in an annual of this type. It has also served to emphasize the great differences in the ontogenetic drifts of this ratio in the different organs, from the leaves which are hardly affected, through the progressive changes of stem and root, to the flower heads themselves, where at an early developmental stage a characteristically high dry weight percentage develops, differing from that of the parent stem.

Thus we see that for some plants, and for suitable conditions, determinations of dry weight/fresh weight ratios and other similar functions

Fig. 24.13. (a) Flower weight ratio (left-hand scale) or flower to vegetative weight ratio (right-hand scale) in relation to age in days from planting for the same experiments on *Callistephus chinensis* as in Fig. 24.12. (b) The corresponding figures for dry matter percentage (left-hand scale) or grams of water per gram dry matter (right-hand scale) of the capitula in relation to flower weight ratio. Two experiments at three levels of atmospheric carbon dioxide concentration, 325 (circles), 600 (triangles) and 900 parts per million (squares). Solid symbols, Experiment 1; open symbols, Experiment 2. From Hughes and Cockshull (1969).

can yield much interesting information, relevant to problems of plant development; suggesting connections with the morphological and anatomical state of the various organs; and suggesting further problems for investigation.

On the other hand, the relationships are likely to be complex, and we have here a further example of the utility of having available observations made under controlled conditions in the laboratory when attempting to interpret observations made on plants growing in the field.

24.15 Water content: relevance

We saw [24.1, 24.2] that tissue water content may vary very widely with the plant's hereditary constitution and systematic position, with the type of habitat in which it lives and with its stage of development. Within this general framework, there may in suitable cases also be wide variations with changes in external conditions, either long-term or short-term. We have examined instances where such information could be collected with relatively little extra trouble as part of the organization of an experiment on plant growth. On occasion the collection of the information for checking purposes is an integral part of experimental planning, involving no extra work at all [11.6].

A link is thus formed between the type of study with which this book is mainly concerned, based chiefly on dry weight, and studies of the water relations of plants. Later, in Chapter 29, we shall examine another type of link, concerned not so much with the make-up of the plant at a particular time (such as the observations which we have been discussing earlier in this chapter) as with rates of loss of water and gain of dry weight, estimated simultaneously.

There remains, however, the direct relevance of simultaneous studies of fresh weight and of dry weight to the concepts which we have been considering in the last twelve chapters, derived from the importance of water as a structural element in plant tissues. In studies made with this end in view it is often convenient to consider the absolute water content of plant tissues, rather than the fresh weight, or the dry weight to fresh weight ratio.

24.16 Water content: estimation

If we assume that the water content is the difference between the fresh weight and the dry weight, we must always bear in mind the conditions under which the two observations were made, and the widely differing status of the water molecules in different parts of the whole living plant. Some of the water, as in xylem sap, is connected only indirectly with the metabolic activity of the plant: to a degree the same is true of vacuolar sap, and in many plants the content of both can vary significantly with minimal change in metabolic functions as measured by the overall indices of photosynthesis or respiration. It is gain or loss of this water which often brings about short-term variations in fresh weight, and this makes it important, for comparative purposes, to know something of the water state of the plant at the time a fresh weight determination was made.

At the other extreme, some water molecules are bound in the structure of the plant's colloidal systems, and some of these molecules are so tightly bound that they are not lost when the plant becomes air dry. In a plant such as *Myrothamnus flabellifolia*, where in the dry season the leaves dry up while remaining alive, a water content of 7 per cent has been observed even under the extreme conditions of temperature, humidity and irradiance of a sub-tropical dry season (Thoday, 1921). We have already discussed [11.19] the difficulties of making accurate determinations of dry weight, because of the other changes going on in the biochemical systems of the plant around 100°C. We therefore chose 80°C as a drying temperature: at this temperature some water would still remain bound, although the amount would be small—negligibly so compared to unavoidable errors in nearly all experiments of the type we have been discussing.

In between lies a whole range of aqueous systems, more or less closely connected with the plant's metabolism, more or less essential for its continuance in a normal state. These considerations raise the question— can water content be used as a measure of these systems, and hence of the mass of active protoplasm? This may prove to be possible only in very favourable cases, and it may be necessary to measure water content under carefully determined standard conditions. Even then our attempts might be frustrated if we were dealing with a plant where during ontogeny, or with changing external conditions, there were marked changes either in the proportional size of the vacuoles in relation to the cell contents, or in the proportions of different types of cell in the organ or plant as a whole, both of which unfavourable features would be found in, say, *Impatiens parviflora*.

Many investigations must necessarily be quite unsuited to this approach, and if we seek for further light on the mechanisms underlying our measurements we must use methods such as those considered in Chapter 27; but in a favourable case valuable additional clues may be provided by studies of water content.

## 24.17	Water content and leaf area

Convenient examples are provided by the studies of Hughes, Cockshull and Heath (1970) on the relations between the absolute water content of leaf tissues and leaf area for plants of several species of dicotyledonous herbs growing under a variety of controlled environmental conditions, mostly in cabinets but also (for two species) in glasshouses. A representative example from the data which they present relates to the first of the two sets of experiments on the growth of *Callistephus chinensis* (cultivar 'Johannistag') which we have already discussed [20.4, 23.7, 24.12, 24.13]. It will be recalled that all these observations were made on plants grown in controlled environment cabinets in the same rooting medium with a common regime of temperature and humidity, daylength and daily light total, and with three different concentrations of atmospheric carbon dioxide ranging from 325 to 900 parts per million. Fig. 24.17 shows the relationship between the mean total leaf area per plant and the mean total leaf water content (fresh weight minus dry weight). We see that the relationship is effectively linear, the regression equation for all three experiments lumped together being

$$L_A = 0.496 L_{H_2O} \pm 0.223, \qquad\qquad 24.17$$

where L_A is the leaf area in square decimetres, and L_{H_2O} is the absolute leaf water content in grams. This is shown in the figure, together with the 5 per cent fiducial limits. The plants studied displayed a wide range both of specific leaf area and of specific leaf water content, as shown in Table 24.18.

Similar relations were observed under a wide variety of controlled conditions of growth in 'Queen of the Market', another cultivar of *Callistephus chinensis*; in two cultivars of *Chrysanthemum morifolium*; in the radish, *Raphanus sativus*; and in *Impatiens parviflora*. In all cases but one there was a small, but significant, intercept of the straight line of best fit on the leaf area axis, ranging from 0.014 up to 0.302 dm², the exception being the radish, where the intercept did not differ significantly from zero. In all these plants it would be expected that the ratio of young to mature leaf tissue would fall as the plants become older, and Hughes and his colleagues point out that if, as might be expected, the young leaf tissue has a different characteristic value of water per unit area from the mature leaf tissue, this may well account for the form of the linear regression equation giving the best fit to all the points.

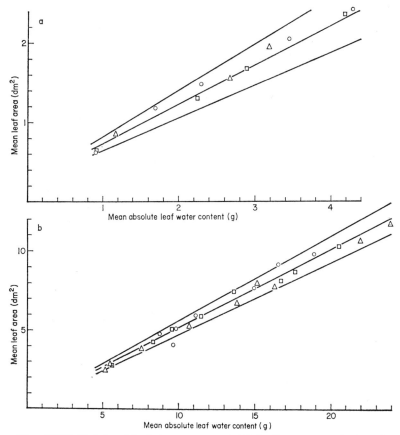

Fig. 24.17. Relationship between mean total leaf area per plant and mean total leaf water content for plants of *Callistephus chinensis* cultivar 'Johannistag' cultured in growth cabinets with three different concentrations of atmospheric carbon dioxide ○ 325, □ 600, △ 900 parts per million. Points are means for three plants in each case, harvested simultaneously (a) for mean total leaf water contents below 5 g and (b) to five times the scale on both axes, 5–24 g. Regression line $L_A = 0.496 L_{H_2O} + 0.223$ with fiducial limits at the 0.05 probability level. Same experiments as in Figs. 20.4 and 24.12 (Experiment 1). From Hughes, Cockshull and Heath (1970).

24.17

24.18 Variations in linear relationships

In spite of the fact that, as shown in Table 24.18, there were marked differences both in specific leaf area, and in specific leaf water content between the three carbon dioxide treatments, statistical analysis confirmed the visual impression from the plot of leaf area against absolute

Table 24.18. Ranges of specific leaf area and specific leaf water content observed during the same series of experiments as in Fig. 24.17 on the growth of *Callistephus chinensis* in controlled environment cabinets with a common rooting medium and a common regime of temperature, humidity, daylength and total daily light, but with three different concentrations of atmospheric carbon dioxide. Means of three plants per harvest. From Hughes, Cockshull and Heath (1970).

Atmospheric carbon dioxide concentration (parts per million)	Range of specific leaf area ($dm^2\ g^{-1}$)	Range of specific leaf water content (grams water per gram dry matter)
325	3·9–7·2	8·0–10·3
600	3·3–6·0	6·4–8·8
900	2·7–5·5	5·5–8·0

leaf water content [Fig. 24.17], that the points for all three treatments fell close to one single straight line, and that no differences between treatments were detectable in this relationship. Other sets of experiments reported by Hughes and his colleagues and also showing single relationships similar to Fig. 24.17 covered variations in daylength, in total daily light, and in plant spacing, although in each case the experimental plants exhibited a wide range both of specific leaf area, and of specific leaf water content, similar to the figures in Table 24.18.

On the other hand, certain changes in conditions including change of rooting medium produced a new straight line, the sets of observations showing similar variability, but the constants of the regression equations changing. On two occasions (one with *Callistephus chinensis*, the other with *Chrysanthemum morifolium*) repetition of an apparently identical

24.18

experiment after a period of time produced relationships with significantly different constants, the reasons for this not being easy to define.

24.19 Leaf water content as a measure of leaf area

Hughes, Cockshull and Heath (1970) point out that so long as one can be sure that a series of experiments falls within the limits of conditions producing a single linear relationship, as in Fig. 24.17, absolute leaf water content (i.e. the difference between fresh weight of leaf material and the corresponding dry weight) can be used as a measure of leaf area in experiments which do not aim at great accuracy (the 5 per cent probability limits of Fig. 24.17 come at roughly ±8 per cent of the leaf area). In such cases much trouble can be saved, particularly when dealing with large quantities of leaves, or when for one reason or another leaf area is difficult to measure.

The condition of a single relationship is most likely to be met when working under controlled environmental conditions. Rather more than half of the instances cited by Hughes and his colleagues were from controlled cabinets, the remainder from glasshouse experiments; but even here their observations on factors causing variations in the constants of the linear relationship [24.18, last paragraph] impose caution. If this method of estimating leaf area is to be employed, it is clearly necessary to include a proportion of checks, in which leaf area is measured, so as to ensure that any deviations from a particular linear relationship are at once detected.

We may also ask whether such a method is likely to be useful under the conditions of field experimentation. Fig. 24.19 shows the relationship for the youngest plants (around 3 weeks old) of *Helianthus annuus* in the experiments over the years 1961–66 and 1968 which we have used already for consideration of dry weight percentages [Table 24.3] and for other purposes [7.6, 14.3, 21.5, 22.4]. In these 7 years the leaf area of these young plants did not exceed 4·0 dm². The relationship appears to pass through the origin, and the mean ratio of leaf area to absolute leaf water content was 0·303 dm² g⁻¹, with 0·05 probability limits of 0·009 dm² g⁻¹ (±3 per cent). Here is a case where it seems that we have a close approximation to one single relationship in spite of the wide range of weather conditions encountered by these plants, growing in the open at fixed dates in seven different years. The variability between years does in fact prove to be

24.19

significantly greater than would be expected from the variability of individual plants within years, and although there is 1 chance in 20 of an individual mean value in Fig. 24.19 differing from the line of best fit for all 7 years by 13 per cent, if we calculate deviations from lines of best fit for each year, the 1 in 20 chance is reduced to 10 per cent. The relationship

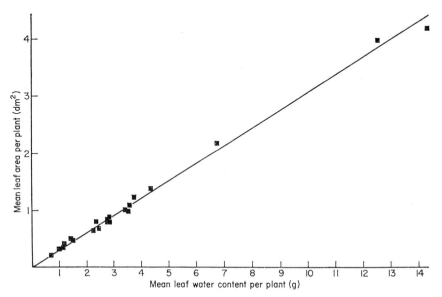

Fig. 24.19. Mean leaf area per plant in relation to mean leaf water content for *Helianthus annuus* grown in the open at the Botanic Garden, Cambridge, in the years 1961–66 and 1968. Points are for means of three plants in each of three size groups (large, medium and small) in each year, all aged around 3 weeks. For other details of these plants see Table 7.5, 7.6, 21.5 and 24.3. Note that none of the plants grown in 1967 had leaf water contents at harvest below 15 g.

could thus be used to estimate leaf area with an accuracy nearly as good as that found in the cabinet experiments of Hughes, Cockshull and Heath. The 8th year, 1967, was, however, an exception. Growing conditions were unusually good, and at the same chronological age the youngest plants were all substantially larger and heavier than any shown in Fig. 24.19. At the same time the mean ratio of leaf area to leaf water content was very significantly higher at 0.360 dm^2 g^{-1}, with 0.05 probability limits based on these nine values alone of 0.018 dm^2 g^{-1}.

24.19

24.20 Effects of age and growing conditions

The same set of data also includes a considerable number of instances of older plants falling within the size limits of Fig. 24.19, aged about 4 weeks (16 cases) and 5–6 weeks (17 cases). Here the slope of the line of best fit increases significantly as the plants become older, the leaf area per unit water content rising to 0.334 ± 0.014 dm^2 g^{-1} at 4 weeks and 0.382 ± 0.010 at 5–6 weeks, the limits being for a probability of 0.05. We see that within this size range the data are not becoming more variable, there being no significant change in the 0.05 probability limits. Nevertheless in using the relationships for estimating leaf area, we should have to bear in mind that we are confined to predetermined limits of plant size, and that within these the slope changes with age. Clearly in any particular instance a check would be necessary to ensure that the appropriate relationship was used.

Similar arguments hold for larger plants, but the picture now becomes more complex. For plants of *Helianthus annuus* all grown side by side in the open in a particular year and harvested at a particular age, the variability of the ratio of leaf area to leaf water content within a sample proves to be small for plants of all sizes up to a leaf area per plant of over 50 dm^2 and an absolute leaf water content for the largest plants of over 130 g.

The years during the period which we are considering (1961–68) in which the older plants had mean total leaf water contents in excess of 15 g at both harvests are 1961, 1962, 1963 and 1967. 1967 was the only year in which the same held for the younger plants also. Table 24.20 compares the mean ratios of leaf area to leaf water content for these ten harvests; and also gives the ages; the mean total leaf water content as an indication of the sizes of the plants (see Tables 7.5, 7.6, 21.5, 22.4 and 24.3 for further details); and the mean relative growth rates during the week between harvests in each year. The variances of the ten mean ratios were reasonably homogeneous, although there appeared to be a tendency for the variances to rise as the plants became larger. (This effect, if real, may well be connected with the physical difficulties encountered by students without previous training in the special techniques of harvesting, when dealing with such large plants.) The ten sets of nine observations then give common fiducial limits for a single mean of 9 of

±0·014 dm² g⁻¹, the least significant differences between two means of 9 being 0·020 at the 5 per cent and 0·026 dm² g⁻¹ at the 1 per cent level. On this basis there are a number of significant differences within the table,

Table 24.20. Mean ratios of leaf area to absolute leaf water content for plants of *Helianthus annuus* grown in the open at the Botanic Garden, Cambridge: same experiments as used for Tables 7.5, 7.6, 21.5, 22.4 and 24.3. All cases during the period 1961–68 when the mean leaf water content per plant exceeded 15 g at both harvests (for the relationship up to 15 g, see 24.19, 24.20). The figures in each year are for two batches of nine plants which were comparable with each other at the time of the first harvest, and the last column relates to the differences which developed during the week between harvests. For comparison, age, mean leaf water content per plant, and mean relative growth rate during the week between harvests. Fiducial limits of each mean ratio, ±0·014 dm² g⁻¹; least significant difference between any two mean ratios ±0·020 at the 5 per cent and ±0·026 dm² g⁻¹ at the 1 per cent level.

Year	Harvest	Age of plants (days)	RGR (week⁻¹)	Mean leaf water content (g)	Leaf area / Leaf water content (dm² g⁻¹)	Difference II–I (dm² g⁻¹)
1961	I	40		79·3	0·335	
			0·834			+0·010
	II	47		119·2	0·345	
1962	I	36		22·1	0·318	
			1·340			+0·017
	II	43		45·9	0·335	
1963	I	36		22·6	0·336	
			1·083			+0·013
	II	43		41·2	0·349	
1967	I	39		78·6	0·364	
			0·615			+0·018
	II	46		119·5	0·382	
1967	I	25		19·8	0·360	
			1·057			+0·041
	II	32		52·0	0·401	

notably between all four sets of plants grown in 1967 and the rest, but the interrelationships between the individual mean values of the ratio of leaf area to leaf water content are clearly complex, and not closely related either to plant size or to relative growth rate in the week preceding harvest.

24.20

There does, however, seem to be a general tendency for the leaf area per unit leaf water content to rise with age in a given population, irrespective of initial size or growing conditions, as shown in the last column of Table 24.20. There the differences are all positive, although only the last one is highly significant ($p \ll 0.001$).

The variances cited in this section can be regarded as maximal values for populations of *Helianthus annuus* grown in the manner described [7.5, 7.6], because the data were collected for checking purposes as part of a class exercise [11.6], by students who were carrying out an experiment of this kind for the first time. No doubt an experienced experimenter could reduce the variability substantially. We conclude that for such plants it would be possible to use leaf water content as a means of estimating leaf area with reasonable accuracy, provided that a value for the ratio of leaf area to leaf water content is determined using a sample from the same population of plants, harvested at the same time.

Our general conclusion is therefore that there do exist instances where measurements of absolute leaf water content could assist in the estimation of leaf area in field experiments. If it seems in the planning stage that the method might be valuable in a particular investigation, it is worth carrying out a pilot study to see whether conditions can be established within which the method can be used with an acceptable degree of accuracy. These conditions might include limitations of age or plant size, or the use of sub-samples to establish the appropriate parameters.

24.21 Significance of these linear relationships

Hughes *et al.* (1970) noted linear relationships between leaf area and absolute leaf water content under a variety of artificial conditions of growth in several species of flowering plants [24.17]. To these we have added *Helianthus annuus*, growing under a variety of natural conditions of the aerial environment. Three of all these species were members of the Compositae, the others belonged to families widely separated in classification. The linear relationships were of two kinds: those observed by Hughes, Cockshull and Heath were in effect time progressions along a pathway which appeared to be fixed for a particular species by the particular, more or less artificial, conditions of growth. Certain variations in conditions of growth merely altered the rate at which the plants progressed along this fixed path (as happened with the three concentrations of atmospheric carbon dioxide in Fig. 24.17); other variations altered the path itself (for example, marked changes in rooting conditions).

Where such a linear progression is present, this implies that any ontogenetic drift of specific leaf area must be reflected inversely in an ontogenetic drift of leaf dry weight percentage, for if (neglecting the small constant term, if any)

$$L_{H_2O} = L_F - L_W = wL_A \qquad 24.21.1$$

where w is constant and L_{H_2O}, L_F, L_W and L_A are respectively the water content, fresh weight, dry weight and area of the leaves, then

$$\frac{1}{\dfrac{L_W}{L_F}} - 1 = w\frac{L_A}{L_W}. \qquad 24.21.2$$

On the other hand the linear relationship of sunflowers in Fig. 24.19 was not a time progression of the same kind, being formed by plants all of roughly the same chronological age. Their size at an age of around 3 weeks varied over a fifteen-fold range because of differences in rate of growth during those 3 weeks. If these were all the facts one might suggest that this is another instance of progression at different rates along a pre-determined path, similar to the instances of Hughes, Cockshull and Heath but involving more drastic changes of growth rate. Against this explanation must be set the further observation of similar linear relationships at around 4 and 5–6 weeks, but with different slopes. Why should the form of the relationship change at 3 weeks in seven different years, in which the plants exhibited wide variations in growth rates and in stage of development? And what happens later on, for larger plants than those of Fig. 24.19, when the pathway broadens out until it includes a range of mean ratios of leaf area to leaf water content extending roughly from 0·3 to 0·4 dm^2 g^{-1}, a range of roughly 15 per cent on either side of the grand mean; and yet for the mean of any one batch of plants, of a particular age in a particular year, the fiducial limits are only ± 4 per cent?

The simplest explanation of the closely linear progressions would be that both leaf area and leaf water content are equally good measures of units of leaf structure, which are increasing in number as the plant grows. On this view these units would be associated with the predominantly aqueous phases of the leaf, the protoplasts, rather than with the walls, which make a relatively large contribution to the dry weight and a relatively small one to the water content, especially in the case of tissues with a high proportion of thick-walled cells, such as are commonly found in the main veins and the midrib. If this hypothesis should prove to hold good, we should be well on the way to the goal outlined in sections 24.15 and 24.16.

A combination of anatomical, biochemical and physiological studies would be needed in order to determine whether such a simple view can be substantiated, or whether the observed linear relationships are the consequence of more complex, compensating, systems. As we have seen, there are difficulties, and at this stage there is little point in speculating on whether or how they might be reconciled with the simple hypothesis; but consideration of the difficulties themselves may well provide useful ideas as the investigation proceeds.

24.22   A source of useful clues

Our consideration of water contents has thus led to the construction of a train of speculative argument involving hypothesis and tentative conclusions which can readily be tested by independent means. It has met two of the aims which we set before ourselves when beginning these analytical studies: on the one hand, quantitative limits have been set within which a particular facet of plant activity must be confined under particular conditions; on the other hand, constructive suggestions have emerged linking this facet with a number of others, spread over several aspects of plant science. In these ways it can be set alongside our other studies as a possible source of clues to situations in the whole living plant. Of course, this is only a beginning on the road to elucidating a problem of relating structural and functional indices. In other chapters, notably 12 (on the making and utility of measurements of respiration) and 27 (anatomical studies in relation to indices of growth) we examine independent means of shedding light on the particular problems which we have been discussing in the last seven sections. Not only in this particular instance, but more generally in the context of this work such further observations are essential for an understanding of the problems posed by the activities of whole plants growing in their natural surroundings. Clues such as those which we have just been discussing, even when they lead to alternative possible explanations, can be a guide to further studies. The ramifications of these further studies, spreading as they do into almost every corner of botanical science, will be the subject of the last Part of this book.

PART IV

PROBLEMS POSED BY
THE GROWING PLANT

INTRODUCTION: PROBLEMS OF THE LIVING PLANT AS A WHOLE

25.1 Structure and function

In our introductory discussions we established a distinction between the structure of a higher plant at any one moment, and the host of functions which are proceeding within that structure. The structure itself is a product of the past activities of the plant within the framework of the past environments within which it has lived; and the distinction is possible because much of a higher plant's total stock of carbon compounds is irreversibly tied up in structural elements, and removed from the total of labile materials which can become involved in future metabolic activities. Furthermore, much of the remaining stock is also tied up in a less permanent but nevertheless enduring way in intracellular structural elements, within which go on the elaborate chains of biochemical change involved in cellular metabolism. A gross example of this is the chloroplast, which may persist as a photosynthetic organ in a leaf for prolonged periods. It may then during senescence of the leaf undergo changes of composition and structure (often leading to transformation into the chromoplasts associated with 'autumn colours'), and finally break down, while part of the material of which it was composed is translocated into the stem. In the case of chloroplasts such changes are obvious and easy to follow, but less obviously during senescence similar changes may go on in other intracellular organelles, and also in the ground cytoplasm itself. In the living state this may be optically empty and capable of flowing, but it must nevertheless be highly organized, as is made clear by studies of fixed and dried preparations in the electron microscope.

25.2 Levels of structure

To understand the life of the plant at any moment, it might seem that ideally we ought first to have available information about its structure at various levels. The first level is that of gross morphology—the number, shapes and dimensions of parts and their relations to each other, numbers and positions of growing points, lengths of translocatory paths, and so on. The second level is that of anatomy—the numbers, shapes and dimensions of the various kinds of cells making up particular parts, and their relations to each other. The third level is that of structural biochemistry, of the non-labile parts of the plant body, principally cell walls; and finally of the quasi-permanent, but nevertheless ultimately labile biochemical structures—intracellular organelles, cytoplasmic structure, and so on.

We see at once that such an ideal programme is impracticable: such a detailed description has never been assembled for any species of plant. Even if it had been assembled, it would contain too much detail to handle. As we have done on previous occasions, we must bear in mind the complications of the detailed picture, while fixing our attention on those particular features of importance to a particular problem.

25.3 Aggregation of activity

Furthermore, the structural picture is only the first step towards understanding the life of a specific plant. It represents, as it were, an immensely elaborate piece of clockwork, wound up but not going. We also need information about the functions which are going on within the structure. Here once again the question arises of the scale of our description of these functions; a full description would be out of the question, even if there were no purely biochemical problems left unsolved. Even within a single cell the number of simultaneous biochemical processes may be very great (especially if the cell contains chloroplasts); and if it were possible to give a full description for every cell, the total picture produced would be hopelessly overdetailed and confusing. We must therefore arrange to aggregate the activities of numerous cells into totals which become comparable with measurements of whole plants and their parts such as those discussed in the previous two sections. This aggregation may either be arranged organ by organ, or function by function.

25.4 Limitations

Unfortunately, at the present time we are rarely in the happy position of knowing too much about the plant, and finding it necessary to cope with excessive detail. For all too many problems, essential data are lacking

25.4

and often unobtainable, so that for the time being we must be content with partial solutions, or none at all. Many of the reasons for this are inherent in the nature of the plant and the present state of natural science as a whole. Thus in certain respects, such as the structure of their cell-walls, plant cells are easier to study than those of animals, but mostly this is not so. The very cell-wall structure makes study of living cells by phase-contrast microscopy more difficult. The relatively low metabolic rate and low concentrations of many metabolites make biochemical studies of plant cells more difficult than those of animals; so does the relative lack of organs exhibiting biochemical specialization; and so does the greater biochemical diversity usual in plant cells. Important compounds such as growth substances, normally present in very low concentrations, may as yet be impossible to identify and measure in preparations from living plants making normal growth. In consequence physiological studies must often be made in ignorance of many details of the biochemical processes involved [26.1, 28.1]—to resume our earlier analogy, we have to study the operation of the elaborate clockwork without being able to open the case.

Moreover, when making physiological studies of whole plants or parts of whole plants, we are frequently hampered by the necessity of using techniques which do not interfere destructively with the system being studied, or at the very least, techniques whose interfering effects can be estimated in quantitative terms. These restrictions rule out many standard laboratory methods which have proved their value in the study, under controlled conditions, of partial systems which do not involve whole plants. The same restrictions can also make it difficult to apply the results of these latter investigations directly to the interpretation of studies of whole plants, unless some independent means can be found of relating conditions in the plant itself to those present in the laboratory studies. Many instances of all these difficulties have been encountered in Part II, notably in Chapter 12 on the measurement of respiration.

Furthermore, the limitations mentioned in the last two paragraphs apply equally to all studies of whole plants, whether in the laboratory or not. When in addition the complexities of the natural environment must be considered simultaneously, the difficulties mount up substantially (see, e.g., Chapters 9 and 10), leading to the complexity of plant/environment interrelations mentioned in 7.1 and 17.1. But there is nothing new in all this. Increasing information and improved methods of investigation

25.4

may have made us more conscious of some of the difficulties, but at the same time they have opened up new lines of investigation and new ways of solving old problems. They hold out the promise of a much more complete understanding of the life of the plant in its natural surroundings than would have been possible a decade or two ago, provided that the old difficulties and limitations are not forgotten in the excitement of applying new techniques.

25.5   The inevitability of ontogenetic drift

In the general terms of sections 25.1, 25.2 and 25.3 we see, when observing a particular plant, that the totality of its functions *now*, operating within the structure *now*, determines the type and rate of change of structure *now*, including both the laying down of new structures and the liquidation of old ones. In turn this brings about alterations both in the structure and pattern of working of the same plant *in the future*.

Proceeding thus step by step, the structure and pattern of working of the plant inevitably pass through a grand progression of ontogenetic drift. The changes at any one moment are brought about by the combined operation of effects of environment and the activities of the plant at that moment: and these in turn are a complex integration of such responses throughout the life of the plant up to that moment. Individual plant organs also pass through analogous progressions of ontogenetic drift, and these individual drifts need not be in phase, as happens, for example, in *Helianthus annuus* and many other plants, where new leaves are being continually produced at the top of the stem, while lower down, after a region of maturity, comes a zone where the leaves are becoming senescent and dying away. For some purposes it is useful to regard the ontogenetic drift of the plant as a whole as a summation of the drifts of the individual organs, but this is not the whole story. There are numerous correlation mechanisms which connect the activities of one part of a plant with that of others, and by variations in the rate of production or consumption of metabolites, by the release of growth regulating substances, and in other ways, the ontogenetic drifts of the different parts are interlinked.

25.6   The total activity of the whole plant: its analysis

The overall aim of our investigations has been to take a few first steps towards an understanding of this total activity of the whole plant. Our method of doing so has been the classical one of analysis. We have

25.6

started from the broadest possible view of the plant as a whole and its stock of carbon compounds: we have gradually broken this totality down, considering groups of organs and broad groupings of specific functions such as overall photosynthetic gain. We have examined methods of measurement of what can be measured directly, and of estimating the rest. In this way we have been able to assemble a broad framework of quantitative description of the structure and activity of the plant as a whole; and if our methods have been sound, any more detailed descriptions of structure and functioning must fit within this framework. We can now go on to examine means by which studies along the classical lines of morphology, anatomy, and the biochemistry and physiology of particular plant processes fit into the general framework, as further steps on the way to understanding the life and activities of the plant as a whole. We can then consider the plant's relations to the ecosystem in which it lives, thus leading finally to questions raised by the study not so much of individual plants, as of entire communities.

25.7 The synthesis of methods of studying plants

From what we have been saying it will be clear that in our context these various studies are ways of examining different facets of a single gem. In time it should be possible to link them all together into a single view of the structure and activities of the plant as a whole. As yet, however, no such synthesis can be made for any single species of higher plant. The studies which have been begun cover less than 1 per cent of the possible species, and those have almost entirely been chosen from plants readily accessible in temperate regions of the globe. Yet we have already seen that this small sample has revealed great diversity of behaviour, so that great care must be taken in endeavouring to put together the jigsaw puzzle out of parts relating to different plants. Members of a group of plants may show substantial qualitative uniformity of many biochemical mechanisms and yet may exhibit great quantitative diversity of physiological behaviour.

We shall continue to examine specific examples, and to endeavour to work out quantitative solutions to specific problems, even if only partial ones; but our consideration of them will be found to raise many more questions than can be answered at the present stage of our knowledge of whole plants. The subsequent chapter headings therefore lay emphasis on problems rather than on solutions.

25.7

SOME PROBLEMS OF MORPHOLOGY: GROWTH AND FORM

26.1 Restrictions of available information and techniques

Many volumes have been written about plant form: at first these served the necessary descriptive purpose of natural history, exemplifying the very great diversity of forms, and relating these in a general way to taxonomy, geographic situation, climate, plant community, and so on. With this as a basis necessary relations have been and are being worked out between the overall forms of plants and the mechanical and other physical properties of the plants themselves and their environment. At the anatomical level of investigation the genesis of particular plant forms can be followed by developmental morphology, by observing the activities of meristems and experimenting with the factors controlling both these and the development of the organs arising from meristematic activity. These observations and experiments are linked with biochemical studies of plant hormones and of the normal and abnormal nutrition of plant organs (we have examined an example of this, in relation to flowering behaviour, in 23.11), and with studies of the movement of labelled assimilates in the plant [23.6, 28.16, 32.8].

These last studies, at the biochemical and biophysical level of investigation, represent an important and rapidly expanding field of investigation, which is already assuming a remarkable degree of complexity. Except for ethylene, for which see Burg (1962) and Mapson (1969), it is at present most conveniently entered through review articles in Wilkins (1969), with its 78 pages of bibliographies containing more than 1800 entries (and for a fuller account of phytochrome see Furuya, 1968). As yet, however, relatively little of this great body of information has been brought into relation with the type of quantitative studies of whole plants with which we are concerned, and as the complexity of the interrelated mechanisms in the actively growing green plant is more fully appreciated, it becomes increasingly clear that there are great difficulties in the way of doing so. Interesting discussions bearing on this will be found in Wilkins's book, notably in Cleland (1969), pp. 62–71, on the role of endogenous gibberellins; Fox (1969), pp. 92–94, on naturally occurring cytokinins; Phillips (1969), pp. 185–189, on environmental effects on apical dominance, and on the relation of apical dominance to the forms of trees; Furuya (1968), pp. 388–391, and Siegelmann (1969), pp. 500–503, on the *in vivo* properties of phytochrome, and on related physiological responses; and see also Audus (1959) on correlation, Phillips (1969) on hormone-directed transport and the nutrient diversion theory, and Barlow (1970), pp. 36–40, on the forms of trees. The main difficulties arise because the chemistry of the naturally occurring cytokinins is very imperfectly understood, and although more is known of the structural chemistry of the natural gibberellins, reliable methods of extraction and assay of physiological quantities are largely lacking, especially for those gibberellins which do not produce marked effects on stem elongation; because phytochrome is always difficult and frequently impossible to estimate *in vivo* in chlorophyllous tissues; because in consequence little is known either of the concentrations of these substances naturally present in normal growing tissues, or of the physiological effects of changes in them; and because these physiological responses all interlock with each other, and with the natural concentrations of auxin and (it seems likely) of ethylene. Thus a wider application attaches to the remark of Fox (1969), made in a particular context related to the cytokinins, 'To measure respiration rates in various plant tissues before and after cytokinin treatment is probably an exercise in futility in the absence of information about the level of endogenous, natural cytokinins and the requirement of the tissues for these substances. For similar reasons one wonders about the significance of studies in which cytokinins sprayed or otherwise applied to intact plants or plant parts are shown subsequently to affect the metabolism of given substances.' It might be thought that we are in the position of a man groping round in a dark cellar for a black cat which may or may not be there, but the position is not quite as bad as this. In 1959 the Linnean Society held an interesting symposium on, 'Experimental approaches to the problems of growth and form in plants' (collected in the Society's *Botanical Journal*, **56**, pp. 154–302), and more recently Morgan (1968) has reviewed the application of these laboratory investigations to the interpretation of problems arising in field studies, and has cited instances where it seems probable that particular changes in plant form can be related to the activity of particular growth substances. This is a vast field, of which only a tiny corner has as yet been examined, and which offers great opportunities in the future.

We have already mentioned that characteristically the higher plants combine few types of organ with very great diversity of form and structure within any one type [2.1]; and that little is known about the formative influences bringing about this diversity. Up to the present, with good reason, most quantitative investigations along the lines we have been discussing have been made on annual plants of temperate regions. As a general principle, when opening up any new field of scientific investigation it is wise to take the simple cases first; additionally in our particular case the first economic applications related to temperate annual crop plants. We therefore still lack the quantitative data which would be needed for any wide ranging discussion. In particular, in wet tropical regions where the modes of plant growth are many and diverse, the observer is constantly struck by problems of all kinds relating to plant form about which practically nothing is known.

Accepting these restrictions, we can most usefully discuss certain problems of fundamental importance concerned with establishing the quantitative distribution of material between organs of different types, and we can begin to consider some of the factors which may influence the deployment of material within the group of organs of the same type possessed by an individual plant. Such studies, if extended to a wider range of growth forms, would contribute much to an understanding of causal morphology; would form a framework within which detailed biochemical and biophysical studies could be fitted; and would help to identify particular problems to which biochemical and biophysical methods could profitably be applied.

26.2 Value of comparisons

Our previous discussions have revealed the difficulties in understanding the growth of any specific plant under specific conditions in absolute terms. These difficulties are an important element in our problems. Most of our useful conclusions have to be reached by comparisons, and here we have seen the further ranges of difficulty which intervene if the gap between the two plants being compared is too wide. Differences of systematic position of species (with concomitant biochemical differences),

26.2

of plant form, of age, and of cultural conditions can all contribute towards widening the gap; and generally speaking, the narrower the gap, the more meaningful the comparison.

At the same time care is necessary in selecting suitable bases for the comparisons themselves, both to make it possible to distinguish inherent plant characteristics from the idiosyncrasies of particular cultural conditions, and to ensure that the conclusions are not affected by artefacts of the methods of analysis. In Part III numerous dangers in the latter direction have been mentioned, and in the present context perhaps the most likely is the generation of inbuilt correlations. We have considered [14.5, 18.12, 18.13] different examples of cases where the risk could be run with safety, but in general, when we cannot be sure of this, it is wiser to avoid the danger altogether by using independent observations as a basis, and ensuring that when they are combined the same basic observation does not contribute both to the numerator and denominator of a ratio, or both to the x- and y-coordinates of a plotted point. Therefore in the sections which follow [26.4, 26.7, 26.8] we shall consider an example of the use of such a treatment: it may be contrasted with that already used in discussing stem and root weight ratios [21.2–21.7], and the changes in distribution of dry matter associated with flowering [23.5–23.7].

26.3 Distribution of material between organs: an example

We can conveniently open our discussion by considering data on two species differing in plant form but belonging to the same genus, grown side by side under the same cultural conditions, and where we have simultaneously data on plants of each species of different ages. Such data are furnished by the series of experiments on the growth of young plants of two species of sunflower, *Helianthus annuus* and *H. debilis*, already described when we were considering techniques [7.5, 7.6, 11.6, 11.21, 14.2, 24.3, 24.4, 24.5]. We recall that while *H. annuus* is tall and at first unbranched, *H. debilis* begins to branch at an early stage (around 3 weeks old, and at a total dry weight of the order of 1 g), long before flower primordia begin to form. While *H. annuus* goes straight up to a large capitulum at perhaps 3 m, the main axis of *H. debilis* terminates in a much smaller head at a height of 1 m or less; growth then continues in the lateral branches, which behave in the same manner as the main stem, branching and terminating in a flower head. In consequence isolated plants become rounded bushes, where the diameter can exceed the height. With these differences in branching go differences in internode length and leaf size. An isolated plant of *H. annuus* has relatively few and long internodes, and few and (for a temperate plant) large leaves, up to roughly 0·1 m² in area, while the internodes of *H. debilis* are much shorter, and the leaves much smaller (up to around 200 cm² in area) and much more numerous. These differences in the broad architecture of the plant are no doubt connected with the facts that the main stem of *H. annuus* soon becomes very thick, the stem weight ratio rising steadily as the plant becomes larger [21.4, 21.5]; and this thick stem is supported by a massive root-stock. In *H. debilis* the main axis and branches are much thinner: in plants of comparable total dry weight the base of the stem is less thickened, and the main branches of the root system are correspondingly thinner. The overall effect of these differences of growth form on the stem weight ratios of young plants [Table 26.7] might not at first have been expected, but one must remember the progressive effect of the increasingly large leaves of *H. annuus* [Fig. 26.8].

26.4 Proportion of roots: comparison of two related species

It is interesting, however, to find that a difference in root weight ratio is established at a very early stage in growth, before the difference in plant form has had time to establish itself. This is shown in Fig. 26.4.1a, where the root dry weights are plotted against the total top dry weights of individual plants of the two species at the first harvest, when the young and old plants both of *Helianthus annuus* and *H. debilis* are approximately 22 and 29 days from germination, respectively. It will be seen that the difference is very marked even for the smallest plants, of around 0·5 g total dry weight, still unbranched and having about eight leaves in addition to the cotyledons. In Fig. 26.4.1b the large difference can be seen to persist through the whole range up to the largest plants at the second harvest, with total plant dry weights around 10 g.

The first-order regression equations are, for *H. annuus*,

$$R_W = 0·519(S_W + L_W) - 0·086; \qquad 26.4a$$

and for *H. debilis*,

$$R_W = 0·215(S_W + L_W) - 0·028; \qquad 26.4b$$

26.4

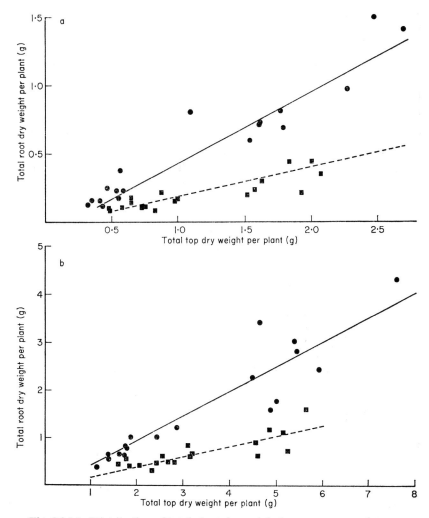

Fig. 26.4.1. Distribution of total plant dry weight between roots and tops: individual young plants of *Helianthus annuus* compared with *H. debilis*. For experimental arrangements, see 26.3. For mean root to top dry weight ratios of the same plants, see Table 26.4. *Helianthus annuus,* ●, ——— regression line, equation 26.4a. *H. debilis* ■, ----- regression line, equation 26.4b. (a) First harvest, 24 July 1968, plants aged 22 and 29 days from germination. (b) Second harvest, 31 July 1968, plants aged 29 and 36 days from germination.

26.4

where R_W, S_W and L_W stand for the dry weights of the roots, stems and leaves respectively, all measured in grams. We see that for a given value of $(S_W + L_W)$, R_W for *H. annuus* has on average a value more than double that for *H. debilis*. The distribution of points shown in Fig. 26.4.1 does not suggest that either relationship shows significant departures from linearity in the region considered [however, see 26.7, 26.8]. Neither relationship passes through the origin, but for all the larger plants we can neglect the effect of this on the values of root weight ratio. It is therefore worth comparing the weight ratios directly, and if we take the age groups separately, as set out in Table 26.4, the differences between

Table 26.4. Root to top dry weight ratios of two species of sunflowers compared for plants of different ages (in days) harvested on 24 and 31 July 1968. Brackets connect values of samples from the same initial population: each value is a mean for nine plants, with standard error. Same series of experiments as in Tables 7.5, 7.6, 14.3, 24.3, and cf. Table 14.4 and Fig. 11.23a and b.

Helianthus annuus		*H. debilis*	
Age	Ratio	Age	Ratio
⌠ 22	0.423 ± 0.040	⌠ 22	0.194 ± 0.014
⌡ 29	0.432 ± 0.019	⌡ 29	0.240 ± 0.016
⌠ 29	0.491 ± 0.038	⌠ 29	0.162 ± 0.015
⌡ 36	0.492 ± 0.042	⌡ 36	0.193 ± 0.017

species are highly significant in each case, and the cumulative significance of the difference is so great that a statistical demonstration is hardly necessary.

Equally, in this particular instance the magnitude and consistency of the difference are such that we do not require a more elaborate analysis, taking account of age groups and harvest sequences; but this is not necessarily so (see, e.g., 11.21–11.23 and 14.5, 14.6), and in many cases the more elaborate treatment would be essential.

However, we have already seen that in *H. annuus* the fraction of the increment of new dry matter going to roots out of that added in a particu-

26.4

lar week can vary within wide limits (the data for seven years and two age
groups given in Table 21.5 ranging from 19 to 46 per cent) and that this
may be connected with the stage of the ontogenetic drift [21.4.] We must

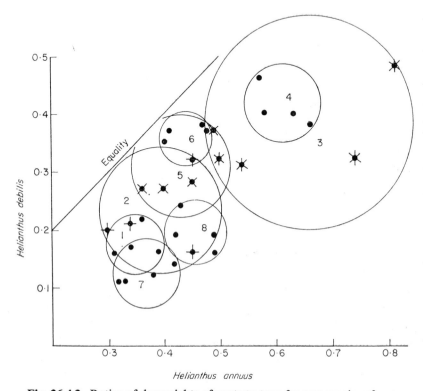

Fig. 26.4.2. Ratios of dry weights of roots to tops for two species of sunflowers compared. Four groups of plants of *Helianthus annuus* [21.5,
22.4] grown side by side with *H. debilis* during June and July, 1961–68 at the
Botanic Garden, Cambridge. Each point compares a mean of nine even-aged
plants of each species. The four points for each year enclosed in a circle
marked with the last digit of the year number, those for 1962, 1963 and
1965 separately distinguished. Equality of the two mean ratios is shown
by a line.

therefore ask—were the differences shown in Table 26.4 also the product
of particular cultural conditions? In Fig. 26.4.2 the mean root to top
ratios for *H. debilis* are plotted against the corresponding ratios for *H.
annuus*, at four harvests in each of the years 1961–68. Each point compares

26.4

means of nine even-aged plants of each of the two species. The four points for each year are enclosed in a circle, marked with the year number in the 1960s, and the points for the years with the largest variations—1962, 1963 and 1965—are separately distinguished. We see that the mean ratio for *H. debilis* is in every case smaller than that for *H. annuus*. The smallest difference, of around 10 per cent of the values of the ratios themselves, is for the younger plants in 1966, and all the other differences are much larger. We also see that the small variations in the year 1968, already noticed in Table 26.4, are not exceptional, the points in 5 out of the 8 years having a similar small spread, but distributed over a wide range of the ratio for each of the two species. Each year appears to have its own idiosyncrasies, and while there is a tendency for the ratios for the two species to rise or fall together, this is by no means invariable. The figure is a further demonstration of the value of repeating such field experiments under a variety of weather conditions.

26.5   The comparison continued: relative growth rates

A marked difference in form between the two species of *Helianthus* is thus established. The plants of the two species were grown side by side under the same cultural conditions, and the observations covered substantial ranges of total size, of weather conditions and of growth rates. We can therefore regard the difference as a genetically determined interspecific one. It is interesting to ask whether this difference in form is associated with a difference in overall plant performance under the specific cultural conditions of this experiment, as expressed by relative growth rate in dry weight. The figures for 1968 are set out in Table 26.5. We see that the relative growth rates of the two species are practically identical at both ages, although the age groups differ from each other. What is more, the differences between the age groups are not merely a matter of overall plant size. We saw from Fig. 26.4.1 that the largest plants at the first harvest overlapped in total dry weight the smallest plants at the second harvest, yet the differences in relative growth rate here are just as pronounced. We have considered a similar phenomenon before [15.2]. With this experience in mind it seems likely that the exactitude of the agreement is partly coincidental, the product of the particular weather of the year acting upon plants of a particular size. In general, if the experiment were repeated in another year quite a different pattern of results would probably be found (see, e.g., for *H. annuus*, the relative growth rates in Table 22.4).

Table 26.5. Relative growth rates, in week⁻¹, of two species of sunflowers compared. Plants of two age groups harvested simultaneously on 24 and 31 July 1968, as in Table 26.4. Each sample divided into three sub-samples, of matched large (L), medium (M), and small (S) plants. For similar figures for *H. annuus* in 1966 and 1967, see Table 15.2.

Helianthus annuus			H. debilis		
Age range		RGR	Age range		RGR
22–29	L	1·25	22–29	L	1·28
days	M	1·36	days	M	1·29
	S	1·19		S	1·26
	All	1·27		All	1·28
29–36	L	0·93	29–36	L	1·02
days	M	1·14	days	M	1·03
	S	1·00		S	0·94
	All	1·01		All	1·01

26.6   Comparable experiments in 7 other years

Fortunately we have available data for comparable experiments made at the same time of year (harvests around the end of July) in the years 1961 to 1967, and the corresponding figures for relative growth rates of the two species in the two age groups are given in Table 26.6. We see an identity as close as that observed in 1968 also in 1962, and one almost as close in 1965, making in all 3 years out of 8. Of the remaining 5 years, *H. annuus* had a higher relative growth rate in both age groups in 1966, and *H. debilis* in 1961 and 1964. In 1963 and 1967 the young plants of *H. annuus* grew relatively faster than those of *H. debilis*, whereas for the older plants the reverse was the case.

We also notice that while the relative growth rates of the older age group are always below the corresponding figure for the younger one, the fall for *H. annuus* is greater on five occasions out of eight (1961, –63, –64, –65, –67), equal on two (1962, –68), and less on one only (1966).

26.6

Table 26.6 Relative growth rates, in week^{-1}, of two species of sunflowers compared, as in Table 26.5. Plants of two age groups, harvested simultaneously around the end of July, 1961–67. Age ranges in days from mean date of germination.

Year	*Helianthus annuus*		*H. debilis*	
	Age range	RGR	Age range	RGR
1961	19–26	1·43	12–19	1·57
	37–44	0·83	30–37	1·06
1962	18–25	1·21	18–25	1·21
	32–39	0·96	32–39	0·95
1963	18–25	1·57	18–25	1·38
	32–39	1·08	32–39	1·26
1964	18–25	1·42	18–25	1·59
	32–39	1·00	32–39	1·31
1965	21–28	1·06	21–28	1·03
	35–42	0·62	35–42	0·68
1966	21–28	1·06	21–28	0·89
	35–42	0·89	35–42	0·53
1967	21–28	1·06	21–28	0·92
	35–42	0·61	35–42	0·70

We now begin to get an inkling of the mechanism underlying the apparent identity. We have already seen [Fig. 13.6] that at this stage in its life cycle the relative growth rate of *H. annuus* is falling very rapidly from an early peak. Kidd, West and Briggs (1920) pointed out, when examining the data of Noll's pupil Gressler, that for *H. cucumerifolius* (which was almost certainly a variety of *H. debilis*, App. 3.5) the peak of RGR is lower, and the subsequent decline less rapid. In consequence, in the specific year and for the two particular taxa which they examined, shown in Fig. 26.6, the two curves of RGR against time crossed at an age of about 6 weeks, and thereafter the RGR of *H. cucumerifolius* continued higher than that of *H. macrophyllus*, which behaved similarly to *H. annuus* in respect of ontogenetic drift of RGR, for the rest of the growing season. We are obviously working close to the mean age at which the curves cross for the variants of the two species which we are using. Identity for two age groups is achieved in those years when some combination of past and present weather either causes the RGR of *H. debilis* to decline more

26.6

rapidly than usual, or that of *H. annuus* less so. Fig. 26.4.2 has shown that climatic differences from year to year can have systematic differential

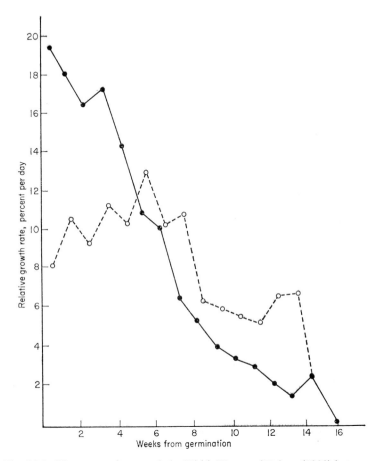

Fig. 26.6. The comparison made by Kidd, West and Briggs (1920) between the ontogenetic drifts of relative growth rate of two species of sunflowers studied by Gressler (1907). ——, *Helianthus macrophyllus*, which behaved similarly to *H. annuus*, and -----, *H. cucumerifolius*, which was probably a variety of *H. debilis*.

effects on the growth of the two species; that these differential effects are usually very consistent within any one year; and that they vary considerably from one year to another.

26.6

26.7 The comparison continued: stem weights

In the years in which this identity occurs (1962, 1968), we have the interesting example of two sets of plants of different species, showing marked and consistent differences in gross morphology [26.3, 26.4], and yet growing at the same relative rate for two different age groups, and therefore providing a particularly suitable example for comparative purposes. Later on it will be worth while to extend the comparison to the other elements, both anatomical and physiological, which contribute to the overall relative growth rate. In this chapter, however, we are concerned with morphological elements, and we must ask whether the difference in root weight ratio between the two species has been compensated for by changes in stem weight of opposite sign, or whether there is an interspecific difference in leaf weight ratio also.

We can conveniently examine this question in a preliminary way by constructing Table 26.7, similar to Table 26.4, giving the ratios of stem

Table 26.7. Ratios of stem dry weight to the total of leaf plus root dry weight of two species of sunflower compared, for plants of different ages (in days) harvested on 24 and 31 July 1968. Brackets connect values of samples from the same initial population: each value is a mean for nine plants, with standard error. Same experiments as Tables 26.4, 26.8, and cf. Table 14.4.

Helianthus annuus		*H. debilis*	
Age	Ratio	Age	Ratio
⌠ 22	0.156 ± 0.010	⌠ 22	0.237 ± 0.017
⌊ 29	0.210 ± 0.004	⌊ 29	0.296 ± 0.010
⌠ 29	0.208 ± 0.010	⌠ 29	0.294 ± 0.010
⌊ 36	0.269 ± 0.010	⌊ 36	0.382 ± 0.020

dry weight to the total of leaf plus root dry weight for the same eight sets of plants. These figures show the early development of the interspecific difference, *Helianthus debilis* already having a higher mean stem weight ratio than *H. annuus* for the youngest plants at the first harvest. We also see that for both species the ratio increases substantially as the plants become older and larger. The form of the relationship between stem weight and the weight of the rest of the plant is obviously of interest, and we should not conclude from Table 26.7 that it is necessarily non-linear.

The most convenient way of examining this question is to extend the method of Fig. 26.4, and plot stem dry weight against the total of leaf and root, shown in Fig. 26.7a for the first harvest, and Fig. 26.7b. for the second harvest. In Fig. 26.7a we see for the younger age group of both species a partially overlapping distribution of dry matter between the stem and the rest of the plant. Above a total for leaf and root of more than 1 g the interspecific difference becomes more marked and continues in all the older and larger plants shown in Fig. 26.7b.

The other important point to emerge is the difference between the relationships for the two species. If we calculate second order regression equations we obtain, for *H. annuus*,

$$S_W = -0.084 + 0.264(L_W + R_W) + 0.00068(L_W + R_W)^2; \quad 26.7a$$

and for *H. debilis*,

$$S_W = -0.048 + 0.285(L_W + R_W) + 0.0252(L_W + R_W)^2, \quad 26.7b$$

where S_W, L_W and R_W are the dry weights of stems, leaves and roots respectively, all measured in grams.

We see that for *H. annuus* the relationship is effectively linear; the coefficient of the quadratic term does not reach significance, and the whole term is in any case negligibly small within the range in which we are working. For *H. debilis*, on the other hand, the relationship is significantly curved.

Both regression lines show a positive intercept on the $(L_W + R_W)$ axis, that for *H. annuus*, at 0·32 g, being much larger than that for *H. debilis*. This does not, of course, mean that the young plants of *H. annuus* reach a total dry weight of around 0·3 g before any stem develops; Fig. 26.7 shows that seven of the nine points for the first harvest are above the line of best fit through all the data for this species in this year, and obviously the relationship must be curved in the early stages, with the proportion of new dry material going to the stems slowly rising, up to a total dry weight of rather over one gram. (The disadvantages of polynomials for fitting curves to data with a pronounced change of curvature at one end have already been discussed in 20.4, and this is another example of the same difficulty.) At greater dry weights it appears that, taking all these observations together, a roughly constant proportion of the increment is being

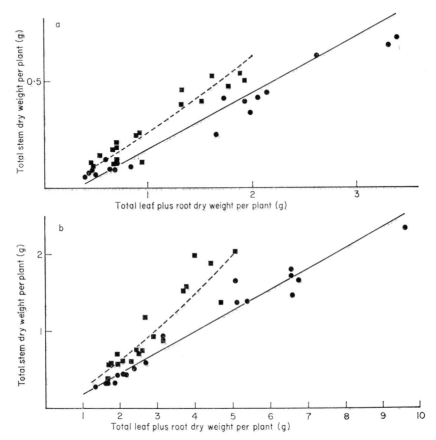

Fig. 26.7. Distribution of total plant dry weight between stems and the rest of the plant: individual young plants of *Helianthus annuus* compared with *H. debilis*, as in Fig. 26.4.1. For mean ratios of dry weight of stem to that of the rest of the plant, see Table 26.7. *Helianthus annuus* ●, ── regression line, equation 26.7a. *H. debilis* ■, ----- regression line, equation 26.7b. (a) First harvest, 24 July 1968, plants aged 22 and 29 days from germination. (b) Second harvest, 31 July 1968, plants aged 29 and 36 days from germination.

added to the stems. However, this relationship was shown in one particular series of observations, under one particular set of environmental con-

26.7

ditions, and as we have several times seen, even for plants of the same overall size, different growth conditions may well produce widely different results. In Table 14.4, for example, we saw that in 1966 for plants of *H. annuus* falling within the same size range (1·5 to 4·5 g total dry weight), 17·7 per cent of the increment during the experimental week was added to the stem. The point was driven home by the wide variations from year to year shown in Table 21.5, covering the much larger range of total dry weights up to 550 g, and affording several comparisons between sets of plants of comparable sizes. We conclude that the approximately linear relationship shown for *H. annuus* in Fig. 26.7, above a (leaf plus root) dry weight of around 1 g is fortuitous.

In fact, as we noted above [26.4], more generally we might expect to have to treat separately the two harvests and the two age groups. If we look at the data again with this idea in mind, we see that in Fig. 26.7a a regression line calculated for the plants of *H. annuus* at the first harvest alone would have a smaller slope than the line for the two harvests combined (the six lowest points are all above the line, seven of the nine highest points below it). At the same time the line for the second harvest would have a slightly greater slope. In the context within which we are working, these differences are small enough to neglect, but in a wider context they are worth noting as an example of effects which may on occasion be very important.

26.8   The comparison continued: leaf weights

Our examination of the proportion of the total dry material in stems has made clear that in this respect the interspecific differences are considerably smaller than they were for roots. It follows that there must also be a marked interspecific difference in the amount of dry matter in the form of leaves, and this expected difference is seen in Fig. 26.8. As for roots, the difference is marked in the youngest and smallest plants, and persists to the largest ones. It would also be expected that the relationship would be a curved one. Fig. 26.4 shows a close approximation to linearity, but the straight line of best fit had a positive intercept on the *x*-axis, reflecting a tendency of the root to shoot ratio to rise in the early stages, as could be seen in Table 26.4. This age and size difference would thus tend to reinforce the substantial difference which we have just noted for stems: if there is an increase in the proportion of dry matter in the roots (most marked in the early stages) and in the stems (more marked later on), then *a fortiori* there must be a decrease all along in the proportion of dry matter in the leaves. This we see in Fig. 26.8. The magnitude of the

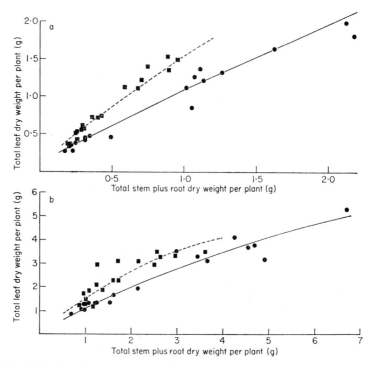

Fig. 26.8. Distribution of total plant dry weight between leaves and the rest of the plant: individual young plants of *Helianthus annuus* compared with *H. debilis*, as in Figs 26.4.1 and 26.7. For mean ratios of dry weight of leaves to that of the rest of the plant, see Table 26.8. *Helianthus annuus* ●, —— regression line, equation 26.8a. *H. debilis* ■, ----- regression line, equation 26.8b. (a) First harvest, 24 July 1968, plants aged 22 and 29 days from germination. (b) Second harvest, 31 July 1968, plants aged 29 and 36 days from germination.

interspecific difference is best brought out by examining the ratios, set out in Table 26.8. It can be clearly seen for both age groups and both harvests that *H. debilis* has a substantially higher proportion of its dry matter in the form of leaves than has *H. annuus*. The even more marked fall of the ratio with age and size in both species is also obvious.

26.8

Table 26.8. Ratios of leaf dry weight to the total of stem plus root dry weight of two species of sunflowers compared, for plants of different ages (in days), harvested on 24 and 31 July 1968. Brackets connect values of samples from the same initial population: each value is a mean for nine plants, with standard error. Same experiment as in Tables 26.4, 26.7, and cf. Table 14.4.

Helianthus annuus		H. debilis	
Age	Ratio	Age	Ratio
$\lceil$ 22	$1{\cdot}362 \pm 0{\cdot}080$	$\lceil$ 22	$1{\cdot}851 \pm 0{\cdot}045$
$\lfloor$ 29	$1{\cdot}119 \pm 0{\cdot}042$	$\lfloor$ 29	$1{\cdot}407 \pm 0{\cdot}068$
$\lceil$ 29	$1{\cdot}018 \pm 0{\cdot}048$	$\lceil$ 29	$1{\cdot}752 \pm 0{\cdot}059$
$\lfloor$ 36	$0{\cdot}872 \pm 0{\cdot}050$	$\lfloor$ 36	$1{\cdot}312 \pm 0{\cdot}082$

It is most convenient to proceed as before [26.4, 26.7], working out a common regression for each species, including both age groups and both harvests, when we have, for *H. annuus*,

$$L_W = 0{\cdot}131 + 0{\cdot}992(S_W + R_W) - 0{\cdot}041(S_W + R_W)^2; \qquad 26.8a$$

and for *H. debilis*,

$$L_W = 0{\cdot}124 + 1{\cdot}543(S_W + R_W) - 0{\cdot}139(S_W + R_W)^2, \qquad 26.8b$$

where L_W, S_W and R_W are the dry weights of leaves, stems and roots respectively, all measured in grams. These lines are plotted in Fig. 26.8. They correspond to a much more rapid increase of leaf weight for *H. debilis* than for *H. annuus* at low values of $(S_W + R_W)$. At the same time for *H. debilis* the quadratic term is much more important, giving a more curved relationship, so that at values of $(S_W + R_W)$ between 3 and 4 g the two curves are already converging. This interpretation agrees generally with the disposition of points in Fig. 26.8; in particular, around 3–4 g the points for the two species intermingle. This particular instance thus reinforces the importance of extending morphological comparisons of this kind over a substantial range of plant size.

A further point emerges from Table 26.8. We see that there is a large and highly significant difference in the ratio of dry weight of leaf to that of the rest of the plant for the 29-day-old plants of *H. debilis* at the two harvests ($0.002 > p > 0.001$). We see from Fig. 26.8 that this is not due simply to plant size—the ratio is declining with total dry weight, yet the plants having the larger ratio (at the second harvest), are also themselves on average larger. There is a strong probability that we have here a continuing effect of past climate, before the harvests began, showing how difficult it is, in field experimentation, to avoid the interlocking effects of the complications which we considered in 3.5, 7.1 and 17.1.

26.9 The morphological comparison of two related species summed up

Differences such as those which we have been discussing cannot but affect overall plant performance. Other things being equal, the plant which can deploy a higher fraction of its total dry matter in the form of leaves would have a higher relative growth rate, with all the advantages which this confers. In the comparison which we have just been making, *Helianthus debilis* started with a markedly higher leaf weight ratio, and retained this advantage throughout. Yet, as we saw in Table 26.5, for both age groups in 1968 it had the same relative growth rate as *H. annuus*: the advantage conferred by the morphological difference must have been counterbalanced elsewhere, mainly, it turns out, in unit leaf rate (on average 12 per cent lower in *H. debilis*), and to a lesser extent in specific leaf area (on average 3 per cent lower). An examination of the values of unit leaf rate for the 8 years and two age groups (similar to that in Table 26.6) shows that this main difference is also systematic, *H. debilis* having a lower unit leaf rate than *H. annuus* in every one of the sixteen comparisons, the mean ratio of the two being roughly 0.8. Table 26.6, in which over a number of years, and within a limited age range, neither species had on average a markedly higher relative growth rate, thus offers an example of compensatory differences between species, comparable to the intraspecific instances which have already been noticed [17.6–17.10]. Such cases are not uncommon, and in the context of the plant's natural environment may be of considerable biological importance. The further investigation of this particular example is, however, made more difficult by additional differences between the two species of *Helianthus*, in the forms of ontogenetic drift of the various derivates, which together combine to produce the differences in overall drift of relative growth rate already discussed [26.6]. Complications of this kind

26.9

are always liable to arise when comparing one species with another, and they must be tackled by sequences of harvests long enough to elucidate the forms of the drifts. We shall investigate one of the simpler examples of intraspecific differences later [26.13, 26.14, 27.11].

Meantime four general points emerge from our discussions. Firstly, *H. debilis* had a root system, for a given size of top, only half the weight of that of *H. annuus*: in some circumstances, such as a shortage of water or mineral nutrients, this reduction in material in the root system might have very disadvantageous effects, a suggestion which could be followed up by field observation and experiment, including investigations of root distribution and activity, in the natural habitats of each of the two species. An interesting parallel is afforded by the work of van Dobben (1967), which, however, was unfortunately confined to a single growing season. He was concerned to explain the complete, but short-lived, dominance of *Senecio congestus* in the developing vegetation of a newly drained polder. This species has an excellent distribution mechanism, a well-developed stand being estimated to produce 1.6 tonne ha^{-1} of seed in a single season. Its dominance soon after draining can be related to this, and also to its low root weight ratio, which contributes to its rapid growth in the conditions of high concentrations of mineral nutrients, especially nitrogen, and relative absence of competitition. The nitrogen content of the upper layers of the soil then rapidly falls (it is calculated that 60 kg ha^{-1} is lost to the local ecosystem during one season in the seed alone), and the plant is replaced by deeper rooted species.

Secondly, the architectural pattern of *H. annuus*, with its larger, fewer leaves, longer internodes and less branching causes it to grow in height more rapidly than *H. debilis*. In a situation of competition with other plants growing rapidly in height (say for the colonization of open ground), slower growth in height might be disadvantageous. On the other hand, the same architectural difference between the two species offers a possible countervailing advantage to *H. debilis*, which, because of its cymose branching system and early flowering, produces a small quantity of mature seed much earlier in the life cycle than *H. annuus*. Consequently, in places where the growing season is liable to be cut short, say by drought, it might well be able to reproduce when *H. annuus* could not. The plants which we have been comparing were growing in an experimental garden, well spaced out and well supplied with both water and mineral nutrients: but it is, of course, essential to view the whole plant, not against such an artificial background, but against a background of the environment in which it normally lives, and in which it must survive in competition with, and subject to interference by, other components of the ecosystem.

Thirdly, Table 26.6 showed that the agreement in RGR for two different age groups in the same week only occurred in two years out of eight. In other years sometimes both age groups of one species had the advantage, sometimes the other, and sometimes one age group only. We were able to explain this behaviour in part on the basis of displacement of the ontogenetic drift of RGR in one species relative to that of the other— generally speaking the RGR of *H. annuus* is falling more rapidly than that of *H. debilis* at the stage of the life cycle with which we are concerned. We have here an example of the combined influence of environment and ontogenetic drift.

Fourthly, there is the question of the combined influence of past and present environments. In the comparison which we have been considering, the two species had characteristic ontogenetic drifts which differed substantially in form. But even within a single variety of a single species, at a given size as measured by total dry weight, behaviour may vary very widely as a result of the combined effect of past environment (as influencing the form, structure and functioning of the plant at the beginning of the experimental period) and present environment, with its influences on the functioning and development of this machinery during the experimental period. We have noted examples of this, but for our present purposes it is convenient to look back to Table 21.5, dealing with the same set of experiments. A number of morphological indices of a uniform genetic strain of *H. annuus* are there related to (i) age, (ii) plant size at the beginning of the experimental period (which together provide a measure of the influence of past environment) and (iii) increment during the experimental period (which together with size and shape at the beginning of the experimental period, provides a measure of the influence of present environment). If we consider only the five largest sets of plants, with mean initial dry weights within the range 45–250 g, and with mean increments in dry weight in the range 75–325 g, we still see very great differences in the development of plant form during the experimental week. In 1967, plants having an initial dry weight of 233 g at 39 days gave a ratio of stem increment to leaf increment during a week of 0·56. In 1961, plants having an initial dry weight of 240 g at 40 days (and therefore initially rather

26.9

similar) gave a ratio of stem increment to leaf increment of 2·05, nearly four times as large. Similar variations were shown in the development of stems and roots. In 1963 26 per cent of the increment during the experimental week went into stem; and 46 per cent into root. In 1961 the position was reversed, 49 per cent going into stem and 29 per cent into root. 1967 was intermediate, with 37 per cent to stem and 36 per cent to root.

26.10.   Conclusions on experimental design

Such variations emphasize the importance of taking three practical precautions when comparing the performance of two different species under natural or semi-natural conditions. It is essential that the plants to be compared should be growing side by side under identical cultural conditions; and the comparisons should embrace both more than one age group and more than one set of cultural conditions. In a climate such as the English one this last requirement can be met simply by repeating the experiment in a different year or even in a different week: in a more stable and predictable climate it may be necessary to work during different seasons or alter the cultural conditions in some other way. Variations in relative response can then be detected, and their extent and direction can be determined.

When this has been done we are still a long way from being able to specify causal mechanisms relating the differences in age, ontogeny, past and present environment to the observed changes in plant form. For example, differences in response which we have noted in the preceding paragraphs cannot as yet be explained in this way (26.1, and see Morgan, 1968, especially the latter part of the discussion on p. 213). We will therefore conclude this chapter by examining one or two examples of work under controlled environmental conditions, when it may be possible to relate specific changes in plant form to specific ontogenetic and environmental changes, even if the intervening mechanism cannot be positively identified.

26.11   Comparisons of growth under controlled environmental conditions

In the comparisons between species just discussed we have seen the particular interest of comparisons where the overall performance of the two sets of plants, as measured by relative growth rate, is identical, although the plants being compared are different. There is no difficulty in finding similar instances where the prime difference is not genetic, but lies in a difference in ontogeny or environment.

If we are concerned to establish specific relationships between changes in plant form and changes in any of these factors, there are great attractions in working under controlled environmental conditions. We have seen in earlier chapters how we can thus largely avoid the inextricable complex of effects which made our comparisons earlier in this chapter so difficult to explain; and there is the particular advantage to investigators both of physiology and of causal morphology, of being able to return at will to the investigation of a specific plant form, a facility impossible in field experiments such as we have been discussing (but see Chapter 5 on sources of variability of seed or propagule). Yet problems as complex as these cannot be solved by laboratory experimentation alone, however ingenious. The differences between the complex of the plant and artificial conditions in the laboratory and that of the plant and its natural conditions are so numerous that the intellectual gap must be bridged by observations under a variety of intermediate, semi-natural conditions, and advances must thus take place simultaneously on a broad front, the different sets of experiments under varied conditions being planned to support each other. With this caution in mind, let us turn to a few instances of experiments under largely or wholly artificial conditions.

26.12 Leaves versus the rest of the plant

The effects of a number of environmental factors on leaf weight ratio provide good examples, which we have already discussed in detail in Chapter 18, the effects of the various factors being summarized in 18.15. It will be recalled that of the dozen or so aspects of the aerial environment investigated, only two, daylength and low temperature, were shown to affect the proportion of total dry weight found in the leaves of *Impatiens parviflora*. Five factors were shown to have no effect under controlled conditions, and there was a strong probability that the complex of other factors associated with the change from controlled conditions in the laboratory to natural conditions in the field also produced no effect.

Daylength changes are necessarily involved with changes either in radiant intensity or in total daily light; and although, as we have seen, the leaf weight ratio at a given plant size is not affected by intensity of radiation in the absence of a daylength change, nevertheless intensity or total daily light have marked effects on relative growth rate, acting through other components of the overall growth system. What is more, as

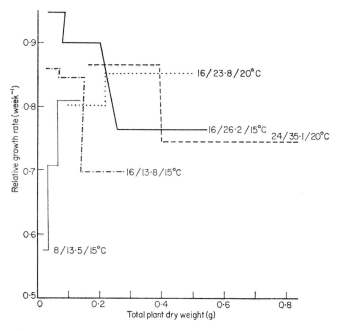

Fig. 26.12. Relationships between relative growth rate and total plant dry weight for means (usually of eight plants) of *Impatiens parviflora*, harvests being at weekly intervals and plants aged between 17 and 45 days. Five experiments, each under constant conditions in growth cabinets, but differing in daylength, total daily visible radiation, and temperature, indicated in this order, in hours, cal cm^{-2} day^{-1}, and °C. From Hughes and Evans (1963).

relative growth rate in this species also shows a marked ontogenetic drift, any comparisons must allow for this aspect also. Comparative figures are shown in Fig. 26.12 for plants grown at two different temperatures and otherwise under comparable conditions in all respects except daylength and total daily light. For the smallest plants, of total dry weight below 0·1 g, daylength appears to have a marked effect, but otherwise it seems that under these conditions the relative growth rate is much more

26.12

affected both by plant size and by total daily light than it is by daylength, in spite of the fact that the latter alters the proportion of dry matter finding its way to the leaves, while plant size and total daily light do not [18.9, 18.13].

Once again we have found that attempts to take account of the effects of particular changes in environment on plant form have to consider also ontogenetic changes, even under controlled conditions. We now see also the further point that a change in an aspect of plant form as important as the proportion of dry matter finding its way to the leaves may produce less effect on overall plant performance (as measured by changes in relative growth rate) than do ontogenetic drifts or environmental changes not connected with changes in this aspect of plant form (and see 27.2, 27.8).

26.13   Roots versus the rest of the plant

In growth studies carried out under controlled environmental conditions, and primarily concerned with the effects of changes in the aerial environment, we have already come across instances where the root system is affected. One effect of reducing irradiance at a fixed daylength, for example, is to increase the proportion of dry weight in the stem of *Impatiens parviflora*, and decrease that in the root [21.3, and Fig. 21.3]. It appears that the form of the ontogenetic drift is not affected—there is a slight tendency for root weight ratio to increase with age—but the whole drift is displaced with changes in irradiance. For all the factors of the aerial environment which were investigated in this particular series of experiments, such effects were small when compared with certain effects of changing the root environment. In Table 18.9 we saw that much the largest effect on leaf weight ratio was produced by changing the root environment from loam and sand, with added mineral nutrients, to vermiculite and culture solution; this was accompanied by a large change in root weight ratio. Such changes in a variety of species have been well known for a long time, but most of the instances in the extensive literature deal with agricultural, horticultural or other semi-natural situations where conditions around the growing root system cannot be specified with any accuracy [10.1–10.5, 10.8; Rorison, 1969].

The problem was therefore investigated further by Lewis (1963), who grew *Impatiens parviflora* from seed on a substratum of coarse crushed flint with a frequent intermittent spray of culture solution. This had the

26.13

effect of continually renewing the nutrient solution round the roots, while at the same time maintaining good aeration of the root system. Conditions were therefore in these respects similar to one of the natural habitats of the species, on shingle banks in rivers.

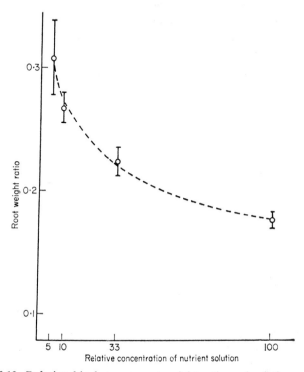

Fig. 26.13. Relationship between root weight ratio and relative concentration of nutrient solution for four sets of plants of *Impatiens parviflora* cultured from seed for 5 weeks in growth cabinets on a medium of coarse crushed flint with constant renewal of four concentrations of the nutrient solution. 100 = full strength of a comprehensive nutrient solution, having the following concentrations of particular ions in milliequivalents per litre: NO_3^- 16·6; PO_4^{---} 3·4; K^+ 6·6; Ca^{++} 11; Mg^{++} 4·4. Points are means of six plants, with standard errors. From Lewis (1963).

Using a standard culture solution, Lewis was able to demonstrate that in this system the form and functioning of the plant fell within the naturally occurring range; and that only small changes were produced by diluting the whole culture solution to one-third of its normal concentration. The effects became more marked with further dilution, as shown in

26.13

Fig. 26.13, which gives the relationship between root weight ratio and concentration of nutrient solution at the time of the fourth harvest of these plants at approximately 29 days from germination. It will be seen that there is a progressive change in root weight ratio, which at a dilution to one-twentieth has become large, having risen to 0·31 as compared with 0·18 in the full strength nutrient solution. Yet even at such a dilution the plants showed no visible signs of deficiency apart from slower growth overall, and a relatively larger root system, as compared with plants grown in the full strength nutrient.

However, in spite of the progressive change in root weight ratio, the relative growth rate in dry weight fell only at the greatest dilution (one-twentieth), to 0·82 g/g/week as compared to 0·99 at one-tenth and 0·98 in the full strength solution. Considerable interest therefore attaches to a comparison between these last two sets of plants: once again we have an example of plants differing markedly in form—the root weight ratio at one-tenth dilution being 0·27 as compared with 0·18 for full strength— yet having the same relative growth rate.

26.14   Effects of differences early in development

As in the previous instances which we have examined it turns out that the higher root weight ratio is accompanied by a lower leaf weight ratio, which, other things being equal, would reduce the relative growth rate. However, as we shall see later [27.11], other changes compensate for this, so that by 15 days after germination the overall effect of the various changes is to leave the relative growth rate unaltered.

Nevertheless, as we mentioned above, the plants grown in the lower concentration are smaller. At 15 days from germination their mean weight was 81 per cent of that of the controls in full strength; at 22 days 81 per cent; and at 29 days 81·5 per cent. The relative growth rate being about 14 per cent per day, this means that the plants growing in the more dilute solution had been delayed in development by nearly 2 days. This delay presumably corresponds to the time needed to effect the changes in plant form and functioning which when complete (as they must be by 15 days from germination) restore the relative growth rate to equality with that of the controls.

Further investigations of growth during the first few days after germination (Hegarty, 1968) have suggested that there is no specific compensating mechanism, the countervailing changes being produced by the operation of the normal developmental processes, as affected by supplies of mineral and organic nutrients.

We have already come across other example [17.6, 17.7] of a similar phenomenon, where plants growing at closely similar rates when a particular dry weight is once reached have, nevertheless, taken widely differing lengths of time between germination and the attainment of this weight. Further investigation of the early developmental processes involved may well show that these are other instances of changes in normal developmental mechanisms producing growth changes of opposite sign which, when combined, leave the relative growth rate almost unchanged.

As investigation proceeds, it seems likely that substantial differences will be found between different ecological types in the quantitative operation of mechanisms of this kind. For example Hunt (1970) observed relatively small changes in the root/top ratio of seedlings of *Carex flacca*, *Deschampsia flexuosa* and *Scabiosa columbaria* under conditions of lowered mineral nutrient supply which brought about large changes in *Holcus lanatus*, *Rumex acetosa* and *Urtica dioica*.

26.15 Vegetative and reproductive phases

In Chapters 23 and 24 we have already examined a number of instances of the profound effects on plant form, and on the distribution of materials between the various organs, associated with the change from the vegetative to the flowering phase, in species where this transition is a sharp one. We have also noted that in some species the flowering phase starts early, is very prolonged, and overlaps the vegetative phase for a large part of the total life cycle [23.3]. Here we need only refer back to the figures there examined, and note that in drawing conclusions from them the cautions discussed earlier in this chapter apply with equal force.

26.16 General conclusions

It will have been noticed that although we began this chapter with the intention of considering growth processes in particular relation to plant form, as our discussions have developed they have involved us in considerations of changes of form in relation to the ontogeny of the plant, in relation to environment at various stages in this ontogeny, and in relation to the detailed structure and functioning of the plant, all of which make their own contributions to overall growth performance, which in turn has its effects on plant form. At the end of a long series of detailed discussions we are, in fact, coming back full circle to the complex of problems which we considered in broad general terms at the outset [3.5, 7.1, 17.1]. In the coming chapters we shall carry this process further, taking in turn the various aspects of detailed structure, functional activity, responses to environment, and so on. Each will lead off into some different branch of the overall study of plants, yet it is already clear that as in this chapter the study of the growth of whole plants will inevitably link them all together. Granted that the problems which arise must first be analysed into these component elements, nevertheless solutions can only be reached by a subsequent synthesis of the conclusions reached in the different specialized studies.

We have reached these and a number of other general conclusions [26.9] with reference to a very small range both of species and life form, but we have seen enough to make clear how much the quantitative expression of any particular morphological index depends on ontogeny and environment, so that any comparison between species and varieties must be broadly based, both in range of environmental conditions, and in respect of the main phases of the life cycle. In particular, if such comparisons are to have relevance to field situations, which are likely to include competition with other plants, they must include observations made on plants growing under such conditions. Here success or failure may depend, not on the growth form as it appears under good growing conditions in a garden, but on the plant's ability to grow under relatively adverse conditions, which in turn may involve adaptive changes in plant form [10.5].

26.16

SOME PROBLEMS OF ANATOMY: GROWTH AND STRUCTURE

27.1 Anatomy and plant physiology

We have already seen [13.8] how the connection between studies of plant anatomy and plant growth was established over 80 years ago, when the first edition of Haberlandt's *Physiologische Pflanzenanatomie* (Leipzig, 1884) gave prominence to Weber's pioneer studies on growth, and combined with them anatomical studies of Haberlandt's own on chloroplast numbers in leaves of the same species. It is unfortunate that in the first decades that followed this pioneer conjunction little progress was made, but it was perhaps inevitable. It could be held that as pioneers both Weber and Haberlandt were before their time, and that, in particular, solutions to the grand problems which Haberlandt posed on the relations of structure and function could not be developed until a much wider range of physiological experimentation could be viewed against a much more extended background of physical chemistry, biochemistry and biophysics.

In the light of modern knowledge and aided by modern techniques, anatomical studies could throw much light on the types of growth problem which we have been studying. It will be recalled that we broke down relative growth rate, which we have been using as an overall index of plant performance, into three components, and that, instantaneously,

$$\frac{dW}{dt} \cdot \frac{1}{W} = \frac{dW}{dt} \cdot \frac{1}{L_A} \times \frac{L_A}{L_W} \times \frac{L_W}{W} \qquad 13.16.1$$

Relative Growth Rate = Unit Leaf Rate × Specific Leaf Area × Leaf Weight Ratio,

where W is the total dry weight of the plant, L_A its leaf area and L_W its leaf dry weight. Of these three components, leaf weight ratio is an index of plant form [13.16; 18.2–18.16; 26.8, 26.9, 26.12]. Unit leaf rate is a physiological index of some complexity [13.11, 28.2], and specific leaf area is an anatomical index, related to the expansion of the plant's leaf material in space, and hence to leaf structure and development [13.16;

19.2–19.12]. Here again, one would wish that there had been more investigations relating differences in specific leaf areas to the underlying anatomical variations—variations not only between one species and another, but also between plants of a particular species grown under different cultural conditions and at different phases of the life cycle, and between different leaves on a particular plant at any one time.

27.2   Specific leaf area and development of individual leaves

It must first be remembered that the term specific leaf area as we have hitherto used it without qualification refers to the average leaf area per unit leaf dry weight of the whole plant. As new leaves develop and old leaves die off, the contribution made by any individual leaf to the overall figure would change, even if there were no change in the leaf itself; but the specific leaf areas of individual leaves may also go on changing over considerable periods, firstly while processes of growth and expansion up to full area are going on, and thereafter as processes of maturation, such as increase of cell wall thickness, continue.

Not all the leaves of a plant necessarily respond in the same way to environmental changes, as is shown for young plants of *Impatiens parviflora* in Fig. 27.2. The paper from which this is reproduced is a good example of the application of anatomical studies to problems posed by quantitative analysis of the growth of other members of the same populations of plants. The plants were grown individually in pots in a glasshouse, and at 23 days from germination were sorted, graded, and divided into a number of comparable samples, which were taken out into the field and allotted at random to a number of cultural conditions involving artificial shade. The pots were plunged at an ample spacing, some in the open field ('100 per cent daylight'), and others under artificial neutral screens transmitting 67 per cent, 42 per cent, 25 per cent and 7 per cent of daylight. Under each screen the development in area of individual leaves up the axis was followed from 23 days after germination up to 40 days, the time of the final harvest.

In Fig. 27.2 O_1 and O_2 represent the first pair of opposite leaves, A_1, A_2, etc. being the subsequent alternate leaves in the order of their production (for the phyllotaxy of the young plant see 27.9). It is apparent that at all light levels successive leaves produced on any one plant have higher maximum rates of growth in area, and that although no leaves had ceased

27.2

to grow in area at the end of the period of observation, nevertheless the leaves inserted at successive nodes would be larger at maturity. Increasing shade produces progressive accentuation of both these trends. As the percentage of daylight transmitted by the screen decreases, the rate of

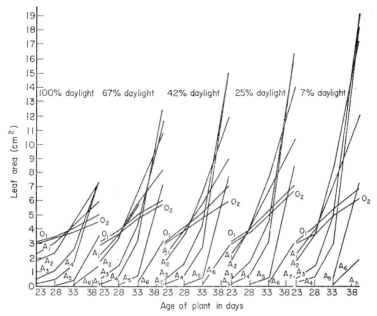

Fig. 27.2. Growth in area of individual leaves of plants of *Impatiens parviflora* aged between 23 and 40 days growing in the open (100 per cent daylight) and under neutral screens of four different transmissions. O_1, O_2, opposite leaves; A_1, etc, alternate leaves. From Hughes (1959b). (For corresponding stem anatomy see Fig. 27.9; for other aspects of the growth of the same batch of plants see Figs. 3.3.1, 19.6, 21.2 and 24.7).

growth of a leaf at any individual insertion increases, and these increases are observable at all stages, including the highest observed rates of growth in leaf area. Secondly, at any particular degree of shading, this highest rate of growth itself increases for successive leaves; and as a corollary, the final mature leaf size must also increase similarly.

This implies a very complex interrelationship of specific leaf area with ontogeny and with environmental conditions. At any particular ontogenetic stage, however this is defined, the proportional contributions of each of the individual leaves to the overall specific leaf area will vary

27.2

with the degree of shading; and owing to the varying changes of slope, the whole pattern will alter progressively from one stage of the ontogeny to another. We are seeing once again the interweaving of the different threads which we are endeavouring to separate; and we conclude that when describing the anatomy of such leaves one must specify not merely the stage of development, but the cultural conditions also.

27.3   Leaf growth and development

With this broad pattern of development in mind, we can now consider the anatomical changes going on within an individual growing leaf. The development of the venation is shown for a half-expanded leaf in Fig. 27.3.

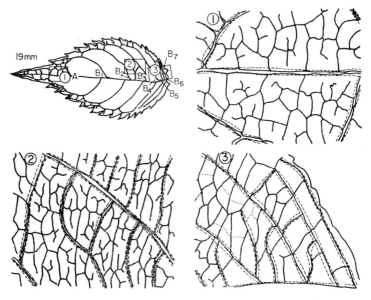

Fig. 27.3. Development of venation in a half-expanded leaf of *Impatiens parviflora*, 19 mm in length. Dotted lines indicate procambial strands, broken lines show veins with bundle sheaths. Approximate magnifications: entire leaf, ×3; regions 1–3, ×240. From Hughes (1959b).

At the base, to the right of region 3, there are veins of the second order, but only a few tertiary cross-veins. To the left of region 3, progressing towards the apex, first quaternary and then quinary veins form a continuous closed matrix. Throughout this development procambial strands at

27.3

various stages of their differentiation can be recognized, and they are shown as dotted lines in the top left of region 3. In region 2 vacuolation and expansion are going on. Disruption of the single files of tracheids in the veinlets of the ultimate order occurs, giving rise to blind vein endings. The rupture was found to be usually at the junction between veinlets, and, with few exceptions, was in single tracheid veinlets only.

In region 1, near the apex, the lamina is fully mature. Thus this sequence in space represents the sequence in time at any one place. The type of leaf development is therefore basiplastic with a ground mesophyll from which procambial strands of higher order differentiate successively. A consequence of this pattern of development is that for a considerable period, and after the tip of the leaf has become mature, the middle and base retain the capacity to respond to changes of environment by changes in development.

27.4 Shading and leaf anatomy: tissues

Investigation showed that this basic system of meristematic activity applied to leaves in all the light climates studied, but that the lower the total daily light the greater the expansion of the lamina. The effect of this difference in expansion is seen in Fig. 27.4, which compares at two positions the mature venation of a leaf developed in full daylight with that of a leaf developed in 7 per cent of daylight, under the densest neutral screen. Early in the differentiation of the young lamina crystal cells are formed, containing bundles of raphides (see also Fig. 27.5,3). These persist throughout the phase of vacuolation and expansion, and can readily be seen in the mature lamina. Their positions are also indicated in Fig. 27.4.

We see that in both positions on the leaf the general pattern is the same (with regard not only to the vein arrangement, but also to the disposition of the crystal cells) but the spacing is much wider in the shade leaf. The lengths of the leaves differed by a factor of 1·8 and the areas by 3·2, the latter figure corresponding to the differences in frequency both of the crystal cells and of the blind-ending veins.

It therefore seems that, over this wide range of light climates, there is a common broad organization of the mature leaf in the plane of the leaf lamina, and that the effect of shading is to cause a unit of this organization to be spread over a larger area.

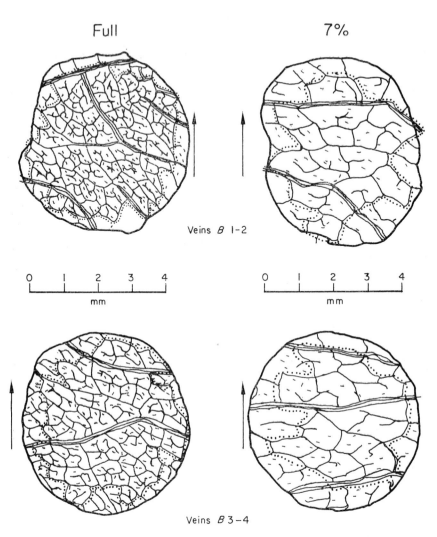

Full   7%

Veins B 1-2

Veins B 3-4

Fig. 27.4. Venation of the mature intercostal region of the second alternate leaves of plants of *Impatiens parviflora* grown in full daylight and under a neutral screen transmitting 7 per cent of daylight. Above, between veins B_1 and B_2 (as in Fig. 27.3); below, between veins B_3 and B_4. Arrows indicate the orientation of the midrib. Broken lines enclose complete islets. Crystal cells indicated —. From Hughes (1959b).

27.5 Shading and leaf anatomy: the transverse section

We must now examine the consequences of this on the leaf as seen in transverse section. Before expansion there is little difference between the

27.5

sun and shade form. Fig. 27.5,3 shows a section of a young shade leaf, having a dense palisade layer with a crystal cell towards the left, two layers of ground mesophyll with a vein and its sheath differentiating towards the right, and below these the larger cells of the arm mesophyll with dense contents, making with the upper and lower epidermis six layers of cells in all. The existence of this common structure before vacuolation and expansion begin accounts for the common structure after expansion, which we have just noted. It also helps to account for the completeness

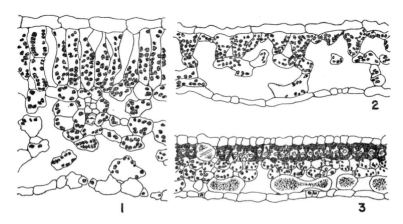

Fig. 27.5. Transverse sections of laminas of leaves of *Impatiens parviflora*. (1) Mature, grown in full daylight (thickness 270 μm). (2) Mature, grown in 7 per cent daylight (thickness 100 μm). (3) Unexpanded part of young leaf grown in 7 per cent daylight (thickness 90 μm). All to same scale. From Hughes (1959b).

with which the specific leaf areas of young plants can adapt to changes of light climate, both upwards and downwards (as we have seen in Fig. 19.7).

A transverse section of the mature lamina grown in full daylight is illustrated in Fig. 27.5,1. The well developed columnar palisade and the loose spongy mesophyll are characteristic of a sun leaf. The corresponding section of a mature leaf grown in 7 per cent daylight is shown in Fig. 27.5,2. The individual palisade cells are much shorter, and although their lower ends, abutting on the very diffuse spongy mesophyll, have a similar diameter to those of the sun leaf, by contrast the upper ends are funnel-shaped. This particular leaf was 100 μm in thickness, as compared with 270 μm for the sun leaf of Fig. 27.5,1.

The lateral extension of 3·2:1 thus corresponds to a reduction in thickness of 1:2·7, a close correspondence of ratio in view of the differences in cellular shape and arrangement which we have just seen, and of the difficulties of measurement mentioned below. The corresponding difference in specific leaf area was about 3·6:1, somewhat greater, as we should expect from the further observation that the cell walls of the shade leaves were thinner than those of the sun leaves. Indeed this, coupled with the large intercellular spaces seen in Fig. 27.5,2, makes it by no means easy to secure a satisfactory preparation of the shade leaf for sectioning, and complicates the measurement of leaf thickness.

27.6 Shading and leaf anatomy: the epidermis

Thus a study of leaf anatomy goes a long way towards explaining observed differences in specific leaf area brought about by shading: but an interesting problem remains. Seeing that in the shade form the upper ends of the palisade cells are funnel-shaped, and the spongy mesophyll so diffuse, the expansion of the leaf in shade cannot be effected by the pressure of the interior tissues of the leaf on a passive and plastic epidermis. It must rather take place by active growth of the epidermis in area. It is therefore worth looking at the epidermal cells in surface view, as shown in Fig. 27.6. Here the upper and lower epidermal cells of both sun and shade leaves are compared, all to the same magnification. On both sides the shade leaf has larger cells than the sun leaf, but what is more remarkable is the difference in their outline. Speaking generally, the margins of the upper epidermal cells are less convolute than those of the lower, but in both cases shading causes a great increase in the degree of convolution. This finally reaches the state shown in the lower right of the figure, where three of the four cells illustrated have 16 or more lobes.

It would be expected that the decrease in photosynthetic activity accompanying shading would lead to a general lowering of the availability both of labile carbohydrate and of the necessary chemical energy for the synthesis of new cell materials, including cell walls. This expectation would itself account for the thinner cell walls. The increase of cell size in shade can also be accounted for qualitatively by taking account of the facts that in general vacuolation, extension and maturation of the

27.6

cell involve two sets of processes, one concerned with the generation of osmotically active substances within the cell, and the other with the plasticity and growth of the cell wall. But the increase in convolution of the epidermal cells as shading increases implies that the mechanisms of cell wall growth are not acting equally all over the cell surface at one and

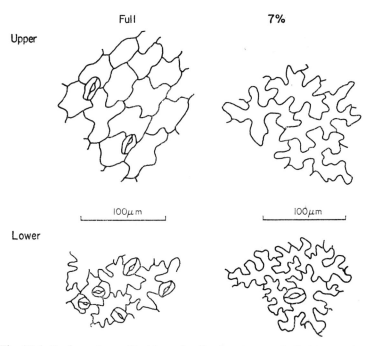

Fig. 27.6. Surface views of epidermal cells of mature parts (between veins B_1 and B_2) of the fourth alternate leaves of plants of *Impatiens parviflora* grown in full daylight (to the left) and under neutral screens transmitting 7 per cent of daylight (to the right). Above, upper epidermis; below, lower epidermis. All to same scale. From Hughes (1959b).

the same time, and this in turn raises further interesting questions, to which we shall return when we are considering primarily physiological problems [28.10].

27.7 Growth and other aspects of leaf anatomy

We have examined in detail one series of problems connecting the development and structure of leaves with specific leaf area. Our earlier studies

27.7

17

have shown this to be an important component of overall plant perform-ance [13.15, 13.16; 19.2–19.12], and these anatomical studies therefore assist our overall task of elucidating the normal activities of a higher plant. But as the plant functions as a whole organism, many other features of its detailed anatomy have a bearing on one or another of the aspects which we have considered when analysing the grand problem.

Other aspects of leaf anatomy are closely connected with questions of the paths of diffusion of gases into and out of the leaf, and hence with the important functions of photosynthesis, respiration and transpiration. We shall return to these matters in Chapter 29, and here we need notice only that the scale of cell measurement needed in connection with the

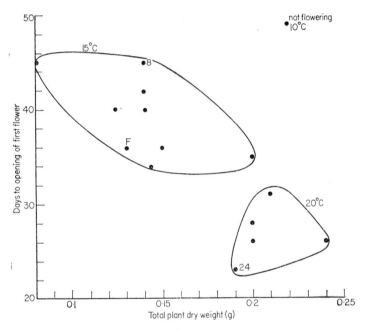

Fig. 27.8.1. Mean number of days from emergence of seedling to opening at first flower in relation to total plant dry weight for *Impatiens parviflora*, grown under a variety of constant, controlled conditions, except for one field experiment marked F. Three groups of treatments with temperatures of 20°C, 15°C, and 10°C, the last not flowering at all. Mean temperature for the week during which flowering occurred in the field, 15·2°C; corre-sponding mean daylength 16¼ hours. All other periods of illumination 16 hour per day, except for two experiments having 8 and 24 hours, labelled accordingly. Redrawn from Hughes (1965c).

27.7

solution of some of these problems exceeds the theoretically possible limit of resolution of a microscope utilizing visible radiation, so that electron microscopy is here essential [29.7, and Plate 29.7].

27.8 The problem of flowering in *Impatiens parviflora*

Impatiens parviflora is a day-neutral plant, flowering equally in 8-hour or 16-hour days or in continuous light. The problem of what factor or factors induced flowering was obscure: no simple correlation was found with age, with number of hours of exposure to light, or with total plant dry weight, even if one grouped plants which had in many respects grown under identical conditions in controlled environments. Fig. 27.8.1, redrawn from Hughes (1965c), illustrates the problem. Points represent the mean onset of flowering in a population of plants grown from seed under controlled environmental conditions in the laboratory, the age in

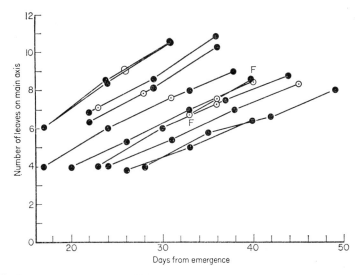

Fig. 27.8.2. Mean number of leaves on main axis as a function of days from emergence for plants of *Impatiens parviflora* grown under a variety of constant, controlled conditions, together with one field experiment marked F. Mean day of opening of first flower in each experiment marked ⊙. No flowering in the two slowest growing populations, one reaching eight leaves at the lowest temperature (10°C), the other reaching 6·5 leaves under the lowest daily total of visible radiation (11·2 cal cm^{-2}). Redrawn from Hughes (1965c).

days from emergence to the opening of the first flower being plotted against the mean dry weight on that day, derived by interpolation in the growth curve of the population. Lines enclose all the experiments carried out at 15°C and at 20°C. The single point in the top right-hand corner indicates a non-flowering population of plants grown at 10°C. Daylength in hours is indicated for 8- and 24-hour days, all the other points representing 16-hour days. F represents a field experiment carried out with a daylength near 16 hours and a mean temperature of 15·2°C. If in addition account be taken of the intensity and spectral composition of the radiation, and of rooting conditions, the problem remains equally obscure.

Fig. 27.8.2, on the other hand, represents a morphological approach. Here for twelve experiments the number of leaves on the main axis is plotted against days from emergence, a point with circle indicating the onset of flowering in each case. The two slowest growing sets of plants did not flower during the course of the experiment. One of these (grown at a low intensity of radiation) did not grow beyond a mean number of seven leaves. The other, which reached eight leaves, was grown at 10°C; we noted this non-flowering population in Fig. 27.8.1. All the others flowered between seven and nine leaves, most of them within a day or two of the production of the eighth leaf.

27.9   Flowering and stem anatomy

Fig. 27.9 illustrates the primary vascular anatomy of a stem of *Impatiens parviflora*. The particular plant shown had a counter-clockwise genetic spiral, the pattern for a clockwise spiral being the mirror image of this. The plant first produces a pair of opposite leaves (1, 2) and the spiral sequence then begins with leaf 3. The first flower to open in this series of experiments was nearly always on a branch in the axil of leaf 3. This is also usually the case for this species in the field, but by no means invariably so —it seems probable that often in the field the production of both the first flowers and the first fruit is complicated by coincident falling irradiance (Coombe, 1966, Fig. 2) with its effects both on changing plant form and lowering carbohydrate status. However, none of these complications are present in the observations with which we are now concerned.

Here the time of opening of the first flower effectively coincides with the onset of expansion of leaf 8, which is a member of the same orthostichy. It seems likely that this is more than a coincidence, and that a causal relationship exists, although the exact form of this is not clear [cf. 23.11, 26.1]. Hughes suggested that the increase of vascularization associated with the expansion of leaf 8 may enable the developing bud in the axil of leaf 3 to effect connection to the bundles flanking it as they increase in size with the transport demands of the expanding leaf. More recently, the establishment by Sheldrake and Northcote (1968) of a strong probability that growth substances may be produced during differentiation of the vascular tissue provides a possible biochemical linkage which is not necessarily at variance with Hughes's suggestion.

In spite of the unanswered questions, we have in this investigation a good example of the contribution which can be made to the elucidation of a growth problem by combining observations on the morphology of the shoot with a study of the vascular anatomy of the stem.

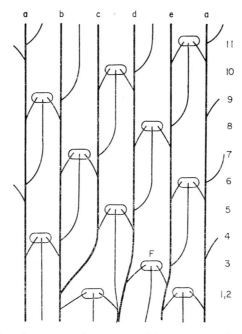

Fig. 27.9. The primary vascular system of the stem of a plant of *Impatiens parviflora* having a counter-clockwise genetic spiral (the pattern for a plant with a clockwise spiral being the mirror image of this). a–e, main stem traces; 1–11, leaf traces; F, position of first flower to open. From Hughes (1965c).

27.9

27.10   Mineral nutrition and root anatomy

Studies of root anatomy can play a part in elucidating other problems of plant growth, and the ones which immediately suggest themselves are those connected with the effects of changes in mineral nutrition on root morphology to which we have already referred [18.9 and 26.13]. We have there seen that using dilutions of nutrient solution which would not cause any visible pathological symptoms of deficiency, Lewis (1963) showed that the root weight ratio of *Impatiens parviflora* would rise with increasing dilution to 50 per cent above its value in a normal nutrient solution. This growth in weight of roots was accompanied by an increase in extent of the root system, and anatomical investigation showed that the structure of this was normal. There was no obvious change in cell size or in the general appearance of a cross-section of root, for plants growing in the nutrient solutions used by Lewis. These root systems can therefore be regarded as having a normal anatomical structure, and the first observable effects of diluting the nutrient solution are a relative extension of the root system accompanied by a retardation in the development of the young shoot.

At much greater dilutions (down to one-thousandth of the normal concentration: Lewis had gone down only to one-twentieth) Hegarty (1968) observed both reduced growth in length of the first formed roots, and inhibition of the formation of laterals. Differential effects on these two processes were observed when the ionic balance of the dilute nutrient solutions was altered.

Morphological and anatomical studies here demonstrate the existence of a series of effects upon the root system, partly dependent upon the dilution of the nutrient solution as a whole, and partly on the concentrations of specific ions.

27.11   Mineral nutrition and the assimilatory system

We have already mentioned [26.14] Lewis's observations that the increased proportion of dry weight in the form of root produced by plants in nutrient solutions diluted to one-tenth of normal was accompanied by a decrease in the proportion of dry weight in the form of leaves, reaching 8 per cent at the last harvest. At this level of dilution specific leaf area appeared to be affected little, if at all. The values at three out of the four harvests fell within the range covered by the normal solution and the same diluted to one-third, where the effect of mineral deficiency had not begun to show and the plants were closely similar to the normal ones. At this dilution it appears that leaf structure has not been affected, and that any compensation for the decrease in leaf weight ratio must come by way of an increase in unit leaf rate, which proves in fact to be the case.

On the other hand, working at much greater dilutions, down to one-thousandth of that of the normal nutrient solution, Hegarty (1968) found marked effects on the expansion of the photosynthetic organs. His plants being young, and the main part of the photosynthetic system being the cotyledons, which differ markedly in structure from the foliage leaves, he distinguished 'specific cotyledon area', as applying to these organs alone, leaving out both the weight and area of the developing leaves. As one would expect, when the seedlings emerged the cotyledon area was at first low, as expansion had hardly begun. Under normal nutritional conditions, cotyledon area then rose steadily, as shown in Fig. 27.11a. On the other hand, at great dilution of the nutrient solution the rise of the cotyledon area was slow and slight, and very little expansion had taken place at 14 days, when these observations ended. At one-tenth normal, on the other hand, cotyledon area had expanded thirty-fold, and specific cotyledon area rose from 0.2 to 0.5 cm^2 mg^{-1} between days 8 and 14. Anatomical examination showed that this lack of expansion in very low nutrient concentration was not due to inhibition of cell division, which continued at much the same rate as in the control [Fig. 27.11b], but to an inhibition of cell expansion. This was accompanied by reduced formation of cell wall materials, and biochemical studies showed that both the absolute amounts per cell and the relative quantities of the various monosaccharide constituents of the cell wall were affected.

Thus the effect of deficiency of mineral nutrients can be traced from its effects on the different aspects of our analysis of growth, through changes in the form of the plants to the underlying processes of growth and development in individual cells of the organs concerned.

27.12 Anatomical problems and others

This brief survey has sufficed to show by examples involving leaves, flowers, cotyledons, stems, roots and meristems how a great variety of anatomical studies can be used to throw light on problems arising in the

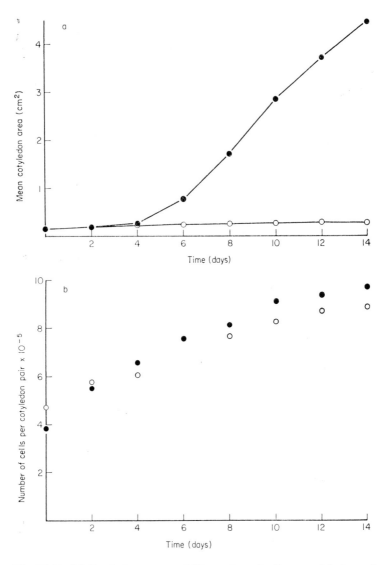

Fig. 27.11. (a) Cotyledon area, and (b) number of cells per cotyledon pair for young seedlings of *Impatiens parviflora*. Batches of stratified seed whose radicles had just begun to emerge were selected for uniformity and put in the feeding apparatus at time 0. Points are means of samples of five in each case, harvested at intervals of 2 days. (i) Solid points, each seedling supplied with nutrient solution, flowing steadily at 350 cm^3 day^{-1}, and containing major nutrients K, 0·67; Ca, 0·55; Mg, 0·22; N, 1·65;

27.12

course of the quantitative analysis of plant growth. It has also reinforced the view put forward at the end of the last chapter, that to obtain full value from the information of anatomical investigations, account must at the same time be taken of morphology, ontogeny, physiology, environment, and other aspects of the life of the plants concerned. At the same time we have seen how little has as yet been done in the way of utilizing anatomical studies to elucidate quantitative problems of plant growth. There is here an almost unexploited field of great potential value and interest.

P, 0·11; and S, 0·22, all in mg atoms per litre, together with a full complement of micronutrients. (ii) Open points, similar in all respects except that all the major nutrients were present at one-hundredth of the former concentrations. Fiducial limits: (a) 19 per cent of each individual value; (b) (i), 0·88, and (b) (ii), 0·64 × 10⁵ cells per cotyledon pair. At 6 days the two points in (b) coincide. Data of Hegarty (1968).

27.12

SOME PROBLEMS OF PHYSIOLOGY:
THE PROCESSES OF GROWTH

28.1 Physiological processes and the plant as a whole

All the physiological and biochemical processes going on in a plant have some effect on the growth of the plant as a whole. Such processes are very numerous and a very great deal is known about them. A mere glance at the row of volumes of Ruhland's Encyclopaedia indicates the magnitude of the task which awaits the eventual synthesis. Meantime, however, many obstacles hamper the application of this great accumulation of knowledge to quantitative problems of the growth of plants in nature. They arise not from how much is known, but from how little. We do now know enough of plant physiology to be sure that our present knowledge, great though it is, corresponds to no more than the outer skin of a very large onion. Beneath lies layer after deeper layer. As in all scientific fields this great structure of ordered knowledge normally grows by accretion, the successful solution of one problem suggesting others, to be explored in turn. Progress is never equal over the whole field of enquiry. It is therefore not surprising that to the great majority of questions proposed from outside the ordered system, no detailed reply can as yet be given. One may well be able to propose the lines of a qualitative solution in general terms, but such an answer may be of little value when a problem has to be solved quantitatively.

We can also be sure that the very great diversity of form and structure of the higher plants corresponds to an equal diversity in the quantitative operation of the physiological and biochemical mechanisms underlying form and structure, and the techniques for a quantitative characterization are often lacking (see, e.g., 26.1). As yet morphology, anatomy, physiology, and biochemistry cannot be brought together to produce a working model of the life of even one species out of the quarter of a million or so available.

What is more, the values of quantities observed in a physiological experiment carried out on some organ of a particular species of higher plant depend not only upon the environmental conditions imposed by the experimental technique, but also upon the form, structure and physiological state of the material in question at the time that the experiment begins. These in turn depend upon the past history of the experimental plant. This point is of general importance and by no means confined to results such as those which we have already seen in surveying series of similar field experiments carried out in successive years (e.g., Tables 7.5, 21.5, 22.4, 24.3, 26.6; and Fig. 26.4.2).

When these obstacles are considered together it is clear that if we wish to explain the growth of a particular plant in quantitative physiological terms, we must initiate a programme of physiological research on plants of the same genetic constitution, grown and growing under the same conditions. Accumulated physiological knowledge will help us in building up a qualitative model of the kind of system with which we are working, but the quantitative detail must be supplied by study of the plants themselves. Such studies are as yet only beginning, but as they continue they will bridge the gap which now exists between our knowledge of the lives of plants growing in the field, and our knowledge of the physiological processes of other plants derived from laboratory experimentation. At present such bridges are tenuous and scattered, but we can examine a few which spring from studies already discussed [25.4,26.1]. As time goes on they will become more numerous, solid, and broader based; but meantime where they are absent no ready-made answers present themselves, in spite of the bulk of physiological knowledge.

28.2   Unit leaf rate: its meaning in physiological terms

Our analysis of relative growth rate as an overall, but necessarily exceedingly complex, index broke the concept down into three parts [13.16, and Chapters 16, 18, 19, 20]. We have dealt with leaf weight ratio together with other morphological indexes in Chapter 26; and with specific leaf area together with other anatomical matters in Chapter 27; now we must turn to unit leaf rate. But whereas the first two are relatively simple concepts and therefore relatively easy to handle, unit leaf rate is more complex, as we saw when considering terminology in 13.11. Indeed, in physiological terms it is exceedingly complex. If we modify the symbols of 13.11

slightly so that the total leaf area of the plant appears as such in our expression for unit leaf rate and if

A′ = daily rate of gain in dry weight of the whole plant due to photosynthesis;

B′ = the rate of loss of dry weight due to respiration by the whole plant over a 24-hour period;

C′ = the daily rate of gain in dry weight of the whole plant due to mineral uptake;

D′ = the daily rate of gain of dry weight due to metabolic processes not comprised in A, B, and C;

L_A = leaf area; then

$$\text{unit leaf rate} = \frac{A' - B' + C' + D'}{L_A}. \qquad 28.2\,(\text{cf. }13.11)$$

Having gone so far, we promptly find that every one of these five components is in turn complex. Thus A includes all sources of photosynthetic gain, not just by the leaves, but by the whole plant; and even supposing that the leaves happen to be the only photosynthetic organs, they are likely to be of different ages and to have inherently different rates of photosynthesis per unit area under uniform conditions. But conditions on the growing plant are not uniform; leaves differ in their position on the plant, in their orientation and angle of inclination, and hence in their interception of radiation from sun and sky; they differ in their optical properties, and hence in the proportion of incident radiation absorbed; they differ in their transpiration rates and rates of advective heat loss per unit area; hence their temperatures differ; this in turn affects the rates of respiration and other physiological processes; and so on. What is more, these complex relations shift as external conditions shift—as the apparent position of the sun in the sky changes, as the wind changes in speed and direction, as air temperature and humidity change, and so on.

The quantity which we have called A′ is therefore itself a complex integration, over time and over a series of non-uniform physiological systems which are themselves distributed in space. The problem of translating this integration into physiological terms is not made easier by the fact that measurements are rarely made over a 24-hour period, but usually over much longer periods of around a week or 10 days. What is more, we do not measure A′; instead we estimate unit leaf rate. Before we can begin to consider the problems involved in interpreting A′ in

28.2

physiological terms, we have to consider the other terms in the relationship also. A simple example will show what is likely to be involved in practice.

28.3   A problem in changing conditions

One of the problems tackled by Coombe (1952), and later examined in more detail by Evans and Hughes (1961), was that of adaptation of the unit leaf rate of a plant to the conditions in which it has grown. When a plant is grown in a forest clearing and then transferred into dense shade, is its unit leaf rate in its new environment as high as that of a plant which has grown from seed in this dense shade? And what happens for the reverse transfer from dense to lighter shade? The evidence from both sets of investigations was assembled and discussed by Evans and Hughes (1961, p. 175). We have already looked at this in another context (11.20, where column A represents the adapted plants, and column B the transferred ones), and here we can summarize it by saying that of three pairs of comparisons involving the downward transfer, two showed negative differences and one a positive one, the probabilities of such differences appearing by chance being 0·1, 0·2 and 0·5. Of the three pairs of comparisons involving the upward transfer, again two were negative and one positive, but here the probabilities were even higher, 0·7, 0·9 and 0·6. The verdict so far was clearly 'not proven', but with a good indication that any real difference must have been small.

   The problem was obviously a suitable one for experimentation under closely controlled environmental conditions, and the results of these experiments are given in Table 28.3 (from Evans and Hughes, 1962, and Hughes and Evans, 1962). Two light regimes were provided, giving high and low light values of 26·2 and 11·2 cal cm^{-2} day^{-1} respectively, with a 16-hour day and all other conditions of the aerial environment controlled and constant. The final adapted values of leaf area ratio of 2·20 dm^2 g^{-1} in high light and 4·59 in low show that the difference in light regime brought about a substantial change in structure. When the transfer experiment began, an initially adapted population of plants around $4\frac{1}{2}$ weeks old was in each case divided into three comparable samples. One was harvested, one remained in the light regime to which it was adapted, and one was transferred to the other light regime. At the end of a further week both adapted and transferred samples were harvested. The figures for leaf

28.3

Table 28.3. Effects of transfers between two light regimes on the mean relative growth rate, leaf area ratio and unit leaf rate (with fiducial limits) of
populations of plants of *Impatiens parviflora* cultured from seed in growth
cabinets. Other conditions constant at 15°C and 2 mmHg saturation deficit,
with a 16-hour day; experimental period 7 days. From Hughes and Evans
(1962), and Evans and Hughes (1962).

	Relative growth rate week^{-1}	Leaf area ratio dm^2 g^{-1}		Unit leaf rate g m^{-2} week^{-1}
		Initial	Final	
(a) *High light, 26·2 cal cm^{-2} day^{-1}*				
Adapted	0·76	2·58	2·20	32·0 ± 0·9
Transferred from low light	1·09	3·74	2·88	33·1 ± 2·9
(b) *Low light, 11·2 cal cm^{-2} day^{-1}*				
Adapted	0·61	3·74	4·59	14·5 ± 1·2
Transferred from high light	0·46	2·58	3·48	15·2 ± 1·4

area ratio show how large were the differences in structure between the
two sets of plants. Connected with this are the large differences in relative
growth rate. In high light transferred plants have a relative growth rate
43 per cent higher than the adapted ones: in low light the reverse is the case,
and the adapted plants have a relative growth rate 35 per cent higher than
the transferred ones. Yet the unit leaf rates are closely similar: in both
cases the transferred plants have slightly higher values (3 per cent in high
light, 5 per cent in low), but the differences are nowhere near significance,
both probabilities lying between 0·3 and 0·4. For all practical purposes
they are identical. What has happened on the transfer? At first sight it
might appear that the simplest explanation is that the effect of transfer
on unit leaf rate is a more or less instantaneous change to the new value
appropriate to adapted plants. Are we to accept this explanation?

28.4 Mineral content

In tackling a problem of this kind are we justified in concentrating on
carbon metabolism and ignoring the plant's content of minerals? Granted
that the primary effect of shading will be upon the photosynthetic system,
yet as we have seen, the secondary effects on plant form, structure, and

28.4

functioning are profound, and they may well extend to mineral uptake and mineral content. Even supposing that mineral uptake and mineral content are maintained in both adapted and transferred plants at the same level in relation to the living protoplasmic systems, differences in the mineral content of the plant as a whole may arise through changes in the proportion of dead cells, and in the thickness of cell walls generally.

It so happens that *Impatiens parviflora* is a plant with an unusually high mineral content, totalling around 20 per cent of the dry weight. If this content remained unchanged throughout an experiment, when considering the metabolism of carbon compounds alone we must subtract this proportion of unit leaf rate, so that for a unit leaf rate of 30 g m^{-2} week^{-1}, the carbon dry-weight rate [13.11] would be 80 per cent of this, 24 g m^{-2} week^{-1}. On the other hand, if the mineral content changes during the course of an experiment, the size of the change in rate of mineral uptake required to produce a given change in mineral content depends on the ratio of the final to the initial dry weight, $_2W/_1W$. If this ratio equals Y, then to produce a 1 per cent change in mineral content requires a change in rate of mineral uptake of $Y/(Y - 1)$ per cent of the rate of increase of dry weight. For these experiments Y ranges from 2·97 (for the plants transferred from low light to high) to 1·58 (for the reverse transfer); and $Y/(Y - 1)$ from 1·51 to 2·72 respectively.

Let us consider the case involving the largest effect; if during the week's growth by the plants transferred from high light to low their mineral content increased from 20 per cent to 21 per cent of the total dry weight, then this would be the effect of an increase in rate of mineral uptake from 20 per cent to 22·7 per cent of the rate of increment of total dry weight. Therefore, if both adapted and transferred plants had the same unit leaf rate of 15 g m^{-2} week^{-1}, and the adapted plants maintained throughout a mineral content of 20 per cent, the carbon dry-weight rate of the adapted plants would be 12 g m^{-2} week^{-1}, and that of the transferred plants 11.6. In other words, without any difference in carbon dry-weight rate, increase of one-tenth in the mineral content of the transferred plants, from 20 per cent to 22 per cent of the total dry weight, would produce a difference in unit leaf rate of around 5 per cent, roughly equal to that shown in Table 28.3. As we have seen, this was not significant, and indeed it is difficult, in growth experiments, to attain a significance level of 5 per cent. Such a large change in mineral content in a single week in the middle of the main phase of vegetative growth is unlikely, but

28.4

could easily happen at other phases of the life cycle. Thus the result demonstrates that during the main phase of vegetative growth, even of a plant with a rather high mineral content, significant differences in unit leaf rate are in general not likely to be due mainly to changes in mineral content, and that this can be neglected in our present discussion. More generally, however, it is wise to keep a check on the mineral contents of plants in the course of experiments of this kind.

28.5   Effects of changing leaf area ratio

Furthermore, of the remaining three components discussed [28.2], together making up the total carbon metabolism, we can in this instance neglect changes in the proportions of D′, the term concerned with metabolic balance. We are here dealing with plants in the middle of their vegetative growth, neither juvenile nor senescent; they were amply supplied with mineral nutrients; even in the low light regime they received a total daily light quantity higher than in many natural habitats of the species; for all four sets of plants both increase in dry weight and unit leaf rate were substantial, and there can be no question of stresses due to near-starvation.

If we then treat D′ as negligible, and further assume that the mineral content of the plants is constant at 20 per cent, we have

$$0\cdot8 \text{ (unit leaf rate)} = 0\cdot8 \text{ E} = \text{Carbon dry-weight rate}$$

$$= \frac{A'}{L_A} - \frac{B'}{L_A}, \qquad\qquad 28.5$$

where L_A is total leaf area, and A′ and B′ represent the mean rates of photosynthetic gain and respiratory loss by the plant as a whole [28.2].

Observations on the respiration of complete, intact, undisturbed plants of *Impatiens parviflora* (Rackham, 1965) have shown that the respiration rate is a nearly constant fraction of the dry weight over the whole age range with which we are concerned, and for plants grown over an even wider range of light climates extending from 9 cal cm^{-2} day^{-1} of visible radiation up to full daylight in the open (say 200–300 cal cm^{-2} day^{-1}), all the observed values lying within 15 per cent of the mean. Correcting to the temperature of the transfer experiments, the mean rate, on a dry weight basis, was 2 mg g^{-1} h^{-1}. To illustrate the consequences of a constant respiration rate, let us assume for the moment that the whole plant respiration rate is the same in the light as in the dark, when the mean value would become 336 mg g^{-1} week^{-1} or, on a weekly basis $\overline{B} = \overline{W}/3$. Then the respiration term per unit leaf area becomes $\frac{1}{3} \cdot (\overline{W}/\overline{L_A})$, taking means for the week. But L_A/W is the leaf area ratio, and this changed very substantially more in the course of the experiment for the transferred plants than it did for the adapted ones [Table 28.3], inevitably resulting in differential changes in the respiration term.

As an example: the plants transferred from low light to high had initially a value of L AR approximately $\frac{2}{3}$ that of the adapted plants. Their respiration term would thus be approximately $\frac{2}{3}$ of that of the adapted plants, assuming that both sets were respiring at the same rate per unit dry weight. The adapted plants had a mean leaf area ratio for the week of about 2·40 dm^2 g^{-1} (Hughes and Evans, 1962, Fig. 3), so that their mean respiration rate, on a leaf area basis, was approximately 14 g m^{-2} week^{-1}. The non-adapted plants would have initially a rate approximately $\frac{2}{3}$ of this, or 9 g m^{-2} week^{-1}, 5 g m^{-2} week^{-1} less than the adapted ones. Allowing for mineral uptake [28.4], the carbon dry-weight rate [13.11] for these adapted plants was $\frac{4}{5} \times 32 = 26$ g m^{-2} week^{-1}, so that the decrease in the respiration term for the transferred plants amounts to nearly 20 per cent of the net rate of addition of new carbon compounds. If the net figure is to remain the same, this would suggest the existence of a corresponding initial short-fall of photosynthetic gain, of about 5 g m^{-2} week^{-1}, from a total for the adapted plants of $26 + 14 = 40$ g m^{-2} week^{-1}, or around 12 per cent. The state of the transferred plants immediately after transfer would then be, on this hypothesis,

Carbon dry-weight rate = Photosynthetic gain − Respiratory loss per
                                     per unit leaf area   unit leaf area
                          =   35   −   9
                          = 26 g m^{-2} week^{-1};

while, when they have become fully adapted, we should have

                          =   40   −   14
                          = 26 g m^{-2} week^{-1}

We see that in consequence of the observation that respiratory loss is almost a constant fraction of the dry weight, the assimilation rate could initially be as much as 12 per cent below the fully adapted rate per unit area, rising as adaptation went on. If the shortfall of photosynthetic

28.5

rate were initially not quite so large as this, then the result over a week would be to give the transferred plants a higher unit leaf rate. What is more, if the low light plants before transfer had a lower respiration rate per unit dry weight [28.14], the effect would be accentuated. However, in this particular instance the rate of photosynthesis per unit area might well have been lower, because of the thinner leaf of the non-adapted plants.

We need not follow through the corresponding reasoning for the transfer from high light to low; but in both cases we can conclude that the apparent constancy of unit leaf rate probably covers processes of adaptation in which both assimilation and respiration per unit leaf area are changing. It will be seen that this conclusion does not depend on the exact form of the relationship between the whole plant respiration rate in the light and in the dark [and see 29.4].

28.6 Unit leaf rate and specific leaf area

We can now examine more closely the relationship between unit leaf rate and specific leaf area, which we have already mentioned in passing [17.4]. Fig. 28.6.1 shows the results of several series of experiments on *Impatiens parviflora*, some in the field and some under controlled conditions in the laboratory. The field grown plants had at first been grown as a large population in a glasshouse, and from this a series of comparable samples had been selected to put out under a number of neutral screens in the field. Lighting conditions in the glasshouse would lie between those in the open field (100 per cent), and those under the screen of highest transmission (67 per cent). The plants put under these screens would therefore be nearly adapted to field conditions when the observations began. On the other hand, the plants placed under the screen of lowest transmission (7 per cent) had initially a form and structure far from those appropriate to an adapted plant, and large changes in specific leaf area ensued. These appear among the field experiments to the left of the diagram in the form of a long line joining the point representing the specific leaf area at the beginning of the experimental period to that at the end (enclosed in a circle). Thus the points indicated by circles represent the more nearly adapted end of the line. As we have already seen [19.7], adaptation of specific leaf area of young plants to substantial changes of light climate is largely complete at the end of a week's exposure to the new conditions.

28.6

Our discussion in the last sections [28.3 to 28.5] now enables us also
to reach some conclusions about the probable extent of the variation in
unit leaf rate during the period of adaptation. We have seen the existence
of a compensatory mechanism, such that the initial deviations from the
adapted value are likely to be small, of a few per cent only, and that in the

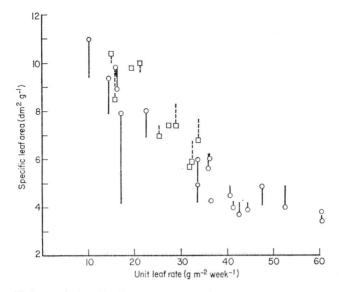

Fig. 28.6.1. Relationship between specific leaf area and unit leaf rate
for plants of *Impatiens parviflora* grown in the field under various levels
of shading (○), and under controlled conditions in the laboratory (□).
Mean values for samples of six or eight plants. Lines join values of specific
leaf area for the beginning and end of the period of growth, symbols at the
end. From Hughes and Evans (1962).

case of a transfer from high light to low, the initial deviation is likely to
be positive.

When we allow for both these sets of changes, we see that Fig. 28.6.1
provides evidence of one single broad relationship between unit leaf rate
and specific leaf area, in spite of the very wide differences between conditions under neutral screens in the field, and those under controlled
conditions in the laboratory (summarized in 18.15, ix–xii).

We also see that between unit leaf rates of about 15 and 40 g m^{-2}
week–^{1}, and specific leaf areas of 10 to 4 dm^2 g^{-1}, the relationship between

28.6

the two is almost a reciprocal one. We have already seen [18.6, 18.15] that in young plants of this species leaf weight ratio is almost constant, and little affected by total daily light, or, indeed, by all the other changes involved in the transition from the field to controlled conditions in the laboratory. Here, therefore, we have the mechanism which brings about a compensation in relative growth rate, so that the relative growth rate

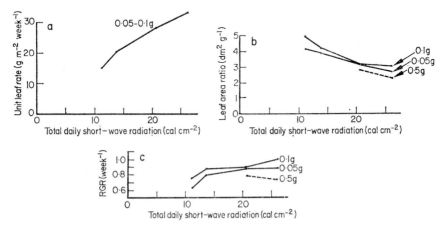

Fig. 28.6.2. Relationships between growth and total daily light for plants of *Impatiens parviflora* of different sizes grown under constant, controlled conditions. (a) Unit leaf rate. Points apply to total plant dry weights of both 0·05 and 0·1 g. For the two highest radiation totals, the unit leaf rate of 0·5 g plants was probably about 5–10 per cent lower. Fiducial limits, ±9 per cent. (b) Leaf area ratio. Values for the three plant sizes derived by interpolation. Least significant difference, 0·39 dm² g⁻¹. (c) Relative growth rate. Derived by multiplying together the appropriate values of (a) and (b). Least significant difference, 0·11 week⁻¹. From Hughes and Evans (1962).

of adapted plants of a given size changes comparatively little over a wide range of light climates. This can be seen in operation for plants of *I. parviflora* of different dry weights in Fig. 28.6.2. Here we see unit leaf rate (for all the relevant plant sizes) rising with total daily radiation; leaf area ratio falling, in a family of curves depending on plant size; and relatively little change in relative growth rate over the range of light climates which bring about the bulk of the change from the high to the low light forms [27.3–27.6].

28.6

28.7 Problems of causal relationship

We have now to ask how the causal relationship operates. Is unit leaf rate determined by leaf structure, of which specific leaf area is a measure? Or, is the particular value of specific leaf area assumed by adapted plants determined by the rate of production of new assimilates, of which unit leaf rate is a measure? Or are both determined independently by the light climate? Or is some more complex system of interlinking at work?

The transfer experiments discussed earlier in this chapter enable us to rule out the first of these possibilities. Our examination of the data has shown that immediately on transfer to a new light regime unit leaf rate must assume a value within a few per cent of the fully adapted rate, and remote from the value associated, in fully adapted plants, with the value of specific leaf area at the time.

The second and third possibilities, in the simple form stated above, are ruled out by the facts that, in a controlled climate where lighting conditions do not change from day to day, unit leaf rate drifts slowly downward while specific leaf area passes through a complex ontogenetic drift. But is this drift simply inherent, its general form unaffected by external conditions? We first observed the relationship between unit leaf rate and specific leaf area when considering figures for plants of a fixed dry weight (for *Impatiens parviflora*, 100 or 150 mg). Fig. 19.8 showed the specific leaf areas plotted against the total dry weights at harvest of individual plants of *I. parviflora* grown from seed under controlled environmental conditions. For all the plants shown, air temperature was maintained at 15°C, vapour pressure deficit at 2 mmHg, with a 16-hour day of constant illumination from fluorescent tubes combined in the proportion of 3 De Luxe Warm White to 1 Blue. The five different series of experiments shown differed, as shown in the legend to the figure, in daily total of visible radiation, and in rooting medium (either a mixture of vermiculite, sand and gravel with culture solution, or 1:1, soil:sand with added artificial fertilizers).

Influences of rooting medium are evident. The plants grown at 11·2 cal cm^{-2} day^{-1} in soil show clearly the rising phase of specific leaf area at very low dry weights (this experiment ended before the subsequent falling phase was established). This rise is not shown by the plants grown at 7·8 cal cm^{-2} day^{-1} in vermiculite, and it may be that initial leaf expansion is facilitated in these rooting conditions. If growth in vermiculite did not affect specific leaf area, one would expect the points for plants growing at 16·0 cal cm^{-2} day^{-1} in vermiculite to fall between those for plants grown at 13·8 and 20·6 cal cm^{-2} day^{-1} in soil, and nearer the former than the latter. In fact their scatter largely overlaps that of the plants grown in 20·6 cal cm^{-2} day^{-1}, and is well clear of the points for plants grown in 13·8 cal cm^{-2} day^{-1}. For our present purposes it is not necessary to quantify these differences [19.8]: it suffices to note that qualitatively the effects of growth in vermiculite and culture solution are not confined to the very early stage of leaf expansion, but appear throughout.

Thus in its general form the falling phase appears to offer a family of curves, falling more steeply the higher the initial values, but maintaining a roughly constant relationship to each other. It seems that our practice of making comparisons at a particular fixed dry weight [17.2, 19.2–19.4] is justifiable, qualitative conclusions drawn from such comparisons being valid at other parts of the range of plant weights under observation.

28.8 Experiments under a variety of conditions

Fig. 28.8 shows such a set of comparisons for plants of 100 mg total dry weight grown under a variety of conditions, each point being an interpolation in the appropriate specific leaf area/total plant dry weight curve derived from samples of six or eight plants per harvest.

The experimental conditions differed in total daily light but the spectral composition did not differ widely, nor did vapour pressure deficit. In Fig. 28.8 the interpolated values are plotted both against total daily light and against unit leaf rate.

1 To simplify consideration of a complex mass of data we can first divide the points into families based on temperature—constant temperatures of 10°C, 15°C and 20°C. As plotted against total daily light we see that the families do not overlap.

2 We then divide the temperature families by rooting medium, and we see that the points relating to vermiculite always lie to the left: in every case of culture in vermiculite a given specific leaf area is attained at a lower value of total daily light or of unit leaf rate, as compared with culture in soil.

3 As plotted against unit leaf rate, the vermiculite series of the temperature families are again distinct, with an even more marked separation of

28.8

the 10°C point than is shown in the plot against total daily light. The change with temperature is in the same sense for both plots, a higher temperature giving a higher specific leaf area for a given value of total daily light or unit leaf rate.

4   As plotted against unit leaf rate, for the soil series there is a coincidence of one of the 20°C points with the 15°C line. The other 20°C point is distinct.

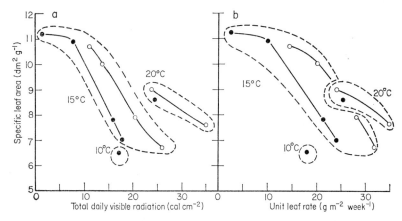

Fig. 28.8. Mean specific leaf areas of plants of *Impatiens parviflora* of 0·1 g total plant dry weight, grown under constant controlled conditions, in relation to (a) daily total of visible radiation, and (b) unit leaf rate. Experiments in three different temperatures, of 10°C, 15°C and 20°C, and two substrata, sandy loam, ○, and vermiculite, sand and gravel with added culture solution, ●. Adapted from Hughes (1965c).

From this consideration we can conclude that, although at a given stage of ontogeny specific leaf area is related both to unit leaf rate and to total daily light, the whole relationship between either pair can be displaced by other factors, including both temperature and rooting conditions; and also [from 28.7] that the whole complex of relationships can be expected to change quantitatively with ontogeny, while in all probability maintaining its qualitative form.

28.9   Influence of spectral composition of radiation

Either temperature or rooting medium will have profound and complex effects on the metabolism of the plant as a whole: it may be suggested that if we wish to distinguish between the direct influence of light on leaf expansion, on the one hand, and its indirect influence via the supply of metabolites (of which unit leaf rate is a measure) on the other, we should do better to examine the effects of changing spectral composition. Fig. 28.9 shows the results of two such experiments, against the relevant background of points from Fig. 28.8. One was carried out in wholly blue

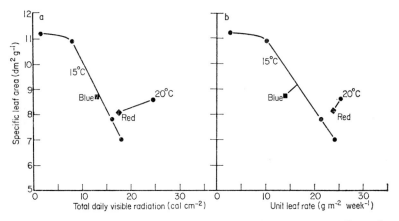

Fig. 28.9. Mean specific leaf areas of plants of *Impatiens parviflora* of 0·1 g total plant dry weight, grown under constant conditions, in relation to (a) daily total of visible radiation and (b) unit leaf rate. Experiments at 15° and 20°C, with a substrate of vermiculite, sand and gravel. ● (points as in Fig. 28.8.), illumination from fluorescent tubes, either Daylight:Warm White 1:1, or De Luxe Warm White:Blue 3:1. ■, illumination from Blue fluorescent tubes only, temperature 15°C. ♦, illumination from Red (magnesium arsenate) fluorescent tubes only, temperature 20°C. Adapted from Hughes (1965c).

light from fluorescent tubes [Fig. 18.10], and otherwise with the conditions of the 15°C, vermiculite, experiments. Plotted against total daily light, this point falls almost exactly on the line through the four comparable points. Plotted against unit leaf rate, it is well below.

The second experiment was carried out under fluorescent tubes with a magnesium arsenate phosphor, having a sharp peak of emission around 660 nm, but by no means devoid of blue light, because of the breakthrough of the mercury lines [Fig. 18.10]. We see that on a total daily light basis it is well to the left of the comparable point. On a unit leaf rate basis it is much closer.

28.9

Thus the situation is complex, with no straightforward agreement between the different spectral compositions of the incident radiation on a unit leaf rate basis; with an agreement between 'white' radiation and blue, but not with red, on a total incident energy basis; and consequently with no agreement in either case on a quantum basis.

28.10   Physiological control of leaf expansion

The anatomy of leaves of *Impatiens parviflora* grown in different light regimes [27.3–27.6] showed that leaf expansion was probably largely controlled by growth of the epidermal cells; and that this cellular growth was not a simple process of expansion of a given structure to occupy a greater area, but that it was accompanied by a remarkable increase in the convolution of the walls separating adjacent epidermal cells. As the plant grows, cells are becoming mature all the time, and adding to the fixed element in the leaf area, the remainder consisting of cells still capable of expansion. At any moment the specific leaf area therefore embodies an integration of past conditions.

Thus the reciprocal relationship between specific leaf area and unit leaf rate for plants grown in the field and under controlled conditions in the laboratory, which we saw in Figs. 17.4b and 28.6.1, was real enough, but limited both as to temperature, rooting conditions, and spectral composition. All the plants were grown in a 1:1 mixture of loam and sand: for the laboratory grown plants, temperature was fixed at 15°C, and for the field grown ones the average mean temperature was near this: and the normal fluorescent lighting was chosen to give a reasonable approximation to daylight.

The causal relationship was there, but deeply buried. We now know that leaf expansion is influenced by many factors including the radiation climate itself; its indirect effects via unit leaf rate; and a number of other factors, both external and internal to the plant. Among these we can list the spectral composition of the incident radiation, temperature, and rooting conditions.

The problem of how the controlling mechanism or mechanisms operate is an interesting and important one, which awaits investigation. It seems likely that part of the mechanism is concerned with the supply of new materials for growth—this is where the relationships with unit leaf rate and with temperature come in; and that another part of the mechanism is concerned rather with some morphogenetic process, which may be affected by the spectral composition of the incident radiation. We may instance the comparison of Fig. 28.10 between Coombe's

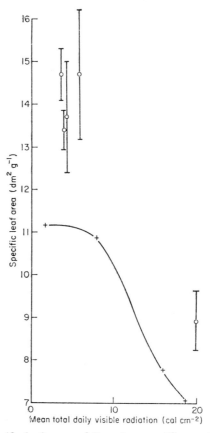

Fig. 28.10. Specific leaf areas of *Impatiens parviflora* at 0·1 g total plant dry weight, as a function of the mean daily total of visible radiation. ○, observations of Coombe (1966) on plants growing in Madingley Wood, with 19:1 probability limits; +, observations of Hughes (1965c) on plants grown under artificial conditions in the laboratory. From Evans (1969).

observations on the growth of *Impatiens parviflora* in Madingley Wood, near Cambridge, and the 15°C vermiculite data of Hughes. We see that whereas Hughes's figures appear to have an asymptotic value, at very low light, of not much over 11 dm² g⁻¹, the corresponding value for Coombe's

28.10

woodland plants is much higher, around 14 dm^2 g^{-1}. Indeed, in all Hughes's experiments the mean specific leaf area for a sample of six or eight plants at 100 mg total dry weight (which is what we are comparing) never reaches a value above 12 dm^2 g^{-1}, although individual plants have higher values. Yet conditions below the ground were the same for both; and of the aerial environmental conditions, much the most likely one to account for the difference is the spectral composition of the incident radiation.

A direct solution of the question of whether phytochrome plays an important part in bringing about the above discrepancies and those of sections 28.8 and 28.9 is made difficult by the paucity of information about the balance between the two forms of the pigment induced by exposure of chlorophyll-containing tissues to radiation of a mixture of wavelengths, and about the physiological effects associated with changes in this balance. It is clearly desirable to have more information, both on these matters, and on the actual mean differences in spectral composition of the natural light climates themselves [9.6, 9.7]. However, at the present stage of investigation it seems unlikely that all the discrepancies we have noted in the last three sections can be reconciled on the basis of a phytochrome-controlled system alone, and other pigment systems may be effective, as, for example, in the experiments of Klein *et al.* (1965).

Meantime, there are immediate practical applications of the general conclusions, without waiting for a definitive solution of the problem [9.10]. Twenty years ago Millener (1952, 1961) noticed that the light-demanding seedlings of the gorse, *Ulex europaeus*, would not make normal growth in a cabinet supplied wholly with fluorescent lighting—it was necessary also to include some tungsten filament bulbs (and see Wassink and Stolwijk, 1956; and Canham, 1966). These observations have recently been confirmed in more detail by Blackman and his co-workers for beans, cotton, maize and sunflowers (Rajan *et al.*, 1971).

28.11   Respiration of plants in deep shade

The observations on the growth of *Impatiens parviflora*, made in deep shade in woodlands, and under comparable conditions in the laboratory, also raise interesting questions about the respiration of plants growing under these conditions, as it may affect growth and, in the case of very dense shade, survival itself. We saw in Fig. 3.2.1a evidence that there is, near the origin, an almost linear relationship between unit leaf rate and the daily total of visible radiation, for laboratory experiments using a 16-hour day of constant irradiance. Fig. 28.11 shows that an essentially

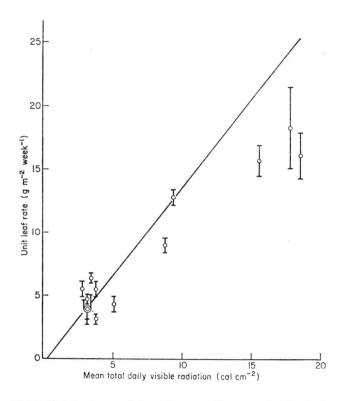

Fig. 28.11. Unit leaf rate of *Impatiens parviflora* growing in shady conditions, as a function of the mean daily total of visible radiation. ○, observations of Coombe (1966) on plants growing in Madingley Wood, with 19:1 probability limits. —— line of Hughes from Fig. 3.2a, assuming mean visible radiation in the wood to be 38 per cent of the mean total short-wave radiation. From Evans (1969).

similar relationship exists for plants growing under natural conditions of the aerial environment in dense woodland shade. The points are those of Coombe (1966), based on the same set of observations carried on

28.11

throughout the summer of 1963 in Madingley Wood as were used in Fig. 28.10. The straight line is that of Fig. 3.2.1a, based on observations made under the very different spectral composition of radiation in growth cabinets. In order to plot both on a single axis of abscissae, Coombe's total short-wave radiation measurements have been converted to a visible radiation basis by assuming that total visible radiation in the wood was 38 per cent of the mean total short-wave radiation (the reasons for this assumption are considered below, 30.6–30.8).

We have just discussed the approximately reciprocal relationship between unit leaf rate and specific leaf area, extending down to unit leaf rates between 10 and 20 g m^{-2} week^{-1}, and seen that when combined with an almost constant value of leaf weight ratio, the relationship of leaf area ratio and unit leaf rate leads to very small changes in relative growth rate over a wide range of total daily light [28.6].

The same approximately reciprocal relationship between leaf area ratio and unit leaf rate also affects the contribution made to the latter by the respiration of the plant as a whole. If the respiration rate per unit dry weight has a constant value, W_R; then the respiration term per unit leaf area [13.11 and compare 28.5], $B = W_R/LAR$, will be approximately proportional to unit leaf rate. Thus as the rate of apparent assimilation per unit leaf area falls with decreasing daily light, so does the absolute value of the respiration term, maintaining a roughly constant fraction of the daily gain per unit leaf area.

28.12   Effect of the maximum value of specific leaf area

The reciprocal relationship between unit leaf rate and specific leaf area thus helps to explain both the linear relationship between unit leaf rate and total daily light, and the very low value of the total daily light at which the extrapolated compensation point for unit leaf rate is reached: but the range over which these relationships operate is limited. We have seen in 28.8 that for *Impatiens parviflora* under any particular set of cultural conditions there exists a maximum value for specific leaf area, no doubt set by leaf structure, and this value begins to be approached at a total daily light of around 12 cal cm^{-2} day^{-1}, corresponding to a unit leaf rate around 15 g m^{-2} week^{-1}. Figs. 3.3.2 and 28.10 show that at around 9 cal cm^{-2} day^{-1}, with a unit leaf rate of about 11 g m^{-2} week^{-1}, the asymptote is almost reached. As the total daily light decreases further, changes in leaf area ratio can have only a small effect on the absolute contribution of the respiration term to the total unit leaf rate: yet the straight line relationship between unit leaf rate and total daily light continues almost to the origin.

The respiration measurements of Rackham (1965) included one set made on plants grown under controlled conditions with a total daily light of 9 cal cm^{-2} day^{-1}; this gave a value about 10 per cent below the mean of all the measurements made on plants grown in a range of light climates up to full daylight. The difference from the mean was not significant, but the variability was high, due to the small size of the plants, and the correspondingly small volume measurements. Let us, however, assume that this measurement was identical with the mean value, for plants grown in all light climates, of 2 mg g^{-1} hr^{-1} ($\equiv$ 336 mg g^{-1} week^{-1}, 28.5), and investigate the consequences of assuming that the rate of respiration per unit dry weight was the same for plants grown in even deeper shade.

28.13   A quantitative example

In Fig. 28.13 for experiment 19, plants grown in 16 cal cm^{-2} day^{-1} had a leaf area ratio of 3·70 dm^2 g^{-1} at 100 mg total plant dry weight, so that the respiration rate per unit leaf area of the whole plant (B of 13.11) would work out at 9 g m^{-2} week^{-1}. At 7·8 cal cm^{-2} day^{-1}, with 4·70 dm^2 g^{-1}, it would be 7 g m^{-2} week^{-1}, and at 1·5 cal cm^{-2} day^{-1} (5·50 dm^2 g^{-1}), 6·1 g m^{-2} week^{-1}. The corresponding unit leaf rates were 21, 10 and 2 g m^{-2} week^{-1}, giving carbon dry-weight rates [13.11] of 17, 8 and 1·6 g m^{-2} week^{-1}, so that the photosynthesis rate per unit leaf area, A (=unit leaf rate + B) would have been 26, 15 and 8 g m^{-2} week^{-1}. But this is absurd. At these very low intensities of radiation it would not be unreasonable to suppose that photosynthesis might increase linearly with intensity, following a straight line passing through the origin. Alternatively, as intensity increases, photosynthesis might fall short of the value of this linear relationship (as we know to be the case for higher intensities), and the relationship would curve down. What is not believable is that as light decreases from 16 cal cm^{-2} day^{-1} there should be increasing positive departures from linearity, which would cause the efficiency of photosynthetic energy conversion to rise rapidly until at 1·5 cal cm^{-2} day^{-1} it had reached a value in excess of the theoretical maximum. What is more, at this point the curve would have to have a pronounced discontinuity in

28.13

order for it to reach the origin—if extrapolated it would give a value for photosynthetic gain of over 4 g m^{-2} week^{-1} in total darkness.

We may well ask—does this arise from our assumption that mineral uptake is a constant fraction of unit leaf rate, maintaining the mineral

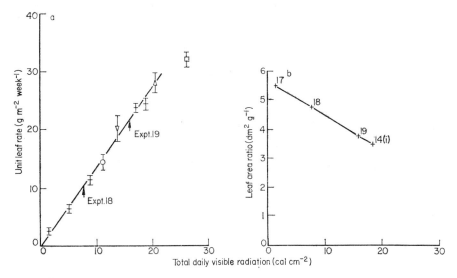

Fig. 28.13. Dry weight increase and leafiness in plants of *Impatiens parviflora* growing under constant controlled conditions, as functions of daily total of visible radiation (cal cm^{-2}). (a) Unit leaf rates (g m^{-2} week^{-1}). Fiducial limits at the 5 per cent probability level. Highest and lowest points excluded from the regression line calculation. Unit leaf rates for Experiments 18 and 19 by interpolation on this line, as shown. (b) Leaf area ratios (dm^2 g^{-1}) for plants of 100 mg dry weight. Values determined by interpolation in the relationships between leaf area ratio and total plant dry weight, for plants growing under four different daily totals of visible radiation. From Hughes (1965a). (See also Figs 3.2.1a and 28.10.)

content of the plant constant at around 20 per cent? But consideration shows that this cannot be so. To reconcile these figures on a basis of increase in dry weight through mineral uptake would require a rate of between 4 and 5 g m^{-2} week^{-1} in total darkness, and at least 4 at 1·5 cal cm^{-2} day^{-1}. But here unit leaf rate itself is around 2 g m^{-2} week^{-1}, which would require a decrease in non-mineral dry weight of 2 g m^{-2} week^{-1}, so that the carbon content of the apparently growing plants would be decreasing and the mineral content increasing twice as fast, leading to the observed unit leaf rate, which is also absurd.

28.14 Conclusions on respiration rates in deep shade

There remains only one explanation which can be regarded as probable—that there was a substantial decrease in the respiration rate per unit dry weight for plants growing in low light. As we have seen [28.12], this may have begun at 9 cal cm^{-2} day^{-1}, although if so the effect was small. To fit in with this hypothesis, and maintain a linear fall in rate of photosynthesis per unit leaf area (A), would require the respiration rate per unit dry weight for the plants growing in 5·1 cal cm^{-2} day^{-1} to be rather less than half that for plants growing in light regimes above 16 cal cm^{-2} day^{-1}.

Granted that some uncertainty attaches to the exact value of the unit leaf rate of plants growing in 1·5 cal cm^{-2} day^{-1}, both because they were very small and tenuous, and also because they were growing so near the compensation point that photosynthesis during the procedure for harvesting and photographing the leaves may have introduced a systematic positive error; nevertheless, while they were undisturbed in the cabinets the plants were definitely growing: they were increasing in dry weight and in leaf area, and they were obviously above the compensation point. But even to bring unit leaf rate to zero would, on the above assumptions, require the respiration rate per unit dry weight to fall to about 40 per cent of its value for medium and high light.

No such uncertainty, however, attaches to the values of unit leaf rate at 5·1 cal cm^{-2} day^{-1}. For plants 23 or 24 days old at the beginning of the experimental week, we have three values, from separate populations, of 6·5 ± 0·9, 6·7 ± 0·6 and 7·0 ± 0·8 g m^{-2} week^{-1} (all means from samples of 8, with fiducial limits); for plants initially 30 days old a further value, of 5·9 ± 0·2. It seems difficult, if not impossible, to reconcile these consistent values with the rest of the available data on any other basis than a reduction in respiration rate per unit dry weight.

Further observations are clearly desirable; if such a reduction in respiration rate can be proved to occur in consequence of shading, this may be a further important mechanism in the make-up of a successful shade plant, supplementing in its operation the consequences of the large increase in specific leaf area at low light also found in facultative shade plants [28.5, 28.11, 28.12]. Further investigation is likely to show that the

28.14

quantitative operation of these mechanisms relative to each other varies from one species of shade plant to another, and that consequently the form of the relationship between unit leaf rate and the daily total of visible radiation also varies. Where, as in *Impatiens parviflora*, there is an approximately linear portion at low light [Fig. 3.2.1a], and where also unit leaf rate reaches a maximum value at radiation totals well below the average for summer days [Fig. 3.4.1], a plot of unit leaf rate against the logarithm of the daily radiation total is bound to be of a sigmoid form similar to that shown in Fig. 3.3. Consequently, variations in the quantitative operation of these mechanisms may well underlie the variations in responses of different species to shading observed by Blackman and Wilson (1951a, b) and Blackman and Black (1959). In the former cases the apparently linear relationships of unit leaf rate with the logarithm of the daily radiation total may prove to represent segments around the points of inflection of a family of sigmoid curves generally similar to that of Fig. 3.3; in the latter cases the corresponding quadratic relationships would approximate to portions of sigmoid curves below the point of inflection.

28.15   Other physiological problems

Obviously such problems of leaf physiology arise immediately from a quantitative analysis of the growth of whole plants, and they have occupied the bulk of this chapter. However, the methods which we have been discussing lend themselves to a variety of other investigations. To pursue these in equal detail would exceed our available space, and we must merely glance at them in passing.

There are first considerations of general utility for physiological studies of higher plants. Since such studies began, variability of plant material has been recognized as a major obstacle. This has led to the development of the techniques of drawing conclusions from the comparison of the behaviour of experimental material with a control; either using comparable samples, or taking the material through a cycle of control conditions—experimental conditions—control, and so on. Until the development of apparatus for growing plants under uniform, specified, conditions, it was rarely possible either to relate the behaviour of one control with that of another remote in place or time, or to return to identical material later for a further experiment. As we have seen, this is still an occupational hazard of field experimentation. Controlled environment equipment now makes it possible to generate a particular plant for experimental purposes, and to return to the same plant at will. It is still, however, not possible to relate such a plant to others grown in different places and using different equipment, by relying on physical measurements of the controlled environment itself. Light climates, in particular, are difficult to characterize in this way. But if we can relate the two controlled environments to each other approximately, a simple quantitative analysis of growth enables us to use the plants themselves as biological integrators [30.3–30.9], and puts the comparison on a sound basis. We are also enabled to compare the plant grown under controlled conditions with its counterpart growing in the field [e.g., 28.10, 28.11], and thus to decide how far conclusions reached by the study of laboratory-grown plants are relevant to the solution of problems arising under natural, agricultural, or horticultural conditions.

28.16 Translocation

It will have been noticed that our analysis of plant growth provides incidentally a picture of the movement of material from organ to organ throughout the plant. Some elaboration in detail, together with respiration studies, would make it possible to convert this into a balance sheet of production and consumption organ by organ, and of the net movement of materials. Data would thus be available on translocation in the plant as a whole. Much work has been done on translocation into and out of particular organs—into fruits and tubers, for example, and out of leaves, but much remains to be found out about translocation in the plant as a whole. In time studies using labelled assimilates [23.6, 26.1] may be expected to make major contributions in this field, although there are still substantial difficulties to be overcome [32.8].

28.17 Ion uptake: unit root rate

The methods of analysing growth which we have discussed lend themselves equally to the investigation of ion uptake. Welbank (1961, 1962, 13.14 and 13.20) studied competition between two higher plant species at two levels of water supply and two levels of nitrogenous manuring. He put forward the concept of unit root rate—rate of uptake of mineral nutrients per unit of root—as a quantity analogous to unit leaf rate. This approach, although obviously a valuable one, has been little followed

28.17

because of the sheer practical difficulty of making the necessary measurements, particularly for plants growing in a natural soil.

There is the added complication, especially in studies of plant competition, of knowing what are the actual concentrations of the various ions in the soil solution in immediate contact with the absorbing surfaces of the root. As we have seen, it is next to impossible to estimate these concentrations in a natural soil [10.8], and this has led to the devising of artificial systems for maintaining known conditions just outside growing root systems.

28.18   Ion uptake: standard conditions

It is possible to grow certain woody plants with the roots not in soil, but supported in a chamber filled with saturated air and constantly sprayed with a nutrient solution of known concentration. Apple trees have been grown in this way at the East Malling Experimental Station (Roach *et al.*, 1957, and see 10.10). The root systems of the majority of herbs, and of the seedlings of most species, must, however, be supported in some rooting medium. We have already discussed at some length the difficulty here [10.10]; if the medium is wettable, giving good support to young and tenuous roots, and also an easy spread of the flowering nutrient solution, it is also inevitably polar, and will adsorb either anions or cations or both. At once it becomes impossible to state the effective concentrations of the adsorbed ions in the immediate vicinity of the root. This is tolerable if we intend to concentrate our attention on the effects of varying concentrations of either cations or anions. We can then choose a polar medium which does not adsorb them, as did Lewis (1963), who used crushed flint (a pure form, effectively amorphous SiO_2), and concentrated on the effects of reducing the concentrations of phosphate and nitrate around the roots.

The technical difficulties of supporting the root system in a non-polar (and therefore non-wettable) medium, and at the same time ensuring that it is adequately bathed in a thin layer of a flowing nutrient solution without being waterlogged, are very great. They have been solved for small seedlings by Hegarty (1968). This makes it possible to establish with precision the concentrations in the 'soil' solution immediately outside the root of a seedling which is entirely confined within an artificial environment, and yet is making normal growth. Studies can be made both of general and particular mineral starvation and their effects on the

28.18

metabolism and growth of the seedling, and effects can be followed at the level of organ, tissue or individual cell [27.10, 27.11]. By making mineral analyses of seedlings the operation of ion pumps can be studied, and conditions of mineral starvation created such that the ion pumps themselves cease to operate. The effects on the operation of the whole pumping system of increasing the concentration of a single ion can then be studied, on seedlings of various ages and more or less developed.

In fact, we have here a very powerful tool for the study of the ionic relations of whole plants: granted that these are seedlings, but in one way this is an advantage as the detailed morphology and anatomy can more readily be described and measured—for example, the exact length and area of the root system at any particular stage of development.

28.19 Small seedlings: their special advantages

The introduction of rapid balances reading to 1 μg [11.19] has made it possible to extend the type of analysis we have been discussing to young seedlings (Williams, 1960). This development opens up a number of new possibilities in tackling the physiological problems of whole plants, particularly as regards ion uptake (e.g., Pitman, 1965, 1966). The small scale makes it easier to specify external concentrations and to characterize the plant organs concerned with accuracy (Hegarty, 1968). In a system so relatively simple (as compared with the average higher plant) the whole geometry of the organism can be described—numbers of growing points, their distances from storage and photosynthetic organs, and so on. This is advantageous when the synthesis of new systems is being studied, and the detailed specification also makes it possible for the first time to tackle the problems of constructing mathematical models of growing plants, which are now beginning to interest control engineers on both sides of the Atlantic (e.g., Curry and Chen, 1970).

28.20 Conclusion

And so, again, we come full circle to the combination of physiological observation with morphology, anatomy, and the study of the environment which we have found inseparable from the quantitative study of plants as whole organisms. All the physiological problems which we have been discussing in this chapter involve one or another or several of these other ways of looking at the study of plants.

28.20

GROWTH STUDIES AND SOME PROBLEMS OF THE PHYSIOLOGY OF GASEOUS EXCHANGE

29.1 An historical purpose

By labyrinthine ways our discussions have almost returned to their historical starting point. We saw [13.8] how 90 years ago Weber was concerned with the capacity of leaves for carbon assimilation, which he called 'Assimilationsenergie', and how he used the growth method as the readiest means of estimating an average value for this over a period of plant growth.

Our discussions so far have related the rate of increment of dry weight (from all causes) of the plant as a whole to the total dry weight itself (relative growth rate) and to the total leaf area (unit leaf rate). When considering the concept of unit leaf rate we have further analysed the overall increment of dry weight by the plant as a whole [13.11]. By allowing for the portion of the increment due to mineral uptake we can calculate the carbon dry-weight rate, the rate of increase of carbon compounds, in the dry state, per unit leaf area. We can further allow for changes in the mean biochemical composition of the carbon compounds of the plant, producing in turn changes in the dry weight equivalent of one gram atom of carbon. We then have the net carbon reduction rate, which can if necessary be converted into the weight of carbon dioxide reduced by the whole plant during a 24-hour period (mainly, of course, in photosynthesis) less the weight of carbon dioxide produced in respiration by the whole plant during the same 24-hour period. This will normally include a period of darkness, and by allowing for the loss of dry weight by respiration of the whole plant during the dark period, we can next derive the light-phase whole-plant carbon balance.

We are now approaching an estimate of mean uptake of carbon dioxide by the most important photosynthetic organs, the leaves, during the light phase, but we are not quite there yet, because the light-phase whole-plant carbon balance gives us, for the light period, the mean rate of uptake by the leaves minus the mean rate of loss by the whole plant during the light phase. We must make yet a further allowance for respiration by the non-photosynthetic part of the plant during the light phase, in order to arrive at the light-phase leaf carbon balance. It will be recalled that by the definitions discussed in 13.11, the last three derivatives, the net carbon reduction rate and the two carbon balances, are all expressed in terms of weight of hexoses. To derive what Rackham (1965) has called 'net carbon dioxide intake' by the leaves in the light, it is necessary to convert the light-phase leaf carbon balance into its equivalent in weight of carbon dioxide by multiplying by the factor 1.47 [13.11].

29.2 Estimates of photosynthesis by whole plants

The method of the previous section is a complex one, involving measurements of many different kinds. Direct measurement of carbon dioxide uptake by the plant would be preferable, if it were practicable to make direct measurements on a plant growing under natural conditions without gross interference either with the plant itself or the environment in which it is growing. The first great advantage of the procedures which we have described is that during an experimental period, of the order of days, the plants being studied can be exposed to the natural environment without interference.

The method is clearly not adapted to studies of changes taking place in short periods of the order of minutes or hours; but it has the counterbalancing advantage that by a reasonable number of repetitions observation can be extended over a whole growing season.

Finally, there is the further advantage that the plant produces its own natural integration of the effects of its own genetic makeup and ontogeny as functioning within its past and present environments [30.3, 30.4]. To produce such an integration by inference from a series of short period observations under specified conditions (of environment, past history and present structure of the plant material) is at its simplest a difficult task, and frequently an impossible one.

29.3 An example

The most comprehensive example of the use of growth studies as applied to gaseous exchanges of leaves is that of Rackham (1965, 1966), and we

29.3

will make use of this in the discussion which follows. The plants used were specimens of *Impatiens parviflora*, grown either under constant, controlled environmental conditions in the laboratory, or as potted plants in the field, where the aerial parts were subjected to the natural aerial environmental conditions.

It was unnecessary, for Rackham's purposes, to go through the intermediate steps of estimating successively carbon dry-weight rate, net carbon reduction rate and the light-phase carbon balances. He used a modification of the simplified procedure discussed at the end of section 13.11, multiplying unit leaf rate by the factor W_C, the quantity of carbon dioxide required to produce unit mass of dry material. This equals the amount of carbon dioxide produced by combustion, which can be measured directly.

His subsequent computations were then based on the quantity $\mathbf{E}W_C$, which is equivalent to the net carbon reduction rate, but converted into weight of carbon dioxide (compare 13.11, end), and all the subsequent allowances for respiration were also made in terms of weight of carbon dioxide.

In Chapter 12 we discussed some of the difficulties in the way of making measurements of the respiration of whole plants, including the effects of wounding and handling [12.4]. With these in mind Rackham measured the mean respiration rates of whole potted plants growing in vermiculite, by the method described in Appendix 1. He then had to measure or estimate the respiration rates of the stems and roots during the light phase. As the whole allowance for respiration in the examples given in Table 29.4 comes to around one quarter of the final figures for net carbon dioxide intake, and as that part of the allowance appropriate to stems and roots during the light phases is at most half of this, Rackham considered (probably correctly) that negligible errors were involved in assuming that the mean respiration rate, per unit dry weight, for stems and roots in the light was the same as the mean rate for the whole plant in the dark, measured as W_R grams of carbon dioxide produced per hour by the amount of fresh plant material having a dry weight of one gram.

29.4 Estimation of net carbon dioxide intake

The measured mean rate of respiration by the whole plant in the dark is then $W . W_R$ g hr^{-1}, and if the length of the light period is H hours out of

24, the total respiration during the dark phase is $(24 - H)W \cdot W_R$ g day^{-1}. During the light phase the rate of respiration by all organs other than the leaves is assumed to be $(W - L_W)W_R$ g hr^{-1}, so that the total respiration of these organs during the light phase is $H(W - L_W)W_R$ g day^{-1}. The total allowance is then $W_R(W(24 - H + H) - HL_W)$ g day^{-1}, and the correction to **E**, on a basis of leaf area measured in square metres [28.2, 28.5] is

$$W_R\left(24\frac{W}{L_A} - H.\frac{L_W}{L_A}\right) \text{ g m}^{-2} \text{ day}^{-1}.$$

But $L_A/W = \mathrm{LAR}$, and $L_A/L_W = \mathrm{SLA}$; if these are measured in appropriate units, the allowance is therefore

$$\frac{24W_R}{\mathrm{LAR}} - \frac{HW_R}{\mathrm{SLA}} \text{ g m}^{-2} \text{ day}^{-1}.$$

E being normally expressed on a time basis of weeks, the mean net carbon dioxide intake is therefore

$$\frac{EW_C}{7} + \frac{24W_R}{\mathrm{LAR}} - \frac{HW_R}{\mathrm{SLA}} \text{ g m}^{-2} \text{ day},$$

or

$$\frac{1}{H}\left(\frac{EW_C}{7} + \frac{24W_R}{\mathrm{LAR}}\right) - \frac{W_R}{\mathrm{SLA}} \text{ g m}^{-2} \text{ per hour}$$

of the light period itself, which is the quantity required. If, as is probable in the cabinet experiments, the difference between the rate of photosynthesis and the rate of respiration of the leaves is effectively constant during the light period, then this mean rate will equal the instantaneous rate at any time during the light period. The same cannot be said of the field experiment, No. II in Tables 29.4 and 29.5, where there is the added difficulty of making an estimate of the effective daylength from the point of view of photosynthesis, which is not the same thing as the length of the astronomical day. Rackham therefore devised an approximate method of estimating the effective daylength for field experiments, based on Evans and Hughes's (1961) data on the dependence of unit leaf rate in this species in total radiation, measured in absolute units, and the figure given for Experiment II in Table 29.4 was derived in this way.

It will be noted that this method of computing the mean rate of net carbon dioxide intake involves no assumptions about the rate of respiration by the leaves in the light, the net rate of carbon dioxide intake corresponding to the apparent assimilation rate.

29.4

Table 29.4. Observations on potted plants of *Impatiens parviflora* growing under a variety of controlled conditions in the laboratory (Experiments O, I, III and IV) and in the field (Experiment II), to illustrate the procedure for the calculation of mean net carbon dioxide intake described in 29.3 and 29.4. The mean measured value of the pure number W_C (the weight of carbon dioxide needed to produce unit dry weight) for the whole series of experiments was 1·55, and this is used throughout. Data of Rackham (1965, 1966).

Experiment	O		I		III		IV		II	
Period	I	II	I	II	I	II	I	II	I	II
Effective daylength, H, hours	12·0		12·0		16·0		16·0		13·0	
Mean daily short-wave radiation, cal. cm^{-2} day^{-1}	11		16		8		28			
Unit leaf rate, E, g m^{-2} week^{-1}	10·6	11·2	14·6	12·4	12·2	8·7	24·1	19·1		43
$10EW_C$, mg dm^{-2} week^{-1}	164	174	226	192	189	135	374	296		666
Respiration rate, W_R, mg g^{-1} hour^{-1}	2·0		2·8		2·1		2·5		2·9	
Leaf area ratio, LAR, dm^2 g^{-1}	4·68	4·49	4·53	4·48	4·90	5·20	3·88	3·80		1·08
Specific leaf area, SLA, dm^2 g^{-1}	9·03	8·51	8·06	7·75	11·36	10·82	7·54	7·03		2·90
Net carbon dioxide intake, mg dm^{-2} hour^{-1}, $\frac{1}{H}\left(\frac{10EW_C}{7} + \frac{24W_R}{LAR}\right) - \frac{W_R}{SLA}$	2·60	2·74	3·59	3·18	2·16	1·61	3·97	3·28		11·3

Table 29.4 sets out all the necessary terms for the calculation of mean net carbon dioxide intake for five experiments, one (Experiment II) under natural conditions of the aerial environment in the open, the other four under controlled environmental conditions in the laboratory. One of those was equivalent to deep shade in a wood, with mean daily visible radiation of 8 cal cm^{-2} day^{-1}, under which *Impatiens parviflora* assumed its typical shade form, with specific leaf areas ranging from 10·8 to 11·4 dm^2 g^{-1}. The other three were intermediate (specific leaf areas ranging from 7·0 to 9·0 dm^2 g^{-1}) between this form and the typical full sun form of the experiment in the open field, with specific leaf areas of 3·0 dm^2 g^{-1} or less. We thus have a range of thirty-fold in mean total daily light, six-fold in net carbon dioxide intake, while the plants show a range of nearly four-fold in specific leaf area, but only about one-third in respiration rate per unit dry weight.

Values of W_C, a pure number representing the number of grams of carbon dioxide equivalent to one gram of dry weight, were obtained by oxidation with chromic acid from separate sets of plants grown over a similar range of conditions. The individual values were apparently not correlated with the conditions of growth, and accordingly the mean value of 1·55 was used throughout.

29.5   Transpiration and intake of carbon dioxide

Having reached this point, much interesting additional information can be gained from studies of transpiration undertaken simultaneously with the growth studies. This is not the place to pursue this subject, abounding with fascinating problems both of measurement and interpretation. Here we will merely give Rackham's methods in the barest outline, together with a representative selection of the types of quantitative inferences which can be drawn from such studies.

He devised a method of weighing a number of the pots in which the young plants were growing in the controlled environment cabinets, without opening the cabinets or disturbing the plants. Evaporation from the upper surface of the black plastic pots was reduced, while the normal aeration of the root system was maintained, by covering the vermiculite in which the plants were growing with alternate layers of expanded polystyrene granules and fine gravel. The residual evaporation was allowed for by weighing blank pots. The course of transpiration was then

29.5

Table 29.5. Observations on the diffusion of water vapour out of, and carbon dioxide into, the leaves of plants of *Impatiens parviflora* growing in a natural aerial environment and under a variety of controlled conditions in the laboratory. Same series of experiments as for Table 29.4, where the figures for mean net carbon dioxide intake are derived (note the change of units). Data of Rackham (1965, 1966).

	O	I		III		IV		II	
Experiment period	II	I	II	I	II	I	II	I	II
Age of plants at end of experiment (days)	54	34	41	37	44	26	33	32	39
Temperature (°C)	14.2	15		14.2		14.2		natural conditions	
Relative humidity (%)	85	75		63		63			
Intensity of transpiration (mg cm^{-2} hr^{-1})	2.6	3.87	4.19	3.04	3.24	5.45	5.12	22.0	23.8
Total water vapour diffusion resistance (cm^{-1} sec)	2.0	2.25	2.08	4.77	4.48	2.76	2.93	0.98	0.79
Total gas phase CO$_2$ diffusion resistance (cm^{-1} sec)	3.5	3.9	3.6	8.0	7.5	4.8	5.1	1.26	1.01
Net carbon dioxide intake (μg cm^{-2} hr^{-1})	27.4	35.9	31.8	21.6	16.1	39.7	32.8	135	113
CO$_2$ concentration difference (ng cm^{-3})	27	39	32	48	34	53	46	47	32
The same, as a fraction of atmospheric concentration (%)	4.6	6.6	5.4	8.2	5.7	9.0	7.9	8.0	5.3

followed in the dark and in the light, and it was shown that steady rates were established very shortly after the lights were switched on or off, and maintained until the next change (Rackham, 1966). Parallel studies of a similar kind were made of the plants growing in pots in the open, transpiration in the light being again distinguished from transpiration in the dark. Leaf temperatures were measured throughout.

From these transpiration figures combined with environmental measurements were calculated values for the resistance to the outward diffusion of water vapour from the surfaces of the mesophyll cells to the bulk of the air, and the path was divided into that element (roughly up to the surface of the leaf) where diffusion through a static gas phase is taking place, and the eddy diffusion path outside this (for the distinction see, e.g., Rider, 1954). It is then possible to convert these into the corresponding resistances to the inward diffusion of carbon dioxide; and then, using the figures for the net carbon dioxide intake which we have just seen, to work out what the difference of carbon dioxide concentration must have been between the bulk of the air and the intercellular spaces immediately adjacent to the mesophyll cell walls.

These broad principles underlying the calculations are easily stated, but the actual calculations themselves depend on very many diverse measurements and an intricate line of reasoning, the details of which need not concern us here. In Table 29.5 are set out the major results, for the experiments from which net carbon dioxide intake was worked out in Table 29.4.

29.6 Conditions in the intercellular spaces of the leaf

We see that the difference between the concentration of carbon dioxide in the air in contact with the mesophyll cell walls and that in the bulk of the outside air lies in all cases between 4·6 and 9·0 per cent of the atmospheric concentration, figures which are of the same order of magnitude as those obtained by, or implicit in the results of, other workers (e.g., Gaastra, 1959, and other authors quoted therein). Even so, the constancy of this figure is remarkable in a series of experiments which covered almost the whole range of form assumed by the plant, between full sun and deep shade. Such variation as is seen is mainly due, not to an association with leaf structure, as expressed in the specific leaf area, or with external conditions, but to an ontogenetic drift. This is seen in Fig. 29.6,

29.6

where the concentration differences are plotted against chronological age between germination and the end of the experimental period. The partial correlation coefficients are:

between fall in concentration and age, allowing for any linear association with specific leaf area, $r = -0.905$, highly significant ($p \simeq 0.002$)

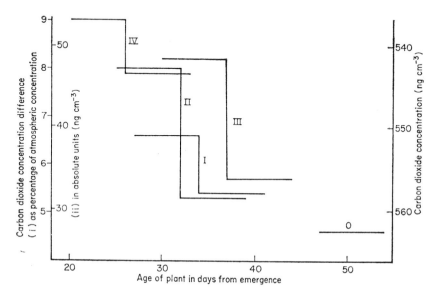

Fig. 29.6. Concentration differences for carbon dioxide between the outside air and the internal surface of leaves of *Impatiens parviflora* as a function of the ages of the plants in days from emergence of the seedling. Same experiments and labelling as in Tables 29.4 and 29.5. At right, ordinates showing also the carbon dioxide concentration in the gas phase at the internal leaf surface. From Rackham (1965).

between fall in concentration and specific leaf area, allowing for a linear association with age, $r = 0.512$, not significant ($p > 0.10$).

Experiments I to IV, in Table 29.4, all showed a decline in net carbon dioxide intake between period I and period II. We now see that diffusion in the gas phase is always a small restriction on the supply of carbon dioxide to the mesophyll cells, and one that progressively decreases with age. It follows that changes in the gas phase diffusion resistance cannot have caused the decline in net carbon dioxide intake, and that the cause must be sought elsewhere in the system.

29.6

29.7 Cellular dimensions and properties

The obvious next step is to follow the inward diffusion of carbon dioxide through the cell walls of the mesophyll to the chloroplasts. To do this we need to know both the dimensions of the various parts of the system and the conditions of the movement of carbon dioxide through it, particularly the resistances to diffusion. In dealing with these latter we are hampered by a lack of direct measurements: the available evidence is mainly indirect. It has been critically reviewed by Raven (1970), and we shall make use of his conclusions as we proceed. In view of the uncertainties, however, it will be best to follow Rackham and calculate the concentration differences on the assumptions that the properties of the cell wall and chloroplast interior are the same, for the diffusion of carbon dioxide, as those of pure water. From these we can infer probable values for concentrations in different parts of the system, and the main sources of uncertainty will become apparent. Rackham assumed that the bulk of the movement through the wall and chloroplast was in the form of molecular carbon dioxide in solution (for the evidence for this assumption see Raven, 1970). The partition coefficient of carbon dioxide between the gas and liquid phases would then be 1·02 at 15°C and 0·88 at 20°C, and the diffusion constant would be about $1·6 \times 10^{-5}$ cm^2 sec^{-1}.

The diffusion constant of carbon dioxide in air is about 0·15 cm^2 sec^{-1} at temperatures between 15 and 20°C, so that 1 cm of diffusion path in air is equivalent to about 1 μm in water. It follows that electron microscopy is needed in order to make the necessary geometrical measurements for estimating the diffusion resistance of a thin cell wall. Plate 29.7 shows one of a series of electron micrographs made from leaf material of a plant of *Impatiens parviflora* grown in conditions equivalent to moderate shade, and having a specific leaf area of 5·56 dm^2 g^{-1}.

We see that the chloroplasts of this species are lens-shaped objects about 3 μm thick (although exceptionally their thickness may attain 6 μm), closely appressed to the very thin walls of the mesophyll cells. These walls are closely lined with chloroplasts where they abut on intercellular spaces, but chloroplasts are less abundant where the cell walls abut on other cells. Close study of the whole series of preparations, which included some stained by permanganate and some by osmium, showed that the intercellular spaces are not lined by any substantial lipoid layer. If present, such a 'cuticle' would impose a finite resistance to dif-

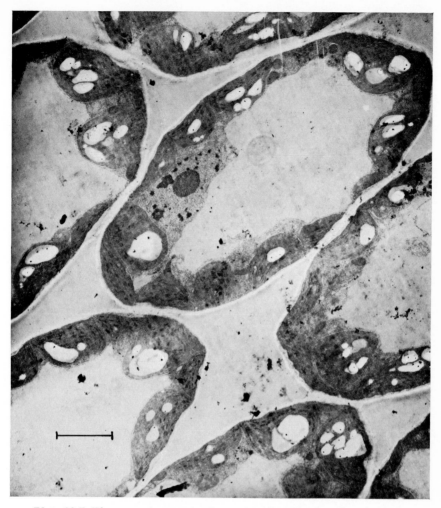

Plate 29.7. Electron micrograph of an approximately paradermal section through the palisade mesophyll of a leaf of *Impatiens parviflora* treated with osmic acid. The elongated appearance of the cells is due mainly to obliquity of sectioning rather than to compression. Scale line 5 μm. From Rackham (1965).

fusion of water, and would probably also interfere with the passage of carbon dioxide. Both permanganate and osmium staining do in fact show an exceedingly thin membrane at or near the outer face of those cell walls which abut on intercellular spaces. It could not be shown that this was an internal cuticle rather than remains of the ruptured middle lamella or some other structure; but its thickness as measured under high resolution was so small, 5 nm, that whatever its structure it is unlikely to have added appreciably to the diffusion resistance to the passage of small, uncharged molecules (Raven, 1970, p. 184).

The measurements of the wall were made on several differently prepared specimens, in order to reduce the errors due to artefacts. They showed a thickness ranging from $0 \cdot 15$ to $0 \cdot 4$ μm, with a mean of $0 \cdot 22$ μm.

Measurements made on leaf sections under the light microscope showed that the total area of mesophyll cell walls abutting on intercellular spaces was approximately five times the area of the leaf itself. Both this figure, and those just quoted for wall thickness, apply to a mature leaf with a specific leaf area of $5 \cdot 56$ dm^2 g^{-1}, a form characteristic of a plant growth in high light and a long day in the laboratory, or under light shade in the field. It is thus intermediate between the sun form of Experiment II and the shade forms of Experiments O, I, III and IV in Tables 29.4 and 29.5.

29.8 Movement of carbon dioxide in the mesophyll cell walls

Rackham pointed out that diffusion in the mesophyll cell walls is not strictly static, in that there is a mass movement of water as well. Water evaporates from the outer surface of these walls, and is replaced from within by a current through the interfibrillar spaces of the wall, inward diffusion of carbon dioxide taking place against this current. His figures enabled him to estimate the rate of backward transport of carbon dioxide by this current for the most favourable cases, those of Experiment II, periods I and II, when the intensity of transpiration was more than four times as high as in any other. It turned out to be rather less than 1 part in 10,000 of the rate of inward diffusion of carbon dioxide, even for these occasions where the effect was maximal, and it can accordingly be neglected.

On the basis of the assumptions mentioned [29.7], together with the measurements and calculations quoted, the mean diffusion resistance to the movement of carbon dioxide through the mesophyll cell walls works out at 1·4 sec cm^{-1} on the basis of unit area of the mesophyll cell walls themselves. To bring this to the same basis of unit leaf area as in Table 29.5 we must divide by the ratio of area of mesophyll cells abutting on intercellular spaces to leaf area, which we have seen to be about 5. If the cell wall had the same properties as water, the diffusion resistance would thus be 0·3 sec cm^{-1} for the leaf studied.

By interpolation in the results given in Table 29.5, the value of the total gas phase diffusion resistance would be about 3·5 sec cm^{-1}, while the concentration of carbon dioxide in the gas phase immediately outside the mesophyll cell walls would be 47 ng cm^{-3} less than the atmospheric concentration, i.e., close to 543 ng cm^{-3}. Taking the partition coefficient of carbon dioxide between the liquid and gas phases as 0·9, this means that the concentration of carbon dioxide in the liquid phase at the external surface of the mesophyll cell walls would be close to $0·9 \times 543 = 489$ ng cm^{-3}.

The concentration difference across a layer of water of the same thickness as the cell wall would then be $47 \times 0·3/3·5 = 4$ ng cm^{-3}, a fall of less than 1 per cent. Raven (1970, p. 184), after reviewing the available evidence, concluded that the resistance to the diffusion of molecular carbon dioxide in solution 'for the cell wall component of the aqueous diffusion pathway is about twice that of the same thickness of water'. The concentration difference would then be about 8 ng cm^{-3}, a fall of about $1\frac{1}{2}$ per cent of the concentration in water in equilibrium with atmospheric carbon dioxide, and the concentration at the inside of the cell wall would be close to 481 ng cm^{-3}. It is clear that for this plant the resistance to the inward diffusion of carbon dioxide through the mesophyll cell walls is an almost negligible factor in limiting the supply to the chloroplasts. It is also clear that this conclusion does not rest on the exact validity of the assumptions we have made [29.7] about the state of the carbon dioxide in the cell walls, and the resistances to its movement.

29.9 Carbon dioxide supply to the chloroplasts

Plate 29.7 showed that the layer of cytoplasm between the outer face of a chloroplast and the adjacent cell wall is exceedingly thin, or non-existent. Permanganate staining showed a unit membrane as the plasmalemma, but no other structure was ever seen between the outer surface of the chloroplasts and the adjacent cell wall. Raven (1970, p. 184) gives figures suggesting that the resistance to diffusion of carbon dioxide through a lipid bilayer 'can be as low as 6×10^{-2} sec cm^{-1}'. Two such membranes would have together a resistance of approximately one-fifth of that which we have just postulated for the cell wall, and the concentration difference across them would lie between 1 and 2 ng cm^{-3}. The carbon dioxide concentration would then be around 480 ng cm^{-3} at the outer edge of the interior of the chloroplasts in the plant we are considering.

In Plate 29.7 and the other preparations, the grana and intergranar lamellae (or, to use a more modern terminology, the thylakoids) are more or less evenly distributed throughout the chloroplasts, and arguing from the extremely large size of the carboxydismutase molecule and from what is known of its distribution, it is to be supposed that the incorporation of carbon dioxide goes on throughout the volume of the chloroplast; although as yet there is no information on whether the rate of the reaction varies from one part of the chloroplast to another (Raven, 1970).

The form of the actual fall in concentration of carbon dioxide across the chloroplast depends, of course, upon the rate per unit volume at which carbon dioxide is being consumed in the different parts of the chloroplast. Rackham was able to show that whatever this distribution, the maximum possible value of the fall in concentration is obtained by assuming that the whole consumption of carbon dioxide is taking place close to the surface of the chloroplast remotest from the cell wall. For a chloroplast 3 μm thick, having the same diffusive properties as water, this would correspond to an effective diffusion resistance of up to 4 sec cm^{-1}, and to a concentration difference of carbon dioxide across the chloroplast of 50 ng cm^{-3}. Raven (1970, p. 184) concludes that the resistance to the diffusion of the molecular carbon dioxide within the chloroplast is likely to lie between the value for pure water and that for the cell wall (twice the value for pure water). This would raise a concentration difference of 50 ng cm^{-3} to around 75 ng cm^{-3}, although we must remember that the basis of estimation produced a maximum figure.

The maximum concentration of carbon dioxide in the chloroplasts of this plant was thus shown to be 480 ng cm^{-3}, or 90 per cent of the concentration in water in equilibrium with the concentration in the bulk of the atmosphere. In the average chloroplast in a part of the cell abutting on the intercellular spaces the concentration would not be below 400 ng cm^{-3}, or about 75 per cent of the atmospheric equilibrium

29.9

concentration, although the analysis showed that the concentration was unlikely to be as low as this. Lower concentrations would, however, be expected to occur in the largest chloroplasts, in those where the whole of one side of the chloroplast did not abut on the intercellular spaces, or in those having above-normal photosynthetic rates. In the first two cases the path of aqueous diffusion would be longer than average, in the third the concentration gradient would be steeper. On the other hand it must be remembered that we have been dealing with average rates of carbon dioxide intake, and that we have calculated concentration differences as if we were dealing with linear diffusion through a stratified system having five times the leaf area, i.e. an area equal to the areas of mesophyll cells abutting on the intercellular space. In so far as the effective area for diffusion becomes larger as the diffusion pathway proceeds inwards from the cell wall adjacent to the intercellular space, a less steep gradient would be needed to maintain the same rate, and the concentration differences would be correspondingly less. Making all allowances for the variations mentioned it seems unlikely that in this plant under the conditions of these experiments the carbon dioxide concentration anywhere in the chloroplasts ever fell much below 70 per cent of the concentration in water in equilibrium with the normal atmospheric carbon dioxide.

29.10 Conclusions

Even if the above calculations are regarded as tentative, Rackham's attention to the possible sources of error may be held to have established that, in a form of *Impatiens parviflora* growing in conditions of light shade to which it has become adapted, the mean daytime concentration of carbon dioxide in the chloroplasts is at any rate a substantial fraction, and probably at least 80 per cent, of the maximum possible (i.e. that in water in equilibrium with air). This conclusion is of interest in view of the frequent (but unproven) assumption in the literature that the concentration in chloroplasts approaches zero except in low illumination.

It follows that, in this form of *I. parviflora*, almost certainly in other forms of the same species, and probably in other species as well, diffusion resistance, in the true sense of the term, cannot be regarded as the most important limitation on the rate of photosynthesis under the conditions in which the plants normally grow. We may thus conclude that the explanation of the quantitative relations between photosynthesis, external carbon dioxide concentration, and illumination must be sought in the cellular organization itself—either in the properties of the photosynthetic mechanism and its associated enzyme systems, or, perhaps less likely [23.7 and Fig. 23.7], in the capacity of the cellular organization for disposing of the products of photosynthesis in such a way that they do not interfere, osmotically or otherwise, with the mechanism.

Such conclusions are of interest in a purely physiological context. They are of particular interest in having been reached by observations on plants all of which were making natural growth undisturbed, and some of which were making natural growth in a natural aerial environment quite free from interference. Similar conclusions of physiological relevance can be reached by purely laboratory experimentation, and not necessarily on intact plants, but to apply these conclusions to intact plants growing in the field would have required comparable field studies, and it is difficult, if not impossible, to see how these could have been carried out without gross interference either with the plant, or with the natural environment, or both, by any other method than using the type of quantitative study of growth with which we have been concerned.

The conclusions also have substantial relevance in interpreting the quantitative relations between certain aspects of growth and the environment in which the plants grew. We shall turn to this in the next chapter.

It is to be hoped that in the near future similar investigations on a range of species, of varying growth forms, ecological preferences, anatomical structure and biochemical make up (Loomis *et al.*, 1971), will make it possible to present a more generalized view [32.4, 32.6, 32.7].

29.10

SOME PROBLEMS OF EXTERNAL RELATIONS: GROWTH AND ENVIRONMENT

30.1 Inescapable problems

We saw in our preliminary consideration of the conditions of plant life that problems connected with the growth of plants are not easily separated from the environment in which the plant has grown and is growing, and our subsequent discussions have produced numerous instances of how much the structure and functioning of a plant are the products of both past and present environments. Not surprisingly our discussions in the last four chapters, centred as they have been on aspects of the plant, have quite failed to eliminate consideration of the environment. Indeed, it has played an important part throughout.

We have already delimited a number of environmental problems, and considered many of the methods available for solving them—statistical, experimental, and so on—in connection with specific instances. In certain cases [17.5–17.10; 18.15] we have been able to compare the effects of a considerable number of specific changes in the external environment on particular aspects of the growth of particular plants. It is not necessary to retrace our steps, but there are certain matters on which we have not yet touched.

30.2 Inferences from the study of gaseous exchange

It was possible to show [29.5] for the particular case of *Impatiens parviflora*, that the carbon dioxide concentration in the intercellular spaces of actively growing plants in light is within 1 or 2 per cent of 92 per cent of the concentration in the bulk of the surrounding air, when this is close to the normal concentration of around 0.6 g m^{-3}. This relative constancy holds for adapted plants over a wide range of light climates, from full natural daylight to artificial days of 12 or 16 hours in controlled environment cabinets, with daily totals of visible radiation down to 8 cal cm^{-2}.

It almost certainly follows that concentrations of carbon dioxide within the chloroplasts of this plant, growing in nature, seldom fall below 80 per cent of the concentration in water in equilibrium with the atmosphere. Five of the nine instances which we examined in Table 29.5 were based on observations made on plants grown with daily light totals between 16 and 8 cal cm^{-2}. We had earlier shown that from 20 cal cm^{-2} day^{-1} downwards, the unit leaf rate of adapted plants approximates closely to a linear relationship with the daily total of visible radiation [Fig. 3.2].

We have already examined [28.11–28.14] the reasons for believing that this linear relationship for unit leaf rate corresponds to a linear relationship between rate of photosynthesis and total daily light, at least to a first approximation, and possibly closer still. Now the points on the linear relationship corresponding to experiments made on plants growing in controlled environment cabinets were all determined with a constant radiant intensity and all with the same period of illumination each day. It follows that over this intensity range (up to at least 2 cal m^{-2} s^{-1} of visible radiation) photosynthetic rate is almost certainly linearly related to light intensity, on an instantaneous basis. It also follows that this linear relationship is very unlikely to be disturbed by the normal exigencies of gaseous exchange, such as variations in wind velocity, or small changes in stomatal aperture. Indeed, very large changes in stomatal aperture would be needed to cause a significant departure in a specific case.

Temperature poses a somewhat more complex case. There is good reason to believe that under the conditions of these experiments the temperature coefficient of respiration is substantially higher than that of photosynthesis, so that the temperature coefficient of unit leaf rate would depend on the proportion which the two processes bore to each other in a particular case. For constant conditions in growth cabinets, with a daily total of visible radiation around 20 cal cm^{-2} day^{-1}, the net result for this species is a temperature coefficient less than 1 [Fig. 17.6]. We have already seen [28.11] that for the range from this radiation total down to 6 cal cm^{-2} day^{-1} in growth cabinets the rates of respiration and photosynthesis bear a roughly constant proportion to each other, so that here the temperature coefficient of unit leaf rate would be expected to be roughly constant. But there is no reason why this should be so under the more complex temperature conditions of Coombe's woodland experiments [28.10, 28.11], where day and night temperatures would have been different, and where different parts of the same plant would have

30.2

been at different temperatures. Variations in the temperature coefficient of unit leaf rate would then be expected, and these would be particularly significant at low light, where unit leaf rate is small. It will accordingly be necessary to examine the data for evidence of this effect.

Finally, the rate of the photosynthetic process is little affected by carbon dioxide supply under daily light totals below 20 cal cm^{-2}. In one experiment an increase in carbon dioxide concentration by a factor of 2·77 brought about an increase in unit leaf rate of between 20 and 25 per cent only. Although the carbon dioxide concentration around plants growing on the floor of a woodland is liable to fluctuate substantially, the mean differences in concentration between different 10-day periods are likely to be small, and their effects on mean unit leaf rates to be negligible.

We thus have a system where total photosynthesis over a period in a fluctuating environment would not be expected on average to depart substantially from the linear relationship with total visible radiation, so long as the variations in radiant intensity do not cause the instantaneous values to exceed 2 cal m^{-2} s^{-1}, which appears to be about the limit of the linear range, often enough to add up to a substantial fraction of the day. As we have seen from Fig. 28.11, this proves in fact to be the case.

This is an interesting train of thought. We have already seen much data leading to the belief that changes in the light climate are among those having the most profound effects on many aspects of plant growth [e.g., 17.5–17.9]. We have also seen that the light climate is one of the most difficult aspects of the whole environment to characterize with precision. Could we use the plant as a measuring device? Specifically, could we use it as a biological integrator in competition with, or to supplement, physical methods?

30.3   Integrators of aspects of the environment: phytometers

The use of biological systems as integrators of aspects of the physical environment can serve several purposes in the investigation of effects of environment on the organism. As it grows, the organism accumulates the consequences of past responses to its environment, and so performs a natural integration as it goes along. Usually these integrations involve elaborate chains of mechanisms, non-linear responses, complex interactions, and other features which make the detailed disentangling of

30.3

relationships impracticable. What we usually have to do is to accept the result of the integration, measure it, and endeavour to establish broad causal relationships with aspects of the environment, without attempting any detailed understanding of the actual mechanism itself.

The idea of plants as integrators of environmental experience is an old one in agriculture and horticulture: it was with ideas of this kind in mind that Kreusler and his co-workers made regular climatological measurements in parallel with their quantitative studies of the growth of maize (Kreusler *et al.*, 1877a, b, 1878, 1879). Our previous studies (e.g., 7.1, Chapter 13) have shown why during the last 90 years progress in this direction has been slow. On the other hand, ecologists have had substantial success in using plants as indicators of particular conditions, especially catastrophic events such as high winds or frosts of a particular intensity; and as integrators of such occurrences over periods of time. As a good example of this latter, Malloch (1970) was able to show experimentally that *Asplenium marinum* has a precise sensitivity to frosts below $-0.2°C$, so that the existence in the spring of undamaged plants in particular cliff-top situations could be used as an index of practically frost-free localities. In this instance the fact that the organs affected, the leaves, are long-lasting, the plant having no leafless season, means that an integration of past unfavourable experience persists, and if needs be a few occasional observations can provide information on the frosts, or lack of them, of an entire year. This is not so for ephemeral organs such as the frost-sensitive flowers of *Fagus sylvatica*. Similarly, after the work of Pigott and Taylor (1964), normal growth of *Urtica dioica* in the field can be used as an index of the absence of phosphate deficiency. An extensive literature has grown up along these lines, and examples will be found in Goodman *et al.* (1965, for industrial wastes), Rorison (1969, for mineral nutrition in general) and Walter (1960, a wide range of examples).

But these mainly qualitative cases of an all-or-none reaction pose relatively few difficulties of interpretation as compared with the use of specific plants as quantitative integrators of environmental experience. Clements endeavoured to use such a system for ecological investigation, and characteristically coined the word 'phytometer' for a plant used in this way (Clements and Goldsmith, 1924). During the subsequent years he and his co-workers published many studies along these lines, with limited success. This was not only due to the inherent complexities of the system studied, but also to a neglect of important aspects, particularly of the problems posed by ontogenetic drift, so that sequential harvests were few and when they occurred, far between. Further difficulties were introduced by the extensive use as phytometers of medium-sized plants of *Helianthus annuus*, a species difficult to grow under artificial conditions at anything approaching its full possible vigour. In consequence, in many series of experiments the most vigorous, 'control', plants were depauperate (see, e.g., the plates illustrating Clements and Long, 1934, 1935, and compare with Plate 24 of Clements, Weaver and Hanson, 1929). This amounts to using the phytometers themselves as a means of demonstrating that growth conditions in the 'controls' fell far short of optimal, and, combined with the general difficulties already mentioned, makes quantitative interpretation of this great body of data difficult and often impossible.

The phytometer method has recently been revived in greatly improved form by Leach and Watson (1968), for the purpose of studying conditions at different levels inside the canopy of a number of annual crops. They used seedlings of an in-bred strain of sugar beet (*Beta vulgaris*, cultivar KLT 50, derived from cv. Kleinwanzleben at the Plant Breeding Institute, Cambridge) of 50–100 mg dry weight, growing in specially designed containers of culture solution. The samples were selected for uniformity on three occasions during the preparatory phase [8.3]. At the last selection one representative sample was harvested, and the other plants were transferred, each to an individual container of culture solution, which were set up in position in the field. They remained undisturbed for one week, and were then harvested. Unit leaf rates were calculated using equation 16.8.2. For calibratory purposes, and as a test of reproducibility, six series of observations were made in the open and under neutral screens of known transmission. For the crop observations, the whole experiment was repeated several times (usually, 3, 4, or 5) during a growing season. The seedlings were small and compact, of a rosette form, and therefore convenient for making measurements at a particular level in a crop.

However, in spite of these improvements in planning and the meticulous attention to practical detail in their work, the difficulties in the way of this approach are underlined by the results of the series of experiments under artificial shades. For these they plotted unit leaf rates against the logarithm of the total daily visible radiation measured in absolute

30.3

units, and worked out the straight line of best fit [28.14, last paragraph]. Taking this as datum, rather over a quarter (6 out of 23) of the mean values of ULR fell outside a range of -14 to $+10$ g m^{-2} week^{-1} (or a range of from 60 per cent up to 200 per cent of a given value for total daily visible radiation). These large deviations among sets of plants placed under standardized conditions of artificial shade greatly complicate the interpretation of the field observations, when it is desired to characterize an unknown condition within a crop, and it will be necessary to make a further substantial reduction in the variability of behaviour of the plants under standardized conditions in order to make the phytometer method one of wide utility for field investigations. The variability observed in these experiments might have many causes among those which we have discussed, but there is internal evidence in favour of one specific suggestion concerning choice of species. It has been pointed out (D. E. Coombe, personal communication) that both the leaf bases and hypocotyls of the seedlings of *Beta vulgaris* are brittle, and thus subject to the kind of almost invisible damage already considered in other connections [11.20, 28.3], leading to negative deviations in the values of growth parameters for the individual damaged plant, lowering the value of the mean of the population of which it forms a part, and increasing the variance. Clearly both choice of species and of substrate are important in a particular investigation, and as only a limited range of species make normal growth in unaerated culture solution, when trying out species not previously studied it is in general better to use coarse vermiculite, already treated with a suitable culture solution, and if necessary to provide an automatic device to maintain a level of water or culture solution near the bottom of the pot.

However, the phytometer method was devised to investigate the combined effects of a group of ecological factors: the problem which was proposed at the end of the last section is much more limited in scope, although formally similar—can we in suitable cases use a specific plant as a biological integrator of one particular aspect of the environment, in competition with, or to supplement, physical methods?

It is always possible to make use of a measuring device of proved reliability without necessarily understanding the underlying mechanism —the hair hygrometer, for example, was a common instrument at a time when very little indeed was known about the physico-chemical processes involved in the change in length of the hair. Thus provided that enough is

30.3

known about its reliability it is possible to use a biological integrator as a substitute for physical measurement of an environmental factor.

If we go further, and compare under the same environmental conditions the data from a biological integrator with those from a physical one, we may in favourable cases be able to gain additional information of two kinds.

30.4 Inferences about biological integrators

The first concerns the formal nature of the integration being performed by the biological system. We may have reason to believe that this is basically linear, but suspect that there may be significant departures on occasion, perhaps caused by aspects of the environment other than the one with which we are concerned. If we make a number of comparisons with a physical integrator known to have linear operation, and find significant coincidence, this can be taken as a demonstration that the departures from linearity, from whatever cause, are confined within the observed statistical variability. But this conclusion cannot, of course, be extrapolated to the use of the integrator outside the limits of the range covered by the comparisons. In a really favourable case we should then have been able both to check our hypothesis of the linear nature of the integrator, and to show that the suspected interference is negligible, or at least that it falls within acceptable limits.

In theory this line of reasoning is capable of extension to non-linear integration, or even to complex integration involving several variables. Here there will usually be no standard physical integrator available (for obvious reasons these nearly all aim to approach linearity). On the basis of our knowledge of the plant system concerned, we must construct a physical simulation containing all the most important items with their cross connections, and then, having set up the necessary physical sensors, use a computer to perform the actual integration. Once again, coincidences between the physical and biological integrators produce evidence in favour of the particular pattern of physical simulation.

Such investigations are only in their beginnings. So far they have been used mainly in the study of transpiration, where physical models are not difficult to construct, and even here comparatively little use has as yet been made of them. In future they are likely to become increasingly important.

30.4

30.5   Inferences about physical integrators

Comparisons of physical with biological integrators can also produce valuable evidence when the validity or relevance of the physical measurements themselves is in doubt. This is liable to happen for environmental measurements in the soil, inaccessible as it is to investigation without damaging interference [10.1]. Much the most difficult aspect of the aerial environment to characterize adequately by physical measurements is the light climate (Bainbridge, Evans and Rackham, 1966, passim).

This is because most of the basic work on the physical measurement of visible radiation has been done using point sources or parallel light: for biological systems extended sources are the rule rather than the exception. These sources themselves may vary simultaneously in intensity, spectral composition, and spatial distribution; in any one place these three variables may change independently with time; and there may also be spatial variation from place to place, say over the leafage of an individual plant.

Faced with all this complication, it may be difficult to demonstrate that any particular form of physical measurement and integration is the relevant one in a particular investigation; and uncertainty hangs particularly over problems connected with the spectral composition of natural radiation, where recording and integration by physical methods in the field are not yet practicable (Evans, 1969).

Here in a favourable case comparison with a biological integrator may enable us to cut the Gordian knot of complication, and demonstrate that the physical integration is relevant to the biological problem.

30.6   Biological and physical integrators: an example

Both these lines of inference can be illustrated by an example involving plants growing in low light in the laboratory and in the field. We have already made several references to the work of Hughes (1965a) on the growth of *Impatiens parviflora* in low light in the laboratory [3.2; 28.10–28.13]. As was shown in Fig. 3.2 he found a linear relationship between unit leaf rate and the daily total of visible radiation. We have just seen [30.2] that this finding fits in well with inferences from studies concerned with gaseous exchange, and it might at first be thought that it provides a comparison of a biological and physical integrator of the type that we need. However, this is not so. In these experiments the spectral composition of the radiation was constant, and so was its intensity over a constant daylength of 16 hours. With these restrictions it was legitimate to use the data, as we did, to postulate that under these conditions the instantaneous unit leaf rate would be related linearly to radiant intensity, but this is no test of the power of the biological system as an integrator, or of the general suitability and relevance of the physical measurement. To test these we need a further set of comparisons with growth of comparable plants under conditions where spectral composition and intensity are constantly changing, and preferably where daylength also alters. Such data are furnished by the observations of Coombe (1966) on plants of the same species growing under natural conditions of the aerial environment in Madingley Wood, near Cambridge, throughout the summer of 1963 [28.10, 28.11].

Hughes's daily totals of visible radiation were arrived at by measuring the instantaneous intensity of visible radiation in his controlled environment cabinets using a Kipp solarimeter, calibrated against a sub-standard derived from the Meteorological Office's standard instrument, and assuming that all the short-wave radiation from the fluorescent tubes that he used lay within the limits of the visible spectrum (see the discussion in Hughes and Evans, 1962, p. 157; it is likely that of the total short-wave radiation measured something between 1 and 5 per cent fell outside the limits of the visible spectrum, in the near infra-red). There was close control of the supply voltage, combined with checks of the decay of the phosphors. Apart from the small uncertainty about the proportion of near infra-red radiation, Hughes was therefore on safe ground in converting his instantaneous readings of intensity of visible radiation to daily totals by multiplying by the daylength.

Coombe's figures for radiation in his woodland sites, on the other hand, were derived by integrating the records of a Robitzsch-type actinograph (Blackwell, 1953, 1966) the calibration of which was also checked against a Meteorological Office sub-standard. This gave readings of total short-wave radiation, and it was assumed that the visible component of this was a constant fraction of the total. We shall discuss two possible values for this fraction shortly [30.7, 30.8].

30.7 Possible reasons for disagreement

Leaving aside the effects of different conditions of temperature [30.2], there are at least three other reasons why the observations of Hughes and those of Coombe should not agree when plotted together on a common basis of daily totals of visible radiation derived in these two quite different ways (Evans, 1969). Firstly, in a wood it is not possible to put the instrument measuring radiation in the same position as is occupied by the population of plants. As there is spatial variation in the amount of radiation reaching the woodland floor, our estimates of the radiation incident on the plants may be higher or lower than the actual radiation itself. Furthermore, as the apparent track of the sun passes over different parts of the canopy, the average pattern in one week differs from that in the next, so that we cannot expect our estimates of radiation to be consistently either high or low. If this were the only factor operating we should accordingly expect the Madingley Wood observations to bracket the line of Hughes.

The second reason for expecting a discrepancy is connected with the proportion of infra-red radiation in the total short-wave. Deciduous woodland canopies are well known to be comparatively transparent to the near infra-red. As the comparisons must be made on the basis of total visible radiation, it will be necessary to apply some factor to Coombe's non-selective measurements of total short-wave radiation. In the open, when sunlight forms a large fraction of the total short-wave radiation, an estimate for the visible of 42 per cent proves not to be far out (see, e.g., Szeicz, 1966). However, in a wood, unless most of the radiation reaching the woodland floor is coming unfiltered through holes in the canopy, we should expect a higher proportion of infra-red (Evans, 1966b). A 42 per cent correction would then overestimate the amount of visible radiation, and the woodland figures would fall below the expected line.

The third reason concerns the daily fluctuations in natural radiation. Hughes's plants growing under artificial conditions had a 16-hour day of uniform radiation. Under natural conditions this cannot be so. The total radiation for the day includes periods of relatively low and periods of relatively high intensity [Fig. 7.3]. But we have already seen that the relationship between unit leaf rate and irradiance is a curved one, except near the origin [Fig. 3.4.1], so that as the radiation increases there is less than a proportional increase in unit leaf rate. For comparisons made on a basis of daily totals, we should accordingly expect the increase of dry weight per unit leaf area under natural radiation to fall below that under artificial. Experiment shows that this is so (Hughes and Evans, 1962, Fig. 8). We notice that the last two causes would both bring about a

30.7

decrease in unit leaf rate at a given figure for daily light total. One would, however, expect the last mechanism, non-linearity of response, to produce a minimal effect at very low intensities, where the fluctuations in irradiance may not be large enough to take the response of the plant much off the linear portion of the relationship.

30.8   The woodland observations

In Fig. 30.8a the observations of Coombe, made under dense shade in Madingley Wood during the summer of 1963 are compared with the regression line of Hughes (Fig. 3.2), based on laboratory observations. The abscissae have been converted from the measured total daily short-wave radiation to the visible component by making the assumption mentioned above [30.7], that the visible is 42 per cent of the whole.

We notice that at the very lowest radiation totals the unit leaf rates observed in the wood do in fact bracket the Hughes line. Four sets of observations do not differ significantly from it, two are significantly above and two significantly below. It would therefore seem at first sight that possible departures due to high infra-red transmission of the woodland canopy do not have a large overall effect in this region, where we expect a minimum departure due to non-linearity. At higher daily light totals, where a larger departure would be expected, the unit leaf rates of the woodland plants are consistently below the Hughes line.

There is another possible interpretation. Suppose that there really is a substantially greater transmission of infra-red radiation through the woodland canopy, so that for a given amount of visible radiation the amount of infra-red radiation reaching the woodland floor is 20 per cent above that in the open. The proportion of visible radiation would then drop from 42 per cent to 38 per cent. This would alter all the abscissae of the Coombe observations on a scale of visible radiation, and, using the lower abscissa scale (b), the Hughes line must now be replotted as the broken one. We see that in this plot the fit is just as good, and possibly even better. (It was for this reason that it was chosen for the simplified version in Fig. 28.11.) In the absence of more information on the proportion of infra-red in the radiation in the woodland and in the cabinets we have no means of deciding which possibility to choose, although we see that the possibilities do not diverge widely.

Even with this uncertainty we may therefore say that there is remarkably good agreement between these observations of unit leaf rate under natural conditions (the biological integrator) and those under fully artificial conditions (the calibratory standard), when the basis of compari-

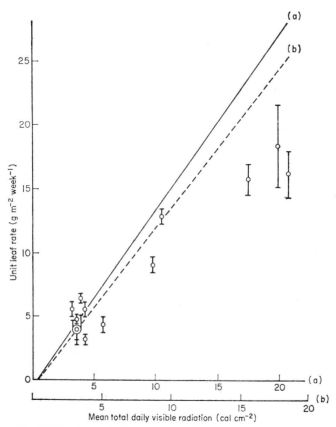

Fig. 30.8. Unit leaf rate of *Impatiens parviflora* as a function of mean daily total of visible radiation. o, observations of Coombe (1966) on plants growing in Madingley Wood, with 19:1 probability limits. (a) and —— line of Hughes from Fig. 3.2.1a assuming visible radiation to be 42 per cent of total short-wave radiation; (b) and – – – – the same, assuming 38 per cent. The scale and line (b) were those used in Fig. 28.11. From Evans (1969).

son is recorded measurements of total short-wave radiation (the physical integrator). The agreement is particularly good when one remembers that in the nature of the woodland observations, which required the plants to be left exposed without interference to all the natural aerial environmental factors of the wood, no attempt could be made to control the aerial environment, and in consequence such important aspects as average temperature [30.2] were liable to differ between one experimental period and another, the daily temperature fluctuations differing even more. This is no doubt a further reason why, around the lower radiation totals, half the points differ significantly from the Hughes line (the limits given being those of 19:1 probability).

30.9 The performance of the integrators compared

Fig. 30.8 therefore provides a favourable case for applying the arguments of 30.4 and 30.5. As far as the physical integrator is concerned, we are hampered by an uncertainty (though admittedly not a large one) as to the scale to be used: but despite this we get a good approach to a linear association with unit leaf rate in the lower part of the range, up to around 10 cal cm^{-2} day^{-1}, where such an agreement might have been expected from the nature of the biological system. It therefore seems that the daily total of visible radiation is indeed a useful and relevant quantity to use in conjunction with the interpretation of the biological data. The uncertainty we have spoken of was confined to the question of the proportion of infra-red radiation. For the purposes of these comparisons the considerable differences in spectral composition of the visible radiation itself between the laboratory and the field do not appear to be of great importance: equal totals in the two cases, neglecting differences in spectral composition, produce closely similar values of unit leaf rate.

Where the biological integrator is concerned, we should not expect it to give a correct integration when the fluctuations in the intensity of natural daylight carried it for substantial periods above the linear region of the relationship between instantaneous values of unit leaf rate and instantaneous values of intensity. This accounts for the negative deviations from the Hughes line of the unit leaf rates observed in the wood with daily light totals between 15 and 20 cal cm^{-2} day^{-1}. Otherwise we have the same good linear agreement. Now that we are looking at it from the biological side, we can make both inferences which we discussed previously [30.4]. We have confirmation that under the whole range of conditions extending from woodland shade to artificial conditions in the laboratory (summarized in 18.15), our biological system is indeed performing a linear integration, so long as we do not go outside the specified range of radiant intensity.

30.9

We have also seen that in half the cases observed, all the various possible interfering causes combined did not bring about a significant deviation from the expected line—their combined effects lay within the limits of the combined experimental errors. In the other half of the observed cases, the deviations were significant, and the experimental errors remained small. This suggests that, while the idea of using this particular system as a biological integrator is basically sound, further work is needed to identify the most important causes of deviation, so that if possible their effects can be predicted and allowances made for them.

We conclude that in favourable cases the use of plants as biological integrators of the effects of aspects of their natural environment can yield useful information; and that this is particularly so when there are difficulties in making the appropriate measurements by physical means. Here comparisons of the two methods can both tell us whether the physical methods are appropriate and relevant to the biological problem, and also provide information about the biological system itself.

30.10   How to bridge the gap between natural and artificial environments?

We have already discussed at length the difficulties which arise when one endeavours to experiment with plants growing in a natural environment [3.2–3.5; 7.4–7.13; 9.1–9.9; 10.1–10.8], and also the advantages which accrue from the use of controlled environmental conditions in the laboratory [3.6; 7.2, 7.3; 9.10–9.12; 10.9 and 10.10], and we have discussed numerous examples of both types of experimentation. There exists, however, a very wide gap between the two sets of conditions. The two environments differ from each other fundamentally, and in very many respects: some of the differences are related to each other, some are quite unrelated and can vary independently.

In Sections 18.6, 18.7, and 18.15 we considered a particularly simple instance, confined to aspects of the aerial environment, and to a transition from the laboratory to open ground, where no account had to be taken of shading or any other aspect of plant competition, or, indeed, of any other biotic factor. Even so we were able to list seven major changes in the aerial environment, some of which were themselves of considerable complexity. In this particular instance we were able to show that the aspect of growth concerned, the distribution of dry material between the leaves and the rest of the plant, was little altered by the overall effect of all these changes. These observations lead to two possible conclusions; either none of these aspects of the environment had any marked effect; or some compensation mechanism was at work, leading to the cancellation of one effect by another.

In this case the results were such that we were able to draw useful inferences, but more usually such a complex environmental change would be accompanied by some marked effect on growth, and then how is the problem to be disentangled? The environmental changes produced by the simple transition just mentioned are too many and too subtle for it to be practicable to consider adding these features one at a time to a controlled environmental installation, even if it were technically possible to reproduce natural conditions in the laboratory.

The way out of the difficulty is implicit in the instance from Chapter 18, which we have just considered. It will often be found that of the very numerous and complex changes involved in passing from the laboratory to the field, only relatively few produce significant effects upon the particular aspect of growth which we are investigating in a particular experiment. Our experiments must be planned so that there are a number of experimental conditions intermediate between the most carefully controlled conditions in the laboratory, and natural conditions, as free as possible from interference, in the field. These intermediate conditions must be disposed so that each gap is an understandable one—we must be able to specify the changes in the environment involved, and if possible we should know which of them are involved in the changes produced on the experimental plants. Where changes in the environment are many and complex, it is best if we can arrange to clear at a bound a group of environmental alterations which together have no effect upon the aspect of growth being considered, so as to have a new firm basis from which to proceed to investigate the next step.

30.11   An example from Chapter 18

The investigation of environmental effects upon the leaf weight ratio of *Impatiens parviflora* provides a good instance of the use of this principle. We were able to show [Fig. 18.6] that there was no net effect of the change from laboratory conditions of constant temperature, humidity, wind speed; an on-off pattern of lighting, of constant intensity; and

30.11

the spectral composition, spatial distribution, and level of irradiance characteristic of fluorescent tubes: to natural conditions of daily march and short-term fluctuations of temperature, humidity, air movement and irradiance, together with the other characteristic features of natural spectral composition, level of irradiance, and changing spatial distribution of sources. This gap, inherently of immense complexity [18.15], could be crossed with confidence in one step, although under less favourable circumstances elaborate investigations might have been needed to elucidate the consequences of so many simultaneous changes, all of such complexity.

The next step was a relatively simple one—changes in daylength appeared to produce marked effects on the distribution of dry material between the leaves and the rest of the plant in the field [18.12]. This could then also be demonstrated to happen under controlled conditions in the laboratory [Fig. 18.12].

These experiments were carried out using standard rooting media, all more or less artificial. They showed that change in rooting medium produced an even more profound effect on the distribution of dry material than did changes in daylength [18.10]. We have thus been able to take two steps in changing aerial environment, leaving the root environment unchanged; and in the process we have been able to elucidate one of the controlling influences in dry matter distribution.

30.12   Aerial and root environments separately considered

The instance we have just considered is one among many where independent changes of both aerial and root environments can be shown to have similar effects upon a particular aspect of plant growth.

Such observations on the plant reinforce the mechanical reasons for treating aerial and root environments separately when we are attempting to bridge the gap between wholly artificial, and wholly natural, environments. What we require, in fact, is not a simple chain of environments, but a network, where changes can be made in the aerial environment, leaving the root environment unchanged, and vice versa. Only in this way can independent effects be measured separately, and interlocking effects be detected and investigated.

The particular form which the network of intermediate environments should take must, of course, be dictated by the nature of the problem being investigated. In practice we set up a small number of intermediate conditions, covering the whole range of transition, and then further steps will be interpolated only when necessary to elucidate some complex change.

30.13 A network of environments

Fig. 30.13 gives such a network for the investigations on the growth of *Impatiens parviflora* which have furnished us with several examples. There must necessarily be a considerable element of the arbitrary in the

		THE AERIAL ENVIRONMENT			
		Wholly Artificial	Partly Artificial	Wholly Natural	
		I. Laboratory	2. Neutral screen in the field	3. Full exposure	4. Under a woodland canopy
THE ROOT ENVIRONMENT	**Wholly Artificial** a. Non-polar medium +culture solution	+		✕	✕
	b. Crushed flint + culture solution	+		✕	✕
	c. Vermiculite + culture solution	+		+	+
	Partly Artificial (Pot culture) d. Soil + sand	+	+	+	+
	e. Soil + sand + root competition			+	
	Wholly Natural f. Undisturbed soil	✕			+

Fig. 30.13. A network of intermediate environments used in investigations on *Impatiens parviflora*, in order to bridge the gap between wholly artificial conditions in the laboratory, and wholly natural conditions in the field. + indicates experiments carried out in this combination of aerial and root environments. ✕ indicates an impracticable combination.

way in which experimental environments are grouped together for presentation in this way. In Fig. 30.13 each transition marked by a line in the diagram corresponds to a large and complex change in conditions, but in most cases it has proved possible to understand what happens, both in the environment and in the plant, when the transition is made. The least understood and most difficult one, as would be expected, is that separating undisturbed soil from the other conditions.

30.13

The division of the aerial environment into four groups has worked well. As our discussions have progressed we have seen numerous examples of investigations under a wide range of artificial environmental conditions in the laboratory; under artificial screens in the field, at different times of year; in open ground, with full exposure; and throughout the growing season under woodland canopies of varying structure and density. We have seen in each case how the effects of changes in particular aspects of the environment can be worked out, and how the different investigations in the whole corpus support each other.

Investigations have so far been made in eleven of the nineteen practicable combinations of Fig. 30.13. The core consists of a block of seven (out of eight possible), extending over the whole range of aerial environments, and covering two root environments, the artificial vermiculite and culture solution, and the semi-natural soil plus sand, with added artificial fertilizers.

From the laboratory end of this central block, investigations extend into increasingly artificial root environments; from the wholly natural aerial environment end, they extend to increasingly natural root environments.

30.14 Special problems of root environments

The difficulties in the way of understanding conditions surrounding a root system growing undisturbed in a natural soil have already been discussed [10.1 et seq.]. In Fig. 30.13 we have proposed six groups of environments for the roots, as opposed to the four groups which have worked well for the aerial environment. This may well be far too few. For an adequate understanding of the problems involved it may eventually be necessary to approach the natural root environment in at least a dozen steps.

At the laboratory end we are dealing with almost wholly inorganic sub-strata. The soil used was a Kettering loam of low humus content, and almost free from macroscopic plant and animal remains. The addition of an equal quantity of sand ensured an open texture and good aeration, essential if normal growth of this experimental plant is to be maintained. The mineral content was brought up to an adequate level by the standard additions used in John Innes compost. In this series of experiments all that was necessary was to avoid both deficiency and excess.

With the use of vermiculite and culture solution we pass to a wholly inorganic substratum, uniform, well aerated, with a very large adsorbing area for water and dissolved substances of all kinds, particularly anions and cations. We see normal growth of the plants, but the complex adsorbing system makes it still impossible to specify conditions round the roots. By changing to coarse crushed flint (a form free from contamination with metallic oxides—practically pure silica) we maintain support of and aeration for the root system, while greatly reducing the adsorbing area and effectively eliminating adsorption of anions; but the low water retention now makes necessary almost continuous supply of culture solution. However, there are still uncertainties about cation concentrations; to be able to specify all ionic concentrations round the root system with accuracy it must be supported on a non-polar medium, with all the attendant complications [28.18].

So far, so good; these steps are not too large to understand. The difficulties come in the other direction. When we make the transition from the inorganic vermiculite to the semi-artificial soil, this should at the first step be sterilized; at the second left alive. In our diagram we have not distinguished this step, which for the Kettering loam used appeared to make no difference. It would probably also be desirable to consider the effects of using instead a richly organic woodland soil, with and without its natural flora and fauna, as modified by their own reactions to the changed conditions of pot culture. Yet even when we have added root competition under conditions of pot culture we are still a very long way from understanding the natural soil; we can only hope that with a wholly natural aerial environment, and having taken so many steps towards simulating the natural root environment the changes in the plant's response to the final step into undisturbed soil will not be too large.

30.15 Biotic factors

Our scheme has included steps involving the flora and fauna of the soil, and also competition from other plants, but we have avoided by side-stepping all the complications attendant on investigations extending to the other biotic components of the whole ecosystem. Under natural conditions we may also have to cope with nibbling, biting, mining or gall-forming insects; with more or less serious fungal parasitism; with

30.15

nematode worms; with birds and mice, and so on. Often one or another of these is of great importance, but this is no place to consider them in detail. The field observations preliminary to the programme of investigation itself should always include the fullest notes, quantitative if possible, of all other organisms which may impinge upon the main problem. Such observations should be kept up, and systematic notes should be made, not only of the presence, but also of the absence of attack by other organisms. It is sometimes possible to rescue information from an experiment which would otherwise have been ruined by some sudden and unexpected attack, by having a note that on a definite date this attack had not begun.

30.16   Plant pathology and growth of the host

Plant pathological problems are a special case of the biotic factor in general, and much light could no doubt be thrown on many of them by quantitative studies of the growth of the host plant. Apart from the pioneer work of Watson and Watson (1953) on the effects of two virus diseases on the growth and yield of sugar beet, little has been done along these lines until the work of Harrison (1968; Harrison and Isaac, 1968; Isaac and Harrison, 1968) on wilt diseases of potatoes caused by infection with *Verticillium albo-atrum* and *V. dahliae*. Previous studies of these diseases had indicated that infected and control potato plants were morphologically indistinguishable during the early stages of growth prior to tuber initiation, after which the infected plants became progressively more stunted, followed by premature senescence and death of the foliage. These symptoms appear earlier, and are more severe, following infection by *V. albo-atrum* than by *V. dahliae*, as can be seen in Fig. 30.16 (reproduced from three figures in Harrison and Isaac, 1968), which shows the progressions of stem height, leaf area index, and a 'symptom and senility index', measuring the visible progress of the disease. At 4 weeks after infection there were no significant differences in stem height or leaf area; at 6 weeks there were still no obvious pathological symptoms, although significant differences in stem height had developed and even greater differences in leaf area; thereafter Fig. 30.16 shows the progress of the disease. We see that both pathogens bring about substantial reductions in maximum leaf area index, and even larger reductions in leaf area duration, owing to the premature senescence of the foliage. The reduction in tuber yield in these experiments was found to be broadly similar to that of leaf area duration, infection by *V. albo-atrum* reducing leaf area duration by 41 per cent and tuber yield by 32 per cent, the corresponding figures for *V. dahliae* being 16 per cent and 27 per cent (Harrison and Isaac, 1968, Table 3).

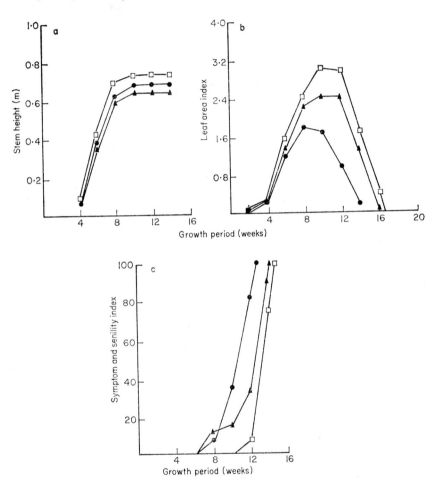

Fig. 30.16. Effects of infection by two species of *Verticillium* on potato plants (cv. King Edward), showing comparisons of diseased and control plants in the progressions of (a), stem height; (b), leaf-area index (green leaf area only); and (c), a 'symptom and senility index' measuring the visible progress of the disease. ●, *V. albo-atrum*; ▲, *V. dahliae*; □, control. From Harrison and Isaac (1968).

30.16

30.17 Elucidation of hidden symptoms

Harrison also investigated the effects of the two pathogens on the growth of young potato plants up to the end of the sixth week after infection, growing uniform single stemmed plants in 20-cm pots well spaced out in the open, grading and sorting into equivalent samples at the end of the second week, and making four harvests at weekly intervals up to the end of the sixth week. The plants were watered artificially during the first 2 weeks, and thereafter left to the natural rainfall. The prevailing weather

Table 30.17. Prevailing climatic conditions during the total harvesting period of potato plants (King Edward) infected with *Verticillium albo-atrum* and *V. dahliae* (from Harrison, 1968).

Harvest interval	Average daily air temperature		Total rainfall (mm)	Mean cloud cover (%)	Relative humidity (%)
	Max. °C	Min. °C			
(a)	18·8	13·1	17·5	85	82·6
(b)	18·1	12·0	37·1	81	83·5
(c)	20·1	10·7	19·6	73	77·1

conditions during the three intervals between harvests are shown in Table 30.17. Of all the aspects of local climate recorded, much the largest difference between the weeks was in total rainfall, 37·1 mm falling in the second interval as against 17·5 and 19·6 mm in the first and third respectively. The only differences between the plants visible to the naked eye before the last week (when the controls became obviously larger) were that the infected plants showed recoverable wilting during the first and third intervals, but not during the second. The control plants did not wilt. Fig. 30.17 shows the mean relative growth rate, a; mean unit leaf rate, b; and mean leaf area ratio, c, of the infected and control plants for the three intervals between harvests. We see a higher relative growth rate in the control plants during the first and third intervals, a lower one during the second. Unit leaf rate is always higher in the controls, but during the second week only marginally so, and its effect on relative growth rate during this interval is outweighed by the infected plants having substantially higher mean leaf area ratios. When we examine, in Fig. 30.17d, the progression of leaf area ratio against plant dry weight, we see similar

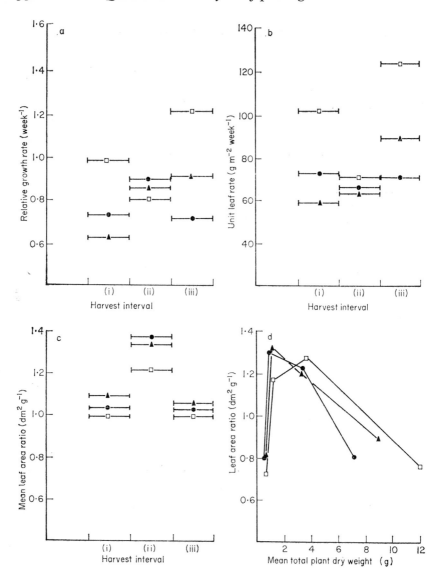

Fig. 30.17. Effects of infection by two species of *Verticillium* on potato plants (cv. King Edward), showing comparisons of diseased and control plants in the progressions of (a), relative growth rate; (b), unit leaf rate; and (c), mean leaf area ratio over three intervals between harvests; together with (d), leaf area ratio plotted against total plant dry weight at harvest. ●, *V. albo-atrum*; ▲, *V. dahliae*; □, control. From Harrison (1968).

30.17

ontogenetic drifts in all three cases, but displaced relative to each other, the infected plants starting higher, rising earlier to a higher maximum and declining earlier than the controls.

It seems most probable that the different behaviour of the three sets of plants as regards unit leaf rate in the three periods is explicable by the visible wilting of the infected plants in the two drier weeks, and it is a pity that there is no information on transpiration of the plants, or on radiation apart from the figures for cloudiness. If one accepted what climatic information is available at its face value one would conclude that transpiration conditions did not differ widely between the three intervals, and that the turgid condition of the infected plants in the second interval was due mainly to a larger supply of water in the soil. However, if one considers the plants as biological integrators [30.3–30.9], the markedly lower unit leaf rate of the controls during the second interval (Fig. 30.17b) suggests a correspondingly lower mean irradiance. If so, then the infected plants would also have been under less water strain from this cause, overall transpiration losses being lower.

30.18   Analysis of differences in leaf area ratio

Fig. 30.18 shows the progressions of leaf weight ratio and of specific leaf area against total plant dry weight corresponding to the data for leaf area ratio given in Fig. 30.17c. Once again we see similar ontogenetic drifts, but whereas from the second harvest onward the leaf weight ratio of the infected plants is always higher than the controls, the specific leaf areas of the infected plants appear to reach their maximum earlier, and fall earlier, than the controls, so that they are at first higher, and then lower. Taken together, these two drifts suggest that the increased leaf weight ratio is due rather to the retention of a higher proportion of assimilates in the leaves of infected plants (possibly due to some inter-ference with translocation) than to the removal of a higher proportion of assimilates to young, developing leaves. If translocation were affected, root growth would be impaired, and this would provide an added reason for the wilting of infected plants during the two intervals when less water was available in the soil.

If retention of assimilates in the foliage were the only phenomenon leading to a higher leaf weight ratio and lower specific leaf area in infected plants at the later harvests, the combined effects should cancel out as far as

30.18

leaf area ratio is concerned. But Fig. 30.17d showed that this was not so. When plotted against plant dry weight, leaf area ratio of the infected plants was lower than that of the controls from the third harvest onward, because infection caused a proportionately larger decline in specific leaf area. Harrison attributed this to early maturation of the foliage, the infected plants having the specific leaf area characteristic of control plants of considerably larger size.

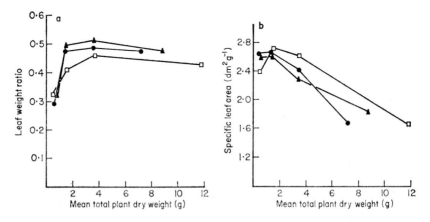

Fig. 30.18. Effects of infection by two species of *Verticillium* on potato plants (cv. King Edward), showing comparisons of diseased and control plants in the progressions of (a), leaf weight ratio and (b), specific leaf area, as related to total plant dry weight, for comparison with Fig. 30.17d. ●, *V. albo-atrum*; ▲, *V. dahliae*; □, control. From Harrison (1968).

Harrison's study is thus an instructive example of the application of the methods of analysing plant growth which we have been considering to a problem in pathology. He was able to demonstrate a whole series of effects of the pathogen on the growth of the host in advance of the appearance of obvious pathological symptoms. His results also exemplify the essential connection between the various lines of attack on the overall problem of plant growth in the field, which we have singled out in the last five chapters. There were present the linking of response to treatment with response to environment as modified by ontogenetic drift, of physiological with morphological and anatomical responses, which our earlier studies have led us to expect. At the same time the quantitative comparison of the growth of infected and control plants in the very early stages of infection has delimited a number of particular problems suitable for further study by the methods appropriate to the branches of botany concerned—problems of early maturity of the foliar system, of translocation and root growth, of water relations and unit leaf rate.

30.19 Natural patterns of activity in a community of organisms

In the last four sections we have been considering problems arising out of the influences of the activities of one species of organism on the growth of another. The influences are, of course, reciprocal, and such paired cases of mutual influence are only part of the larger problems existing in nature, of the mutual interactions of all the organisms comprising the ecosystem. Most of the detailed problems arising there are outside the scope of this study, but it is possible to apply to plant communities as entities some of the concepts which we have been considering in connection with the growth of individual plants, and this will form the subject of the next chapter.

30.19

SOME PROBLEMS OF PLANT GROWTH
AND THE ECOSYSTEM: PRODUCTIVITY

31.1 Productivity: energy flow

Productivity is a term derived from agriculture and forestry, referring originally to some measure of crop-yield in some appropriate unit of time, often a whole growing season or a calendar year. The crop might be of plant or animal material, or of both. In this form the term is related to human activities, so that it is not a very precise or fundamental biological concept in the wider ecological sense, however important it may be for human society. The 'yield' of a plant crop normally implies only a part of the net primary photosynthetic production, and will be a different fraction for different crops, or even for the same crop in different societies, with different patterns of consumption. The yield of a given plant crop in a given locality will, moreover, vary with the agricultural or sylvicultural techniques employed—methods of preparation of the soil, fertilizer treatments, etc. The yield of an animal crop will be similarly variable with variations in the techniques of both crop and animal husbandry.

If the idea of productivity is to be applied to a whole ecosystem, rather than to a selected part of one, questions at once arise. What are the fundamental biological concepts involved? What measurements will be valuable for making relevant comparisons between ecosystems of different kinds, or in different places? (The concepts and methods of measurement involved in these questions are reviewed and discussed in Lieth (1962), and Westlake (1963).)

The photoreduction of carbon dioxide, by converting the photons of visible radiation into the chemical energy of 'reduced-carbon units' [2.8], provides nearly all the basic supply of energy required to maintain the activities of all the living organisms in a particular ecosystem. There is then an analogy to one of Thomas Tompion's year equation clocks, where a pulse of kinetic energy applied to the winder is stored as potential energy in the weights, and then doled out in more than thirty million movements of the escapement and main train, and in tens of thousands of other movements, of the mechanism operating the equation ring, the calendar, the moon dial, and so on. Once a carbon atom has been reduced from the level of carbon dioxide to that of one of the products of photosynthesis, there is no telling how long its subsequent history may be until it returns to the level of carbon dioxide again. It may become part of a fossil and last for hundreds of millions of years. It may be reoxidized in a matter of minutes. The average life in the reduced form will vary very greatly, depending upon the type of system within which the photosynthesis is taking place.

In time, when analysis of the changes in atmospheric carbon-14 following various nuclear tests is completed, it may be possible to give a figure for the average half-life of a molecule of atmospheric carbon dioxide which has been subjected to photosynthetic reduction. This approach to the problem is very complex, because of equilibration with carbon dioxide dissolved as such, and contained in exchangeable buffer systems, in the sea. Indeed, the problem may prove to be insoluble; it seems likely that the mean half-life in the reduced form might be of the order of months or years, and it might not be possible to distinguish the effect from that of atmospheric mixing, on the one hand, and equilibration with the sea, on the other. If so, it will be necessary to estimate the global turnover by less direct means, via estimates of the total standing crop at any one time, which are likely to be very tentative for a long time to come.

On the other hand, we are on much surer ground in dealing with a single ecosystem, where it is often possible by suitable measurements, to estimate both the total dry weight of standing crop at any one time, and the rate of change. In turn, these figures can be converted into energy units [4.3], giving direct estimates of the total energy stored in the ecosystem at any one time, and of energy flow through it, the amount of energy flowing per appropriate time-unit (such as a growing season) being one fundamental measure of the whole activity of the system. There is also a basic biological interest in comparisons between figures of this sort for natural terrestrial ecosystems and similar figures for the maximal energy flow in certain standard crop plants grown under comparable conditions of the aerial environment, but with non-limiting conditions of the root environment. There is an analogy between such comparisons

31.1

and certain of the transitions of an environmental grid such as Fig. 30.13, but in interpreting the data we have in addition to remember the consequences of the large genetic differences involved.

31.2 Carbon dioxide turnover

The energy flow is all in one direction, between the high-energy quanta of short-wave radiation received at the surface of the earth from the sun, and the low-energy quanta of the long-wave 'thermal' radiation from the earth into space. On the other hand there are many cyclic aspects of ecosystems which are of great importance. If the dry weight figures for a particular ecosystem are converted into units of carbon dioxide, as we have several times had occasion to do [e.g., 29.3, 29.4], it becomes possible to consider aspects of the carbon cycle. One way of looking at the matter is to consider the annual production of crop in relation to the carbon dioxide content of the air over the same area. For example, a biennial such as sugar beet growing from seed may produce a crop of plants containing 10–15 tonne ha^{-1} of dry matter by mid-autumn, equivalent to say 4–6 tonne ha^{-1} of carbon. The atmosphere above a hectare contains about 47 tonne of CO_2, equivalent to about 13 tonne of carbon. The autumn standing crop is a substantial fraction of this.

Alternatively, we may use our figures to estimate rates of turnover for shorter periods. Suppose a plant has a unit leaf rate of 25 g m^{-2} $week^{-1}$, and a leaf area index of 3, then the net increase of dry matter per week due to this plant is 75 g m^{-2} $week^{-1}$, or about 30 g m^{-2} $week^{-1}$ of carbon. This is equivalent to about one-fortieth of the carbon (as carbon dioxide) in the atmosphere above one square metre.

Thus in a fertile, well-watered, temperate region the turnover of atmospheric carbon dioxide is substantial: on an annual basis it is higher still in some places, such as a tropical swamp, but globally the figure is far lower. When account is taken of the polar regions, the dry regions of the continents and the deserts of the open oceans (Westlake, 1963, Table 4), the order of magnitude of the annual turnover probably lies between 1 and 10 per cent.

31.3 And net productivity

Compared to such a rate of turnover, the production of really long-lived reduced carbon, which effectively means fossil carbon, is very small on the

31.3

land surface of the globe. Apart from human activity, we can take it that for the whole land surface, over a moderate period, the rate of reoxidation equals the rate of photoreduction to several significant figures. To the same number of significant figures, the net productivity of the totality of ecosystems is therefore zero. The same thing holds for such a plant community as a large area of mature wet tropical high forest. Here again the rate of production of fossil carbon is negligible, and in an area large enough to contain representative numbers of all local species, together with representative phases of the regeneration cycle, there will be no net change in the total content of reduced-carbon units over a period. Indeed, leaving aside ecosystems producing fossil carbon, this steady state criterion is inherent in the idea of a stable community, although in some cases it would be necessary to average not only over a wide area, but also over an extended period of time—as when the vegetation contains a substantial proportion of long-lived, simultaneously flowering, monocarpic plants [23.3].

However, (a) the attainment of the steady state takes time—sometimes a very long time: many interesting ecosystems have never attained it. And (b) even when the steady state condition exists, the period of study may have to be prolonged, in order to go through the progression and return to the position of no net change in the total of reduced-carbon units. Further (c) in any community showing a pronounced mosaic of up-grade and down-grade phases (as in the example of mature wet tropical high forest just mentioned), net accumulation is frequently observed, if the area studied does not include representative proportions of all the phases. Consequently, the literature abounds in records of positive net productivity (in effect, the accumulation of reduced-carbon units) due to the abrogation of one or another or all of these conditions (for examples see Lieth, 1962; and Westlake, 1963). This does not make these observations any less interesting: the study of parts sheds much light on the operation of the ecosystem as a whole: but it is necessary to bear these restrictions in mind when comparing certain parts of one ecosystem with parts of another. They may not be exactly equivalent.

31.4   Total primary photoreduction

Let us, however, for the moment suppose that we are dealing with a large enough sample of an ecosystem where there is no net accumulation or depletion, so that we can to a close approximation equate the totality of all the energy-degrading processes of the ecosystem as a whole to the total primary photoreduction of carbon dioxide. If this could be measured, we should thus have a measure of the total energy flow through the system, and the proportions taking the various possible paths could be evaluated. This, then, is the question to which we must apply the results of our discussions.

The simplest way of proceeding is to choose an ecosystem whose characteristics include pronounced up-grade and down-grade phases, separated in space or in time. Ideally, we should make a quantitative study of both, so as to be able to check one estimate against the other; but frequently this is not practicable, and we must for the time being be content with an estimate based on whichever is easier to measure. Let us first consider some consequences of studying the up-grade phase, when reduced-carbon units are accumulating in the ecosystem or in some part of it.

The harvest results which we have been considering have provided data on the accumulation of dry matter by the plant. We have spent much time considering this accumulation in relation to the leaf area produced by the plant. We could just as well have related the accumulation to the equivalent ground area corresponding to a single plant. This is often done for crops of known spacing: it is less easy in natural communities, but is usually possible. To take a simple case, suppose that we are considering a population of seedlings growing in bare ground. Under natural conditions these seedlings are likely to be very thick on the ground, and to compete with each other. Week by week we could count the number of seedlings per unit area, and harvest and measure a representative sample. We should then find that as self-thinning went on, the mean ground area per seedling would increase week by week, and the rate of accumulation of dry matter could be related to this by methods of computation analogous to those which we used for unit leaf rate. We could also make an allowance each week for the contribution of dry matter made to the ecosystem as a whole by seedlings which had died during the previous week. In this way our methods could be converted from a leaf area basis to a ground area basis.

We should, however, still be dealing with *accumulated* dry matter; i.e. total primary photoreduction of carbon dioxide during the photoperiods less respiration of the whole plant during the whole period of light plus dark between harvests, plus additions of minerals and allowing for

31.4

the metabolic balance term [13.11, 28.2]. Even if we determine the carbon dioxide equivalent of the dry weight [13.11, 29.4], so as to be able to convert the dry weight into units of carbon dioxide reduced, we should still need to make an allowance for respiration in order to arrive at the total primary photoreduction. Even if for the time being we set aside the particularly difficult problems of photorespiration [29.4, 32.6], and confine our attention to net carbon dioxide intake [29.3, 29.4], it may still be a most difficult matter to make an allowance for the respiration of the whole plant in the dark, and for the non-photosynthetic part in the light, especially when dealing with plants growing in natural communities. In Chapter 12 we have already examined many of the complications involved in handling the material and in the techniques of measurement, and we shall examine other difficulties later [31.15, 31.16].

31.5   And dry weight accumulation

We must now ask—can we draw a distinction between the reoxidation of that part of the total primary photoreduction which is respired by the plant itself, out of the stock of reduced-carbon units which it is contributing to the ecosystem, and the reoxidation of the remainder by other organisms? We could liken reoxidation by the plant to frictional losses in a heat engine—they are inevitable if the engine is to run at all. If we draw this distinction we then provide ourselves with a revised estimate of productivity—for total primary photoreduction we have substituted net contribution of reduced-carbon units to the rest of the ecosystem.

In some ways this is a simplification: it enables us to make use of our harvest data, without a long and elaborate additional scheme of research into respiration—if we were simple-minded we could say that the contribution of an annual plant was the dry weight of the whole plant at death, less the dry weight of one seed for the next generation. But this is a gross over-simplification in several different ways, and it is worth considering these shortly in order to show the trouble which follows as soon as we depart from the basic notion of total productivity as equated with total primary photoreduction.

31.6   Some difficulties: above-ground parts

Firstly, everything produced by the plant at every stage of its life cycle must be collected when the particular part dies: because, as long as the plant is alive it is respiring, and this respiration we have agreed to neglect. But under natural conditions, as plant organs become moribund they are liable to attack by other organisms, notably fungi: how is the respiration of these attackers to be distinguished from that of the moribund primary producer?

Secondly, as we have already mentioned, we cannot just write off at this stage all seeds except the one which will replace this season's plant, because the others, which germinate, grow some way, and then die, may each make their own additional contribution of dry matter over and above the dry weight of the seed, and this future contribution must clearly be assessed [26.9].

Thirdly, under natural conditions many plants are intimately associated with other organisms, which are supported in one way or another by the plant's total primary photoreduction. Some of these may operate in ways easy to assess—for example, the removal of material by leaf-cutting bees can be measured at the same time as leaf area; others, such as the weak parasitism of rusts, may be difficult; and some, such as nematode worms, impossible.

31.7 Some more difficulties: roots

If these are awkward problems for the above-ground parts of the plant, they are a great deal more awkward for the parts below the ground. We have already considered the difficulties of extracting root systems for dry weight determinations [11.14, 11.15]. All these difficulties apply in a much more acute form to the question of assessing the magnitude of the contribution which the plant has made to the total reduced carbon pool of the soil, in addition to that associated with the dry weight of the root system which we have dug up.

Firstly, there are the difficulties of assessing the amounts of dry matter lost in deciduous roots, and attacks by various organisms on the root system. Here the problems are essentially similar to their counterparts for the above-ground parts, but complicated by the peculiar inaccessibility of the root system of a plant growing in a natural soil [10.1, 10.2, 10.4].

Then there is the question of exudations from root systems, which may be of sufficient magnitude to exercise a profound influence on the species composition and populations of many different kinds of organisms in the soil immediately around the roots, giving rise to the concept of the rhizosphere.

Finally, what are we to do about symbionts? They often form an important, and at times essential part of the whole natural system associated with the plant below-ground—essential, that is, for healthy growth of the plant. The growth of many leguminous plants is much restricted in the absence of root nodule bacteria; some conifers will not grow in the absence of mycorrhizal fungi; and so on. These last instances make clear how difficult it can be to draw a line when once we depart from our earlier concept of productivity as total primary photoreduction.

31.8   A further restriction

So great are the difficulties associated with assessing what is going on in the soil that many workers effectively neglect them, and for many practical purposes this is clearly justified, although we are now departing very far from our original concept. The harvest of many crops is confined to the above-ground parts of the plants in question. When studying the production of the particular organ which is of economic importance in relation to the rest of the plant body, it is obviously easier to cut the plant off at ground level, and abandon the below-ground part. Can we then apply a simple correction to allow for the roots? Our discussion of the changes of root weight ratio with various factors, both of the environment and of the plant [21.1–21.5] suggests that the application of any simple proportional correction inevitably involves the acceptance of a substantial margin of error.

Similar considerations apply to forestry considered as the production of a woody crop. The proportion which the commercially valuable woody stem bears to the plant as a whole varies very widely from species to species, but in some the proportion of stem is so large that errors in estimating the actual dry weight of the remainder are small when considered against the dry weight of the plant as a whole. Also, in many timber trees the main stem is so regular in shape that its dry weight can be assessed with considerable accuracy by measurement, and comparison with the known relationship between measurement and dry weight of other, harvested, stems (for an example of the information which can be gained in favourable cases by sufficiently elaborate measurement, see Tischendorf, 1926, a portion of whose data was used in the construction

31.8

of Fig. 2.9). Here there is no argument about the commercially important basic datum—the rate of production of marketable timber. However, with less certainty it is possible to assess the dry weight of the whole plant by an extension of these procedures; for accounts of methods which are in use for this purpose see the general description in Newbould (1967); and the applications to particular communities with examples, in Ogawa *et al.* (1965b); Bunce (1968); and Whittaker and Woodwell (1968). Difficulties proliferate as the attempt is made to argue from these figures along the line leading back to total primary photoreduction.

When we come to consider retracing our steps in this way, it can easily be argued that the problems posed by the below-ground parts are so much more difficult than the rest, that the attempt to solve them has the effect of 'muddying the wells of enquiry': that there are so many things to consider, and the errors associated with each attempt at estimation are so large, that when they come to be combined the overall answer is almost meaningless. When we are trying to think back to total primary photo-reduction, what is needed is some method of integrating automatically everything which is going on in the soil, without the necessity of consider-ing any part in detail.

We could do this if we had any means of estimating the net rate at which carbon compounds are being transferred into the soil across the soil/air interface. Broadly speaking, this transfer takes two forms—firstly, the fall of solid matter of all kinds from the above-ground part of the eco-system, dead animals, dead leaves, shed stems, flowers, seeds, caterpillar frass, etc. etc. The rate of this fall can be estimated. Secondly, there is the translocation of organic compounds down the stems of the plants, and across the notional ground-level interface between the stem and the root. The direct estimation of this is not at present possible.

31.9   Overall estimates of soil activity: soil respiration

It is suggested, therefore, that the integration must be in the other direction. If we cannot simply integrate the processes by which carbon compounds are finding their way into the soil through living plant parts, let us try to measure the rate at which the final product of their reoxida-tion, carbon dioxide, is coming out. Lundegårdh (1924, p. 144) took the view that soil respiration is almost equal to carbon assimilation by the plant cover, and consequently paid considerable attention to methods of

31.9

measuring it; and he has had many followers (see, e.g., Haber, 1962; and the general account in Lieth, 1962, p. 108). However, this is one of those simplifications which are very attractive at first sight, but which dissolve again into complexity on close examination. Soil respiration is usually measured by establishing a closed boundary across the soil-air interface, enclosing the air above this, and estimating the rate at which carbon dioxide is liberated into the enclosed volume. If there is a closely-spaced ground cover, the first question to arise is of access to the soil surface for measurement—how is this to be done without interference with the above-ground vegetation? And if this is easy, as it would be, say, in a pinewood without a ground flora, the next question is, where is the soil surface? Does the 'soil' include the litter of fallen needles and small parts? fallen branches? fallen trunks? Granted that all this is no longer part of the living, above-ground plant system, and is in process of slow degradation and ultimate incorporation of the residues into the soil, nevertheless the larger pieces represent accumulations of reduced-carbon compounds made by living plants in years long past: and the respiration of the organisms bringing about the degradation of these large pieces is difficult to measure. Yet if we take the view mentioned above of the equation of upgrade and downgrade processes in general, all these ought to be measured.

However, our purpose was more restricted, being confined to the use of soil respiration studies as a means of estimating the totality of the flow of carbon compounds into the soil by movement in living parts of plants. To obtain this we need to deduct from figures of overall soil respiration that element due to the breakdown of residues of all form of litter. The estimation of this last, in a particular community, requires a choice of the particular level across which it is most convenient to estimate the rate of downward movement of litter residues into the soil, which can then be taken as the 'soil surface' datum for the respiration studies.

When all these problems have been solved it is then necessary to consider spatial and temporal variations. Spatial variations may be due to the variable distribution of plants and of organic residues, or to variations in the soil and its populations of organisms: here some ecosystems are relatively uniform, others very variable over wide areas, requiring an extended survey. There are similar differences in the pattern of temporal variations: short-term variations in soil respiration rates have been observed with time of day and with weather changes: there are longer-term variations in seasonal climates, and for many, and perhaps most soils here, it is necessary to make estimates on a yearly basis, to cover the complete cycle of change: and so on. Unless we happen to be confronted with a particularly simple case, an investigation of this kind inevitably involves a complex system of measurements.

Next there are the problems of the measurements themselves, with considerable formal similarity to the problems discussed for plant respiration in Chapter 12. Referring to the argument developed around Fig. 12.5, there are similar reasons for desiring to establish a steady state, which is most readily attained by gas-current methods, although in the practice of soil respiration studies these have been used comparatively rarely. Of the available methods the simplest is probably that of Walter (1952), which in its simplest form consisted of a yogurt-pot 5·5 cm diameter by 12 cm high with a mouth 3·5 cm wide, containing 10 cm³ of half normal KOH, covered by an empty jam tin 22 cm diameter by 25 cm high, pushed a short distance into the ground. This method, which is not as unsophisticated as would appear at first sight, was subjected to intensive study by Haber (1958), whose calibrations showed that for moderate weather in shaded situations the fiducial limits for a single observation were around 6 per cent of the value measured: for more extreme conditions—frost, water under the cover, exposure to sun— around 20 per cent. Thus carefully used under suitable conditions even this very simple device gives adequate reproducibility without very large replication of observations. More sophisticated field methods for estimating carbon dioxide in air have been discussed by Macfadyen (1970), and their application to soil respiration studies reviewed by Chapman (1971), who describes a small conductimetric device which makes possible several readings of carbon dioxide absorbed on a single filling of absorber.

Unfortunately, in all too many cases in the literature simple figures are given for soil respiration; and these have little meaning unless attention has been given to estimating the importance of each of these numerous complexities in the particular case studied. We stand, in fact, on the edge of a little-explored field of study, which is outside the scope of the present work. Nevertheless, substantial though the difficulties may be, there is no doubt that the study of the respiration of the soil as a whole promises a much easier road to the answer to our overall question than any amount of study of the individual components of the complex system existing below the surface of a natural soil.

31.9

31.10 A practical example

With the ideas developed earlier in this chapter in mind we can turn finally to a practical example, which will indicate the very wide range of measurements and studies needed to make estimates of productivity in a complex ecosystem. It will also give us some specific figures indicating the magnitude of the major effects which we have been discussing, and hence the importance of the various sources of uncertainty which we have considered. A very convenient work to consider to this purpose is that of Müller and Nielsen (1965) on tropical high forest in the Ivory Coast. Not only does this include one of the most comprehensive pieces of work to date on this particular type of plant community, which is one of peculiar interest, but the work of the same authors, using the same methods, on beechwoods in Denmark makes possible a comparison between forests of the wet tropical and northern temperate zones. Particular interest also attaches to comparison with the very elaborate study of tropical rain forest made by the Joint Thai-Japanese Biological Expedition to South East Asia, 1961–62, in Thailand (Ogawa et al., 1961, 1965a, b; Yoda, 1967; Kira et al., 1967).

Samples plots (of 30 × 30 and 40 × 40 m) were established in undisturbed forest near Anguédédou about 25 km west of Abidjan. Rainfall measurements taken at a station approximately 1 km away gave annual totals of 245, 197, 205, 288 and 247 cm in the years 1959 to 1963 respectively. However, as is usual on the West Coast of Africa, there is a marked rainy season from May to July, the rainfall at Abidjan for June reaching 50 cm, and a marked dry season from December to April inclusive, the rainfall from mid-December to mid-March not exceeding 10 cm per month [Fig. 31.12.1]. The climate is thus much more seasonal than in the classical rain forests of South America and Malesia, but nevertheless there is sufficient distributed rain to maintain mature high forest essentially similar in structure to that described by Richards (1939) and Jones (1955; 1956) in Nigeria.

31.11 Measurements of intact plants

All the trees on the sample plots were numbered and measured in 1955. They were also identified, itself quite a task in these species-rich forests when the rules of the game do not permit one to fell the tree, which is the usual way of making sure in doubtful cases. All the trees were then measured again in 1957 and in 1960. Comparable trees covering the same species range were felled in similar forests, but far enough away not to disturb the sample plots. From the measurements of these felled specimens and the weights of their parts the weights of the standing trees and their parts could be estimated. As is usual in wet tropical forest, the stems formed the bulk of the stand, having on average three-quarters of the fresh weight of the whole above-ground parts of the trees. The roots were assumed to represent one-sixth of the whole, a reasonable assumption (Bray, 1963) which would make the fresh weight of the roots about four-fifths of that of the branches.

These measurements gave a mean rate of increase for the measured, above-ground parts of the trees of 7·5 tonne ha^{-1} year^{-1}; and assuming the roots to increase at one-fifth of this, the total annual increment of dry matter would be 9 tonne ha^{-1} year^{-1}. This figure is very closely similar to their earlier estimate for a beech plantation in Denmark (55°N latitude; Fig. 31.14) of 9·6 tonne ha^{-1} year^{-1}, but higher than the figure of 6·5 tonne ha^{-1} year^{-1} obtained by Kira *et al.* (1967) for the Khao Chong rain forest in Thailand.

31.12 Respiration measurements: stems

Measurements of intact plants were, however, the easiest part of the work. Attention was then turned to the estimation of respiration. For the larger pieces of stem use was made of the apparatus shown in Fig. 12.9, of tin-plate coated with paraffin wax. The pieces of trunk were cut in the forest approximately 20 cm longer than the final specimen, brought into the laboratory, sawn to length, and the freshly cut surfaces covered with lanoline. Measurements of respiration, measured as carbon dioxide output, were then made as soon as possible thereafter. The authors discuss the problems posed by wound respiration [12.4]; they do not mention the possible changes in aeration of the central tissues consequent on preparation of the specimen, but the two effects would, of course, go on together. In long-term trials they found that respiration increased steadily for several days after cutting; for this reason they made their measurements as soon as possible. There is here a possible source of error, but probably not a large one, because of the large fraction of the total tissue which was undamaged at the time that the apparatus was assembled. Respiratory quotients were measured, and found to be 0·96 for a trunk of *Strombosia postulata* 25 cm in diameter, and 0·93 for branches of the same species less than 1 cm in diameter. A piece of the trunk of *Combretodendron africanum* 30·4 cm in diameter had an RQ of 0·95. All the respiration figures were corrected to the mean temperature for this part of the Ivory Coast, 26·9°C, using an assumed correction.

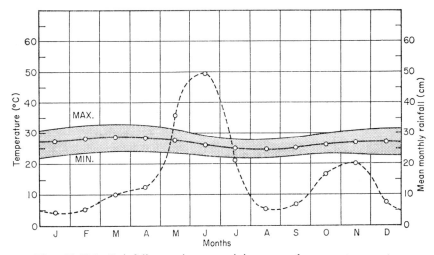

Fig. **31.12.1.** Rainfall, maximum, minimum and mean temperature throughout the year at Abidjan, Ivory Coast: means of monthly figures for the years 1943–53. Mean annual rainfall for the same years 196 cm. For comparison, mean annual rainfall for the years 1959–63, measured 1 km from the sample plots in la forêt de l'Anguédédou (25 km west of Abidjan) was 245, 197, 205, 288 and 247 cm respectively. From Müller and Nielsen (1965).

Once again, this is an example of a calculated risk—to have measured temperature coefficients for all the species and parts studied would have been an enormous task; the errors involved are likely to be small, because the annual range of mean temperature is a few °C only, and the daily range rarely exceeds 10°C, as shown in Fig. 31.12.1. This makes it a very convenient climate for physiological experimentation.

These respiration studies were made on large and small specimens of three of the commonest species, and on pieces of trunk and branches of various size. The results are set out in Fig. 31.12.2, as carbon dioxide produced per unit fresh weight. It will be seen that the respiration rate per

31.12

unit fresh weight shows very large changes with diameter; that the rate
of respiration of both large and small pieces of trunk varies considerably

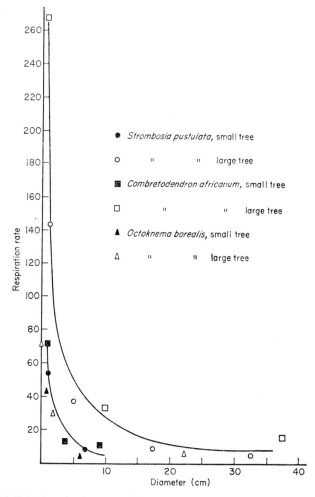

Fig. 31.12.2. Respiration rates (in μg CO_2 evolved per g fresh weight per
hour) of sections of trunks and branches of large and small specimens of
three species of trees grown in la forêt de l'Anguédédou, Ivory Coast.
From Müller and Nielsen (1965). For apparatus used, see 12.9.

and systematically from species to species; that respiration by small
branches is likely to be important; and that the respiration rates of parts
of large trees are consistently higher than those of parts of small ones of

31.12

comparable diameter. Altogether this presents a complex picture, and much calculation is needed to produce from these figures estimates of the respiration rates of the measured trees standing on the same plots. When this is done the estimated respiration of trunks and branches was 18·5 tonne ha^{-1} year^{-1}, as compared with 4·5 for the Danish beechwood. We thus have a very interesting comparison—the trunk and branch respiration of the tropical forest being double the net annual accumulation of dry material, while that of the Danish beechwood was only half of the corresponding quantity. In view of the large difference in mean annual temperature, this is not surprising. On the other hand the estimate of respiratory losses in the trunks and branches of the trees in the Khao Chong rain forest in Thailand was very much higher still, at 32·1 tonne ha^{-1} year^{-1} (Kira *et al.*, 1967). We shall examine possible reasons for the discrepancy later [31.15].

31.13   Respiration of leaves

Simplifying assumptions are also necessary when considering the respiration of leaves. Leaf respiration is known to have a marked ontogenetic drift; there are marked differences between sun leaves and shade leaves; and we have some 35 species to contend with, some of which grow entirely in the shade, while others have individuals of several sizes, from tall emergent trees down to small saplings. Clearly to evaluate the whole of this problem in detail would be an impossibly large task. Four species were chosen by Müller and Nielsen, and respiration measurements were made on mature leaves of trees of varying heights. In every case measurements were made on leaves from the upper parts of the crown, and, for trees over 10 m in height, also from the centre of the base of the crown. From these measurements there were derived average figures of respiration per unit area of sun leaves in general (135 mg CO_2 m^{-2} hour^{-1} corrected to 26·9°C) and of shade leaves in general (59 in the same units). It was further assumed that the whole leaf area of all trees less than 15 m in height consisted of shade leaves (1·04 ha of leaves per ha of forest) while all the leaf area of trees above 15 m consisted of sun leaves (2·12 ha per ha of forest). A mean figure for the carbon content of the leaves was then determined at 49·1 g carbon per 100 g of leaf dry matter (mean of four determinations on two species: *Strombosia pustulata*, sun 51·0, shade 47·4; *Combretodendron africanum*, sun 49·6, shade 48·3). Combining these figures the annual respiration of the leaves then worked out at 3·0 tonne dry matter per hectare for the shade leaves, and 13·9 tonne ha^{-1} for the sun leaves, 16·9 tonne ha^{-1} in all. We shall examine the validity of these estimates in 31.16.

This figure is 3·8 times as large as the corresponding one which they had earlier determined for Danish beechwoods. They point out that this is a reasonable difference, as the leaves at Anguédédou respire for 365 days at a temperature of 27°C, while the leaves of the beechwood in 55°N latitude respire for only 160 days at temperatures around 10–16°C. On the other hand, the estimate for total yearly leaf respiration in the Khao Chong rain forest in Thailand was 57 tonne ha^{-1}, more than three times the estimate for leaf respiration in the Ivory Coast forest, and indeed exceeding the total estimated gross production there. This discrepancy will be discussed later [31.16].

31.14   Other losses: the overall balance sheet

There is very little information on the average longevity of leaves of tropical trees, and the estimation of this for an individual species is complicated by irregular production of flushes. The authors assume that all shade leaves have a life of 1 year, and all sun leaves one of 2 years, and that then the whole dry weight of the mature leaf is shed. This gives an annual loss in dry weight to the plants of 2·1 tonne ha^{-1} (but see below). Dry weight loss of stems and roots was estimated at 2·3 tonne ha^{-1} year^{-1}. However, these are only estimates, based in the former case on assumptions about the length of life of leaves. Measurements of litter fall were made in the Khao Chong rain forest for two 3-week periods, and the results show good agreement between litter traps in two localities between mid-January and late February, 1962. Leaves formed almost exactly half of the litter, and the mean rate of fall, 6·36 g m^{-2} day^{-1}, is equivalent to 23·2 tonne ha^{-1} year^{-1}, more than five times the total estimated loss for the Ivory Coast forest. As Kira and his co-workers were well aware, the main question mark hanging over the estimate for the Thai forest is—how representative were the six observational weeks of conditions the year round? This is not only a matter of possible seasonal shedding of leaves and twigs: it depends also on weather conditions, and particularly on wind. The high winds associated with tropical storms usually tear off quantities of leaves and twigs from the upper layers of the forest, but where there are marked seasonal changes in the pattern of air movement, much lighter winds can affect litter fall at the quieter times of year. This is because, in a canopy of such complexity, a leaf dropped from the upper part of the crown of an emergent tree is unlikely to reach the ground for a long time, and much decomposition goes on in the upper layers of the forest itself. Anyone who has spent any length of time in one of these forests is familiar with the fall of leaf mould which ensues when a light breeze stirs the upper crowns during one of the quieter times of year. This has two implications for the estimation of litter fall: the amount reaching the ground is bound to be an underestimate of the amount detached; and even if the rate of detachment were constant, the rate at which the litter reached the ground would vary with seasonal changes in air movement.

After an examination of the various factors involved, Kira and his co-workers concluded that their extrapolation of the litter collections during January and February (the short dry season at Khao Chong) to the whole year led to an overestimate of annual litter production, and the overestimation may well be substantial. In their compendious account of records of litter fall in different parts of the world, Bray and Gorham (1964, Table IV) give a digest of the extensive unpublished data collected by Mitchell in a variety of plant communities in Malaya over several years. His mean annual totals for litter fall of all kinds in forests (excluding plantations) extending from the lowlands up to an altitude of 600 m ranged from 5·5 up to 14·4 tonne ha^{-1} year^{-1}, and taking all their figures together Bray and Gorham (Fig. 1) concluded that the mean rate of litter fall in wet tropical forests was about 11 tonne ha^{-1} year^{-1}. The observations of Nye (1961) in mature secondary forest in Ghana agreed almost exactly with this (10·5 tonne ha^{-1} year^{-1}), and his observations show a substantial increase in the rate of fall during the dry season. Such a seasonal effect may well account for the high values observed at Khao Chong. The general conclusion remains, however, that Müller and Nielsen's estimate for loss of leaves, twigs and branches, of around 4 tonne ha^{-1} year^{-1} was probably too low, and that a figure between two and three times this is more likely.

The whole picture is then set out in Fig. 31.14, in which the Ivory Coast forest is compared with a Danish beech forest. We see a very close similarity of forest increment above and below ground (9·0 against 9·6

31.14

tonne ha⁻¹ year⁻¹). The big differences are in respiration, four times as large overall for the Ivory Coast forest, and consuming three-quarters of

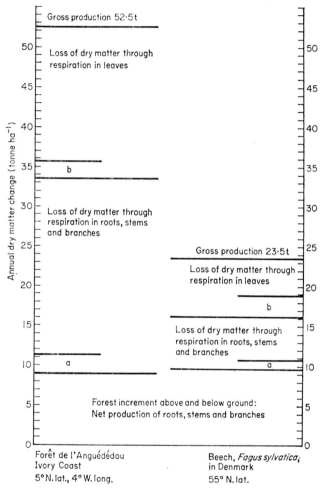

Fig. 31.14. Annual dry weight changes per hectare in a lowland wet tropical high forest and in a 46-year-old cultivated Danish beech forest compared. (a), estimate annual losses of roots and branches; (b) annual loss of leaves. From Müller and Nielsen (1965).

the 'gross production' of 52·5 tonne ha⁻¹ year⁻¹, whereas in the Danish beechwood respiration of the beeches themselves consumed less than half of a total itself less than half that for the Ivory Coast.

31.14

31.15   Sources of uncertainty: respiration of stems

We now have to ask ourselves—how accurate are these estimates of respiration? On the credit side we have the relatively non-seasonal climate and the relatively invariant temperature, shown in Fig. 31.12.1. If we were working in a seasonal climate we should have to investigate the seasonal changes in physiological activity of the living cells of the stem, as well as the changes in temperature of the stem itself. This latter also involves the complication of temperature gradients inside the stem in response to wide fluctuations in the external environment. The majestic progress of the annual temperature cycle at Anguédédou makes it possible to correct all the respiration measurements to a constant 26·9°C without risk of serious error.

On the debit side are several uncertainties, the first being connected with wounding, which we have mentioned. This is likely to lead to too high a value for respiration of the stems and branches, but the error is probably not serious with the precautions which the authors describe.

More serious is the question of the effect of the species richness of the forest being studied. We have already seen from Fig. 31.12.2 the possibility that the respiration rate of *Combretodendron africanum* may be substantially higher than the other two species studied, especially in large trees. The sample plot trees actually measured belong to 35 species, and in view of the variability shown by the three studied, it is not easy to assess what the behaviour of the other 32 would have been

On the one hand, the species studied were among the most numerous. There were no fewer than 44 specimens of *Strombosia pustulata*, seven of *Octoknema borealis*, and seven of *Combretodendron africanum*, out of an initial total of 132 trees on the 40 × 40 m plot. On the other hand, none of the measured specimens of the first two species were large trees. Only two specimens of *S. pustulata* attained a diameter of more than 30 cm (32·0, 34·7), with heights of 22·3 and 23·6 m. The largest specimen of *O. borealis* was 23·4 cm in diameter and 17·6 m high. *Combretodendron africanum* can become a large tree, and one measured had a diameter of 82·5 cm and a height of 50·2 m (for a photograph of this, see Müller and Nielsen, 1965, Fig. 30). Two other measured trees were more than 40 m high, specimens of *Albizia zygia* and *Piptadenia africana* (now *Piptadeniastrum africanum*) (ibid, Fig. 28). There cannot be room for many emergents on a 40 × 40 m plot, but this means that as far as the respiration studies of the large emergents goes, much rests on a few observations on a medium sized tree of *C. africanum*.

Yet in view of the time and effort which have to be expended, it would not be reasonable to expect more information at this stage. We must be grateful that so much has been achieved in the study of an important plant community about which so little is known. The uncertainty stems from how much one would need to know in order to present a complete picture. We simply have to guess at the margin of error in these respiration measurements—is it 10 per cent?, 20 per cent? There is no means of knowing. We can probably say with fair confidence that the estimate is very unlikely to be out by as much as 50 per cent; and it is not unreasonable that it should be around four times as large as the corresponding figure for a Danish beechwood, when we remember that there is roughly a 20°C difference in mean annual temperature.

The observations of Yoda (1967), made by a method similar to that of Fig. 12.9, in the Khao Chong forest in Thailand [31.12] are more difficult to assess, although there is reason to suppose that the method used to calculate, from the observations themselves, the respiration rate of the roots, stems and branches of the stand as a whole would lead to too large an estimate. For each of the nine tree species investigated, the logarithm of the respiration rate of the trunks per unit fresh weight was plotted against the logarithm of the diameter of the sample, and a curved regression line was fitted. A similar procedure was used for branches, except that here the regression was linear. The respiration rate of whole individual trees of these nine species was then derived by an elaborate algebraic process from these relationships and from measurements of the trees themselves. The largest container used for these respiration measurements had a volume of 0·04 m³, much smaller than the largest used by Müller and Nielsen (0·33 m³), and in consequence the largest trunk section investigated had a diameter of around 25 cm, the largest possible in a container of this size.

Unfortunately, inspection of Yoda's Figs. 2 and 3 shows that at the larger diameters of trunk sections the observations begin to fall below the fitted regression line: for diameters between 10 and 20 cm, taking all nine species together, there are seven points above the line, seven on it (taking the diameter of the figured circles) and 21 below, five of the nine species showing no point above the line in this diameter class. For diameters of 20 cm and above there are 10 points below the line, and none above or

31.15

on it. What is more, the scales being logarithmic, some of the departures by individual measurements are substantial, one measurement giving only about one-third of the fitted value. The figures for branches show a similar tendency. The trunks and branches of full-sized rain forest trees must have gone up to many times the diameters of the experimental pieces, and how large an error would be introduced when one began to extrapolate the fitted relationships there is no means of telling, although the procedure used cannot fail to have led to a substantial overestimate of the respiration rate of a whole tree. Here we must accordingly prefer the estimates of Müller and Nielsen.

31.16   Sources of uncertainty: respiration of leaves

We have already mentioned some of the problems associated with periods of measurement [12.13]. Yet others are created by changes in leaf structure over a wide range. If we consider the figures given for the respiration of leaves of *Combretodendron africanum*, which included some leaves from the crown of the tallest tree on the plot (Table 31.16),

Table 13.16. Respiration rate, in mg of CO_2 per m² of leaf area per hour (corrected to 26·9°C) of leaves taken from specimens of *Combretodendron africanum* of different height (from Müller and Nielsen, 1965).

Height (m)	Respiration rates of leaves from	
	Upper part of crown	Middle of base of crown
<2·0	28	—
10·5	48	40
37·8	172	—
50·2	212	114

we see a continuous change of respiration rate per unit leaf area with height of tree, which presumably corresponds to large changes in structure and hence in specific leaf area. There is a much larger difference between the shade and sun forms of the largest tree than there is between the sun leaves of the two largest trees themselves. All this is not unexpected, but it emphasizes the difficulty of arriving at any simple mean figure for the respiration of the leaves of large trees.

Next there is the question of variability from tree to tree. Two large specimens of *Strombosia pustulata*, one 29·0 and the other 32·2 m in height, and having respectively 295 and 305 m² of leaves each, gave respiration figures of 96 and 164 mg CO_2 m^{-2} hour^{-1} for leaves from the top of the crown, and 96 and 122 for leaves from the centre of the crown base. This may have been partly due to variation from one specimen to another, but it seems more likely to have been due to differences in exposure of the crowns of these trees relative to the large emergents. That such variations are important is emphasised when one considers the large leaf areas involved. Of 79 trees felled and measured on a 30 × 30 m plot, only five had leaf areas of more than 100 m² (105, 126, 151, 226 and 868 m²), and these five had roughly half the whole leaf area on the plot (1476 out of 2847 m²).

The authors raise another point, of interest when we are considering our attempts to work back to gross primary photoreduction. We have already noted that the figure of 16·9 tonne ha^{-1} year^{-1} is obtained by assuming that 100 units of leaf dry matter contain 49·1 units of carbon, the mean of several determinations. But are we justified in using this as a measure of the dry matter equivalent of the CO_2 respired? Almost certainly not. Whatever the respiratory substrate, it can hardly have had the mean composition of the leaves themselves. The respiratory through-put is almost exactly eight times the leaf dry weight (2·1 tonne ha^{-1}) every year; what is being respired by the leaves is almost certainly the current production of photosynthesis—carbohydrate. Now if we converted the respired CO_2 not into average leaf material, but into starch, the annual dry weight equivalent of respiration would be 18·7 tonne per ha^{-1}, almost exactly 10 per cent higher. This particular type of uncertainty is best met, as in all our earlier discussions, by working in terms of carbon units [13.11].

Finally, there is ontogenetic drift. All the foliage studied was mature. There is reason to suppose that young foliage would have had a higher respiration rate, and so might some of the senescent foliage. Both would tend to lead to the estimate of respiration being too low. On the other hand, any extra senescent respiration is compensated for in another way. It has been assumed that what is lost to the tree at leaf fall is the full dry matter of the mature leaf. This is almost certainly too large an allowance —the tree loses to the rest of the ecosystem what remains when both translocation and senescent respiration have removed a good deal of

31.16

material from the mature leaf. In considering the total of respiratory losses plus leaf fall, we can accordingly neglect the senescent respiratory increase, although we must remember that carbon compounds lost in plant respiration are not available to other, non-photosynthetic, organisms in the ecosystem.

Thus for leaf respiration we are faced by many uncertainties, not quite the same ones as those we considered for stem respiration, but equally difficult to estimate with any degree of accuracy. And yet, again as for stem respiration, the figure is an essentially reasonable one, almost exactly four times that for the Danish beechwood used for comparison in Fig. 31.14 and, on balance of probability, too high rather than too low. The estimated leaf respiration per unit ground area of the Khao Chong forest was more than three times as large [31.13], at 57 tonne ha^{-1} year^{-1}, but associated with this figure are substantial uncertainties of two kinds, one associated with the estimation of the respiration rates of the leaf samples, the other with the method of converting these into overall estimates for a complete stand of trees.

The respiration measurements were made as soon as convenient after the felling of a sample tree. '100 grams of fresh leaf samples were taken out at random from the well mixed heap of leaves of a sample tree, put in a small bag of plastic net, and enclosed for two hours in a tin container of 6 litre capacity' (Yoda, 1967, p. 89) together with a dish containing 20 ml of 2 N KOH solution. Using this method the effects of wounding and handling [12.4] must have been substantial, but probably much more serious would be the effects on restricting gas exchange between the respiring leaf tissue and the air in the container [12.5]. The extent of this effect would probably vary from species to species and from sun leaves to shade leaves, depending on the shape, size, flatness and flexibility of the leaf sample. It was in fact suspected that the observed respiration rates were low, and 'careful examinations were later made on the reliability of the method in a rain forest of S W Cambodia in 1965, by enclosing different amounts of leaves in the same container. It was revealed by the examination that the respiration rate had indeed been seriously underestimated ... and that the true rate of leaf respiration could be obtained by multiplying [the uncorrected rate] by a correction factor equalling 2·68' (*ibid*, pp. 103–104) which was accordingly applied to all the original observations. This had the effect of bringing the observations on sun leaves into much the same range of values as in the observations of

31.16

Stocker (1935) in Java and of Müller and Nielsen (1965) in the Ivory Coast, but almost certainly led to a serious overestimate of the respiration of leaves in the undergrowth, giving 0·39 mg carbon dioxide per gram fresh weight of leaves per hour, as against a mean figure of 0·135 for the observations of Löhr and Müller (1968), which fits in well with other considerations [31.17]. However, the respiration of the undergrowth plays a relatively small part in that of the stand as a whole (this section, below), and the combination of these uncertainties is such that there is no inherent means of telling whether the estimated rate per unit fresh weight of leaf is high or low, and if either, by how much. On the other hand, the general agreement with the results of other investigators suggests that the estimated mean respiration rate per unit fresh weight is roughly of the right magnitude, and unlikely to be in error by a factor of, say, 2.

The other source of uncertainty is more serious. Ogawa *et al.* (1965b) had worked out a series of allometric relationships between (a) the logarithm of the diameter of trees at breast height (D) and the logarithm of the height of tree (H); (b) between the logarithm of the total dry weight of the trunk of the tree and the logarithm of D^2H; and (c) between the logarithm of the dry weight of foliage and the logarithm of the dry weight of trunk. An average figure was also arrived at for the dry weight/fresh weight ratio of leaves (0·345), and thus, combining all these relationships, an estimate was made of the total fresh weight of leaves on a tree from the diameter at breast height (Yoda, 1967, p. 125).

Unfortunately, as the first three of these relationships are logarithmic, the uncertainties multiply rapidly, and the most difficult is (c), a curved relationship with an asymptotic value for the total dry weight of the foliage of any tree of 40 kg. The effect of this relationship (Ogawa *et al.*, 1965b, p. 56) is that, for an increase in dry weight of trunks of trees from 200 to 3,500 kg (the largest studied), the fitted line shows an increase in dry weight of foliage only from 10 to 32 kg. At the same time over this range there are eight points on the line, none above it and five below, the largest deviation corresponding to less than a quarter of the fitted value. The effect of this small change in the fitted value of total leaf dry weight over a large range of total trunk dry weight, combined with considerable variability, make this method an exceedingly unreliable guide to the actual amount of foliage on a given large tree; and as we have just seen, the foliage of the small number of large trees is an important fraction of the whole foliage of the stand. What is more, in this particular in-

31.16

stance, the method is one which is likely to lead to a substantial over-estimate.

The likelihood of the estimates of leaf respiration at Khao Chong can, however, be checked by a different line of argument. Yoda (1967, p. 103) gives figures for the mean fresh weight/leaf area ratio of each of the 13 species (including nine emergents) which he investigated from the rain forest stands: of these 11 lay between 1·40 and 1·83 g dm^{-2}, the other two being 2·05 (*Sterculia* sp.) and 3·95 (*Alstonia spathulata*), and the mean for the whole series being 1·787. His mean respiration rate of 0·448 mg carbon dioxide per gram fresh weight per hour would then on average be equivalent to 0·80 mg CO_2 dm^{-2} hour^{-1}, or kg CO_2 ha^{-1} hour^{-1}. His overall figure for respiration by the leaves of trees (derived by the method mentioned above) was 9·62 kg CO_2 ha^{-1} hour^{-1} (Yoda, 1967, Table 13, p. 136), so that the main tree layer would have to have had a leaf area index of 12, and below this are small trees and undergrowth adding a further 1·57 to reach the total leaf respiration in the whole stand of 11·19 kg CO_1 ha^{-1} hour^{-1}. Not only is this a most unlikely state of affairs, but there is a further inconsistency, in that whatever the leaf arrangement and inclination, on average half the tree leaves would have above them foliage with a leaf area index of six or more: in other words they would be in dense shade. But if so, as we have seen, their respiration rate has been much overestimated.

Thus, taking all these sources of uncertainty together, without pursuing the finer points discussed earlier in this section, we must prefer Müller and Nielsen's estimates of leaf respiration in a tropical rain forest. There may have been many more layers of leaves in the Thai forest, with a corresponding increase in leaf respiration, but from the evidence given there is no means of telling.

31.17   Mean values of unit leaf rate

We may now ask ourselves what these figures mean in terms of the analysis which we have hitherto been developing, and how they compare with Coombe's figures for seedlings of the West African tree species *Musanga cecropioides* (Coombe and Hadfield, 1962). As a basis it would seem to be best to take the annual increment, 9·0 tonne ha^{-1}, plus the loss of dry material as shed branches, leaves and roots, 4·4 tonne ha^{-1}, as corresponding to the dry weight increase which we should have measured in our harvesting experiments. The leaf area index was measured at 3·16, giving a mean unit leaf rate of 4·24 tonne ha^{-1} year^{-1}, or 8·2 g m^{-2} week^{-1}. *Musanga cecropioides* gave a unit leaf rate of 5·9 g m^{-2} week^{-1} when grown at a leaf area index of 4·0, near enough to comparable conditions. On the other hand, the total short-wave radiation incident on the plants was almost certainly lower in Coombe and Hadfield's work, having a daily total of about 190 cal cm^{-2}, whereas in Anguédédou one might have expected something in the region of 250 cal cm^{-2} (this is a guess, based on conditions in Southern Nigeria, far to the eastward along the same coast, although in the same general climatic region. It depends much on average cloudiness, and we are given no figures either for this or for total short wave radiation at Angédédou).

In general we may regard the agreement as remarkably close, bearing in mind the differences in the plants and in the methods used; and we notice that it is based mainly on that part of Müller and Nielsen's work which is least uncertain—that is, the actual increments of the trees as measured, and the leaf areas as measured after harvest. We may also remark that on a similar basis the mean unit leaf rate for the Danish beech forest of Fig. 31.14 works out at about 10 g m^{-2} week^{-1}, the leaf area index being higher, but the season of leafage shorter.

It seems likely that the relatively large respiratory losses are an important contributory factor to the low values of unit leaf rate for the larger components of tropical rain forest. The same would no doubt hold for the small understorey trees, shrubs and ground vegetation. Although what few observations are available suggest that these would have low respiration rates as compared at the same temperature with temperate plants from similar situations (e.g., Löhr and Müller, 1968), nevertheless the prevalent high temperatures would bring about substantial respiratory losses by the non-photosynthetic parts of the plants, and by the leaves in the dark, which can be estimated from Löhr and Müller's figures at around 3–5 g m^{-2} week^{-1} for a single layer of leafage in the undergrowth. In consequence all these lower layers contribute relatively little to the dry weight increment of the stand as a whole.

31.18   Root metabolism and soil respiration

We are now in a position to use the information on the particular ecosystem which we have been discussing as a basis for reconsidering the questions—can studies of soil respiration be used in practice to elucidate the obscure question of the mean net rate of downward passage of carbon compounds into the roots and therefore into the soil through the organs of living plants? Can this element of the overall soil respiration be distinguished from respiration associated with the breakdown of litter? How accurate would the measurements need to be to make a useful contribution of information? What particular conditions of measurement would need to be observed?

In important respects the tropical rain forest is a favourable case, its relatively non-seasonal climate corresponding to relatively uniform regimes of temperature and moisture, and a relatively uniform rate of litter fall; even in the comparatively seasonal climate of southern Ghana the most rapid monthly rate of fall recorded by Nye (1961) was less than twice the average rate for a year. On the other hand there are counter-balancing unfavourable features. The annual rate of litter fall is higher than any other of the numerous plant communities all over the world for which Bray and Gorham (1964) collected data: their mean rate for wet equatorial regions being around 11 tonne ha^{-1} year^{-1}, while in Ghana Nye (1961) recorded 10·5. However, the data collected over a considerable period in Malaya by Mitchell (quoted by Bray and Gorham) indicated only small variations in rate of litter fall from year to year.

As against these figures, Müller and Nielsen's estimate of root respiration was around 3 tonne ha^{-1} year^{-1}, and the yearly increment in dry weight of the tree roots around 1·5. Clearly if this is to be distinguished from the much larger quantity of litter fall it will be necessary to make observations of soil respiration over considerable periods in places where litter fall is being closely studied, and preferably in places where it has been closely studied for some time past.

On the other hand, Nye's observations suggested that whereas the freshly fallen litter decomposed very rapidly, the humus fraction of the soil decomposed only relatively slowly. This suggests a possible way out of the difficulty. If by a suitable system of measurement the above-ground decomposition of the litter could be distinguished from that in the soil proper, it might be possible to reduce the basal soil respiration to such a quantity that root respiration and decomposition of shed roots would form a substantial fraction.

Thus spot observations of soil respiration such as those of Wanner

20*   31.18

(1970) in Sarawak cannot be used to elucidate this problem, although they are broadly similar to the figures we have been discussing (around 10 tonne ha^{-1} year^{-1}). On the other hand, they might be valuable in elucidating conditions in the neighbourhood of recently formed forest gaps due to the fall of large trees. We have seen that Müller and Nielsen's estimate for annual rate of root growth in dry weight was about one-tenth of their figure for root respiration, combined with Bray and Gorham's average total of litter fall for an equatorial forest. If, as seems likely, on the death of the tree this material decomposes at least ten times as fast as it was laid down, the rate of respiration on the lower layers of the soil would be expected to increase very substantially. Even spot observations, say along a transect through the gap, would be expected to throw some light on the rapidity of decay, especially if it were possible to repeat them in a series of gaps of known age.

31.19 Particular measurements and general concepts

We must now consider how these measurements fit into the general picture of productivity which we discussed earlier. In the first place, we have seen how very complex are the considerations involved in translating ideas about productivity into practice, both in the field of measurement and in that of converting the measurements into average figures for a stand of vegetation. Secondly, we have had an excellent object lesson in the importance of respiration by the plants themselves, which in this tropical forest consumes some three quarters of 'gross production'. But what relation does this 'gross production' bear to our concept of gross primary photoreduction? There is firstly, the question of units, which we mentioned in 31.16 above. 'Gross production' is in the form of plant dry matter of mean composition, which is not necessarily what is being respired. On the other hand, we were thinking of 'gross primary photoreduction' in the form of carbon units. Secondly, the measurements of the growth of stem, the harvesting and weighing of leaves, and so on, measure what is left when animals, plants, funguses and so on have taken their toll of the living trees. We have no means of estimating how much loss of dry matter all these things entail—anyone who has tried climbing a tree in a wet tropical forest would think the activities of all these organisms substantial, especially those of the insects. What is more

31.19

like respiration, it is going on the year round; there is no season of dormancy for the animals as in north temperate regions. However, large they are, these losses are to be added to the measured and estimated quantities which we have hitherto considered, making the total turnover larger still.

Next we may view this turnover in terms of the total carbon dioxide content of the atmosphere above the forest. On an annual basis, of around 100 tonne CO_2 ha^{-1}, it is around double the whole CO_2 content of the air, but even on a weekly basis it is sufficiently startling, at about 1·8 tonne ha^{-1} week^{-1}, or about one twenty-fifth.

Finally, how do these measurements fit into the general idea which we discussed at the beginning of this chapter, that net productivity of a climax community is zero? Tropical forest is a mosaic of growth and destruction. The overmature emergent tree dies, and in due course falls; or it is ripped down in a violent storm, characteristic of the climate. With it an area of the surrounding forest is also brought crashing down, and a gap appears. Rapidly the leaves and small branches are destroyed by organisms of all kinds; more slowly the larger parts disappear, but within 5 years little but the bark would remain of all but a few species of trees reaching a metre or more in diameter. In the gap light-demanding seedlings of all kinds spring up, and the cycle begins. Early growth may well be speeded up by the mineral nutrients released in the breakdown of the fallen vegetation. Soon the canopy closes again, growth rates of plants at ground level drop to low values, and the larger trees grow steadily on. This is the phase delineated in the figures which we have discussed. But there is no doubt that if one took, say, a square kilometre instead of 40×40 m, all phases of the cycle would be included, and net productivity would be zero. It seems likely that the figures of Kira and his co-workers for the Khao Chong forest in Thailand are a step in this direction. There the increase in biomass of living trees was 6·5 tonne ha^{-1} year^{-1}, but death of standing trees during the period of the observations led to a loss of living material at an average rate of 1·2 tonne ha^{-1} year^{-1}, reducing the mean rate of biomass increase in the stand investigated to 5·3 tonne ha^{-1} year^{-1}. Müller and Nielsen's observations went a stage further: during the five years of their study the third largest tree on their 40×40 m plot died, and this, together with the death of some smaller trees, reduced the mean rate of increase of living material to 4·2 tonne ha^{-1} year^{-1}.

31.20 A problem in interpretation

These results pose a nice problem in interpretation. The high forest zone of the wet tropics has long been notorious as a region of rapid vegetative growth—what Gilbert Carter used to call a 'vegetative frenzy'. At one time it seemed obvious that one or more of three major systems might contribute to this. Firstly, the wet tropical forest might have had very many layers of leaves; but we have seen that, with a leaf area index little over three, the large trees of the Ivory Coast sample have considerably less than those of a Danish beechwood, and any lower layers could not have contributed enough increment in dry matter to alter the broad picture. Secondly, the rate of production of new dry matter per unit leaf area might have been high; but this is not so. Coombe and Hadfield's (1962) *Musanga cecropioides*, which we have seen to be comparable with the forest of Anguédédou, had at a leaf area index of $1 \cdot 0$ a unit leaf rate of $10 \cdot 25$ g m^{-2} week^{-1} as compared with $42 \cdot 7$ for *Helianthus annuus* grown under the same conditions. Thirdly, the effect might have been due to the long growing season; but as compared with the Danish beechwood, we have seen that 365 days of photosynthesis as compared with around 160 only just compensated for the much higher respiration loss associated with the higher temperature and the longer period of leaf respiration.

Individually, all the elements which we should expect to find in a system showing relatively more rapid vegetative growth appear to be missing. As far as above-ground increment is concerned there is as yet no reason to suppose that the data from the Ivory Coast are not typical; the increment in the Khao Chong forest was less. Then how are they to be reconciled with the widespread subjective conviction of the exceptionally rapid growth of vegetation in the wet tropics? It seems that there may be two complementary explanations, one associated with plant form, the other psychological.

The first concerns resistance to the unfavourable season. Temperate plants growing in a highly seasonal climate can be divided into two very broad categories; those which perennate by structures noticeably above ground level, which includes all trees, shrubs, and plants with woody stems, and those which do not, which includes all annuals and plants which perennate at or below ground (or water) level. The permanent structures of the former category must be capable of resisting the unfavourable season. The above-ground vegetative structures of the

31.20

second category do not have to resist the unfavourable season, but every year they must start again from ground level, and their opportunity for gaining height is limited to perhaps 6 months out of each year.

It may be that the necessity of surviving the unfavourable season imposes very marked restrictions upon the possible structures of woody plants. Certainly all successful self-supporting woody plants in a climate such as the British Isles have, at least initially, a low height growth rate. Although many can grow at a metre a year later in life, when fully established, it is very exceptional for one to grow as much as 1 m in height in its first year from seed; and this in spite of the fact that some of them have large, or even very large seeds.

On the other hand, if such restrictions exist, they do not apply to annuals and the various herbaceous perennials. A great many of these will grow from seed to 1·5 m in a single season, and some exceptional ones to 3 or 4 m.

But in the wet tropics the restrictions of the unfavourable season do not apply. There is no obvious reason why those forms which grow more rapidly early in life should not go on and become shrubs or trees. There are many forms which are relatively much more common in the wet tropics than in more seasonal climates, including some with long internodes and very large leaves. The relative prevalence of some of these forms may be associated with generally low wind speeds. Aerodynamically considered, the rain forest with its irregular surface, is very 'rough'. A great deal of momentum transfer takes place in the upper layers, so that even if a breeze is blowing, the air in a gap in the forest is relatively still. High rates of air movement are thus no bar preventing relatively delicate forms from forming a part of the group of early colonists of forest gaps; and hence of the colonists of abandoned farm land within the forest.

So much for the question of plant form; coming to the psychological explanation, anyone who has established a new shrubbery, or watched the growth of a stand of young trees, is aware of an effect of eye level. The increment of 20 cm, from 80 cm to 1·0 m, is just an increment of 20 cm. The increment from 1·8 m to 2·0 m produces an additional effect on the observer. Suddenly the distant view is shut out, and he is hemmed in by vegetation. But just when temperate vegetation, growing up from the ground, has reached this critical point—critical, that is, in its effect on the observer—it dies down and in the next season must start again from

31.20

ground level. Or, if it consists of woody plants, years elapse before the critical point is reached. To the vegetation of the wet tropics neither of these restrictions apply. The traveller walks out of an abandoned rest-house on to what was a lawn 18 months before, to find self-sown saplings of *Tectona grandis* 5 m high. They have long internodes, and leaves the size of dinner plates. He feels like Alice in Wonderland. Or he notices the weeds growing on abandoned farm land, including such objects as *Vernonia conferta*, a Composite with vegetative parts like a rosette of *Taraxacum officinale* 3 m across, on a stalk up to 4 m high. But usually he does not look behind the facade: beside the road he sees the strange shapes of bananas, gingers and giant bamboos, but he does not attempt to hack his way through the tangle of dull shrubs and creepers on the abandoned farm: he sees the luxurious vegetation bordering the forest river, with its spectacular palms and dangling lianes, but does not traverse the green gloom behind, with its spindly undergrowth like the dullest and most depauperate Victorian shrubbery, its sparse ground vegetation and barely growing seedlings. The effect, on someone used to the vegetation of the temperate regions, is of a 'vegetative frenzy'. But such an effect can be produced by plants growing in all respects more slowly than many annuals of the temperate zone. Why they grow more slowly is another matter. Which brings us back to the point at which our discussions started. Plants are organisms very inaccessible to the human mind.

31.20

32

FUTURE DEVELOPMENTS

32.1 Varieties of development

In considering the ways in which these quantitative studies may develop
it is convenient to distinguish between the use of the techniques which
we have discussed to investigate practical problems involving growing
plants, mostly in the field—the problems of the ecologist, the plant
pathologist, the agriculturist, the horticulturist and the forester; the
use of the same techniques for investigating more purely scientific problems, involving many other plant sciences; improvements in the techniques of investigation themselves, including contributions from other
branches of knowledge; and the use of all these methods in teaching.

32.2 Ecology: experimental taxonomy

The solutions of many ecological problems hinge upon particular
critical conditions of the environments, or critical phases of life cycles,
which determine qualitatively the success or failure of the organisms
concerned in particular circumstances. The solutions of other problems
depend rather upon differential quantitative responses of different
species. The methods of investigation which we have discussed have a
particular relevance to problems of the latter type, providing the only
means at present available of characterizing many physiological activities of the plants concerned while they are growing undisturbed in their
natural surroundings; at the same time furnishing information on form
and structure as they develop during the plant's ontogeny; and being
readily adaptable to work in remote places. We have discussed one
investigation along these lines at some length (Coombe, 1966; 6.4;
7.8–7.13; 8.9; 27.9; 28.10, 28.11; 30.6–30.8), but as yet such investigations
have been relatively few, so that the opportunities for making quantitative
comparisons between the behaviour of different species are limited. It is
therefore often desirable to plan investigations to include comparative
studies of two or more species or varieties, related either taxonomically,

or in their ecological behaviour, or in both. The last type of investigation is often a particularly fruitful one, as limiting the number of variables, and thus reducing the overall complexity of the problem. Such were the studies of Whitehead and Myerscough (1962) on *Epilobium* spp., and of van Dobben (1967, 26.9) on *Senecio* spp.

Advance in this field is likely to come slowly but steadily by the accumulation of detailed studies of particular species, species pairs, and groups, thus building up detailed quantitative knowledge, and providing a framework of reference for future extensions of the work. Advance cannot be rapid, because such field studies have to take account, not only of the circumstances of an individual year or growing season, but also of the differences between one season and another. The mere fact that a particular result of high significance statistically has been obtained in one particular year does not mean that the same thing will happen in the next year, and we have encountered a number of examples of this [e.g., Fig. 26.4.2].

Ecology and experimental taxonomy are so closely linked, representing as they do different ways of approaching the same natural systems, that the relevance of these lines of investigation to particular taxonomic problems hardly needs stressing. However, as in ecology, the available effort is small in relation to the range of problems which present themselves, so that here also there must be selection of particular examples (selection, this time, on taxonomic grounds, rather than ecological ones), and the utility of these methods will grow with the increase in ordered knowledge in the form of accumulated, fully comparable, examples.

If one were to presume to offer advice to the young investigator it would be to select, from the very great number of both types of problem, one which could yield valuable information in the fields both of ecology and of experimental taxonomy.

32.3 Field studies of cultivated plants

Such studies represent a specialized form of the more general problems which we have just been discussing, but they are of such importance as to command many times the resources applied to the study of ecology and experimental taxonomy combined. It is very proper that this should be so. That all animals are ultimately dependent for their continued existence on photosynthesis by plants may be a truism often forgotten by politicians, economists and financiers, but even in a country such as Great Britain those whose memories extend back thirty years will recall that not much stress is needed to bring it home. In many countries less fortunately placed, problems involving cultivated plants are often of importance. Here the methods which we have been discussing can be of great value, both in the simple forms suited to field work in places remote from technological facilities, and also as adapted to the more complex forms of agricultural research. Had this been an agricultural text rather than an ecological monograph, it could have been furnished with a set of agricultural examples: and indeed, it will have been noticed that many of the examples used have been derived from agricultural situations. A great advantage of these methods when starting work on a little-explored agricultural problem is their generality—they start from a wide view of the whole plant, and enable the important facets of the problem to be identified broadly, as a guide to later work. A good example of how these methods can be used to rough out the details of a problem in crop growth, and to provide a series of suggestions for further research, is provided by the problem in plant pathology investigated by Harrison (1968: 30.16–30.18).

It will have been noticed that the advances discussed in the last two sections depend on the application to field problems of methods of investigation which are already available. They do not depend on advances in these and other techniques which we shall discuss in 32.9.

32.4 Morphology and anatomy

When methods have been devised for determining the concentrations of all the principal growth substances in the cells of a plant making normal growth [26.1], very great advances in our knowledge of causal morphology and anatomy will ensue. It will then be possible to determine the effects of changes in the concentrations of these substances on the patterns of growth, and to relate changes in these concentrations to other changes inside and outside the plant. These advances depend on researches outside the scope of our present enquiries, and on present indications they seem likely to be slow in coming to fruition, but meantime substantial advances are possible within the more restricted ambit which we have been considering. Furthermore, quantitative relationships such as can now be elucidated are likely to be of great value when it becomes possible to apply increased knowledge of the activities of growth substances. In Chapters 18, 19, 21, 22, 23, 24, 26 and 27 we have already considered a range of examples of the kinds of relationship which can readily be investigated using existing techniques, and it is unnecessary to recapitulate.

It seems, however, that particularly interesting information is likely to be gained from two interlinked groupings of research—those concerned with young seedlings, and those concerned with the activities of meristems and the patterns of development of the products of their divisions, as in the work of Williams (1960 and subsequent papers) and Hegarty (1968). These, combined with investigations into translocation [32.5] seem to offer the readiest means of shedding light on the problems of correlation between the different parts of growing plants, which at present are very obscure.

The old ideas of correlation mechanisms as based mainly on nutrient flow, on gradients and absolute concentrations of quantitatively important compounds, and on relative concentrations of these, have of late years been largely cast into the shade by investigations into the profound effects of minute concentrations of growth substances. It seems likely, however, that the overall correlation mechanisms of the growing plant will be found to involve all these various systems, and that information on concentrations and bulk movements of solutes will be just as necessary for an understanding of development as information on concentrations and movements of growth substances. This is the kind of complex of problems most easily tackled in a young growing seedling, where the number of meristems involved is small and the whole developing system can be described quantitatively without the description becoming overloaded with detail. The combination of techniques now available makes precise work on small seedlings much easier than in the past.

Such investigations, carried out on a range of species, would have the further attraction of shedding light on the general biology of the plants concerned. Some of the highest relative growth rates ever recorded are from the early stages of development of plants with very small (and very numerous) seeds. For example, Shamsi (1970) found that, between 20 and 35 days from germination, seedlings of *Epilobium hirsutum* increased in mean dry weight from 0·29 to 24·7 mg, an increase of 85-fold in 15 days, corresponding to a mean R G R over the period of 2·07. It seems likely that these very high R G R s are associated with the exceedingly short translocatory pathways both for the intake of mineral nutrients, between the

32.4

knowledge of respiration rates has been essential for interpreting the results of quantitative growth studies; yet we have also seen in Chapter 12 a number of the obstacles to the accurate measurement of respiration, whether of whole plants or of their parts. These obstacles resided partly in difficulties of handling the plant material, and partly in the non-availability of suitable apparatus in the places where the measurements would have to be made. However, although there may be difficulties, and although in specific instances great care may be needed, Chapter 12 also showed that these obstacles are not insurmountable, using the techniques available at the present day.

So important are these measurements for an understanding of the processes underlying changes in dry weight of higher plants [4.7, Fig. 13.11] that it seems likely that major advances in knowledge will accrue from researches combining overall quantitative growth studies with measurements of respiration. Briggs, Kidd and West were aware of this in 1920, and as it turned out their respiration studies were the only part of their corpus of work on *Helianthus annuus* to be published at the time (Kidd, West and Briggs, 1921), but they have had few followers in the last 50 years, and here an important field of research lies open to investigation.

On the other hand, as far as field work is concerned, it seems unlikely that much progress will be made in the near future in elucidating the problems of photorespiration and generally of the respiration of leaves in the light. However, as we have seen [29.3–29.6] much useful interpretation of relationships can be done on the basis of mean rates of carbon dioxide intake by leaves in the light, without needing the further analysis of this into the various individual processes concerned with the uptake and production of carbon dioxide in the cells of the leaf.

Respiratory studies also promise to shed much light on a variety of related problems, such as those associated with stomatal movement, and with the concentration of carbon dioxide in the intercellular spaces of respiring organs [12.16]; on the role of changing respiration rate as an adaptive mechanism in plants growing in deepening shade (as, for example, when a deciduous woodland canopy opens in the spring) [28.14]; or as an assistance in the identification of units of leaf structure in conjunction with other methods for the investigation of the linear progression of leaf water content with leaf area, apparently widespread among a number of species [24.22].

Substantial advances are also to be expected in techniques for measuring soil respiration under field conditions in a variety of plant

32.6

communities, and hence in the available information on spatial and temporal variations in soil respiration rates. This will be of great assistance in assessing the year-round scale of activity of the whole biological complex of the soil [31.9].

Further in the future lies another possible development which harks forward to the next two sections—that of the use of labelled carbon dioxide, assimilated in a short burst by the plant using its normal photosynthetic machinery, translocated for a fixed time, located and measured. It is then possible to obtain information not only about photosynthesis, translocation, and incorporation into the plant body but also, by collection as part of the whole quantity of carbon dioxide respired, to determine the contribution which this particular period of photosynthesis has made to the subsequent substrates of respiration.

32.7 Photosynthesis

Chapter 29 has provided an example of what can be done by way of inferring rates of photosynthesis from estimates of mean values of unit leaf rates. The extension of such studies to a much wider range of species under a variety of conditions will provide a great deal of useful information, not only about the photosynthetic rates themselves in relation to leaf structure and environment, but also about the effects of the changes in leaf structure associated with changes of environment. For *Impatiens parviflora*, under a range of lighting conditions extending from full sunlight to the artificial equivalent of deep woodland shade, we saw that these changes in leaf structure and in the operation of the photosynthetic mechanism had the effect of holding the carbon dioxide concentration in the intercellular spaces of the leaves adjacent to the mesophyll cells almost constant [Fig. 29.6]. Is such a phenomenon widespread? Is it connected with the ecological relationships of the plants concerned? If so, does it make a contribution to the ability of certain plants to succeed in deep shade? Evans and Hughes (1961, Table 7) showed that the unit leaf rate of *Helianthus annuus* fell very much more rapidly with increase of shading than did that of *I. parviflora*. Is there a connection here? At the same time it was possible to infer from Fig. 29.6 that the efficiency of the photosynthetic machinery of this particular plant fell with age. This is confirmed by suggestions from other lines of evidence and observations on other species. It would be interesting to know how widespread the phenomenon is, and whether it plays a part in the biology of plants of different growth forms and biochemical pathways [32.4].

However, definitive answers to these last questions, and the related ones connected with the development of photosynthetic activity by young leaves, depend on developments in methods for the direct measurement of the rate of photosynthesis by individual attached leaves. To be of value in elucidating ecological problems such methods would have to be simple enough and robust enough for use in the field in places remote from normal laboratory facilities, while at the same time causing minimal interference with the leaf itself and its immediate environment. We have already seen in Chapter 12 how difficult are field measurements of respiration: measurements of photosynthesis under an approximation to natural conditions are much more difficult, because of the effects of interference on both radiation and air movement, with consequent effects on transpiration, the heat balance, and leaf temperature. (Indeed, curiously, it is probably very much simpler, in this respect, to work on water plants than on land plants.) The development of adequate field methods here would be a great benefit for comparative studies, both of the behaviour of individual leaves on a particular plant, and as between one plant and another. They would also give a further lead into the problems of the unit of photosynthetic machinery [13.13]. Their application would at first necessarily be confined to spot observations made under particular, specified, conditions; and it would be practically impossible to integrate these measurements alone, so as to give a picture of photosynthesis by the whole plant over any extended period. This would, however, be of little consequence, because the integrations could be based on the established methods for the analysis of the growth of comparable whole plants.

One fruitful line of development here probably lies in the exposure of individual attached leaves to labelled carbon dioxide for a short period, followed by harvest, fixation and subsequent assay in the laboratory. This leads into the next series of developments which we must consider.

32.8   Translocation

Many possible fates await the newly formed products of photosynthesis [Fig. 2.8, 4.3]. These may be roughly classified into (a) entry into one of the respiratory pathways, leading back to carbon dioxide; (b) synthesis into some part of the cellular structure; (c) temporary storage; and (d) translocation to other parts of the plant, where the same four broad possibilities await. The types of analysis of the growth of whole plants which we have examined have told us something about leaf growth and translocation, but in themselves they do not discriminate between one reduced-carbon unit and another. When a molecule of carbon dioxide has once been incorporated into the plant's stock of reduced-carbon compounds, it loses its identity and cannot be followed further by the methods which we have been discussing. Granted that there are times when particular parts of the overall stock can be followed from one part of the plant to another, nevertheless these opportunities depend on particular favourable circumstances, and more usually there is no means of distinguishing, which creates problems at times [e.g., 23.5]. Nevertheless, much could be done, using the analytical procedures which we have discussed, to study problems of translocation, particularly bulk movements. Translocatory balance sheets are readily drawn up, and these, combined with anatomical studies of the relevant pathways and biochemical studies of the concentrations of important compounds, could throw much light on the organization of the plant as a whole. Of particular interest in this connection are studies of the growth and failure of meristems, which play so important a part in the genesis of plant form, in such matters as root systems, branch systems, tillering in grasses, development of inflorescences, and so on. This may, however, lead at once into considerations of the number and capacity of alternative sinks for translocates, and hence to the necessity for the measurements of the concentrations of particular compounds in these very small cell complexes. This also may call for refinements of technique.

Both generally and particularly we thus come back to the desirability of being able to identify a particular group of reduced-carbon units among the plant's whole stock—in fact to label some of it. We have noticed [23.6] the value of studies involving feeding $^{14}CO_2$ in determining the pattern of distribution of assimilates throughout the plant. It might at first sight appear that such studies would rapidly resolve all problems of this kind, but this is not so. Studies such as those of Ryle (1970a) depend on feeding a particular leaf for, say, 30 minutes, and determining the short-term distribution of assimilates after, say, 24 hours. Clearly to build up a picture of the immediate destinations of the assimilates from the various leaves by this means over a whole growing season would

32.8

involve many hundreds of very time-consuming experiments (Ryle, 1970b). There would still remain all the questions connected with the later movements of these assimilates, which in many cases may be the more important fraction of the whole picture.

On the other hand, experiments such as those of Stoy (1963) examine the ultimate destination, at final harvest, of all the assimilates generated during a short period of exposure to $^{14}CO_2$. Clearly what is needed is a combination of the two techniques, feeding $^{14}CO_2$ at various developmental stages and following the products through the following stages; but if enormous proliferation of experiments is to be avoided, it will be necessary to select the most interesting stages for close study. Here, the various types of investigation discussed in Chapter 23 can be very useful.

32.9   Improvements in basic techniques

So far the developments which we have been discussing have all concerned either the applications of existing techniques, of measurement or analysis, or the development of techniques ancillary to the main theme of these discussions. As far as analysis goes, it is not easy to see where the next important development will come, in the absence of foreknowledge of developments in measurement. We have spent several chapters discussing the measurements which can already be made on whole plants and their environments; we have noted limitations on existing methods and gaps in our knowledge, and we need not now recapitulate; but perhaps the sphere where limitations are most pronounced and frustrating is that of the root environment. It is to be hoped that the next few years will see marked improvements in the techniques of characterizing root environments, of investigating the extent of root systems in natural and semi-natural soils, and in identifying the absorbing areas in relation to the root system as a whole. It would then be possible to put the tentative suggestions of section 13.20 on to a much sounder basis, and rapid advances would be possible. On the other hand, the root and its environment together present a system so obscure and difficult of study that one would need to be sanguine indeed to expect major advances in a short time.

Of the aerial part of the plant, one of the most important and elusive problems has always been that of quantifying the photosynthetic mechanism—the identification if possible of a 'unit of photosynthetic machinery' of which we have spoken [13.13]. The total activities of photosynthesis in supplying the plant with reduced-carbon compounds and maintaining its energy supply are of basic importance, and it is very regrettable that there is no single, simple, unified way of characterizing in quantitative form the totality of the machinery in the plant which absorbs the quanta of short-wave radiation and uses them to carry out these functions. In their original analysis Briggs, Kidd and West (1920a, b) used area of leaf lamina as a rough index, and as a rough index it has never been bettered: but stems, leaf sheaths, floral parts, and so on, all contribute, sometimes in very important ways, and if only the 'unit of photosynthetic machinery' could by some means be identified and characterized, the whole analysis could be put on a much more unified basis. Meantime there remains the awkward and tedious problem of the measurement of the total area of leaf laminas—we have examined many methods [11.7–11.12], none of them ideal, and it is difficult to think of a single advance of technique which would be of more value to those involved in the time-consuming job of collecting the basic data on which this whole structure of growth analysis rests than the devising of a quick and reliable means of measuring the area of a batch of leaves of any size and shape and at the same time preserving a record of their shapes and areas in permanent form.

Otherwise the most notable developments in technique are likely to come by way of recent improvements in the recording of very large numbers of instantaneous values of a great number of variables, in our case mainly microclimatic data, and their processing and application to much longer-term problems by means of computer programmes. This is a large field of development which is only just beginning to open up, and one in which the next few years are likely to see great advances, particularly in dealing with specific, circumscribed problems. In the background, for the longer term, are the problems associated with the computer-simulation of whole growing plants, which is already arousing interest [31.1, 28.19], and which must be based on quantitative studies such as we have been discussing.

32.10 Uses in teaching

Studies of the kind discussed have many attractions for the teaching of the advanced student, because they provide a central theme, drawing together the detailed knowledge which he already possesses of many facets of plant life, into the growth of plants as whole organisms. The great differences in behaviour between species, and the paucity of our quantitative information about them, make it easy to devise programmes of investigation in which the student can feel that he is himself helping to push forward the frontiers of knowledge. It will have been noticed how many of the practical examples which we have discussed, often illustrating important principles, have been taken from classwork carried out by students at the end of their second university year, demonstrating both the utility of studies of this kind, and also the fact that no extended apprenticeship is needed to acquire special skills, so long as the investigations are carefully planned (itself involving useful exercises embodying many important principles) and the execution is adequately supervised. Finally, because of the generality of the methods and the width of their applications, students who are going to work with plants in later life are likely to find the experience useful to them in the future, in ways which cannot necessarily be foreseen.

There are, however, also attractions in these methods for the elementary teaching of those with negligible biological knowledge. Here we are using the growth of a whole plant as a starting point, and it is desirable that it should be a local plant which is already a familiar object to the pupil. The experience gained in handling plants, and the insight which it brings into the relative rates of the various processes involved in growth, are an admirable corrective to the sectional nature of many courses based on textbooks: anyone with extended experience of school examining is familiar with the watertight compartments which they engender in all too many young minds. There is the further point that even in cities which are centres of high technology, school laboratories are often ill-equipped for carrying out effective practical work in many branches of science, and, in places remote from such centres, facilities are often minimal. While some of the more elaborate methods of investigation which we have discussed require extensive laboratory facilities, nevertheless the basic quantitative methods do not; and in particular, we have considered methods which are suited to the ecologist working in places very remote from technological facilities of all kinds.

32.11   Conclusion

Five general points have emerged from our discussions. Firstly, these quantitative studies can provide information on the activities of plants growing undisturbed in their natural environments, which could be obtained in no other way. Granted that the great bulk of our examples have involved more or less artificial environments, yet many of these, particularly in agriculture and horticulture, have their own intrinsic interest and furnish their own important problems.

Secondly, although great sophistication of methods is possible, nevertheless the basic investigations described can be carried out with the minimum of expensive apparatus and trained assistance. They are therefore particularly suited to ecological field work, which in its very nature may well have to be carried out in places remote from civilization, its amenities and less desirable influences. The same features make them valuable tools of agricultural research in regions remote from centres of technology.

Thirdly, the basic observations can conveniently be used to identify the most important aspects of problems connected with plant growth. Using them as a centre, lines of investigation radiate into many fields of plant science and involve a great variety of distinct disciplines.

Fourthly, the basic observations can be used to provide a framework of quantitative information on the activities of the growing plant, into which can be fitted the results of independent, and especially laboratory, investigations into parts of the whole system.

And finally, while these four characteristics make the methods of investigations discussed powerful tools of ecological research, the last three also make them well adapted for general botanical teaching, both in school and at the university. This is one reason why in the Part devoted to techniques, so much attention has been paid to the simplest means of obtaining useful information; no school is so ill-equipped that nothing could be started along these lines. Further along the educational road, growth studies have in practice proved to be convenient centres, round which to group other investigations, and at the same time to point to the solution of problems of practical utility.

Now that our discussions are concluded the author would like to thank once again those numerous colleagues and friends who have contributed so much to its completion. One of its aims has been to be available during the working period of the International Biological Programme, and this has limited the time available, while since the book's inception its progress has been hampered by several serious and unforeseeable difficulties, only remotely connected with the study of plants.

32.11

Through these difficulties the help and encouragement of friends and colleagues has been invaluable: it is doubtful whether the book could have been completed in time without their aid. Its very deficiencies (for which the author is solely responsible) may serve a useful purpose, if they encourage young readers to go one better. For the whole future of this, as of other ventures of the human mind, rests in the interest of the young.

32.11

APPENDICES

APPENDIX 1

THE THODAY RESPIROMETER

App. 1.1 Principles

If circumstances make it necessary to use a closed system for measuring the respiration rates of higher plants or their parts, there are good reasons for using an arrangement in which the carbon dioxide produced is taken up by a strong absorber, and the decrease in volume measured at constant pressure [12.9]. The measurement is then of the volume of oxygen taken up in a given time, readily converted into mean rate of uptake of oxygen (which we will call for convenience Rate X). If it is also necessary to determine the carbon dioxide released, the observations are repeated without the absorber, giving the net rate of volume change due to release of carbon dioxide and uptake of oxygen (Rate Y). In principle the rate of release of carbon dioxide (Rate Z) is then given by subtracting rate X from rate Y, taking account of sign, as rate X involves a decrease in volume, and is therefore negative.

Provided that suitable precautions are taken, the measurement of volume changes at constant pressure can readily be made with an accuracy equal to that attained in the more familiar measurement of pressure changes at constant volume, and fundamentally the method has certain advantages (Dixon, 1951). Certainly it is very convenient in use, and, because vigorous agitation is not necessary to maintain equilibrium between liquid phases and a gas phase altering in pressure, it is much easier to alter the scale if desired, and also to improvise the necessary apparatus from parts available in the average laboratory. Thoday put the method forward for school use in 1932, and the apparatus which he described was developed by Maskell (unpublished) between 1940–45. In this form it has been used for university practical class experiments over a period of more than 20 years, and has proved remarkably reliable in un-accustomed hands. The account which follows is mainly based on Maskell's development, but because this was adapted to the requirements of a practical class using a particular design of apparatus under standardized conditions, it has been necessary to make certain changes for the sake of generality. In the theoretical sections the magnitudes of a number of minor corrections have been worked out, mainly as an indication of when and why they may be safely neglected, but those readers who are familiar with Maskell's treatment will find the main differences in the handling of the solubility correction [App. 1.2, Table App. 1.2.1, App. 1.17] and the correction for water vapour [App. 1.15].

Table App. 1.2.1. Solubility of carbon dioxide in water, expressed as cm^3 dry gas, measured at a pressure of 760 mmHg and at water-bath temperature, dissolved in 1 cm^3 of water at water-bath temperature, when the partial pressure of carbon dioxide is 760 mmHg.

Water-bath Temperature °C	Solubility	Water-bath Temperature °C	Solubility
10	1·238	21	0·920
11	1·200	22	0·896
12	1·166	23	0·872
13	1·135	24	0·850
14	1·104	25	0·829
15	1·075	26	0·808
16	1·043	27	0·789
17	1·016	28	0·771
18	0·989	29	0·754
19	0·965	30	0·738
20	0·942		

When carbohydrates are completely oxidized to carbon dioxide and water, one molecule of carbon dioxide is released for every molecule of oxygen taken up. Rate Y is zero; rate Z equals rate X; and the ratio of the two, the respiratory quotient (RQ), is unity. Under these circumstances a measurement of rate X can stand for one of rate Z. If, however, other oxidations or reductions are going on simultaneously (part of the metabolic balance term, D of Fig. 13.11), the respiratory quotient can depart substantially from unity, and a measurement of rate Y is essential [App. 1.12]. Experience of the system being investigated will indicate how far the double measurement is necessary in particular cases.

21

App. 1.1 Table App. 1.2.1

App. 1.2   Units and corrections

The changes in the volumes of gas in the apparatus to be described [App. 1.5] will be of the order of fractions of a cubic centimetre of wet gas per hour, and will apply to the enclosed plant material at the prevailing temperature and pressure. For comparison with other plants or plant organs, it will be necessary to characterize the plant material, for which purpose unit fresh weight or dry weight are the usual standards, and also to apply two corrections.

The first correction allows for the appreciable solubility of carbon dioxide in water and in the aqueous phases of the experimental material. It need only be applied to the measurement of rate Y, where the alkaline carbon dioxide absorber is absent or replaced by water (the 'water set'), and where we are concerned with the measurement of the difference between carbon dioxide released and oxygen taken up. In this case there may be a considerable accumulation of carbon dioxide in the system, and the amount dissolved in the various aqueous phases becomes important [12.5], which is not the case in the 'alkali set', where rate X is being measured, and the carbon dioxide concentration remains uniformly low. Solubilities as a function of temperature are given in Table App. 1.2.1, and they are used to correct rate Z, as in the example below.

The second correction reduces the measured volume change to dry gas at s.t.p.: multiply the volume or rate to be corrected by the factor in Table App. 1.2.2 appropriate to the air temperature and barometric pressure at the time that the experiment begins (for the theory behind this practice, see App. 1.15).

If, in unit time interval,

X = the volume change, in cm³, in the 'alkali set' (a negative quantity),
= the volume of oxygen taken up;
Y = the volume change, in cm³, in the 'water set',
= $Z + X$; where
Z = the volume change, in cm³, due to carbon dioxide given out; and if
V_1 = the total volume, in cm³, of water available for the solution of carbon dioxide in the experimental chamber of the 'water set';
V_g = the total volume, in cm³, of the gas phase in the experimental chamber of the 'water set' and its associated connecting tubes; and

App. 1.2

Table App. 1.2.2. Multiply by the factors below to correct the volume of gas, saturated with water vapour at an initial barometric pressure of $_0P$ mmHg and an initial air temperature of $_0\theta_a$°K to the corresponding volume of dry gas at 760 mmHg and 273°K (s.t.p.).

$_0\theta_a$	$_0P$													
	650	660	670	680	690	700	710	720	730	740	750	760	770	780
273° + 10°	0·813	0·826	0·839	0·851	0·864	0·877	0·890	0·902	0·915	0·928	0·940	0·953	0·966	0·978
+ 11°	0·810	0·822	0·835	0·848	0·860	0·873	0·886	0·898	0·911	0·924	0·936	0·949	0·961	0·974
+ 12°	0·806	0·819	0·831	0·844	0·856	0·869	0·882	0·894	0·907	0·919	0·932	0·945	0·957	0·970
+ 13°	0·802	0·815	0·827	0·840	0·853	0·865	0·878	0·890	0·903	0·915	0·928	0·940	0·953	0·966
+ 14°	0·799	0·811	0·824	0·836	0·849	0·861	0·874	0·886	0·899	0·911	0·924	0·936	0·949	0·961
+ 15°	0·795	0·807	0·820	0·832	0·845	0·857	0·870	0·882	0·895	0·907	0·919	0·932	0·944	0·957
+ 16°	0·791	0·803	0·816	0·828	0·841	0·853	0·866	0·878	0·890	0·903	0·915	0·928	0·940	0·953
+ 17°	0·787	0·800	0·812	0·824	0·837	0·849	0·862	0·874	0·886	0·899	0·911	0·923	0·936	0·948
+ 18°	0·783	0·796	0·808	0·820	0·833	0·845	0·857	0·870	0·882	0·894	0·907	0·919	0·931	0·944
+ 19°	0·779	0·792	0·804	0·816	0·829	0·841	0·853	0·865	0·878	0·890	0·902	0·915	0·927	0·939
+ 20°	0·775	0·788	0·800	0·812	0·824	0·837	0·849	0·861	0·874	0·886	0·898	0·910	0·923	0·935
+ 21°	0·771	0·784	0·796	0·808	0·820	0·832	0·845	0·857	0·869	0·881	0·894	0·906	0·918	0·930
+ 22°	0·767	0·780	0·792	0·804	0·816	0·828	0·840	0·853	0·865	0·877	0·889	0·901	0·914	0·926
+ 23°	0·763	0·775	0·787	0·800	0·812	0·824	0·836	0·848	0·860	0·872	0·885	0·897	0·909	0·921
+ 24°	0·759	0·771	0·783	0·795	0·807	0·820	0·832	0·844	0·856	0·868	0·880	0·892	0·904	0·916
+ 25°	0·755	0·767	0·779	0·791	0·803	0·815	0·827	0·839	0·851	0·863	0·875	0·887	0·899	0·912
+ 26°	0·751	0·763	0·775	0·787	0·799	0·811	0·823	0·835	0·847	0·859	0·871	0·883	0·895	0·907
+ 27°	0·746	0·758	0·770	0·782	0·794	0·806	0·818	0·830	0·842	0·854	0·866	0·878	0·890	0·902
+ 28°	0·742	0·754	0·766	0·778	0·790	0·801	0·813	0·825	0·837	0·849	0·861	0·873	0·885	0·897
+ 29°	0·737	0·749	0·761	0·773	0·785	0·797	0·809	0·821	0·833	0·844	0·856	0·868	0·880	0·892
+ 30°	0·733	0·745	0·757	0·768	0·780	0·792	0·804	0·816	0·828	0·840	0·851	0·863	0·875	0·887

S = the solubility of carbon dioxide in water at the prevailing temperature (effectively the temperature of the water bath in which
the experimental chamber A is immersed, θ_b of App. 1.13);

then, within the limitations discussed in App. 1.17,

$$\text{the total carbon dioxide output} = Z\left(1 + S.\frac{V_1}{V_g}\right)$$

$$= Z' \text{ cm}^3. \qquad \text{App. 1.2.1}$$

$$\text{the respiratory quotient} = \frac{Z'}{X} \text{ (a dimensionless quantity)};$$

App. 1.2.2

the rate of carbon dioxide output, in cm³ dry gas at s.t.p. per unit time,

$= Z'$ (corrected by the appropriate factor from Table App. 1.2.2);

and the rate of oxygen uptake, in the same units,

$= X$ (corrected by the appropriate factor from Table App. 1.2.2).

A fuller account of the theory of the apparatus is given in App.
1.13–App. 1.17.

App. 1.3   The apparatus

The Thoday Respirometer encloses the plant material in a glass vessel,
the experimental chamber A of Fig. App. 1.3. Small volumes of plant
material are supported in the middle of the chamber in a loosely fitting
wire cage, and kept moist by wet cotton wool above and below. The
space below the cage may then contain water (for the measurement of
Y), or alkali (for the measurement of X). The size of the chamber is
limited by the time taken for the establishment of thermal equilibrium.
If too large a chamber and too much plant material are used, the time for
thermal equilibration of the whole may become long compared with
drifts in respiratory activity (App. 1.18). The experimental chamber is
connected to a reference chamber, the thermobarometer B, through a
manometer C, which has a capillary bore and is filled to the datum line D
on the inclined arm with a suitable manometric fluid (such as 'Brodie's
Fluid', 23 g sodium chloride + 5 g sodium tauroglycocholate + a few
grains of thymol dissolved in 500 cm³ of distilled water). The inclination
of the arm increases the sensitivity of the manometer in proportion to the

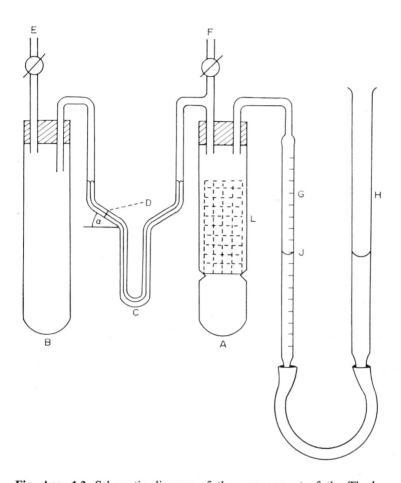

Fig. App. 1.3. Schematic diagram of the arrangement of the Thoday Respirometer. The experimental material is enclosed in a cage of stainless steel gauze L, which slips easily into the experimental chamber A, coming to rest on three projections, of which two are shown. The experimental chamber opens to the air through a tap F, and is connected to the thermo-barometer B through the manometer C, having a capillary bore, and a datum line D on an arm inclined at an angle α to the horizontal. The thermo-barometer also opens to the air through a tap E. Changes in volume of the gas in chamber A are measured by changes in the level of water J in a graduated pipette G, when the level of liquid in the manometer C has been adjusted to the mark D by raising or lowering the water reservoir H. For further details of operation, see App. 1.3 and App. 1.4; for diagrams of two practical forms of the apparatus, see Figs. App. 1.5 and App. 1.18.

App. 1.3

cosecant of the angle of inclination to the horizontal, α. Both the thermo-barometer and the experimental chamber open to the air through taps, E and F. Changes in the volume of gas in the experimental chamber are measured by changes in the level of water in a graduated pipette G, and the necessary pressure adjustment to maintain the level of the fluid in the manometer C on the datum line is made by raising or lowering the water reservoir H (if necessary a fine adjustment can be provided in the form of a side arm with a screw clamp pressing on rubber tubing, but for most purposes the necessary accuracy can be attained without this). In practice the mountings and connections are so arranged that A and B are in a plane far enough behind C, G and H for A and B to be immersed in a water-bath while C, G and H are accessible outside the bath for reading and manipulation.

App. 1.4   Operation

Supposing that we are dealing with a 'water set' (no alkali in chamber A), with the intention of measuring rate Y, the plant material is first loaded in (with the precautions of App. 1.6), and the apparatus stands with chambers A and B in the water bath, and both taps E and F open, long enough for the attainment of thermal equilibrium. Then as both chambers are at atmospheric pressure, the manometer meniscus stands at D. The water reservoir H is adjusted so that the meniscus J is about halfway up the pipette G. The tap E is closed, and a check made that the liquid in the manometer C stands on the datum mark D. As chamber A still communicates with the atmosphere through the open tap F, this means that the gas in the thermobarometer B is at atmospheric pressure ($_0P$) at that time, $_0T$. Tap F is closed; the atmospheric pressure $_0P$ is read (preferably on a good vernier barometer); if necessary the reservoir H is adjusted to bring the liquid in the manometer back to the datum D, and the level of the meniscus J in the graduated pipette G is read. Let this volume be $_0V_p$ at time $_0T$. As the observations proceed, whenever the reservoir H is adjusted so that the manometer meniscus returns to D the volume of gas in B will be the same as at zero time (within the limits discussed in App. 1.15), and while the temperature is unchanged, its pressure will be $_0P$. Therefore the pressure in A will also be $_0P$. Any change of volume in A at this constant pressure will register as a difference in level of the meniscus J, which can be read off the scale of the graduated pipette, giving volume readings $_1V_p$, $_2V_p$, etc. at times $_1T$, $_2T$, etc. Then, unless the respiration rate of the

App. 1.4

plant material is changing rapidly, a plot of V_p against T will give a straight line whose slope is Y.

For the set with alkali in the space at the bottom of the experimental chamber we know that the gas volume must decrease with time, and the experiment is started with J at the bottom of the pipette. Here the slope of the plot of V_p against T is X.

Atmospheric pressure changes subsequent to the closing of the tap E are of no consequence, because the reference pressure has been trapped in the thermobarometer B. It will be seen in App. 1.14–App. 1.16 that no correction need be applied for temperature fluctuations during the course of the experiment, provided that two conditions are met. The first is that that ratio of gas volume at bath temperature to gas volume at air temperature should be roughly the same in each of the two closed systems on either side of the manometer C. The most convenient way of arranging this will be considered in App. 1.5 and App. 1.15. The second condition is that during the course of the experiment the difference between the air and bath temperatures should not change by too large an amount, the limit for moderately accurate work being around 1°C. For smaller changes in the difference, the correction factor would change the recorded volume by less than four parts per thousand, comfortably within the reading error of the graduated pipette in the form of the apparatus next to be described. It follows that if need be a water-bath without a thermostat can be used, a great convenience when working in the field. It is, however, essential to record both bath and air temperatures regularly throughout the experimental period. For larger changes in temperature difference, for substantial changes in water-bath temperature, or for work of higher accuracy, the corrections given in App. 1.15 are applied.

App. 1.5 Construction (i)

In turning the schematic diagram of Fig. App. 1.3 into a practical apparatus it is necessary (a) to decide on the scale to be used; (b) to select a level of tolerable expense; and (c) to take account of the theoretical requirements already considered (App. 1.4, and for more detail see App. 1.15). We will describe first a form where chambers A and B each have a volume of roughly around 50 cm³ (actually close to 60 cm³ in the example discussed later in App. 1.9, but such a deviation is well within the tolerance limits of this type of design). This size is quick to reach

App. 1.5

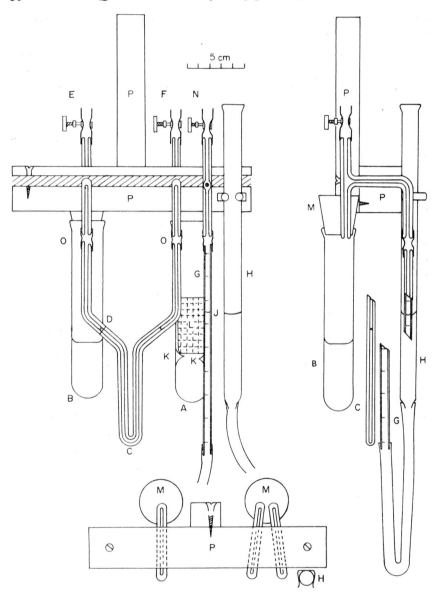

Fig. App. 1.5. Scale diagram of an arrangement of the Thoday Respirometer [Fig. App. 1.3] adapted for measurements on seedlings and small plant parts. A, B, strong boiling tubes, volume around 50 cm³, closed by rubber bungs M, M, carrying wide capillary glass tubing connections to the rubber tubing and screw clip closures E, F and N. Each of these glass

App. 1.5

thermal equilibrium, and therefore convenient for seedlings and plant parts small enough to accommodate without incurring the disturbing effects of wounding and handling [12.4, and App. 1.6]. It would be possible, but costly, to make it all of fused glass and standard joints and taps, and such a form would have advantages in ease and certainty of operation; but the form drawn in Fig. App. 1.5 works satisfactorily, with a little extra care in operation (mainly to avoid leaks), and has the advantages that numbers can be made at small cost, while the rubber joints reduce the risks of breakage when used in the field. The chambers A, B are made from strong boiling tubes (volume around 50 cm³), the chamber A having three constrictions, K, K, pushed into it to hold the cage L. This is preferably made of stainless steel gauze, and designed so that the plant material to be used can be put in without deformation and with the minimum of handling. For some purposes a cylinder open at the top suffices; but for the more awkward material, such as leaves [Fig. 12.4.1], it may be better to use two hemicylinders, one fitted with two circular ends, held together by rubber bands. In any case the cage should slide easily into the chamber and come to rest on the projections.

For respiration measurements on tissues capable of photosynthesis, the chamber A is wrapped in black paper or black polythene sheet held on by rubber bands (as in Fig. App. 1.18).

The chambers A and B fit tightly on rubber bungs, M, M, which in turn are a tight fit on the tap tubes. Because of variations in diameters of commercial boiling tubes it is well to number the units when first assembled, to keep a particular pair of tubes for their matched bungs, and, where field work is involved, to provide matching spares. If capillary glass taps present any problem of supply, the taps E, F and N are conveniently made by screw clamps (with a wide diameter screw head) constricting thick-walled rubber tubing. Two taps, F and N, communicate with the

---

connections has a horizontal arm, supported by the felt and wood clamp P, two of them connected by rubber junctions O, O, to the capillary manometer C and the third to the top of the graduated pipette G, the bottom of which is connected by rubber pressure tubing to the water reservoir H. In use the wooden clamp P rests on, or is supported near, the front edge of a water-bath, in which A and B are immersed; C, G and H are suspended outside and in front of the water-bath. They are coplanar, but in the elevation on the right G and C are shown broken and displaced to the left so as to show their relation to each other.

App. 1.5

experimental chamber A, so that if required during an experiment this chamber can be flushed out with fresh air or with a gas mixture of any desired composition. In subsequent descriptions the tap N will be considered permanently closed unless needed for change of atmosphere. It will be seen that in the upper left-hand diagram of Fig. App. 1.5 the connection to this tap is shown in the same plane as the pipette G, in front of the mounting P (see below), so as to give the maximum separation of the taps F and N, for ease of operation. The volume-measuring tube G is a standard 2 cm^3 graduated pipette, connected at the top to the tap N by a glass-to-glass joint in thick-walled rubber tubing. A length of this is also used to connect the bottom of the pipette to the water reservoir H, which can be slid up and down in a spring clip screwed to the wooden mounting P. In order that the manometer C can be readily cleaned and refilled it is connected to the chambers A and B through glass-to-glass joints (shown open for clarity in the figure) in short lengths of thick-walled rubber tubing.

When this glassware has all been made, the volume of gas in air (as opposed to that in the water-bath) is determined for both sides of the manometer C, reckoning from the datum D to the bung M on the side of the thermobarometer B (V_a' of App. 1.13); and for the experimental chamber A including the corresponding tubing from the other side of the manometer to the other bung M, plus the connection to the tap N and as far as the top of the graduations on the pipette G (V_a). The design requirement of having a similar ratio of gas volumes in the water-bath to those in air on both sides of the manometer C [App. 1.4, App. 1.15] is then most conveniently met when operating the apparatus by adding a calculated volume of water to the thermobarometer B, reducing the gas volume in the water bath on this side of C (where the volume of tubing in air is necessarily less than on the side of A) so as to give the same ratio as on the other. For this reason a volume of water, not necessarily to scale, is shown in B in Fig. App. 1.5.

The whole assembly is held in a simple mounting, P, of wood, with felt or sponge-rubber packing to hold the three horizontal glass connections, and a vertical bar by which the whole is supported, by a retort stand and clamp, over the water bath. This support can be dispensed with, and the mounting arranged to fit over the front edge of the water bath, if the apparatus is to be used personally rather than by a class of students, and particularly for field work where weight may be important. The bungs

App. 1.5

M, M, carrying the chambers A and B, being at the back of the mounting, and the manometer C, the pipette G and the reservoir H in front, the mounting is not far off balance when resting on the front edge of the water bath, and a simple clip suffices to hold it rigid. For class purposes the water-bath need be no more than a jar filled with water and allowed to stand for a day in the laboratory. The temperature of such a simple bath changes little in the course of an experiment lasting an hour or two, and in the same period many rooms not directly exposed to the sun have an air temperature change within the limits discussed in App. 1.4, App. 1.15. The apparatus should not, of course, stand in sunlight or in an obvious draught. It is important that the water in the bath should come right up to the bungs of the chambers A and B so that all the gas space of the chambers themselves is at the uniform bath temperature.

App. 1.6 Precautions

(a) Brief consideration of Fig. App. 1.5 will show that it is easy by clumsy handling to drive the manometer fluid into the connecting tubes or into chambers A or B. It is well to make it a rule always to make sure that both taps E and F are open before putting on or taking off chamber A or B, and always to open the taps when the apparatus, containing respiring material, is to be left unattended. In case of accidents during field work it is prudent to keep a small stock of spare manometers, ready filled and sealed with rubber tube and glass rod, as replacements.

(b) The commonest sources of error in measurement with an apparatus to this design are either a lack of thermal equilibrium, or leaks. The length of time needed for the effective attainment of thermal equilibrium with a particular apparatus, a particular amount of plant material, and a particular initial disequilibrium of temperature, may be tested by setting up a blank apparatus, without plant material, or with the cage loaded with a comparable weight of wet filter paper, and ensuring that there are no leaks by the procedure of App. 1.7. With taps E and F closed (N should be closed in any case), readings of bath temperature, air temperature, and the level of the meniscus J (when the manometer C stands on its datum D) are then taken at intervals of a few minutes. Provided that changes in the two temperatures considered together fall within the limits mentioned [App. 1.4, App. 1.15], the meniscus J should remain steady when once thermal equilibrium is attained. As long as the meniscus is

App. 1.6

moving, the apparatus is detectably out of thermal equilibrium. A simple plot of the movement of the meniscus against time enables the time of attainment of effective equilibrium (a deviation not larger than the reading error) to be evaluated, as the change in volume would be expected to follow Newtonian cooling, i.e., the rate declines exponentially with time.

(c) Because of the lethal nature of the alkaline absorber, great care must be taken not to wet the upper walls when introducing it into chamber A. Also, because the alkali will continue to absorb carbon dioxide even if greatly diluted, very thorough rinsing is needed when changing from an 'alkali set' to a 'water set'. Next to leaks [App. 1.7] and lack of thermal equilibrium (b, above), the commonest source of spurious respiration readings with this apparatus is alkali out of place.

(d) The extent to which the respiration rate of a particular plant tissue is affected by wounding or handling varies a great deal with the species, with the particular organ concerned, and with its physiological state [12.4]. The sensitivity of the particular plant or organ studied can be tested by an experiment involving various controlled degrees of wounding or mechanical stimulation, as a guide to the planning of the investigation and the interpretation of the results.

App. 1.7   Testing for leaks

When the apparatus has been assembled, with the plant material in chamber A, and while both chambers are coming to thermal equilibrium with the water bath, the following routine should be carried out to ensure that there are no leaks. Each chamber is tested in turn with the other open to the air.

(a) With both taps E and F open lower the reservoir H to the bottom of its traverse.

(b) Close tap F and raise H to the top of its traverse, creating a positive pressure in A. The fluid in the manometer C rises in the arm towards B, and should then remain stationary. If it sinks back, there is a leak in A.

(c) From this position, test B by closing tap E and then opening tap F. The manometer fluid drops part of the way back to its rest position and then stops, holding the air in B under tension. If it continues to drop back there is a leak in B.

(d) If a leak is detected it must be rectified at this stage and a further period allowed for thermal equilibrium. The most likely place is at the bungs, which can be wetted to improve the seal, but the taps should be checked also.

(e) If chamber A or the plant material in it are initially at a temperature well above that of the bath, cooling during the above procedure may produce the appearance of a small leak [App. 1.6b]. If this effect is suspected, it can be checked by repeating the above procedure, but creating a negative pressure in chamber A by putting the reservoir at the top of its traverse at operation (a), and then lowering it at operation (b). The contraction in volume of the gas in chamber A due to cooling will then cause the liquid in the manometer C to rise, instead of falling as it would if there were a leak.

App. 1.8   Experimental organization

The method of making an individual reading has already been outlined in App. 1.4. As readings are taken and recorded it is nearly always worth while also to enter them on a plot of volume against time. A study of the features of this graph usually makes it possible at an early stage to detect either an initial thermal disequilibrium, or the development of a leak, and to distinguish both from possible drifts in the rate of respiration of the plant material. Many materials provide early evidence of a steady rate of respiration, as in the example below, and observations under particular conditions can be confined to a short period, and conditions then changed. If information on longer term drifts under relatively fixed conditions is required, it may be convenient then to stop taking readings, open all the taps (including tap N), and pass a slow stream of air through chamber A for a period (possibly of one to several hours) before resuming readings. Before embarking on this type of experiment it is well to check beforehand that the alkali used will retain its effectiveness as an absorber of carbon dioxide at the end of the longest intended period of observation.

App. 1.9   An example

The following example is taken from part of a standard class experiment carried out by students in their second university year: special expertise is thus not necessary to attain this standard of accuracy.

A uniform sample of seeds of *Ricinus communis* were germinated until the radicle had reached a length of about 5 cm, when the following operations were carried out with the greatest care, partly to keep the

App. 1.9

effects of wounding and handling to a minimum, and partly to keep the poisonous protein ricin off the skin of hands or face. The testas were cut round in the plane of the cotyledons, the two halves prized apart, the cotyledons separated from the endosperm and the endosperm from the testa, which was discarded. Two samples of complete embryos around 2 g were weighed and placed in the wire cages, supported top and bottom by plugs of absorbent cotton wool each wetted with 0.5 cm^3 of water. One cage was placed in an experimental vessel which had been rinsed with distilled water, and the drops shaken out but not dried. When assembled, this apparatus formed the 'water set'. The other cage was placed in an experimental vessel already prepared with 5 cm^3 of 4 per cent KOH pipetted carefully into the bottom, which also contained two rolled discs of thick filter paper (Whatman No. 42), of a size not to touch the wire cage. When assembled, this apparatus formed the 'alkali set'. Two samples of the endosperm halves, of similar weight, were set up in the same way. In all four sets the thermobarometer vessel B contained 30 cm^3 of water, to make the ratio of the gas phase volumes surrounded by water to the gas phase volumes surrounded by air approximately the same on both sides of the manometer [App. 1.5, App. 1.14]. Ten minutes was allowed for temperature equilibration, during which time the tests for leaks [App. 1.7] were carried out.

Readings in the water sets were started in the middle of the graduated pipettes, in the alkali sets near the bottom. In each set the tap E on the thermobarometer was closed first, trapping the reference pressure while the manometer fluid stood on the datum D, then the tap F was closed.

If necessary the reservoir was next adjusted to bring the manometer meniscus back to the datum, and the volume reading on the graduated pipette, $_0V_p$, was noted, together with the atmospheric pressure, $_0P$, and the air and bath temperatures, $_0\theta_a$ and $_0\theta_b$, all at time $_0T$. Care was taken to handle the pipette only by the rubber tubing at the bottom, so as to run no risk of warming it up above air temperature.

The whole procedure of the last paragraph was then repeated at 10-minute intervals (except for the reading of atmospheric pressure). The example given in Table App. 1.9 shows that a pair of students, working together, can keep four of these pieces of apparatus going at once, starting and reading them in sequence. The data are plotted in Fig. App. 1.9, and give steady rates of volume change over an hour for both types of material in both water and alkali sets. Had there been

App. 1.9

a shortage of apparatus, in such a case one could with confidence start with the plant material in a water set, and after an hour's observation to determine rate Y, open both taps, remove the cage from chamber A, add KOH and filter papers to convert it to an alkali set, reassemble, equilibrate and test for leaks, and resume observations (starting with the meniscus J near the bottom of the pipette) to determine rate X,

Table App. 1.9. Data obtained on the respiration of embryos and endosperm of germinating seeds of *Ricinus communis*, using four Thoday Respirometers simultaneously. Columns 1, 4, 7, 10: time, minutes from starting the first set of readings; columns 2, 5, 8, 11: graduated pipette reading, cm^3; columns 3, 6, 9, 12: bath temperature, °C. Initial barometric pressure, 764·5 mmHg. Weight of each sample, 2·0 g. Water in each sample, 1·5 g. Total water in each experimental chamber of the water sets, 2·5 g. Air Temperature, times 0 and 9 minutes, 24·5°C; 19 and 29 minutes, 24·25°C; 39, 49 and 59 minutes, 24·0°C.

endosperm						embryo					
alkali set			water set			alkali set			water set		
1	2	3	4	5	6	7	8	9	10	11	12
0	1·51	21·0	2	0·90	21·0	5	1·57	21·0	7	1·60	21·0
10	1·27	21·0	12	0·76	21·0	15	1·40	21·0	17	1·585	21·0
20	1·01	21·25	22	0·62	21·0	25	1·26	21·0	27	1·57	21·0
30	0·77	21·25	32	0·48	21·0	35	1·08	21·0	37	1·55	21·25
40	0·52	21·25	42	0·34	21·25	45	0·90	21·25	47	1·54	21·25
50	0·27	21·25	52	0·205	21·25	55	0·74	21·25	57	1·52	21·25
60	0·01	21·25	62	0·07	21·25	65	0·58	21·25	67	1·50	21·25

returning finally to a water set again if it should be necessary go check for drifts.

At the end of the experiment the plant material was dried in a ventilated oven and weighed. The mean percentages of water were: embryo, 76·8 per cent; endosperm, 73·7 per cent, giving, for the water in a 2 g sample: embryo, 1·536 g; endosperm, 1·474 g. An inspection of the calculation in App. 1.10 shows, however, that if both these figures are rounded to 1·5 g the rounding error in the solubility correction is approximately one part in 2,000, well within the general experimental error. The rounded figure was therefore used in the following calculation.

App. 1.9

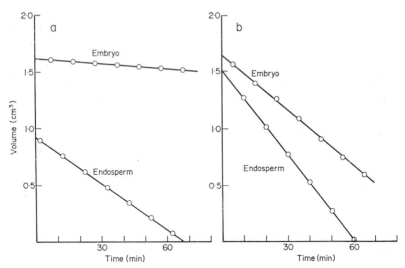

Fig. App. 1.9. Respiration rates of embryos and endosperm of *Ricinus communis*: the volume data of Table App. 1.9 plotted against time. (a), the two water sets, columns 5 and 11, giving rates *Y*, the difference between intake of oxygen and output of carbon dioxide. (b), the two alkali sets, columns 2 and 8, giving rates *X*, the intake of oxygen.

App. 1.10 Calculations: volumes

Fig. App. 1.9 does not give any suggestion that the volumes in the four sets of observations were not linear functions of time, the correlation coefficients lying between 0·998 and 0·99997. Assuming linearity, and taking all four sets together, the fiducial limits of a single observation are ±0·0115 cm³, close enough to the reading limit to indicate that here all sources of random experimental error were small.

From the slopes of the lines of best fit in Fig. App. 1.9 are obtained the mean rates of volume change, giving, for the alkali sets, values of rate *X*, the rate of intake of oxygen; and for the water sets, values of rate *Y*, the net change due to both intake of oxygen and output of carbon dioxide. To obtain values of rate *Z*, the rate of output of carbon dioxide, rate *X* is therefore subtracted from rate *Y*, taking account of sign. We then have the following mean rates of volume increase, all in cm³ hour⁻¹:

Embryo	alkali	X_{emb}	−0·996
	water	Y_{emb}	−0.099
	difference	Z_{emb}	+0.897

App. 1.10

Endosperm	alkali	X_{end}	-1.498
	water	Y_{end}	-0.831
	difference	Z_{end}	$+0.667$

Correcting for the solubility of carbon dioxide in water, the total carbon dioxide output,

$$Z' = Z\left(1 + S.\frac{V_1}{V_g}\right). \qquad \text{App. 1.2.1}$$

Here the value of S (by interpolation in Table App. 1.2.1) is 0.917;

$$V_1 = 2.5 \text{ cm}^3 \qquad \text{(Table App. 1.9)};$$
$$V_g = 55 \text{ cm}^3 \qquad \text{for the apparatus used};$$

therefore the correction factor in the bracket $= 1.042$, and

$$Z'_{emb} = +0.935 \text{ cm}^3 \text{ hour}^{-1},$$
$$Z'_{end} = +0.695 \text{ cm}^3 \text{ hour}^{-1}.$$

$$\text{The respiratory quotient} = \frac{Z'}{X}, \qquad \text{App. 1.2.2}$$

counting oxygen uptake as positive, so that RQ

$$\text{for the embryo} = 0.939,$$
$$\text{for the endosperm} = 0.464.$$

The correction of Table App. 1.2.2 converts rates X and Z' to volumes of dry gas at s.t.p., interpolation giving, for an air temperature of $24.5°C$ and a barometric pressure of 764.5 mmHg, a multiplying factor of 0.895. We then have the following rates, all in terms of volumes of dry gas at s.t.p., and grams fresh weight of plant material:

Embryo

$$O_2 \text{ uptake} = 0.895 \; X_{emb} = 0.891 \text{ cm}^3 \text{ hour}^{-1},$$
$$\equiv 0.446 \text{ cm}^3 \text{ g}^{-1} \text{ hour}^{-1};$$
$$CO_2 \text{ output} = 0.895 \; Z'_{emb} = 0.837 \text{ cm}^3 \text{ hour}^{-1},$$
$$\equiv 0.418 \text{ cm}^3 \text{ g}^{-1} \text{ hour}^{-1}:$$

Endosperm

$$O_2 \text{ uptake} = 0.895 \; X_{end} = 1.341 \text{ cm}^3 \text{ hour}^{-1},$$
$$\equiv 0.670 \text{ cm}^3 \text{ g}^{-1} \text{ hour}^{-1};$$
$$CO_2 \text{ output} = 0.895 \; Z'_{end} = 0.622 \text{ cm}^3 \text{ hour}^{-1},$$
$$\equiv 0.311 \text{ cm}^3 \text{ g}^{-1} \text{ hour}^{-1}.$$

App. 1.10

App. 1.11   Weights of oxygen and carbon dioxide

Finally, volumes of dry gas at s.t.p. are at once convertible into weights of gas by a simple factor, whose value is $1\cdot4276$ mg cm^{-3} for oxygen and $1\cdot9635$ mg cm^{-3} for carbon dioxide. The results of the observations of

Table App. 1.11 Summary of results (to two significant figures) derived from the observations of Table App. 1.9, made using the Thoday Respirometer, on gas exchanges by embryos and endosperm of germinating seeds of *Ricinus communis*, measured separately at temperatures near 21°C. Rates of change of: volume; expressed as cm^3 dry gas at N.T.P. per g fresh weight per hour: weight; expressed as micrograms of the gas per g fresh weight per hour.

	CO$_2$ output		O$_2$ uptake		
	Volume	Weight	Volume	Weight	RQ
Embryo	0·42	820	0·45	640	0·94
Endosperm	0·31	610	0·67	960	0·46

Table App. 1.9 on *Ricinus communis*, expressed in these various ways, are summarized in Table App. 1.11.

How and to what extent these rates of oxygen uptake and carbon dioxide evolution, both expressed in terms of weight of gas, can be used to infer the net rate of change of dry weight of the plant or organ concerned will be considered in App. 1.12.

App. 1.12   Discussion of results

Gas exchange in the endosperm is very different from that in the embryo. As would be expected from its relatively large proportion of meristematic tissue and vigorous metabolism, the embryo has a high respiration rate per unit fresh weight, compared with any of the examples examined in Chapter 12. At the same time it takes in only slightly more than 1 molecule of oxygen for each molecule of carbon dioxide that it gives off, so that its respiratory quotient is close to $1\cdot0$, and the metabolic balance term (D of Fig. 13.11) is likely to be relatively small as compared with the throughput of hexose to carbon dioxide in the respiratory machinery. In consequence of all this it is losing dry weight rapidly, but before we can decide how rapidly, in default of an experimental check it is necessary to

App. 1.12

make some assumptions about the loss or incorporation of the elements of water, in addition to the exchanges of oxygen and carbon dioxide quantified in Table App. 1.11. The oxygen equivalent of the carbon dioxide given off is around 600 μg g^{-1} hour^{-1}, and it would be reasonable to assume that together these correspond to a loss of hexoses at around 570 μg g^{-1} hour^{-1}. Considerable uncertainty attaches to the fate of the extra oxygen absorbed over and above this, but at 40 μg g^{-1} hour^{-1} it is small compared with the hexose loss, and in so vigorously growing a young organ considerable uncertainty also attaches to the rates of the dry weight changes involved in such processes as the condensation of simple carbohydrates into wall materials, which do not directly involve any exchanges of oxygen or carbon dioxide. We may conclude that the net loss of dry weight probably lies between 500 and 600 μg per gram fresh weight per hour, or between 2·2 and 2·6 mg per gram dry weight per hour.

On the other hand, although the rate by volume of evolution of carbon dioxide from the endosperm, on a fresh weight basis, is roughly three-quarters of that from the embryo, the rate of oxygen uptake is 50 per cent higher, leading to an RQ of 0·46, presumably associated with the metabolism of the storage materials, which include much fat and oil. The complete oxidation of fats to carbon dioxide and water would lead to an RQ of around 0·7, accompanied by a loss of dry weight; whereas the conversion of fat to carbohydrate via the fatty acid spiral and the glyoxalate cycle would give an apparent RQ of around 0·35, accompanied by a gain in dry weight roughly equivalent to half the oxygen taken up. The evidence points to the predominance of a process of this latter kind in the metabolism of the endosperm at this stage in the germination of the seed, and it is probably showing a net gain in dry weight, perhaps at around one quarter of the rate of uptake of oxygen by weight, or around 250 μg per gram fresh weight per hour, equivalent to around 1 mg per gram dry weight per hour. This would represent a substantial gain in dry weight over a 24 hour period, and the actual rate could readily be checked by a separate experiment if necessary.

As at this stage in its germination the fresh weight of the endosperm substantially exceeds that of the embryo, we have in the seed as a whole a good example of a considerable metabolic balance term [13.11], and a case where dry weight changes are a very poor guide to gains or losses of carbon units.

App. 1.12

At the same time the example shows the utility of measurements of both carbon dioxide output and oxygen uptake, when it is desired to use respiration measurements to make inferences about dry weight changes in the organs concerned.

App. 1.13 Theory: symbols

V_a The volume of the gas in the connecting tubes, which are in air, between the datum level of the manometer C (Fig. App. 1.3, App. 1.5), the experimental chamber A, and the top of the graduations on the pipette G.

V_a' The same, between the datum level of the manometer C and the thermobarometer B.

V_b The volume of the gas phase in the experimental vessel A in the water-bath.

V_b' The same, in the thermobarometer B.

V_g The volume of the gas phase in the experimental chamber A, together with its associated air spaces $= V_a + V_b + V_p$.

V_1 The volume of water in the experimental chamber A of the 'water set' = water in the plant material plus added water.

$_0V_p, {}_1V_p$, etc.   The volume of gas in the graduated pipette, measured from the meniscus J to the top of the graduations, at times 0, 1, etc.

$_0P, {}_1P$, etc.   The gas pressure in a vessel at times 0, 1, etc., not necessarily equal to atmospheric pressure at the time, but $_0P$ equals the atmospheric pressure at time $_0T$.

$_0P_a, {}_1P_a$, etc.   The total partial pressure of all gases except water vapour in a vessel at times 0, 1, etc.

$_0P_w, {}_1P_w$, etc.   The partial pressure of water vapour in a vessel at times 0, 1, etc. At all times $P_a + P_w = P$.

$_0\theta_a, {}_1\theta_a$, etc.   The absolute temperature of the air at times 0, 1, etc. = air temperature in °C + 273°.

$_0\theta_b, {}_1\theta_b$, etc.   The absolute temperature of the water-bath at times 0, 1, etc.

D   The difference between the temperature changes, from zero time up to the time of reading, in the water-bath and in the air $= ({}_1\theta_b - {}_0\theta_b) - ({}_1\theta_a - {}_0\theta_a)$.

App. 1.13

F The correction factor for water vapour

$$= \frac{{}_0 P_w}{{}_0 P} \left(\frac{{}_1 P_w}{{}_0 P_w} \cdot \frac{{}_0 \theta_b}{{}_1 \theta_b} - 1 \right) \left(\frac{1}{{}_1 \theta_b - {}_0 \theta_b} \right).$$ (App. 1.15.3)

App. 1.14 Theory: note on dimensions

In the particular form of the apparatus used in the practical classes of which an example was given in App. 1.9, both the chambers A and B had a total volume of around 60 cm³, and 30 cm³ of water had been added to B in each set. Allowing for the volumes of cage, sample, and added water or KOH, the gas volumes, in cm³, were therefore approximately as follows:

	Thermobarometer (all sets)	Chamber A Water set	Chamber A Alkali set
Total volume of chamber	60	60	60
Cage, plant material, plus added water	30	5 to 7	10 to 12
Hence, V_b'	30; V_b	55 to 53	50 to 48
V_a'	0·5; V_a	0·8	0·8
V_a'/V_b'	0·017; V_a/V_b 0·015		0·016–0·017

Thus the ratio V_a/V_b, or V_a'/V_b', is always small, and the routine addition of 30 cm³ of water to all the thermobarometers has given values of V_a'/V_b' approximately the same as the corresponding values of V_a/V_b. In App. 1.15 it will be assumed that these two ratios have exactly the same value for the thermobarometer and experimental chamber of any particular set. In App. 1.16 the consequences of deviations from equality will be investigated.

App. 1.15 Theory: changes of volume

In this section it will be assumed that readings are not begun until chambers A and B and their contents are all at the same, bath, temperature, ${}_0 \theta_b$, and their connecting tubes and the contained gas also at the same, air, temperature, ${}_0 \theta_a$. With the precautions detailed in App. 1.6 this is a safe assumption for an apparatus of the size considered in App. 1.5 and App. 1.14, any deviations having undetectably small effects.

App. 1.15

However, this is not necessarily so for other sets of dimensions. An example of the consequences in practice, when the behaviour of the apparatus deviates from these assumptions, will be considered in App. 1.18.

Then since P, P_a, P_w, θ_a, θ_b and also V_a/V_b are the same for both the thermobarometer and experimental chamber of any set at any particular time; and since also the total amount, and hence the total volume at s.t.p., of gases other than water vapour in the control vessel is constant, therefore the total volume (measured at s.t.p.) of gases other than water vapour in the corresponding space $V_a + V_b$ in the experimental chamber is also constant.

The only necessary correction for any change in pressure or temperature is therefore that to be applied to the volume additional to $V_a + V_b$ for the experimental chamber, i.e. the volume V_p of the gas in the graduated pipette, between the meniscus J and the top of the graduations.

$$\text{At time } {}_1T, \; {}_1P_a = {}_0P_a \left(\frac{\dfrac{V_b'}{{}_0\theta_b} + \dfrac{V_a'}{{}_0\theta_a}}{\dfrac{V_b'}{{}_1\theta_b} + \dfrac{V_a'}{{}_1\theta_a}} \right)$$

$$= {}_0P_a \cdot \frac{{}_1\theta_b}{{}_0\theta_b} \left(\frac{1 + \dfrac{V_a' \cdot {}_0\theta_b}{V_b' \cdot {}_0\theta_a}}{1 + \dfrac{V_a' \cdot {}_1\theta_b}{V_b' \cdot {}_1\theta_a}} \right). \qquad \text{App. 1.15.1}$$

With values of V_a'/V_b' around 0·02 and θ_a, θ_b around 290, the expression in brackets in equation App. 1.15 can vary very little indeed from unity. If θ_a gains or loses as much as 10°C on θ_b it will change by only six parts in 10,000. Hence, with quite negligible error,

$$_1P_a = {}_0P_a \cdot \frac{{}_1\theta_b}{{}_0\theta_b}. \qquad \text{App. 1.15.2}$$

However, this relationship applies only to gases other than water vapour. The partial pressure of water vapour, P_w, is a very much more complex function of temperature, but it can readily be simplified over a limited working range. The following treatment is based on the requirement that P_w should be the saturation vapour pressure at water-bath temperature. This will almost inevitably be so for an apparatus such as that of App. 1.5, because saturation with water vapour will take place in

App. 1.15

both chambers A and B at the same time as thermal equilibration, provided that air temperature is not lower than water-bath temperature. If it is, there is a risk of condensation in the tubing in air, with a consequent drop in vapour pressure. This effect will be small provided that $\theta_b - \theta_a$ is small, V_a/V_b is small, and the available area for condensation in the tubes in air is small compared to the evaporating area inside the water-bath. However, any uncertainty here can be eliminated altogether by arranging that the water bath shall always be below air temperature.

Over the temperature range with which we are concerned, say from $10°$ to $30°C$, the saturation vapour pressure of water increases by roughly $6\frac{1}{2}$ per cent for every degree rise in temperature. For small temperature differences the quantity

$$\left(\frac{{}_1P_w}{{}_0P_w} - 1\right)\left(\frac{1}{{}_1\theta_b - {}_0\theta_b}\right)$$

declines slowly from 0.069 per $°C$ at $10°$ to 0.064 at $20°$ and 0.059 at $30°C$. Because ${}_0\theta_b$ has a value around 290, the same holds for the expression

$$\left(\frac{{}_1P_w}{{}_0P_w}\cdot\frac{{}_0\theta_b}{{}_1\theta_b} - 1\right)\left(\frac{1}{{}_1\theta_b - {}_0\theta_b}\right),$$

which, for a temperature difference of $1°C$, also declines from 0.065 per $°C$ at $10°$ to 0.060 at $20°$ and 0.055 at $30°$. Nor, for values of ${}_0P$ around 760 mmHg, are there large changes in the value of the expression

$$\frac{{}_0P_w}{{}_0P}\left(\frac{{}_1P_w}{{}_0P_w}\cdot\frac{{}_0\theta_b}{{}_1\theta_b} - 1\right)\left(\frac{1}{{}_1\theta_b - {}_0\theta_b}\right),$$

which, again for a temperature difference of $1°C$, increases from 0.0008 per $°C$ at $10°$ to 0.0011 at $15°$, 0.0014 at $20°$, 0.0018 at $25°$ and 0.0023 at $30°C$. For any particular experiment which involves only small changes in water-bath temperature, to a close approximation this can be regarded as a constant, which we will call $\mathbf{F}$, and we can write:

$$\frac{{}_0P_w}{{}_0P}\left(\frac{{}_1P_w}{{}_0P_w}\cdot\frac{{}_0\theta_b}{{}_1\theta_b} - 1\right) = \mathbf{F}({}_1\theta_b - {}_0\theta_b). \qquad \text{App. 1.15.3}$$

Then

$$\frac{{}_1P}{{}_0P} = \frac{{}_1P_a + {}_1P_w}{{}_0P_a + {}_0P_w},$$

$$= \frac{{}_0P_a\cdot\dfrac{{}_1\theta_b}{{}_0\theta_b} + {}_1P_w}{{}_0P_a + {}_0P_w},$$

App. 1.15

substituting from App. 1.15.2,

$$= \frac{{}_1\theta_b}{{}_0\theta_b}\left(1 + \frac{{}_0P_w}{{}_0P}\left(\frac{{}_1P_w}{{}_0P_w}\cdot\frac{{}_0\theta_b}{{}_1\theta_b} - 1\right)\right),$$

$$= \frac{{}_1\theta_b}{{}_0\theta_b}(1 + \mathbf{F}({}_1\theta_b - {}_0\theta_b)). \qquad\qquad \text{App. 1.15.4}$$

F could be evaluated from equation App. 1.15.3 in a particular case, but it would be necessary only for work of high accuracy. From the values given above, it can be seen that the water vapour correction can be neglected when working at temperatures below 30°C and air pressures near 1 atmosphere, unless an accuracy greater than 1 part in 400 is aimed at. Even for an accuracy of 1 part in 1000, a rough interpolation in the figures given is all that is needed. However, for the sake of generality the correction will be retained through the next stage.

The change in the gas volume in the experimental chamber and its connected spaces, measured under the initial conditions of pressure and air temperature, ${}_0P$, ${}_0\theta_a$, between times ${}_0T$ and ${}_1T$, is then

$$_0V_p - {}_1V_p\cdot\frac{{}_1P}{{}_0P}\cdot\frac{{}_0\theta_a}{{}_1\theta_a} = {}_0V_p - {}_1V_p\left(\frac{{}_1\theta_b}{{}_0\theta_b}\cdot\frac{{}_0\theta_a}{{}_1\theta_a}\right)(1 + \mathbf{F}({}_1\theta_b - {}_0\theta_b)).$$

$$\text{App. 1.15.5}$$

Since the bath and air temperatures are close to each other,

$$\frac{{}_1\theta_b}{{}_0\theta_b}\cdot\frac{{}_0\theta_a}{{}_1\theta_a} \simeq 1 + \frac{({}_1\theta_b - {}_0\theta_b) - ({}_1\theta_a - {}_0\theta_a)}{{}_0\theta_b},$$

$$\simeq 1 + \frac{\mathbf{D}}{{}_0\theta_b} \qquad\qquad \text{(App. 1.13)}$$

and

$$\left(\frac{{}_1\theta_b}{{}_0\theta_b}\cdot\frac{{}_0\theta_a}{{}_1\theta_a}\right)(1 + \mathbf{F}({}_1\theta_b - {}_0\theta_b)) \simeq 1 + \mathbf{F}({}_1\theta_b - {}_0\theta_b) + \frac{\mathbf{D}}{{}_0\theta_b}.$$

$$\text{App. 1.15.6}$$

At values of ${}_0\theta_b$ around 290°K, **F** is little over 0·001 and $\mathbf{D}/{}_0\theta_b = 0\cdot0034\mathbf{D}$. Since no value of V_p can exceed 2 cm³, the whole correction to be applied to ${}_1V_p$ must almost inevitably be less than ±0·01 cm³ unless air temperature gains or loses more than 1°C on bath temperature.

The effect of the double correction can be illustrated by means of the example in Table App. 1.9. Here bath temperature was around 21°C, so that **F** has a value of approximately 0·0015, while ${}_1\theta_b - {}_0\theta_b = 0\cdot25$°C,

App. 1.15

making this part of the correction approximately 0·0004. Over the experimental period of 60 minutes, $\mathbf{D} = (21·25 - 21·0) - (24·0 - 24·5) = +0·75°C$, so that $\mathbf{D}/_0\theta_b = +0·0025$, and the whole correction is $+0·0029$. The largest value of $_1V_p$ is 1·5 cm^3, for the embryos in the water set, making the largest value of the total correction for the whole 60 minutes of observations $+ 0·004$ cm^3, about one-third of the fiducial limits for a single observation. Here the correction is not worth applying (and as it happens, if it were applied, none of the summary figures of Table App. 1.11 would be altered thereby), but in other circumstances it would be desirable to test the correction for what might be significant systematic effects.

It is also worth enquiring how the readings might be affected by the expansions of water and of glass with increase of temperature, because, although neither is likely to be of significance for an apparatus of these dimensions [App. 1.14] used as in App. 1.9, it might be necessary to take account of them in other circumstances. The coefficient of cubic expansion of water is around 2×10^{-4} per °C, so that, with chambers A and B both of 60 cm^3 capacity, and B half full of water, but relatively little water in A, a 1°C change in bath temperature would change the reading of the graduated pipette G by approximately 0·005 cm^3. So large a change in bath temperature is unlikely in a short experiment (compare Table App. 1.9, where in an hour none of the baths changed by more than 0·25°C), so that this correction is then negligible. However, if the apparatus must be used for experiments where readings must be continued through larger changes of bath temperature, it would be better to reduce the volume of the thermobarometer B to the desired value by cutting a length off the top of the boiling tube, rather than by adding water. The same might also be true if the dimensions of the apparatus were different, and this point should be checked when designing an apparatus of different scale. In either case it would then only be necessary to have enough water present in B to maintain saturation of the gas phase, and the correction for the expansion of liquid water would be negligible.

The coefficient of cubic expansion of glass is an order of magnitude smaller than that of water at around 3×10^{-5} per °C, but its value is almost immaterial so long as chambers A and B both have the same coefficient of expansion and are at the same temperature. This is so whether or not they have the same volume, because a given proportional volume change on both sides of the manometer C will produce the same

App. 1.15

proportional pressure change on both sides, and in the absence of other changes will produce no change in the levels of the manometer C or of the meniscus J.

App. 1.16 Theory: inequality of V_a/V_b

Suppose, instead of the assumed equality, that V_a/V_b is as much as 20 per cent greater than V_a'/V_b' (which is greater than any of the differences given in App. 1.14). This would mean that the readings of the graduated pipette $_0V_p$, $_1V_p$, etc., should all be increased by a volume $0 \cdot 2 \, V_b . (V_a'/V_b')$ before applying the correction of App. 1.15.5 and App. 1.15.6. For an apparatus of the dimensions of App. 1.14, this additional volume would be $0 \cdot 2 \times 0 \cdot 02 \times 50 = 0 \cdot 2$ cm^3, approximately.

The change in volume, measured at $_0P$, $_0\theta_a$, would then be

$$(_0V_p + 0 \cdot 2) - (_1V_p + 0 \cdot 2)(1 + \mathbf{F}(_1\theta_b - _0\theta_b) + \mathbf{D}/_0\theta_b) =$$
$$_0V_p - _1V_p(1 + \mathbf{F}(_1\theta_b - _0\theta_b) + \mathbf{D}/_0\theta_b) - 0 \cdot 2(\mathbf{F}(_1\theta_b - _0\theta_b) + \mathbf{D}/_0\theta_b).$$

For the experiment of App. 1.9, the further correction due to inequality of V_a/V_b would thus be less than $-0 \cdot 0006$ cm^3, while even were $\mathbf{D}$ as large as $10°C$ it would still be less than $0 \cdot 01$ cm^3. For an apparatus of around the dimensions given in App. 1.14, it is therefore unnecessary to correct for small differences between the ratios V_a/V_b and V_a'/V_b'.

App. 1.17 Theory: solubility of carbon dioxide

The output of carbon dioxide to the gas phase is measured in the first instance as cm^3 of moist gas at $_0P$ and $_0\theta_a$, and it is required to correct for carbon dioxide dissolved in the water in the experimental chamber and in the aqueous phases of the plant material. To a first approximation, the solubility in the latter can be taken to be the same as that in water [12.16]. Using the symbols of App. 1.13, and adding

S   the volume of dry carbon dioxide, measured in cm^3 at 760 mmHg pressure and at a temperature of $_1\theta_b$, required to saturate 1 cm^3 of water at $_1\theta_b$;

Z   the volume of carbon dioxide in the gas phase of chamber A and its associated air spaces, measured as cm^3 of moist gas at $_0P$ and $_0\theta_a$, as in App. 1.15;

App. 1.17

then the volume of dry carbon dioxide is $Z.[(_0P - P_w)/_0P]$, and the partial pressure of carbon dioxide in the gas phase is

$$Z.\frac{_0P - P_w}{_0P}.\frac{_0P}{V_g} = \frac{Z}{V_g}(_0P - P_w).$$

At the same time the volume of dry carbon dioxide, measured in cm^3 at $_0P$ mmHg and a temperature of $_0\theta_a$, required to saturate 1 cm^3 of water at $_1\theta_b$ and a partial pressure of 760 mmHg would be

$$S.\frac{_0\theta_a}{_1\theta_b}.\frac{760}{_0P};$$

while if the carbon dioxide were saturated with water vapour the volume would be increased to

$$S.\frac{_0\theta_a}{_1\theta_b}.\frac{760}{_0P}.\frac{_0P}{_0P - P_w}.$$

It follows that the volume of moist carbon dioxide, measured at $_0P$ and $_0\theta_a$, dissolved in 1 cm^3 of water in chamber A, at a temperature of $_1\theta_b$ and a partial pressure of $(_0P - P_w).Z/V_g$ would be

$$S.\frac{_0\theta_a}{_1\theta_b}.\frac{760}{_0P}.\frac{_0P}{_0P - P_w}.\frac{_0P - P_w}{760}.\frac{Z}{V_g} = S.\frac{_0\theta_a}{_1\theta_b}.\frac{Z}{V_g}.$$

But the volume of water in chamber A is V_1 cm^3. Therefore the total volume of moist carbon dioxide in chamber A and its connecting tubes, including that dissolved in water, all measured in cm^3 at $_0P$ and $_0\theta_a$,

$$Z' = Z + V_1.S.\frac{_0\theta_a}{_1\theta_b}.\frac{Z}{V_g}$$

$$= Z\left(1 + S.\frac{_0\theta_a}{_1\theta_b}.\frac{V_1}{V_g}\right). \qquad \text{App. 1.17.1}$$

In an experiment such as we have been considering, the difference between $_0\theta_a$ and $_1\theta_b$ is likely to be small compared to the value of either, so that $_0\theta_a/_1\theta_b$ will be close to 1, and its effect on the correction is likely to be negligible. As an example, considering the figures of App. 1.10, where $S.V_1/V_g = 0.042$, and $_1\theta_b = 294°$K, to change the value of the correction to 0·041 or 0·043 would require $_0\theta_a$ to differ from $_1\theta_b$ by as much as 294/42 = 7°C. Hence, for smaller differences between $_0\theta_a$ and

$_1\theta_b$, or when not trying to work to an accuracy of 1 part in 1000, the correction can be taken to be

$$Z' = Z\left(1 + S.\frac{V_1}{V_g}\right),$$ App. 1.17.2

with an inappreciable error.

App. 1.18 Construction (ii)

The basic theory of the constant pressure method thus does not impose any barriers to a wide range of size of experimental chamber, and the obvious practical difficulties involved in change of scale are small [App. 1.1]. Nevertheless, attempts to scale the apparatus up sufficiently to deal with much larger plants or plant parts have not been uniformly successful, very large increases of scale bringing a different set of problems into view, due to practical difficulties in the way of maintaining the validity of the initial assumption of App. 1.15. Fig. App. 1.18 illustrates an example in which these difficulties were largely overcome (Rackham, 1965). The general arrangement is the same as in Figs. App. 1.3 and 1.5, the main differences being due to the use for the experimental chamber A and the thermobarometer B of cylindrical museum jars, A (of 1,810 cm³) being chosen so that the plant in its plastic pot (capacity 270 cm³) would just fit in conveniently without any part having to be bent. The thermobarometer B was of 1,600 cm³ capacity, close enough to equality of gas volume when allowance has been made for air space within the pot. The jars were closed by ground glass plates, Q, Q, sealed with vaseline and held down by horseshoe-shaped lead weights, W, W. Circular holes drilled in the plates carried rubber bungs, M, M, M, with tubing connections as in App. 1.5, and also sensitive thermometers, R, R, which could be read to 0·025°C, and which had their bulbs halfway up the two chambers. For ease of changing and cleaning, liquids in the two chambers (either both water, or both strong KOH) were contained in beakers, S, S, with crumpled filter papers to increase the liquid/air interface. Chamber A was enclosed in black polythene sheet. The two stirrers, T, T, in the large, lagged, water bath (deliberately without thermostat and heater, see below) were arranged so that the water circulated freely all around and between the chambers A and B. Otherwise the operation of the apparatus was as previously described [App. 1.4, App. 1.9].

To allow free water circulation around such large jars requires the connections (of thick-walled glass tubing, 3 mm bore) to be much longer than in the version of App. 1.5, so that the volume of those attached to chamber A was much larger than in App. 1.14, at approximately 3 cm³.

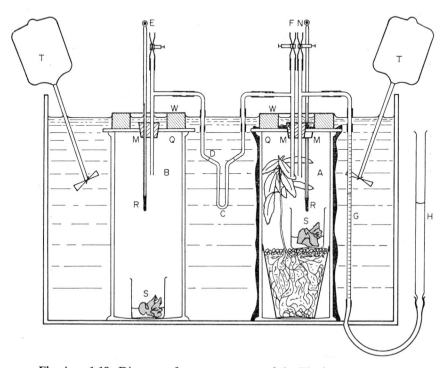

Fig. App. 1.18. Diagram of an arrangement of the Thoday Respirometer adapted for measurements on intact potted plants. A, B, cylindrical glass vessels with the tops ground flat, closed with glass plates Q, Q, sealed with vaseline and held down by lead weights W, W. The plates Q, Q, have circular holes closed by rubber bungs M, M, carrying thermometers R, R, and the usual glass tubing connections to the manometer C, the graduated pipette G, and the rubber and screw-clip closures E, F and N, as in Fig. App. 1.5. Liquids in A and B are contained in beakers S, S, with thick crumpled filter paper to increase the area of the liquid-gas interface. The horizontal glass connections between B, C, A and G are so arranged that C and G (and hence H) are all outside the water bath (adapted from Rackham, 1965).

Even so, V_a/V_b was approximately 0·002, an order of magnitude smaller, so that none of the corrections considered in App. 1.15–1.16 was of any greater importance than for an apparatus of the dimensions of App. 1.14.

App. 1.18

App. 1.19 Further precautions

The apparatus is, however, liable to a source of error which does not trouble users of the smaller version [App. 1.18], that of differential temperature effects between chambers A and B. In Rackham's experiments the rate of change of volume due to plant respiration was usually less than 1 cm^3 $hour^{-1}$, or approximately seven parts per 10,000 per hour. To avoid spurious readings it was therefore necessary that the difference between the temperature of chamber A and that of chamber B should change by less than one part in 10^4, or roughly 0·025°C, per hour. It was not, of course, necessary to control the temperature of the whole water bath system as closely as this, or even to ensure that both chambers were so accurately at the same temperature: the important thing is the rate of change of the temperature difference between them. It was for this reason that no attempt was made to control the bath temperature by means of a thermostat and heater, because of the risk of setting up undesirable temperature gradients.

In observations such as Rackham's, the pot and its contents have a substantial heat capacity. All this inevitably involves a considerable period of equilibration, which must be studied before the observations proper are begun.

There is also the question of the absorption of carbon dioxide in an alkali set. Both chambers A and B, when full of normal air, contain approximately 0·5 cm^3 of carbon dioxide, which is the reason for putting a beaker of alkali in both chambers equally.

In the first few hours after setting up, in a quiet corner of a north-facing room, Rackham found rapid fluctuations in gas volume, sometimes positive, sometimes negative, and initially often many times greater than those observed for the respiration of a plant. The rate of change soon decreased, and steady conditions were approached asymptotically, sometimes with one or two oscillations, in 6 hours or less. These differences in behaviour are no doubt due to differences in the overall temperature distribution at the time of setting up, which would be very difficult to specify with accuracy. Readings at less than 6 hours from setting up were therefore rejected.

Rackham (1965) also states: 'Apart from the transient effects, slow changes in the recorded volume were observed in blank runs. These are doubtless functions of changing temperature gradients in the apparatus,

App. 1.19

Plate App. 1.18. The Thoday respirometer arranged for work on whole potted plants as in Fig. App. 1.18. The datum line D on the manometer C is observed by a horizontal microscope. From Rackham (1965).

though there is some evidence for a correlation with atmospheric pressure. They do not seem to be a periodic function of the time of day. Attempts to correlate them with the recorded temperatures were unsuccessful. It might be possible to account for these fluctuations by more elaborate temperature recording, but this would add greatly to the complexity of the method. The procedure adopted was to treat these changes as errors, partly random and partly systematic, and to assign a mean and limits to them by pooling the observations from 15 days of blank runs.'

Taken together, these observations underline the overriding importance of uniformity of temperature when using an apparatus of this kind with large chambers. If there is available a cellar, which could be closed and left unoccupied in the intervals between readings, the difficulties would be greatly reduced, and the use of a constant-temperature room might well eliminate altogether those connected with long periods of readings.

App. 1.20 Root respiration

Quite apart from the idiosyncrasies of the measuring apparatus, when dealing with any system so complex as plant roots in a pot of 'soil', a relatively long time must elapse before a steady state is established [12.5], even if there is no question of soil respiration. If a whole plant has been enclosed in a chamber such as that shown in Fig. App. 1.18, large changes may then have taken place in the respiration rates of the above-ground parts [12.13] before a measurement can be made of the respiration of the plant as a whole. In such cases it is better to arrange to measure the respiration rates of the above-ground parts separately, by sealing across the top of the pot. It is then necessary either to establish, by calculation or otherwise, that there is an adequate oxygen supply in the sealed-off part to last the root system for the duration of the observations without risk of the establishment of anaerobiosis, or to supply ventilation.

A further reason for the separate study of the respiration of roots and tops is the high ionic exchange capacity of many inorganic soil substitutes such as vermiculite. This provides a very large effective store (E of Fig. 12.5) with which exchanges take place from the air spaces of the 'soil', and the time for the establishment of a steady state may then be greatly extended, to periods of the order of a day. Under these conditions root and top respiration may have to be studied separately if the results are to

App. 1.20

have any meaning, and for the roots it is best to use a gas current method, where (the roots normally being darkened) observations can be extended over many days under conditions approximating to normality. It is also worth ensuring that the storage capacity of the 'soil' for carbon dioxide is reduced as far as possible (e.g. by growing the plant in silver sand, with culture solution), especially if the use of a closed system such as the Thoday respirometer proves unavoidable.

App. 1.20

APPENDIX 2

An expanded version of Table 16.12. Percentage differences between mean values of unit leaf rate calculated on the assumption that $W = h + kL_A^n$ and the corresponding value for the assumption that $W = r + sL_A^2$, $[100(\bar{E}_n - \bar{E}_2)/\bar{E}_2]$, as a function of n and of P, the ratio of total leaf area at the second harvest to that at the first. Note that for any particular value of P the tabulated differences are identical with those for P^{-1}, so that for values of $P < 1$ the column corresponding to P^{-1} is used, i.e. always the ratio of the larger to the smaller leaf area, whether leaf area is increasing with time or not. For details of applications and means of calculating or estimating values of n, see sections 16.12–16.16.

P	1.200	1.400	1.600	1.800	2.000	2.200	2.400	2.600
n								
0.10	0.527	1.802	3.532	5.556	7.773	10.121	12.559	15.059
0.20	0.499	1.706	3.342	5.253	7.344	9.556	11.848	14.195
0.30	0.471	1.610	3.153	4.951	6.917	8.992	11.140	13.335
0.40	0.444	1.514	2.963	4.650	6.491	8.431	10.435	12.480
0.50	0.416	1.419	2.774	4.350	6.066	7.872	9.735	11.631
0.60	0.388	1.323	2.585	4.050	5.643	7.316	9.038	10.788
0.70	0.360	1.227	2.397	3.752	5.222	6.763	8.347	9.953
0.80	0.332	1.132	2.209	3.454	4.803	6.214	7.661	9.125
0.90	0.305	1.037	2.021	3.158	4.386	5.669	6.981	8.305
1.00	0.277	0.942	1.834	2.863	3.972	5.128	6.307	7.495

P	2.800	3.000	3.200	3.400	3.600	3.800	4.000	4.200
n								
0.10	17.601	20.171	22.759	25.356	27.958	30.561	33.160	35.754
0.20	16.578	18.983	21.401	23.825	26.249	28.670	31.085	33.491
0.30	15.560	17.803	20.053	22.305	24.555	26.797	29.031	31.253
0.40	14.550	16.631	18.717	20.800	22.878	24.945	27.001	29.044
0.50	13.547	15.470	17.394	19.312	21.221	23.117	25.000	26.867
0.60	12.553	14.320	16.085	17.841	19.586	21.316	23.030	24.727
0.70	11.568	13.184	14.793	16.391	17.976	19.544	21.095	22.627
0.80	10.594	12.061	13.518	14.963	16.392	17.804	19.197	20.571
0.90	9.632	10.953	12.263	13.559	14.838	16.098	17.340	18.561
1.00	8.682	9.861	11.028	12.179	13.313	14.429	15.525	16.601

P	4.400	4.600	4.800	5.000	5.200	5.400	5.600	5.800
n								
0.10	38.341	40.920	43.488	46.047	48.594	51.130	53.655	56.167
0.20	35.888	38.273	40.647	43.007	45.355	47.689	50.009	52.316
0.30	33.463	35.659	37.841	40.009	42.161	44.298	46.420	48.527
0.40	31.071	33.083	35.079	37.059	39.022	40.968	42.898	44.811
0.50	28.717	30.551	32.366	34.164	35.944	37.706	39.450	41.177
0.60	26.406	28.066	29.708	31.330	32.934	34.519	36.086	37.635
0.70	24.141	25.634	27.109	28.563	29.999	31.415	32.812	34.191
0.80	21.925	23.259	24.573	25.868	27.143	28.398	29.635	30.853
0.90	19.763	20.945	22.106	23.248	24.371	25.474	26.559	27.626
1.00	17.657	18.693	19.710	20.708	21.687	22.647	23.590	24.515

P	6.000	6.250	6.500	6.750	7.000	7.250	7.500	7.750
n								
0.10	58.668	61.776	64.867	67.938	70.991	74.026	77.044	80.043
0.20	54.609	57.456	60.282	63.088	65.872	68.637	71.381	74.106
0.30	50.619	53.212	55.782	58.329	60.854	63.357	65.839	68.299
0.40	46.708	49.056	51.379	53.678	55.954	58.206	60.435	62.642
0.50	42.887	45.000	47.087	49.149	51.186	53.199	55.188	57.155
0.60	39.166	41.054	42.917	44.753	46.564	48.350	50.113	51.852
0.70	35.552	37.228	38.878	40.501	42.099	43.672	45.221	46.748
0.80	32.054	33.529	34.978	36.401	37.800	39.174	40.525	41.853
0.90	28.676	29.963	31.225	32.462	33.674	34.864	36.031	37.176
1.00	25.423	26.535	27.623	28.687	29.727	30.746	31.744	32.721

P	8.000	8.250	8.500	8.750	9.000	9.500	10.000
n							
0.10	83.026	85.991	88.940	91.873	94.790	100.577	106.305
0.20	76.813	79.500	82.170	84.821	87.456	92.674	97.827
0.30	70.739	73.160	75.560	77.942	80.305	84.978	89.581
0.40	64.828	66.993	69.137	71.262	73.367	77.521	81.604
0.50	59.099	61.022	62.924	64.805	66.667	70.332	73.925
0.60	53.569	55.264	56.938	58.592	60.225	63.435	66.571
0.70	48.252	49.735	51.197	52.638	54.060	56.847	59.562
0.80	43.160	44.446	45.711	46.957	48.183	50.582	52.911
0.90	38.300	39.405	40.490	41.556	42.604	44.648	46.627
1.00	33.678	34.617	35.538	36.440	37.327	39.050	40.714

App. 2

P	1.200	1.400	1.600	1.800	2.000	2.200	2.400	2.600
n								
1.10	0.249	0.847	1.648	2.569	3.560	4.591	5.640	6.694
1.20	0.221	0.752	1.462	2.277	3.152	4.059	4.980	5.904
1.30	0.194	0.657	1.276	1.986	2.746	3.532	4.328	5.124
1.40	0.166	0.563	1.092	1.697	2.343	3.010	3.683	4.355
1.50	0.138	0.468	0.908	1.409	1.943	2.493	3.047	3.598
1.60	0.111	0.374	0.725	1.123	1.547	1.982	2.419	2.853
1.70	0.083	0.280	0.542	0.840	1.155	1.477	1.801	2.120
1.80	0.055	0.187	0.361	0.558	0.766	0.979	1.191	1.401
1.90	0.028	0.093	0.180	0.278	0.381	0.486	0.591	0.694
2.00	-0.000	0.000	-0.000	0.000	-0.000	-0.000	-0.000	0.000
2.10	-0.028	-0.093	-0.179	-0.276	-0.377	-0.479	-0.581	-0.680
2.20	-0.055	-0.186	-0.357	-0.549	-0.750	-0.952	-1.152	-1.347
2.30	-0.083	-0.278	-0.534	-0.820	-1.118	-1.418	-1.713	-1.999
2.40	-0.110	-0.370	-0.710	-1.089	-1.482	-1.877	-2.264	-2.638
2.50	-0.138	-0.462	-0.885	-1.355	-1.842	-2.328	-2.804	-3.263
2.60	-0.165	-0.554	-1.059	-1.619	-2.197	-2.773	-3.334	-3.874
2.70	-0.193	-0.645	-1.232	-1.880	-2.548	-3.210	-3.854	-4.471
2.80	-0.220	-0.736	-1.404	-2.139	-2.894	-3.640	-4.363	-5.054
2.90	-0.247	-0.827	-1.575	-2.395	-3.235	-4.063	-4.861	-5.623
3.00	-0.275	-0.917	-1.744	-2.649	-3.571	-4.478	-5.349	-6.178
3.20	-0.329	-1.097	-2.080	-3.148	-4.230	-5.285	-6.294	-7.247
3.40	-0.384	-1.276	-2.410	-3.636	-4.869	-6.063	-7.197	-8.261
3.60	-0.438	-1.453	-2.736	-4.113	-5.488	-6.811	-8.060	-9.224
3.80	-0.492	-1.628	-3.056	-4.578	-6.087	-7.530	-8.882	-10.135
4.00	-0.546	-1.802	-3.371	-5.031	-6.667	-8.219	-9.665	-10.997
4.20	-0.600	-1.974	-3.680	-5.473	-7.227	-8.880	-10.410	-11.811
4.40	-0.654	-2.145	-3.984	-5.904	-7.767	-9.513	-11.118	-12.580
4.60	-0.708	-2.313	-4.282	-6.322	-8.288	-10.118	-11.790	-13.306
4.80	-0.761	-2.481	-4.575	-6.729	-8.791	-10.697	-12.429	-13.991
5.00	-0.814	-2.646	-4.862	-7.124	-9.274	-11.250	-13.035	-14.637

P	2.800	3.000	3.200	3.400	3.600	3.800	4.000	4.200
n								
1.10	7.745	8.787	9.815	10.827	11.821	12.797	13.754	14.691
1.20	6.822	7.730	8.624	9.502	10.362	11.205	12.029	12.834
1.30	5.913	6.692	7.456	8.205	8.938	9.653	10.351	11.032
1.40	5.020	5.673	6.313	6.939	7.549	8.144	8.722	9.285
1.50	4.142	4.675	5.195	5.703	6.197	6.677	7.143	7.595
1.60	3.280	3.697	4.103	4.498	4.882	5.253	5.614	5.962
1.70	2.434	2.740	3.037	3.325	3.604	3.874	4.135	4.386
1.80	1.605	1.805	1.998	2.185	2.365	2.539	2.706	2.868
1.90	0.794	0.891	0.985	1.076	1.163	1.247	1.328	1.406
2.00	-0.000	-0.000	-0.000	-0.000	-0.000	-0.000	-0.000	-0.000
2.10	-0.776	-0.869	-0.958	-1.044	-1.125	-1.204	-1.279	-1.350
2.20	-1.535	-1.716	-1.889	-2.055	-2.213	-2.364	-2.508	-2.646
2.30	-2.275	-2.540	-2.793	-3.034	-3.264	-3.482	-3.690	-3.888
2.40	-2.998	-3.342	-3.670	-3.981	-4.277	-4.558	-4.825	-5.078
2.50	-3.703	-4.122	-4.520	-4.897	-5.255	-5.593	-5.914	-6.218
2.60	-4.389	-4.879	-5.343	-5.782	-6.196	-6.588	-6.958	-7.308
2.70	-5.058	-5.615	-6.140	-6.636	-7.103	-7.544	-7.959	-8.351
2.80	-5.709	-6.329	-6.912	-7.461	-7.977	-8.462	-8.918	-9.347
2.90	-6.343	-7.021	-7.658	-8.256	-8.816	-9.342	-9.836	-10.299
3.00	-6.959	-7.692	-8.380	-9.023	-9.624	-10.187	-10.714	-11.208
3.20	-8.139	-8.973	-9.750	-10.473	-11.147	-11.774	-12.359	-12.905
3.40	-9.253	-10.174	-11.028	-11.818	-12.552	-13.232	-13.864	-14.451
3.60	-10.302	-11.298	-12.217	-13.065	-13.847	-14.570	-15.239	-15.860
3.80	-11.289	-12.350	-13.324	-14.219	-15.041	-15.798	-16.497	-17.143
4.00	-12.217	-13.333	-14.353	-15.287	-16.141	-16.926	-17.647	-18.312
4.20	-13.089	-14.252	-15.310	-16.275	-17.155	-17.961	-18.699	-19.379
4.40	-13.907	-15.109	-16.199	-17.189	-18.090	-18.912	-19.663	-20.353
4.60	-14.675	-15.910	-17.026	-18.035	-18.952	-19.786	-20.547	-21.244
4.80	-15.395	-16.657	-17.794	-18.819	-19.748	-20.590	-21.358	-22.060
5.00	-16.071	-17.355	-18.508	-19.546	-20.483	-21.332	-22.104	-22.809

App. 2

P	4.400	4.600	4.800	5.000	5.200	5.400	5.600	5.800
n								
1.10	15.609	16.508	17.388	18.250	19.094	19.920	20.729	21.522
1.20	13.621	14.391	15.142	15.876	16.594	17.295	17.980	18.651
1.30	11.696	12.343	12.974	13.589	14.189	14.774	15.344	15.901
1.40	9.833	10.366	10.884	11.388	11.879	12.357	12.822	13.274
1.50	8.034	8.460	8.874	9.275	9.665	10.044	10.412	10.769
1.60	6.300	6.627	6.943	7.250	7.547	7.835	8.114	8.385
1.70	4.629	4.864	5.091	5.311	5.523	5.728	5.926	6.118
1.80	3.023	3.173	3.318	3.457	3.591	3.721	3.847	3.968
1.90	1.480	1.552	1.621	1.687	1.751	1.813	1.872	1.929
2.00	-0.000	-0.000	-0.000	-0.000	0.000	-0.000	-0.000	-0.000
2.10	-1.419	-1.484	-1.547	-1.607	-1.665	-1.720	-1.774	-1.825
2.20	-2.777	-2.903	-3.022	-3.137	-3.247	-3.352	-3.452	-3.548
2.30	-4.077	-4.257	-4.428	-4.591	-4.748	-4.897	-5.039	-5.176
2.40	-5.319	-5.548	-5.766	-5.973	-6.171	-6.360	-6.540	-6.712
2.50	-6.506	-6.779	-7.039	-7.286	-7.520	-7.744	-7.957	-8.160
2.60	-7.639	-7.953	-8.250	-8.531	-8.799	-9.053	-9.295	-9.525
2.70	-8.720	-9.070	-9.400	-9.713	-10.009	-10.291	-10.558	-10.812
2.80	-9.752	-10.133	-10.493	-10.833	-11.155	-11.460	-11.750	-12.024
2.90	-10.735	-11.145	-11.531	-11.896	-12.240	-12.566	-12.874	-13.167
3.00	-11.672	-12.108	-12.517	-12.903	-13.267	-13.611	-13.936	-14.243
3.20	-13.416	-13.894	-14.342	-14.763	-15.159	-15.532	-15.883	-16.215
3.40	-14.999	-15.510	-15.988	-16.435	-16.854	-17.248	-17.619	-17.968
3.60	-16.436	-16.972	-17.472	-17.938	-18.375	-18.784	-19.168	-19.530
3.80	-17.740	-18.295	-18.811	-19.292	-19.741	-20.161	-20.554	-20.924
4.00	-18.926	-19.495	-20.022	-20.513	-20.970	-21.397	-21.796	-22.171
4.20	-20.005	-20.583	-21.118	-21.616	-22.078	-22.510	-22.913	-23.290
4.40	-20.987	-21.572	-22.113	-22.614	-23.080	-23.514	-23.919	-24.297
4.60	-21.884	-22.473	-23.017	-23.520	-23.988	-24.422	-24.828	-25.206
4.80	-22.703	-23.295	-23 840	-24.344	-24.812	-25.247	-25.652	-26.030
5.00	-23.454	-24.046	-24.592	-25.096	-25.563	-25.997	-26.401	-26.778

P	6.000	6.250	6.500	6.750	7.000	7.250	7.500	7.750
n								
1.10	22.299	23.249	24.175	25.079	25.962	26.824	27.667	28.491
1.20	19.306	20.106	20.884	21.642	22.380	23.100	23.801	24.486
1.30	16.445	17.107	17.749	18.374	18.981	19.571	20.145	20.705
1.40	13.716	14.251	14.771	15.274	15.762	16.236	16.696	17.143
1.50	11.117	11.538	11.946	12.340	12.721	13.090	13.448	13.795
1.60	8.648	8.965	9.272	9.568	9.853	10.129	10.396	10.655
1.70	6.305	6.529	6.745	6.953	7.154	7.347	7.534	7.714
1.80	4.085	4.226	4.361	4.491	4.616	4.737	4.852	4.964
1.90	1.985	2.051	2.115	2.175	2.234	2.290	2.344	2.396
2.00	-0.000	-0.000	-0.000	-0.000	-0.000	-0.000	-0.000	-0.000
2.10	-1.874	-1.932	-1.988	-2.042	-2.093	-2.142	-2.189	-2.233
2.20	-3.641	-3.751	-3.857	-3.957	-4.052	-4.144	-4.231	-4.314
2.30	-5.307	-5.463	-5.611	-5.752	-5.886	-6.014	-6.135	-6.252
2.40	-6.876	-7.072	-7.257	-7.433	-7.600	-7.760	-7.911	-8.056
2.50	-8.354	-8.584	-8.802	-9.008	-9.204	-9.390	-9.567	-9.735
2.60	-9.744	-10.005	-10.251	-10.483	-10.703	-10.912	-11.110	-11.298
2.70	-11.053	-11.340	-11.609	-11.864	-12.105	-12.333	-12.549	-12.754
2.80	-12.285	-12.593	-12.884	-13.157	-13.416	-13.660	-13.891	-14.110
2.90	-13.444	-13.772	-14.079	-14.369	-14.642	-14.900	-15.144	-15.374
3.00	-14.535	-14.879	-15.201	-15.504	-15.789	-16.059	-16.313	-16.553
3.20	-16.529	-16.897	-17.243	-17.566	-17.870	-18.156	-18.426	-18.680
3.40	-18.297	-18.684	-19.045	-19.382	-19.699	-19.996	-20.275	-20.539
3.60	-19.870	-20.268	-20.639	-20.986	-21.310	-21.614	-21.900	-22.168
3.80	-21.271	-21.677	-22.054	-22.406	-22.735	-23.043	-23.332	-23.603
4.00	-22.523	-22.933	-23.314	-23.669	-24.000	-24.310	-24.600	-24.872
4.20	-23.644	-24.057	-24.439	-24.795	-25.127	-25.437	-25.727	-25.999
4.40	-24.652	-25.065	-25.448	-25.804	-26.135	-26.445	-26.734	-27.005
4.60	-25.561	-25.974	-26.356	-26.711	-27.041	-27.349	-27.637	-27.907
4.80	-26.384	-26.795	-27.176	-27.529	-27.858	-28.164	-28.450	-28.719
5.00	-27.130	-27.540	-27.919	-28.270	-28.597	-28.902	-29.186	-29.453

App. 2

P	8.000	8.250	8.500	8.750	9.000	9.500	10.000	∞
n								
1.10	29.297	30.085	30.856	31.612	32.352	33.788	35.168	450.000
1.20	25.154	25.807	26.445	27.068	27.677	28.856	29.986	200.000
1.30	21.250	21.781	22.298	22.804	23.296	24.248	25.156	116.667
1.40	17.578	18.000	18.412	18.812	19.202	19.954	20.668	75.000
1.50	14.132	14.459	14.776	15.085	15.385	15.961	16.507	50.000
1.60	10.905	11.147	11.382	11.610	11.831	12.255	12.655	33.333
1.70	7.888	8.057	8.219	8.377	8.530	8.822	9.097	21.429
1.80	5.072	5.176	5.276	5.373	5.467	5.646	5.814	12.500
1.90	2.446	2.494	2.540	2.585	2.629	2.711	2.788	5.556
2.00	-0.000	-0.000	-0.000	-0.000	-0.000	-0.000	-0.000	0.000
2.10	-2.276	-2.318	-2.357	-2.396	-2.432	-2.502	-2.567	-4.545
2.20	-4.394	-4.471	-4.544	-4.615	-4.682	-4.810	-4.929	-8.333
2.30	-6.363	-6.470	-6.572	-6.669	-6.763	-6.940	-7.104	-11.538
2.40	-8.194	-8.325	-8.452	-8.572	-8.688	-8.905	-9.106	-14.286
2.50	-9.895	-10.049	-10.195	-10.335	-10.468	-10.719	-10.950	-16.667
2.60	-11.478	-11.649	-11.812	-11.967	-12.116	-12.395	-12.651	-18.750
2.70	-12.949	-13.135	-13.312	-13.481	-13.642	-13.943	-14.219	-20.588
2.80	-14.319	-14.516	-14.705	-14.884	-15.055	-15.374	-15.667	-22.222
2.90	-15.593	-15.801	-15.999	-16.187	-16.366	-16.700	-17.005	-23.684
3.00	-16.781	-16.997	-17.202	-17.397	-17.582	-17.928	-18.243	-25.000
3.20	-18.921	-19.148	-19.364	-19.569	-19.764	-20.126	-20.455	-27.273
3.40	-20.787	-21.022	-21.244	-21.455	-21.655	-22.026	-22.363	-29.167
3.60	-22.421	-22.660	-22.886	-23.100	-23.303	-23.678	-24.018	-30.769
3.80	-23.859	-24.099	-24.327	-24.542	-24.746	-25.123	-25.464	-32.143
4.00	-25.128	-25.370	-25.597	-25.813	-26.016	-26.393	-26.733	-33.333
4.20	-26.255	-26.496	-26.723	-26.938	-27.141	-27.515	-27.853	-34.375
4.40	-27.260	-27.500	-27.726	-27.939	-28.141	-28.513	-28.849	-35.294
4.60	-28.160	-28.399	-28.623	-28.835	-29.036	-29.405	-29.738	-36.111
4.80	-28.970	-29.207	-29.430	-29.640	-29.839	-30.206	-30.536	-36.842
5.00	-29.703	-29.938	-30.159	-30.368	-30.565	-30.929	-31.256	-37.500

App. 2

APPENDIX 3

TAXONOMIC NOTES

App. 3.1 *Camellia sinensis*

The tea bushes of Assam are all inter-fertile, and because of extensive hybridization during the last century and a half, they form a complete series between what is regarded nowadays as the typical Assam plant, with relatively few, very large leaves held almost horizontally, and the typical China plant, with more, relatively much smaller leaves inclined upwards from the petiole to the tip. I have followed Purseglove (1968), who places all these variants in *Camellia sinensis* (L.) O. Kuntze, distinguishing the extreme plants as *C. sinensis* var. *sinensis* (at the China end) and *C. sinensis* var. *assamica* (Mast.) Pierre (at the Assam end).

App. 3.2 *Cannabis gigantea*

Bailey (1914, 1:657) places this under *C. sativa* L., commenting 'In gardens, the form known as *C. gigantea* is commonest; this reaches a height of 10 ft or more.' It seems likely that this was the variant grown by Hackenberg (1909), in which case Table 13.4 illustrates infraspecific variation.

App. 3.3 *Eupatorium adenophorum*

Knight (1922) does not make clear whether he was working on *Eupatorium adenophorum* Kunth (=*E. trapezoideum* Kunth) or *E. adenophorum* Spreng. (=*E. glandulosum* Humb., Bonpd & Kunth, not to be confused with *E. glandulosum* Michaux, which antedates it). However, *E. adenophorum* Spreng. was a well-known conservatory plant, figured by Hemsley (1907) in Curtis's Botanical Magazine (plate 8139), and the probabilities seem strongly in favour of this species as the one used by Knight.

App. 3.4 *Gossypium roseum*

Sir Joseph Hutchinson informs me that a cotton grown in Benares half a century ago under the name *Gossypium roseum* would almost certainly have been a well established annual variant of *G. arboreum* L.

App. 3.5 *Helianthus cucumerifolius* and *H. debilis*

1. Two North American floras which cover both these taxa are J. K. Small's *Manual of the South Eastern Flora* (1933) and M. L. Fernald's 8th Edit. of *Gray's Manual of Botany* (1950). Small treats the two as distinct species, while Fernald treats them as two varieties of *H. debilis*. In any case it seems clear that the two taxa are closely related.

2. However, '*H. cucumerifolius*' is widely grown in Central Europe as a decorative plant, and packets of seed so named are readily obtainable from seedsmen. A packet of such seed was purchased at Winterthur in N. Switzerland and grown in the Botanic Garden at Cambridge, where it proved to be a useful experimental plant. Dr P. F. Yeo examined these plants, and reported: 'In the light of the keys given by Fernald and Small, and of the descriptions given by Small, the characters of the plant used at Cambridge for study of growth physiology which support its determination as *H. debilis* are the unmottled stem, branched from the base, the scarcely swollen bases of the foliage hairs, the broadly cuneate to truncate leaf-bases, and the fact that the petioles are one-third to two-thirds as long as the blades. However, its erect stems and branches clash with Small's description, and its remotely or sinuately but acutely dentate leaf-margins clash with Fernald's. Therefore, though the plant has been determined as *H. debilis* Nutt. sensu stricto (*H. debilis* Nutt. var. *debilis*), it is not entirely typical of that taxon.' Specimens of this plant, which is the one referred to whenever *H. debilis* has been mentioned in the chapters above, are preserved in the herbarium of the Botanic Garden, Cambridge (CGG, 1968).

3. This does, however, raise the question of the taxonomy of the *H. cucumerifolius* grown by Gressler at Bonn in 1907 [Fig. 26.6]. The high reputation of the Botanical Institute and Garden there establishes a probability in favour of correct identification, in which case Gressler's plants would belong to a different taxon from the plants grown at Cambridge; but at the same time, following Fernald, it would nevertheless have been a variety of *H. debilis*.

App. 3.6 *Helianthus macrophyllus*

Dr Yeo notes '*H. macrophyllus* Willd. is usually reduced to synonymy of the perennial *H. strumosus* L. or to a variety of it (var. *macrophyllus*

(Willd.) Britton). Hayek in Hegi's *Illustrierte Flora von Mitteleuropa* (1918) describes an *H. strumosus* var. *willdenowianus* Thellung (*H. macrophyllus* var *sativus* Graebner) as a form cultivated in C. Europe for its edible tubers and as a bee plant. Fletcher and Taylor, in Bailey's *Standard Cyclopedia of Horticulture* (1915), say that *H. macrophyllus* var. *sativus* of gardens might be either *H. strumosus* var. *macrophyllus* or *H. giganteus* var. *subtuberosus* (Bourgeau) Britton.' There is therefore some doubt as to precisely what taxon Gressler was using [Fig. 26.6].

App. 3.6

APPENDIX 4

NOTATION

App. 4.1

One of the difficulties of presenting as a consistent whole a subject with a literature spread so widely in time and space has been devising a notation. This should retain as many as possible of the familiar conventions, for the convenience of established workers; while for the convenience of those new to the field, there should be simple and consistent rules governing the meaning of an individual symbol. As a first step towards conveying as much meaning as possible in this way, each of the main subdivisions of the notation is set in a distinct type.

App. 4.2 Sub-division

	Type
4.2.1. Contractions formed of the initial letters of frequently used names of derivates, such as RGR for relative growth rate.	Small roman capitals.
4.2.2. Measured quantities	Italic capitals.
4.2.3. Derived quantities	Bold capitals.
4.2.4. Parameters of equations	Bold, lower case.
4.2.5. Conventional signs and mathematical symbols.	Roman and lower case italic.

App. 4.3

Consistency within this framework has been maintained as far as possible, and this has necessitated the use of T for time, in breach of a common convention. Even so, some few exceptions are unavoidable. These include the conventional capital Δ for a difference between two similar measured quantities, the only occurrence of a Greek letter in the main text. The difficulty of a symbol for temperature (for which conventionally both t and T are used) was overcome in Appendix 1, where alone it occurs, by the use of θ. Special capitals are used for the notional quantities (A), (B), (C), (D) of 13.11, which are not in practice measurable and cannot be derived simply and directly, but must be inferred by a variety of roundabout routes.

App. 4.4

It is often necessary to deal with series of values, both of measured and derived quantities, and occasionally of parameters of equations. To indicate such a series, subscripts in the prefix position are invariably used. Thus in a series of n harvests, the times of individual harvests would be written $_1T, _2T, \dots _xT, \dots _nT$, where $_xT$ is the time of the xth harvest, and the corresponding total plant dry weights would be $_1W, _2W, \dots _xW, \dots _nW$.

App. 4.5

When a mean value of a derived quantity, say relative growth rate, $\mathbf{R}$, has been calculated over a period of time, it is given a bar, $\bar{\mathbf{R}}$. Members of a sequence of such mean values are similarly distinguished by a prefix subscript, the mean value for the xth period being $_x\bar{\mathbf{R}}$.

App. 4.6

This leaves the subscript suffix position free for distinguishing letters or numerals. (Elaborate suffixes have been avoided, the only exceptions being self-evident cases where the context makes the meaning immediately plain, as X_{end}, X_{emb}, for the measured oxygen intake by endosperm and embryos respectively, in App. 1.10.) These distinctions may be of three kinds:

4.6.1. a simple qualification, thus L, leaf, L_w, total leaf dry weight per plant; L_A, total leaf area per plant.

4.6.2. a qualification involving the amount of some substance, or the rate of some process, as W_C, the carbon dioxide equivalent of unit dry weight, or W_R, the respiration rate per unit dry weight, or A_N the specific absorption rate for nitrogen.

4.6.3. a qualification involving some other type of difference, such as the method of computation used to arrive at an estimate of a derived quantity: thus; $\bar{\mathbf{E}}_n$, $\bar{\mathbf{E}}$ calculated on the assumption that $W = \mathbf{h} + \mathbf{k}L_A^n$; $\bar{\mathbf{E}}_2$, $\bar{\mathbf{E}}$ calculated on the basis that $\mathbf{n}$ in this equation equals 2, and so on.

App. 4.6

App. 4.7

Because it is necessary to combine individual equations in various ways, it seemed that consistency could best be maintained, and confusion avoided, by always using the same letter for a particular parameter in a particular equation. However, the whole alphabet is not available, because for this purpose it is convenient to avoid some letters (e.g. e, i, o, x, y). Although use is made in all of 25 parameters, 8 of them occur in one section only [20.4]. By using z_1–z_8 in this section, it has been possible otherwise to avoid the use of suffixes for parameters.

App. 4.8

The theory of the Thoday respirometer in Appendix 1 involves a complex series of interrelated measurements (of quantities defined in App. 1.13). Here the system of distinctive subscripts of App. 4. 4–7 does not suffice, and the necessary further distinction between closely related quantities is supplied by a superscript, as V_a, V_a'; Z_{emb}, Z_{emb}'.

App. 4.9 Contractions

References are to sections where the concept is defined, or first discussed.

CDR	Carbon Dry-weight Rate	13.11
LAD	Leaf Area Duration	13.19
LAI	Leaf Area Index	13.18
LAR	Leaf Area Ratio	13.7
LLCB	Light-phase Leaf Carbon Balance	13.11
LWCB	Light-phase Whole-plant Carbon Balance	13.11
LWR	Leaf Weight Ratio	13.16
NAR	Net Assimilation Rate	13.10
NCRR	Net Carbon Reduction Rate	13.11
RGR	Relative Growth Rate	13.6
RLGR	Relative Leaf Growth Rate	13.17
RWR	Root Weight Ratio	13.20
SAR	Specific Absorption Rate	13.20
SLA	Specific Leaf Area	13.16
SQ	Substance Quotient	13.4
SWR	Stem Weight Ratio	21.1
ULR	Unit Leaf Rate	13.7

App. 4.10 Measured quantities

The list which follows is confined to those quantities for which symbols are needed in equations used in the text above. These examples show that the system of notation can readily be extended; for example, C is available for cotyledons, and suffixes are available to denote contents of various minerals, as K, N, P (which is why the notional 'total photosynthetic machinery per plant' has been denoted L_p).

Symbol	Instantaneous value at time $_1T$	Mean value over a period	Mean value over the xth period	
H				Number of hours in the light-phase. 29.3
L_A	$_1L_A$	$\bar{L}_A$	$_x\bar{L}_A$	Total leaf area per plant
L_F				Total leaf fresh weight per plant. 24.21.
L_p				Total photosynthetic machinery per plant. 13.16
L_W	$_1L_W$	$\bar{L}_W$	$_x\bar{L}_W$	Total leaf dry weight per plant
M				Number of weeks in the growing season. 13.19
N				Number of plants per m² ground area. 13.18
N^{-1}				m² ground area per plant. 13.18
P	$_1P$			Gas pressure ⎫ In the theory of the Thoday Respirometer, App. 1.13
P_a	$_1P_a$			The same, excluding pressure of water vapour ⎬
P_w	$_1P_w$			Water vapour pressure ⎭
R_W	$_1R_W$			Total root dry weight per plant
S_W	$_1S_W$			Total stem dry weight per plant
T				Time
ΔT				A period of time

Symbol	Instantaneous value at time $_1T$	Mean value over a period	Mean value over the xth period	
θ	$_1\theta$			Temperature, °K ($=°C + 273°$)
θ_a	$_1\theta_a$			Air temperature — In the theory of the Thoday Respirometer, App. 1.13
θ_b	$_1\theta_b$			Water bath temperature
V				Volume. For V_a, V_a', V_b, V_b', V_g, V_1, V_p, see the theory of the Thoday Respirometer, App. 1.13 et seq.
W	$_1W$	$\overline{W}$	$_x\overline{W}$	Total dry weight per plant
$_MW$				Total dry weight per plant at the end of the growing season. 13.19
W_C				Mean carbon dioxide equivalent per unit dry weight of the whole plant. 13.11, 29.4
W_R				Mean respiration rate per unit dry weight of the whole plant. 28.11, 29.3
X, Y, Z, Z'				Volume changes in the operation of the Thoday Respirometer, App. 1.2;
X				in the 'alkali set';
Y				in the 'water set';
Z				due to carbon dioxide given out ($=Y-X$);
Z'				total carbon dioxide output ($=Z$, corrected for solubility).
X_{emb}				Volume changes due to respiration of embryos and endosperm of dissected germinating seeds. App. 1.10
X_{end}				

App. 4.10

App. 4.11　Derived quantities

Symbol	Instantaneous value at time $_1T$	Mean value over a period	Mean value over the xth period	
A				Specific absorption rate. (13.20)
A_N				The same, for nitrogen. (13.20)
D				Difference between changes in water-bath and air temperatures (theory of the Thoday Respirometer, App. 1.13)
E	$_1E$	$\bar{E}$ $\bar{E}_n$	$_x\bar{E}$	Unit leaf rate The same, calculated on the assumption that $W = \mathbf{h} + kL_A^n$
		$\bar{E}_1, \bar{E}_2$		The same, for $\mathbf{n} = 1, 2$
F				Correction factor for changes in partial pressure of water vapour (theory of the Thoday Respirometer, App. 1.13, 1.15)
P				Proportional increase in leaf area between harvests 1 and 2, $= {_2L_A}/{_1L_A}$ (16.10–16.14)
Q				The same, between harvests 1 and 3 $= {_3L_A}/{_1L_A}$ (16.14)
R	$_1R$	$\bar{R}$	$_x\bar{R}$	Relative growth rate
R_L	$_1R_L$	$\bar{R}_L$	$_x\bar{R}_L$	Relative leaf growth rate
S				'Surplus' dry weight. 22.8
Y				Proportional increase in dry weight $= {_2W}/{_1W}$, 28.4 (=Substanzquotient of Noll, 13.4)
Z				$({_3W} - {_2W})/({_2W} - {_1W})$, used in estimating the value of $\mathbf{n}$. 16.14

App. 4.11

App. 4.12 Parameters of equations

	Equation or Section
a	20.2.1.
b	20.2.1.
c	16.8.1; 16.11; 16.15.
d	16.8.1; 16.11; 16.15.
f	20.2.2.
g	20.2.2.
h	16.11; 16.12.1; 16.14; 16.16; 20.5; 22.3.
k	16.11; 16.12.1; 16.14; 16.16; 20.5; 22.3.
m	17.6.
n	16.12.1–16.12.3; 16.13; 16.14.1, 16.14.2; 16.15; 16.16; 20.5; 22.3; 22.5; 22.8.2.
p	16.4; 16.11.
q	16.4; 16.11.
r	16.9.1; 16.11; 16.12.
s	16.9.1; 16.11; 16.12.
u	16.3; 16.11.
v	16.3; 16.11.
w	24.12.1, 24.12.2.
$z_1, z_2, z_3, z_4,$ $z_5, z_6, z_7, z_8.$	20.4.1, 20.4.2.

BIBLIOGRAPHY

As it is hoped that this book may be useful to those interested in plant growth who have not had a formal training in the plant and environmental sciences, marginal letters a–f distinguish a short selection of accounts (many of them introductory accounts) of the following subjects, with an entry to the literature:

(a) genetic make-up and behaviour of higher plant species in nature;
(b) statistics and statistical aspects of experimental design;
(c) soils considered as biological systems;
(d) mineral nutrition, artificial rooting media, culture solutions, etc.;
(e) control of the aerial environment;
(f) meteorological and micrometeorological instruments and measurements.

Marginal letters g–q distinguish a short selection of accounts of aspects of the study of light in relation to vegetation, especially to woodlands and forests:

(g) general review;
(h) instruments and methods;
(i) measurements—daily march;
(j)   —daily totals;
(k)   —spectral composition;
(l)   —sunflecks;
(m)   —spatial variation, horizontal;
(n)   —spatial variation, vertical;
(o)   —seasonal variations;
(p)   —under expedition conditions;
(q) photographic studies of canopies.

References in the right hand margin are to sections, Figures being indicated in bold type, and Tables in italic.

hloq   ANDERSON M.C. (1964a). Studies of the woodland light climate. I. The photographic computation of light conditions. *J. Ecol.*, **52**, 27–41.

gh   ANDERSON M.C. (1964b). Light relations of terrestrial plant communities and their measurement. *Biol. Rev.*, **39**, 425–486.

hjloq   ANDERSON M.C. (1964c). Studies of the woodland light climate. II. Seasonal variation in the light climate. *J. Ecol.*, **52**, 643–663.

jmnoq   ANDERSON M.C. (1966). Some problems of simple characterization of the light climate in plant communities. In *Light as an Ecological Factor* (*Symp. Brit. Ecol. Soc.*, **6**, pp. 77–90). Blackwell Sci. Pubs., Oxford.   6.9, **6.9**

ARBER A. (1934). *The Gramineae. A study of cereal, bamboo and grass.* Cambridge Univ. Press, pp. xvii + 480.   23.3

ip   ASHTON P.S. (1958). Light intensity measurements in rain forest near Santarem, Brazil. *J. Ecol.*, **46**, 65–70.

AUDUS L.J. (1935). Mechanical stimulation and respiration rate in the cherry laurel. *New Phytol.*, **34**, 386–402.   12.4

AUDUS L.J. (1939). Mechanical stimulation and respiration in the green leaf. II. Investigations on a number of Angiospermic species. *New Phytol.*, **38**, 284–288.   12.4, 12.13

AUDUS L.J. (1959). Correlation. *J. Linn. Soc.* (Bot.), **56**, 177–187   26.1

BABINGTON C.C. (1843). *Manual of British Botany.* John van Voorst, London, pp. xxiv + 400.   10.2

BAILEY L.H. (1914). *Standard Cyclopedia of Horticulture*, Vol. I. Macmillan, New York, pp. xx + 602.   App. 3.2

b   BAILEY N.T.J. (1964). *Statistical Methods in Biology.* Second impression (with corrections). English Universities Press Ltd., London, pp. ix + 200.   5.2

f h-q   BAINBRIDGE R., EVANS G.C. & RACKHAM O. (eds.) (1966). *Light as an ecological factor* (*Symp. Brit. Ecol. Soc.*, **6**). Blackwell Sci. Pubs Oxford, pp. xi + 452.   30.5

BARKER J. (1935). A note on the effect of handling on the respiration of potatoes. *New Phytol.*, **34**, 407–408.   12.4

BARLOW H.W.B. (1970). Some aspects of morphogenesis in fruit trees. In *Physiology of Tree Crops* (ed. L.C. Luckwill & C.V. Cutting), pp. xiii + 382. Academic Press, London and New York.   26.1

BARTON L.V. (1961). *Seed preservation and longevity.* Leonard Hill, London; Interscience Publishers, New York, pp. xviii + 216.   5.7

BEAVEN E.S. (1947). *Barley. Fifty years of observation and experiment.* Duckworth, London, pp. xx + 394.   23.4

BEEVERS H. (1961). *Respiratory metabolism in plants.* Row, Peterson & Co., Evanston & New York, pp. xi + 232.   4.7

BINGHAM J. (1966). Varietal response in wheat to water supply in the field, and male sterility caused by a period of drought in a glasshouse experiment. *Ann. appl. Biol.,* **57,** 365–377.   23.10

a.   BIOLOGICAL FLORA OF THE BRITISH ISLES. *J. Ecol.:* for a list of accounts of species published up to 1962, see Index to vols 21–50. Others in subsequent volumes.

BLACKMAN F.F. (1905). Optima and limiting factors. *Ann. Bot.,* **19,** 281–295.   3.4

BLACKMAN G.E. & BLACK J.N. (1959). Physiological and ecological studies in the analysis of plant environment. XII. The role of the light factor in limiting growth. *Ann. Bot.,* N.S. **23,** 131–145.   3.3, 28.14

BLACKMAN G.E. & RUTTER A.J. (1946). Physiological and ecological studies in the analysis of plant environment. I. The light factor and the distribution of the bluebell (*Scilla non-scripta*) in woodland communities. *Ann. Bot.,* N.S. **10,** 361–390.   3.3, 10.5

BLACKMAN G.E. & RUTTER A.J. (1947). Physiological and ecological studies in the analysis of plant environment. II. The interaction between light intensity and mineral nutrient supply in the growth and development of the bluebell (*Scilla non-scripta*). *Ann. Bot.,* N.S., **11,** 125–158.   10.5

BLACKMAN G.E. & RUTTER A.J. (1948). Physiological and ecological studies in the analysis of plant environment. III. The interaction between light intensity and mineral nutrient supply in leaf development and in the net assimilation rate of the bluebell (*Scilla non-scripta*). *Ann. Bot.,* N.S., **12,** 1–27.   10.5

BLACKMAN G.E. & RUTTER A.J. (1949). Physiological and ecological studies in the analysis of plant environment. IV. The interaction between light intensity and mineral nutrient supply on the uptake of nutrients by the bluebell (*Scilla non-scripta*). *Ann. Bot.,* N.S., **13,** 453–489.   10.5

BLACKMAN G.E. & RUTTER A.J. (1950). Physiological and ecological studies in the analysis of plant environment. V. An assessment of the factors controlling the distribution of the bluebell (*Scilla non-scripta*) in different communities. *Ann. Bot.,* N.S., **14,** 487–520.   10.5

BLACKMAN G.E. & WILSON G.L. (1951a). Physiological and ecological studies in the analysis of plant environment. VI. The constancy for different species of a logarithmic relationship between net assimilation rate and light

intensity and its ecological significance. *Ann. Bot.*, N.S.,
15, 63–94. 3.3, 28.14

BLACKMAN G.E. & WILSON G.L. (1951b). Physiological and
ecological studies in the analysis of plant environment.
VII. An analysis of the differential effects of light
intensity on the net assimilation rate, leaf-area ratio,
and relative growth rate of different species. *Ann. Bot.*,
N.S., 15, 373–409. 3.3, 28.14

BLACKMAN V.H. (1919). The compound interest law and
plant growth. *Ann. Bot.*, 33, 353–360. 13.2, 13.4, 13.5

f   BLACKWELL M.J. (1953). *On the development of an improved
Robitzsch-type actinometer.* Met. Res. Cttee. MRP.
No 791. 10 pp. 30.6

fh   BLACKWELL M.J. (1966). Radiation meteorology in relation
to field work. In *Light as an Ecological Factor.* (*Symp.
Brit. Ecol. Soc.*, 6, 17–39). Blackwell Sci. Pubs, Oxford. 30.6

e   BOSIAN G. (1965). Control of conditions in the plant chamber:
fully automatic regulation of wind velocity, temperature
and relative humidity to conform to microclimatic field
conditions. In *Methodology of Plant Ecophysiology* (ed.
F.E. Eckardt), pp. 233–238. Unesco, Paris. 12.2

BRAILSFORD ROBERTSON T. (1908a). On the normal rate of
growth of an individual, and its biochemical significance.
Arch. EntwMech. Org., 25, 581–614. 13.1

BRAILSFORD ROBERTSON T. (1908b). Further remarks on the
normal rate of growth of an individual and its biochemical significance. *Arch. EntwMech. Org.*, 26,
108–118. 13.1

BRAILSFORD ROBERTSON T. (1910). Explanatory remarks
concerning the normal rate of growth of an individual
and its biochemical significance. *Biol. Zbl.*, 30, 316–320. 13.1

BRAY J.R. (1963). Root production and the estimation of
net productivity. *Can. J. Bot.*, 41, 65–72. 31.11

BRAY J.R. & GORHAM E. (1964). Litter production in forests
of the world. *Adv. Ecol. Res.*, 2, 101–157. 31.14, 31.18

a   BRIGGS D. & WALTERS S.M. (1969). *Plant variation and
evolution.* Weidenfeld and Nicolson, London, pp.
0 + 256.

BRIGGS G.E., HOPE A.B. & ROBERTSON R.N. (1961).
Electrolytes and plant cells. Blackwell Sci. Pubs,
Oxford, pp. x + 217. 5.6

BRIGGS G.E., KIDD F. & WEST C. (1920a). A quantitative
analysis of plant growth. Part I. *Ann. appl. Biol.*, 7,
103–123. 13.7, 13.10, 19.1, 32.9

BRIGGS G.E., KIDD F. & WEST C. (1920b). A quantitative
analysis of plant growth. Part II. *Ann. appl. Biol.*, 7,
202–223. 13.7, 13.10, 19.1, 20.6, 32.9

BUNCE R.G.H. (1968). Biomass and production of trees in a mixed deciduous woodland. I. Girth and height as parameters for the estimation of tree dry weight. *J. Ecol.*, **56**, 759–775.   31.8

BÜNNING E. (1948). *Entwicklungs- und Bewegungsphysiologie der Pflanze.* Springer-Verlag, Berlin, Göttingen, Heidelberg, pp. xii + 464.   5.7

BURG, S.P. (1962). The physiology of ethylene formation. *A. Rev. Pl. Physiol.*, **13**, 265–298.   12.6, 12.9, 26.1

BURKILL I.H. (1935). *A Dictionary of the economic products of the Malay Peninsula.* Vol. I. Crown Agents for the Colonies, London, pp. xi + 1220.   23.3

BUXTON P.A. & LEWIS D.J. (1934). Climate and tsetse flies: laboratory studies upon *Glossina submorsitans and tachinoides. Phil. Trans. R. Soc.*, B **224**, 175–240.   9.8

CANHAM A.E. (1957). Lamps for use in controlled environments. Range and characteristics of available lamps. In *Control of the plant environment* (ed. J.P. Hudson) pp. 207–210. Butterworth Sci. Publs, London.   **18.10**, 28.9

e CANHAM A.E. (1966). *Artificial light in horticulture.* Centrex Publ. Co., Eindhoven, pp. vii + 212.   9.10, 28.10

CHAILAKHYAN M.K. (1968). Internal factors to plant flowering. *A. Rev. Pl. Physiol.*, **19**, 1–36.   23.2

CHAPMAN S.B. (1971). A simple conductimetric soil respirometer for field use. *Oikos*, **22**, 347–352.   31.9

CHODAT R. (1911). *Principes de Botanique.* 2nd Edn. J.-B. Baillère et fils, Paris, pp. xi + 842.   13.2, 13.5

e CHOUARD P. & DE BILDERLING N. (1969). *Phytotronique.* Centre National de la Recherche Scientifique, Paris, pp. vii + 111.

CHURCHILL W.S. (1923). *The world crisis, 1911–1915.* Thornton Butterworth, London, 2 vols, pp. 536 + 557.   5.1

CLELAND R.E. (1969). The gibberellins. In *The Physiology of Plant Growth and Development* (ed. M.B. Wilkins), pp. 49–81. McGraw Hill, New York.   26.1

CLEMENTS F.E. & GOLDSMITH G.W. (1924). *The phytometer method in ecology. The plant and community as instruments.* Publs Carnegie Instn 356, pp. vi + 106.   30.3

CLEMENTS F.E. & LONG F.L. (1934). Factors in elongation and expansion under reduced light intensity. *Pl. Physiol., Lancaster*, **9**, 767–781.   30.3

CLEMENTS F.E. & LONG F.L. (1935). Further studies of elongation and expansion in *Helianthus annuus. Pl. Physiol., Lancaster*, **10**, 637–660.   30.3

CLEMENTS F.E., WEAVER, J.E. & HANSON H.C. (1929). *Plant competition*. Publs Carnegie Instn 398, pp. xvi + 340. 23.3, 30.3

CLYMO K.E. (1964). Studies in the autecology of *Calamagrostis epigeios* (L.) Roth and *C. canescens* (Weber) Roth. Ph.D. Thesis, Univ. Cambridge. 10.2

b COCHRAN W.G. & COX G.M. (1957). *Experimental designs*, 2nd Edn. John Wiley, New York, pp. xiv + 611.

COCKSHULL K.E. & HUGHES A.P. (1967). Distribution of dry matter to flowers in *Chrysanthemum morifolium*. *Nature, Lond.*, **215**, 780–781. 23.8

COCKSHULL K.E. & HUGHES A.P. (1968). Accumulation of dry matter by *Chrysanthemum morifolium* after flower removal. *Nature, Lond.* **217**, 979–980. 23.8, **23.9.1, 23.9.2**

COCKSHULL K.E. & HUGHES A.P. (1972). Flower formation in *Chrysanthemum morifolium*: the influence of light level. *J. hort. Sci.* (in the press). 23.8

fh COLLINGBOURNE R.H. (1966). General principles of radiation meteorology. In *Light as an Ecological Factor (Symp. Brit. Ecol. Soc.*, **6**, 1–15) Blackwell Sci. Pubs. Oxford.

COOMBE D.E. (1952). Plant growth and light in woodlands. Ph.D. Thesis, Univ. Cambridge. 28.3

COOMBE D.E. (1956). Biological flora of the British Isles, *Impatiens parviflora* DC. *J. Ecol.*, **44**, 701–713. **18.5**

hk COOMBE D.E. (1957). The spectral composition of shade light in woodlands. *J. Ecol.*, **45**, 823–830.

COOMBE D.E. (1960). An analysis of the growth of *Trema guineensis*. *J. Ecol.*, **48**, 219–231. 9.8, 16.10, *16.10*, 16.12

jo COOMBE D.E. (1966). The seasonal light climate and plant growth in a Cambridgeshire wood. In *Light as an Ecological Factor (Symp. Brit. Ecol. Soc.*, **6**, 148–166). Blackwell Sci. Pubs, Oxford. 3.3, 6.4, 7.8, 8.9, 27.9, **28.10, 28.11,** 28.11, 30.2, 30.6, 30.7, 30.8, **30.8**, 32.2

hpq COOMBE D.E. & EVANS G.C. (1960). Hemispherical photography in studies of plants. *Med. biol. Illust.*, **10**, 68–75. 6.10, 9.14

q COOMBE D.E. & HADFIELD W. (1962). An analysis of the growth of *Musanga cecropioides*. *J. Ecol.*, **50**, 221–234. 6.10, 9.9, 31.17, 31.20

CROCKER W. & BARTON L.V. (1953). *Physiology of seeds*. Chronica Botanica, Waltham, Mass, pp. xv + 267. 5.7

CURRY R.B. & CHEN L.H. (1970). Dynamic simulation of vegetative growth in a plant canopy. *Proc. 1970 Summer Simulation Conf., Denver, ACM/HARE/SCI*, New York, 737–745. 13.1, 28.19

DAINTY J. (1969). The water relations of plants. In *Physiology of Plant Growth and Development* (ed. M.B. Wilkins), pp. 421–452. McGraw-Hill, London.   24.1

DALE I.R. & GREENWAY P.J. (1961). *Kenya trees and shrubs.* Nairobi, Buchanan's Kenya Estates Ltd., pp. xxvii + 654.   23.3

DECKER J.P. (1954). The effect of light intensity on photo-synthetic rate in Scotch pine. *Pl. Physiol., Lancaster*, **29**, 305–306.   12.6

DEVAUX H. (1899). Asphyxie spontané et production d'alcool dans les tissus profonds des tiges poussant dans les conditions naturelles. *C. r. hebd. Séance. Acad. Sci., Paris*, **128**, 1346–1349.   12.4

DIXON M. (1951). Manometric methods as applied to the measurement of cell respiration and other processes. 3rd edn. Cambridge University Press, pp. xvi + 165.   App. 1.1

DONALD C.M. (1962). In search of yield. *J. Aust. Inst. agric. Sci.*, **28**, 171–178.   23.4

fh   ECKARDT F.E. (ed) (1965). *Methodology of plant ecophysiol-ogy.* Proc. Montpellier Symp. Arid Zone Res., **25**, pp. 17 (unnumbered) + 531. Unesco, Paris.   12.7

cd   ECOLOGICAL ASPECTS OF THE MINERAL NUTRITION OF PLANTS (1969). *British Ecological Society Symposium, No. 9.* Ed. Rorison, I.H., assisted by Bradshaw, A.D., Chad-wick, M.J., Jefferies, R.L., Jennings, D.H. and Tinker, P.B. Blackwell Sci. Pubs, Oxford, pp. xxi + 512.   10.8, 26.13

EGGELING W.J. & DALE I.R. (1951). *The indigenous trees of the Uganda Protectorate.* 2nd Edn. The Government Printer, Entebbe, and Crown Agents for the Colonies, pp. xxx + 491.   23.3

ENRIQUES P. (1909). Wachstum und seine analytische Darstellung. *Biol. Zbl.*, **29**, 331–352.   13.1

hiklp   EVANS G.C. (1939). Ecological studies on the rain forest of Southern Nigeria. II. The atmospheric environmental conditions. *J. Ecol.*, **27**, 436–482.   9.8, 12.8, 12.9

hjlmp   EVANS G.C. (1956). An area survey method of investigating the distribution of light intensity in woodlands, with particular reference to sunflecks. *J. Ecol.*, **44**, 391–428.   9.5

e   EVANS G.C. (1959). The design of equipment for producing accurate control of artificial aerial environments at low cost. *J. agric. Sci., Camb.*, **53**, 198–208.   7.3, 9.10, 17.4, 17.9

EVANS G.C. (1966a). Temperature gradients in tropical rain forest. *J. Ecol.*, **54**, 20P–21P.   9.6

ikloq   EVANS G.C. (1966b). Model and measurement in the study of woodland light climates. In *Light as an Ecological Factor*

(*Symp. Brit. Ecol. Soc.*, **6**, 53–76). Blackwell Sci. Pubs, Oxford. 30.7

hk EVANS G.C. (1969). The spectral composition of light in the field. I. Its measurement and ecological importance *J. Ecol.*, **57**, 109–125. **28.10, 28.11**, 30.5, 30.7, **30.8**

hpq EVANS G.C. & COOMBE D.E. (1959). Hemispherical and woodland canopy photography and the light climate. *J. Ecol.*, **47**, 103–113. 6.8, 9.14

EVANS G.C. & HUGHES A.P. (1961). Plant growth and the aerial environment. I. Effect of artificial shading on *Impatiens parviflora. New Phytol.*, **60**, 150–180 **3.4.1**, 3.4, 16.11, 16.12, 18.6, 18.10, *19.6*, **19.6, 21.2, 24.7**, 28.3, 29.4, 32.7

EVANS G.C. & HUGHES A.P. (1962). Plant growth and the aerial environment. III. On the computation of unit leaf rate. *New Phytol.*, **61**, 322–327. **4.5**, 16.12, *16.12*, **16.9.1, 16.13.1**, 28.3, *28.3*

ijlmp EVANS G.C., WHITMORE T.C. & WONG Y.K. (1960). The distribution of light reaching the ground vegetation in a tropical rain forest. *J. Ecol.*, **48**, 193–204.

e EVANS L.T. (ed.) (1963). *Environmental control of plant growth*. Academic Press, New York and London, pp. xvii + 449. 9.10

FAHEEMUDDIN M. (1969). Comparative biology of some fen plants (*Filipendula* and *Iris* spp.). Ph.D. Thesis, Univ. London. 5.9, 16.13

FERGUSON C.W. (1968). Bristlecone Pine: Science and Esthetics. *Science, N.Y.*, **159**, 839–846. 4.6

FERNALD M.L. (1950). *Gray's Manual of Botany*, 8th Edn. American Book Co., New York, pp. lxiv + 1632. App. 3.5

b FISHER R.A. (1966). *The design of experiments*, 8th Edn. Oliver & Boyd, London, pp. 264.

FLETCHER S.W. & TAYLOR N. (1915). *Helianthus*. In *Bailey, L.H. Standard Cyclopedia of Horticulture*, Vol. III, pp. 1201–1760. Macmillan, New York. App. 3.6

FOX J.E. (1969). The cytokinins. In *Physiology of Plant Growth and Development* (ed. M.B. Wilkins), pp. 85–123. McGraw-Hill, New York. 26.1

FURUYA M. (1968). Biochemistry and physiology of phyto-chrome. In *Progress in Phytochemistry* (eds L. Reinhold & Y. Liwschitz), **1**, 347–405. Interscience Publ., London, New York and Sydney. 26.1

GAASTRA P. (1959). Photosynthesis of crop plants as influ-enced by light, carbon dioxide, temperature and stomatal diffusion resistance. *Meded. LandbHoogesch. Wageningen*, **59**(13), 1–68. 29.6

c   GARRETT S.D. (1963). *Soil fungi and soil fertility*. Pergamon Press, Oxford, pp. viii + 165.

GERICKE F. (1908). Experimentelle Beiträge zur Wachstumsgeschichte von *Helianthus annuus*. *Z. Naturw.*, **80**, 321–363. 13.4

GLOCK W.S. (1937). *Principles and methods of tree ring analysis*. Pubs Carnegie Instn No. 486, 1–100. 4.6

GODWIN H. (1935). The effect of handling on the respiration of cherry laurel leaves. *New Phytol.*, **34**, 403–406. 12.4, **12.4.1**

GOODALL D.W. (1945). The distribution of dry weight change in young tomato plants. Dry weight changes of the various organs. *Ann. Bot.*, N.S. **9**, 101–139. 8.10

a   GOODMAN G.T., EDWARDS R.W. & LAMBERT J.M. (eds) (1965). *Ecology and the industrial society*. British Ecological Society Symposium No. 5. Blackwell Sci. Pubs, Oxford, pp. viii + 395. 30.3

GREGORY F.G. (1918). Physiological conditions in cucumber houses. *3rd Ann. Rep., Exptl Res. Stn, Nursery and Market Garden Industries Development Soc. Ltd.* Cheshunt, 1918, pp. 19–28. 13.9, 13.10, 16.2

GREGORY F.G. (1926). The effect of climatic conditions on the growth of barley. *Ann. Bot.*, **40**, 1–26. 13.10

GREIG-SMITH P. (1964). *Quantitative Plant Ecology*. 2nd Edn. Butterworths Sci. Pubs, London, pp. xii + 256. 13.18

GRESSLER P. (1907). Über die Substanzquotienten von *Helianthus annuus*. Inaugural Dissertation, Bonn, pp. 1–25, Tab I–V (quoted by Kidd, West & Briggs, 1920). 13.4, **26.6**

HABER W. (1958). Ökologische Untersuchungen der Bodenatmung. *Flora, Jena*, **146**, 109–157. 31.9

HABER W. (1962). Über Zusammenhänge zwischen der Produktivität eines Pflanzenbestandes und der Bodenatmung. In *Die Stoffproduktion der Pflanzendecke* (ed. H. Lieth), pp. 109–112. Gustav Fischer, Stuttgart. 31.9

HABERLANDT G. (1884). *Physiologische Pflanzenanatomie*. Wilhelm Engelmann, Leipzig, pp. xii + 398. 13.8, 13.13, 27.1

HACKENBERG H. (1909). Über die Substanzquotienten von *Cannabis sativa* und *Cannabis gigantea*. *Beih. bot. Zbl.*, **24**, 45–67. 11.19, 13.2, 13.4, *13.4*, 24.2

HACKETT C. (1969). A study of the root system of barley. II. Relationships between root dimensions and nutrient uptake. *New Phytol.*, **68**, 1023–30. 13.20

HADFIELD W. (1968). Leaf temperature, leaf pose and productivity of the tea bush. *Nature, Lond.*, **219**, 282–284. 9.3

fh HANDBOOK OF METEOROLOGICAL INSTRUMENTS (1965). Part I. Instruments for surface observations. 4th Im-

pression. Meteorological Office, Met. O. 577. London,
H.M.S.O. pp. x + 458.   9.5

c   HARLEY J.L. (1969). *The biology of mycorrhiza.* 2nd Edn.
Leonard Hill, London, pp. xxii + 334.   13.20

HARPER J.L. (1961). Approaches to the study of plant
competition. *Symp Soc. exp. Biol.*, **15**, 1–39.   10.5

HARRISON J.A.C. (1968). *Verticillium* wilt of potatoes.
Ph.D. Thesis, Univ. Swansea.

           30.16, *30.17*, **30.17**, **30.18**, 30,18, 32.3

HARRISON J.A.C. and ISAAC I. (1968). Leaf-area development
in King Edward potato plants inoculated with *Verticillium albo-atrum* and *V. dahliae. Ann. appl. Biol.*, **61**,
217–230.   30.16, **30.16**

HASLAM S.M. (1960). The vegetation of the Breck fen margin.
Ph.D. Thesis, Univ. Cambridge.   10.2

HAYEK A. VON (1918). *Helianthus* L. In *Hegi, G., Illustrierte
Flora von Mitteleuropa* Vol. VI, Pt I, 507–515, J.F.
Lehmann's Verlag, München.   App. 3.6

HEATH O.V.S. and GREGORY F.G. (1938). The constancy of
mean net assimilation rate and its ecological importance.
Ann. Bot., N.S. **2**, 811–818.   20.6

HEGARTY T.W. (1968). Seedling growth and the root
environment. Ph.D. Thesis, Univ. Cambridge.   10.10, 26.14, 27.10,
           27.11, 28.18, 28.19, 32.4, 32.5

HEMSLEY W.B. (1907). *Eupatorium glandulosum. Curtis's
bot. Mag.*, **133**, Tab. 8139.   App. 3.3

a   HESLOP HARRISON J. (1964). Forty years of genecology.
Adv. Ecol. Res., **2**, 159–247.

d   HEWITT E.J. (1966). *Sand and water culture methods used in
the study of plant nutrition.* Commonwealth Bureau of
Horticulture and Plantation Crops, Technical Communication No 22. Commonwealth Agricultural Bureaux,
Farnham Royal, pp. xiii + 547.   10.9, 10.10

HILLMAN W.S. (1969). Photoperiodism and vernalization. In
Physiology of Plant Growth and Development (ed. M.B.
Wilkins), pp. 559–601. McGraw-Hill, London   23.2

HOOKER J.D. (1885). *Chusquea abietifolia. Curtis's bot. Mag.*,
111, Tab. 6811.   23.3

e   HUDSON J.P. (ed) (1957). *Control of the plant environment.*
Butterworth Sci. Pubs, London, pp. xvi + 240.   **18.10**

e   HUGHES A.P. (1959a). Plant growth in controlled environments as an adjunct to field studies. Experimental
application and results. *J. agric. Sci., Camb.*, **53**,
247–259.   3.1, 17.3, **17.4.1**, *17.4*, **17.4.2**, 17.5, **17.6**

HUGHES A.P. (1959b). Effects of the environment on leaf

development in *Impatiens parviflora* DC. *J. Linn. Soc.* (*Bot.*), **56**, 161–165.   **27.2, 27.3, 27.4, 27.5, 27.6**

HUGHES A.P. (1964a). Some modifications of Schwabe's photometric apparatus for leaf measurement. *Ann. Bot.*, N.S. **28**, 473–474.   11.11

HUGHES A.P. (1965a). Plant growth and the aerial environment. VII. The growth of *Impatiens parviflora* in very low light intensities. *New Phytol.*, **64**, 55–64.   **3.2.1.a**, **28.13**, 30.6, 30.7, 30.8, 30.9

HUGHES A.P. (1965c). Plant growth and the aerial environment. IX. A synopsis of the autecology of *Impatiens parviflora*. *New Phytol.*, **64**, 399–413.   **3.4.2**, 19.8, **19.8, 27.8.1, 27.8.2, 27.9, 28.8, 28.9, 28.10**

HUGHES A.P. (1966). The importance of light compared with other factors affecting plant growth. In *Light as an Ecological Factor* (*Symp. Brit. Ecol. Soc.*, **6**, 121–146). Blackwell Sci. Pubs, Oxford.   Plate 7.3

HUGHES A.P. & COCKSHULL K.E. (1969). Effects of carbon dioxide concentration on the growth of *Callistephus chinensis* cultivar Johannistag. *Ann. Bot.*, N.S. **33**, 351–365.   23.7, 23.7, 24.12, **24.12, 24.13**

HUGHES A.P., COCKSHULL K.E. & HEATH O.V.S. (1970). Leaf area and absolute leaf water content. *Ann. Bot.*, N.S. **34**, 259–266.   11.8, 24.17, **24.17**, *24.18*, 24.19, 24.21

HUGHES A.P. & EVANS G.C. (1962). Plant growth and the aerial environment. II. Effects of light intensity on *Impatiens parviflora*. *New Phytol.*, **61**, 154–174.   **3.2.1.b**, **19.4.2, 21.3, 24.8**, 28.3, *28.3*, 28.5, **28.6.1**, **28.6.2**, 30.6, 30.7

HUGHES A.P. & EVANS G.C. (1963). Plant growth and the aerial environment. IV. Effects of daylength on *Impatiens parviflora*. *New Phytol.*, **62**, 367–388.   **26.12**

HUGHES A.P. & FREEMAN P.R. (1967). Growth analysis using frequent small harvests. *J. appl. Ecol.*, **4**, 553–560.   16.19, 16.20, 20.4, **20.4.1, 20.4.2**, 23.7

HUNT R. (1970). Relative growth-rate: its range and adaptive significance in a local flora. Ph.D. Thesis, Univ. Sheffield.   26.14

INAMDAR R.S., SINGH S.B. & PANDE T.D. (1925). The growth of the cotton plant in India. I. The relative growth rates during successive periods of growth and the relation between growth rate and respiratory index throughout the life cycle. *Ann. Bot.*, **39**, 281–311.   13.6, **13.6.2, 18.3**, 18.3, 23.5, **23.5.2**

ISAAC I. & HARRISON J.A.C. (1968). The symptoms and causal agents of early-dying disease (*Verticillium* wilt) of potatoes. *Ann. appl. Biol.*, **61**, 231–244.   30.16

c   JACKS G.V. (1954). *Soil.* Thomas Nelson & Sons, London, pp. ix + 221.

JONES E.W. (1955). Ecological studies on the rain forest of Southern Nigeria. IV. The plateau forest of the Okomu Forest Reserve. Part I. The environment, the vegetation types of the forest and the distribution of species. *J. Ecol.*, **43**, 564–594.   31.10

JONES E.W. (1956). Ecological studies on the rain forest of Southern Nigeria. IV. The plateau forest of the Okomu Forest Reserve. Part II. The reproduction and history of the forest. *J. Ecol.*, **44**, 83–117.   31.10

KEITH LUCAS D.M. (1968). Shade tolerance in *Primula*. Ph.D. Thesis, Univ. Cambridge.   9.15

KIDD F. & WEST C. (1919). Physiological pre-determination: the influence of the physiological condition of the seed upon the course of subsequent growth and upon the yield. IV. Review of the literature. Chapter III. *Ann. appl. Biol.*, **5**, 220–251.   13.2, 13.4

KIDD F., WEST C. & BRIGGS G.E. (1920). What is the significance of the efficiency index of plant growth? *New Phytol.*, **19**, 88–96.   **26.6, 26.6**

KIDD F., WEST C. & BRIGGS G.E. (1921). A quantitative analysis of the growth of *Helianthus annuus*. Part I. The respiration of the plant and of its parts throughout the life cycle. *Proc. R. Soc.*, B **92**, 368–384.   32.6

KILTZ H. (1909). Versuche über den Substanz-Quotienten beim Tabak und den Einfluss von Lithium auf dessen Wachstum. *Bot. Zbl.*, **110**, 455–456.   13.4

KIRA T., OGAWA H., YODA K. & OGINO K. (1967). Comparative ecological studies on three main types of forest vegetation in Thailand. IV. Dry matter production with special reference to the Khao Chong rain forest. *Nature and Life in S.E. Asia*, **5**, 149–174.   31.10, 31.11, 31.12, 31.13, 31.14, 31.19

KLEIN R.M., EDSALL P.C. & GENTILE A.C. (1965). Effects of near ultraviolet and green radiations on plant growth. *Pl. Physiol.*, *Lancaster*, **40**, 903–906.   28.10

KNIGHT R.C. (1922). Further observations on the transpiration, stomata, leaf water content, and wilting of plants. *Ann. Bot.*, **36**, 361–383.   24.1

KOLLER D., MAYER A.M., POLYAKOFF-MAYBER A. & KLEIN S. (1962). Seed germination. *A. Rev. Pl. Physiol.*, **13**, 437–464.   5.7

KORIBA K. (1958). On the periodicity of tree growth in the tropics. *Gdns' Bull.*, *Singapore*, **17**, 11–81.   2.4, 13.21

KRASNOSELSKY-MAXIMOW T.A. (1925). Untersuchungen über Elastizität der Zellmembran. *Ber. dt. bot. Ges.*, **43**, 527–537.   24.1

KREUSLER U., PREHN A. & BECKER G. (1877a). Beobachtungen über das Wachstum der Maispflanze (Berichte über die Versuche vom Jahre 1875). *Landw. Jbr.*, **6**, 759–786.   **2.7**, 11.8, 13.7, 30.3

KREUSLER U., PREHN A. & BECKER G. (1877b). Beobachtungen über das Wachstum der Maispflanze (Berichte über die Versuche vom Jahre 1876). *Landw. Jbr.*, **6**, 787–800.   11.8, 13.6, **13.6.2**, 13.7, **13.7**, **16.3**, **16.4**, **16.8**, **16.9.2**, **16.13.2**, 30.3

KREUSLER U., PREHN A. & HORNBERGER R. (1878). Beobachtungen über das Wachstum der Maispflanze (Bericht über die Versuche vom Jahre 1877). *Landw. Jbr.*, **7**, 536–564.   13.7, 30.3

KREUSLER U., PREHN A. & HORNBERGER R. (1879). Beobachtungen über das Wachstum der Maispflanze (Bericht über die Versuche vom Jahre 1878). *Landw. Jbr.*, **8**, 617–622.   13.7, 30.3

LANG A. (1965). Physiology of flower initiation. In *Encyclopedia of Plant Physiology* (eds W. Ruhland *et al.*), Vol. 15(1), pp. 1380–1536. Springer Verlag, Berlin & New York   23.2

LEACH G.J. & WATSON D.J. (1968). Photosynthesis in crop profiles, measured by phytometers. *J. appl. Ecol.*, **5**, 381–408.   30.3

LEBLOND C. & CARLIER F. (1965). Technique pour la mesure de l'émission de gaz carbonique par les organes végétaux sur pied, en conditions définies de température et d'humidité relative. In *Methodology of Plant Ecophysiology* (ed. F.E. Eckardt), pp. 275–281. UNESCO, Paris.   12.6

LEWIS J.P. (1963). Plant growth and nutritional factors. Ph.D. Thesis, Univ. Cambridge.   **5.11**, 10.10, **13.6.1**, 26.13, **26.13**, 27.10, 28.18

LIETH H. (ed) (1962). *Die Stoffproduktion der Pflanzendecke.* Gustav Fischer, Stuttgart, pp. 156.   31.1, 31.9, 31.3

f h–q   LIGHT AS AN ECOLOGICAL FACTOR (1966). *British Ecological Society Symposium No 6.* Blackwell Sci. Pubs, Oxford, pp. xi + 452.   9.6

LÖHR E. & MÜLLER D. (1968). Blatt-Atmung der höheren Bodenpflanzen im tropischen Regenurwald. *Physiologia Pl.*, **21**, 673–765.   31.16

LOOMIS R.S., WILLIAMS W.A. & HALL A.E. (1971). Agricultural productivity. *A. Rev. Pl. Physiol.* **22**, 431–468.   29.10, 32.4

LUNDEGÅRDH H. (1924). *Der Kreislauf der Kohlensaüre in der Natur.* Gustav Fischer, Jena, pp. viii + 308.   31.9

LUPTON F.G.H. (1966). Translocation of photosynthetic assimilates in wheat. *Ann. appl. Biol.*, **57**, 355–365.   23.6

MACFADYEN A. (1970). Simple methods for measuring and maintaining the proportion of carbon dioxide in air, for use in ecological studies of soil respiration. *Soil. Biol. Biochem.*, **2**, 9–18.   31.9

MALLOCH A.J.C. (1970). Analytical studies of cliff-top vegetation in south-west England. Ph.D. Thesis, Univ. Cambridge.   30.3

MAPSON L.W. (1969). Biogenesis of ethylene. *Biol. Rev.*, **44**, 155–188.   12.8, 12.9, 26.1

MÄRKER M. (1876). Versuche über die Zunahme an Trockensubstanz, Asche und Stickstoff an der Maispflanze. Ausgefuhrt auf der agrikultur-chemischen Versuchs-Station Halle a.S. *Landw. Jbr.*, **5**, 751–753.   **2.7**

MAXIMOV N.A. (1929). *The plant in relation to water*. Edited with notes, by R.H. Yapp. London, George Allen & Unwin, pp. 0 + 451.   24.1

MAYER A.M. & POLJAKOFF-MAYBER A. (1963). *The germination of seeds*. Pergamon Press, London & New York, pp. vii + 236.   5.7

f   THE MEASUREMENT OF ENVIRONMENTAL FACTORS IN TERRESTRIAL ECOLOGY (1968). (Eds Wadsworth R.M., Chapas L.C., Rutter A.J., Solomon M.E. & Warren Wilson J.) *British Ecological Society Symposium No 8*. Blackwell Sci. Pubs, Oxford, pp. x + 314.   9.5

MILLENER L.H. (1952). Experimental studies on the growth forms of the British species of *Ulex* L. Ph.D. Thesis, Univ. Cambridge.   28.10

MILLENER L.H. (1961). Daylength as related to vegetative development in *Ulex europaeus* L. I. The experimental approach. *New Phytol.*, **60**, 339–354.   28.10

MOHR H. (1969). Photomorphogenesis. In *Physiology of Plant Growth and Development* (ed. M.B. Wilkins), pp. 509–556. McGraw-Hill, London.   23.2

f h   MONTEITH J.L. (1972) *Survey of instruments for micrometeorology* (IBP Handbook No 22). Blackwell Sci. Pubs, Oxford.   9.5

MORGAN D.G. (1968). The regulation of growth and development by plant growth substances. *Euphytica*, **17** (1968 Suppl. 1), 189–213.   26.1, 26.10

b   MORONEY M.J. (1951). *Facts from figures*. Penguin Books. Harmondsworth pp. iv (unnumbered) + 472.

MORRIS D. (1886). *Chusquea abietifolia*. *Gdnrs' Chron.*, **26**, 524.   23.3

MÜLLER D. (1924). Studies on traumatic stimulus and loss of dry matter by respiration in branches of Danish forest trees. *Dansk bot. Ark.*, **4** (6), 1–33.   12.4, **12.4.2**, 12.12

MÜLLER D. & NIELSEN J. (1965). Production brute, pertes

par respiration et production nette dans la forêt ombro-
phile tropicale. *Forst. ForsVaes. Danm.*, **29**, 69–160.   12.9, **12.9,**
                    12.11, 13.18, 31.10–.18, **31.12.1, 31.12.2, 31.14,** *31.16*

MUNRO COL. (1868). A Monograph of the Bambusaceae,
including descriptions of all the species. *Trans. Linn.
Soc. Lond.*, **26**, 1–157.   23.3

NEWBOULD P.J. (1967). *Methods for estimating the primary
production of forests.* IBP Handbook No 2. Blackwell
Sci. Pubs, Oxford and Edinburgh, pp. ix + 62.   4.6, 31.8

NOBEL P.S. (1970). *Plant cell physiology, a physicochemical
approach.* W.H. Freeman & Co., San Francisco, pp.
ix + 267.   5.6

NYE P.H. (1961). Organic matter and nutrient cycles under
moist tropical forest. *Plant and Soil*, **8**, 333–346.   31.14, 31.18

OGAWA H., YODA K. & KIRA T. (1961). A preliminary survey
on the vegetation of Thailand. *Nature and Life in S.E.
Asia*, **1**, 21–157.   31.10

OGAWA H., YODA K., KIRA T., OGINO K., SHIDEI T.,
RATANAWONGSE D. & APASUTAYA C. (1965a). Comparative ecological study on three main types of forest
vegetation in Thailand. I. Structure and floristic
composition. *Nature and Life in S.E. Asia*, **4**, 13–48.   31.10

OGAWA H., YODA K., OGINO K. & KIRA T. (1965b). Comparative ecological studies on three main types of
forest vegetation in Thailand. II. Plant biomass. *Nature
and Life in S.E. Asia*, **4**, 49–82.   31.8, 31.10, 31.16

OJEHOMON O.O., RATHGEN A.S. & MORGAN D.G. (1968).
Effects of day length on the morphology and flowering
of five determinate varieties of *Phaseolus vulgaris* L.
J. agric. Sci., Camb., **71**, 209–214.   23.11

OVINGTON J.D. & MACRAE C. (1960). The growth of seed-
lings of *Quercus petraea. J. Ecol.*, **48**, 549–555.   9.15

OXFORD ENGLISH DICTIONARY (1901). A new English dictionary on historical principles. Vol. IV. F & G. Clarendon
Press, Oxford, pp. viii + 532.   1.3

f   PENMAN H.L. (1963). *Vegetation and hydrology.* Commonwealth Bureau of Soils, Technical Communication No.
53. Commonwealth Agricultural Bureaux, Farnham
Royal, pp. viii + 124.   6.6

PETERKEN G.F. & LLOYD P.S. (1967). Biological flora of the
British Isles. No. 108. *Ilex aquifolium* L. *J. Ecol.*, **55**,
841–858.   2.4

PHILLIPS I.D.J. (1969). Apical dominance. In *Physiology of
Plant Growth and Development* (ed. M.B. Wilkins),
pp. 165–202. McGraw-Hill, New York.   26.1

PIGOTT C.D. & TAYLOR K. (1964). The distribution of some
woodland herbs in relation to the supply of nitrogen and
phosphorus in the soil. *J. Ecol.*, **52** (Suppl.), 175–185.   5.6, 30.3

PITMAN M.G. (1965). Sodium and potassium uptake by
seedlings of *Hordeum vulgare*. *Aust. J. biol. Sci.*, **18**,
10–24.   28.19

PITMAN M.G. (1966). Uptake of potassium and sodium by
seedlings of *Sinapis alba*. *Aust. J. biol. Sci.*, **19**, 257–269.   28.19

POORE M.E.D. (1967). The concept of the association in
tropical rain forest. *J. Ecol.*, **55**, 46P–47P.   13.22, 23.2

POORE M.E.D. (1968). Studies in Malaysian rain forest. I. The
forest on triassic sediments in Jengka Forest Reserve.
J. Ecol., **56**, 143–196.   13.22, 23.2

PURSEGLOVE J.W. (1968). *Tropical Crops. Dicotyledons* 2.
Longmans Green, London, pp. viii, 333–719.   App. 3.1

RACKHAM O. (1965). Transpiration, assimilation, and the
aerial environment. Ph.D. Thesis, Univ. Cambridge.   12.13, 28.5,
          29.1, 29.3, *29.4, 29.5*, **29.6**, 29.7, Pl. 29.7, 29.8, 29.10,
          App. 1.18, **App. 1.18**, App. 1.19

RACKHAM O. (1966). Radiation, transpiration, and growth
in a woodland annual. In *Light as an Ecological Factor*
(*Symp. Brit. Ecol. Soc.* **6**, pp. 167–185). Blackwell Sci.
Pubs, Oxford.   29.3, *29.4, 29.5*

RAJAN A.K., BETTERIDGE B. & BLACKMAN G.E. (1971).
Interrelationships between the nature of the light
source, ambient air temperature, and the vegetative
growth of different species within growth cabinets.
Ann. Bot., N.S. **35**, 323–343.   28.10

RAUNKIAER C. (1937). *Plant life forms*. Translated by H.
Gilbert Carter. Oxford, Clarendon Press, pp. vii + 104.   23.12

RAVEN J.A. (1970). Exogenous inorganic carbon sources in
plant photosynthesis. *Biol. Rev.*, **45**, 167–221.   12.16, 29.7, 29.8,
          29.9

RICHARDS P.W. (1939). Ecological studies on the rain forest
of southern Nigeria. I. The structure and floristic
composition of the primary forest. *J. Ecol.*, **27**, 1–61.   31.10

RIDER N.E. (1954). Eddy diffusion of momentum, water
vapour, and heat near the ground. *Phil. Trans. R. Soc.*,
A **216**, 481–501.   29.5

ROACH W.A., NEVE R., VANSTONE F.H., PHILCOX H.J.,
DELAP A.V. & FORD E.M. (1957). A method of growing
apple trees by spraying their roots with nutrient solution. *J. hort. Sci.*, **32**, 85–98.   10.10, 28.18

cd   RORISON I.H. (ed) (1969). *Ecological aspects of the mineral
nutrition of plants. British Ecological Society Symposium*

656 *Bibliography*

No. 9. Blackwell Sci. Pubs, Oxford, pp. xxi + 512. 10.8, 26.13, 30.3, 32.5

c RUSSELL E.W. (1961). *Soil conditions and plant growth.* 9th Edn. Longmans, London, pp. xvi + 688.

RYLE G.J.A. (1970a). Partition of assimilates in an annual and a perennial grass. *J. appl. Ecol.* 7, 217–227. 23.6, 32.8

RYLE G.J.A. (1970b). Distribution patterns of assimilated ^{14}C in vegetative and reproductive shoots of *Lolium perenne* and *L. temulentum. Ann. appl. Biol.,* 66, 155–167. 23.6, 32.8

SALISBURY E.J. (1942). *The reproductive capacity of plants.* London, Bell & Sons Ltd., pp. xi + 244. 13.4, 23.3

SCHWABE W.W. (1951). Physiological studies in plant nutrition. XVI. The mineral nutrition of bracken. *Ann. Bot.,* N.S. 15, 417–446. 11.11

SCOTT J.K. (1949). Respiration in bulky plant tissue. Ph.D. Thesis, Univ. Cambridge. 12.4

SCOTT RUSSELL R. (1970). Root systems and plant nutrition —some new approaches. *Endeavour,* 29, 60–66. 13.20

SEARLE N.E. (1965). Physiology of flowering. *A. Rev. Pl. Physiol.,* 16, 97–118. 23.2

SEIFRIZ W. (1920). The length of the life cycle of a climbing bamboo. A striking case of sexual periodicity in *Chusquea abietifolia* Griseb. *Am. J. Bot.,* 7, 83–94. 23.3

SHAMSI S.R.A. (1970). Comparative biology of *Epilobium hirsutum* and *Lythrum salicaria.* Ph.D. Thesis, Univ. London. 13.4, 23.4, 23.10, 32.4

SHELDRAKE A.R. & NORTHCOTE D.H. (1968). The production of auxin by tobacco internode tissues. *New Phytol.,* 67, 1–13. 27.9

SIEGELMAN H.W. (1969). Phytochrome. In *Physiology of Plant Growth and Development* (ed. M.B. Wilkins), pp. 489–506. McGraw-Hill, London. 23.2, 26.1

SLATYER R.O. (1967). *Plant-water relationships.* Academic Press, London & New York, pp. xii + 366. 6.6, 10.9, 24.1, 24.2

SMALL J.K. (1933). *Manual of the Southeastern Flora.* Univ. North Carolina Press, Chapel Hill, pp. xxii + 1554. App. 3.5

STOCKER O. (1935). Assimilation und Atmung westjavan-ischer Tropenbäume. *Planta,* 24, 402–445. 31.16

STOY V. (1963). The translocation of ^{14}C-labelled photosynthetic products from the leaf to the ear in wheat. *Physiologia Pl.,* 16, 851–866. 23.6, 32.8

SUSSMAN A.S. & HALVERSON H.O. (1966). *Spores: their dormancy and germination.* Harper & Row, New York and London, pp. xi + 354. 5.7

fhk   SZEICZ G. (1966). Field measurements of energy in the
0·4–0·7 micron range. In *Light as an Ecological Factor*
(*Symp. Brit. Ecol. Soc.*, **6**, 41–52), Blackwell Sci. Pubs,
Oxford.   30.7

THAINE R., OVENDEN S.L. & TURNER J.S. (1959). Translocation of labelled assimilates in the soybean. *Aust. J.
biol. Sci.*, **12**, 349–372.   23.6

THODAY D. (1921). On the behaviour during drought of
leaves of two Cape species of *Passerina*, with some notes
on their anatomy. *Ann. Bot.*, **35**, 585–601.   24.2, 24.16

THODAY D. (1932). Apparatus for plant physiology. *Sch.
Sci. Rev.*, **14**, 168–172.   App. 1.1

TISCHENDORF W. (1926). Wuchsgesetze von *Pinus silvestris*.
Forstwiss. ZentBl., **48**, 578–589, 652–662, 689–698,
729–738.   **2.9**, 31.8

TOOLE E.H., HENDRICKS S.B., BORTHWICK H.A. & TOOLE
V.K. (1956). Physiology of seed germination. *A. Rev.
Pl. Physiol.*, **7**, 299–324.   5.7

VAN DOBBEN W.H. (1967). Physiology of growth in two
Senecio species in relation to their ecological position.
Meded. Inst. biol. scheik. Onderz. LandbGewass., **346**,
75–83.   26.9, 32.2

VERNON A.J. & ALLISON J.C.S. (1963). A method of calculating net assimilation rate. *Nature, Lond.*, **200**,
814.   16.16, 16.19, 20.4

f   WADSWORTH R.M. (ed.) (assisted by Chapas L.C., Rutter
A.J., Solomon M.E. and Warren Wilson J.) (1968).
*The measurement of environmental factors in terrestrial
ecology.* Blackwell Sci. Pubs, Oxford, pp. x + 314.   9.5

WALTER H. (1952). Eine einfache Methode zur ökologischen
Erfassung des CO_2-Faktors am Standort. *Ber. dt. bot.
Ges.*, **65**, 175–182.   31.9

WALTER H. (1960). *Grundlagen der Pflanzenverbreitung. I
Teil. Standortslehre.* 2 Aufl. Stuttgart, Eugen Ulmer,
pp. 0 + 566.   24.1, 30.3

WANNER H. (1970). Soil respiration, litter fall and productivity of tropical rain forest. *J. Ecol.*, **58**, 543–547.   31.18

WASSINK E.C. & STOLWIJK J.A.J. (1956). Effects of light
quality on plant growth. *A. Rev. Pl. Physiol.*, **7**, 373–
400.   28.10

WATSON D.J. (1947). Comparative physiological studies on
the growth of field crops. I. Variation in net assimilation
rate and leaf area between species and varieties, and
within and between years. *Ann. Bot.*, N.S. **11**, 41–76.   13.18, **13.18.1**,
    13.19, **13.19**, *13.19*, 20.6

WATSON D.J. (1958). The dependence of net assimilation rate on leaf area index. *Ann. Bot.*, N.S. **22**, 37–54. 13.19

WATSON D.J., THORNE G.N. & FRENCH S.A.W. (1958). Physiological causes of differences in grain yield between varieties of barley. *Ann. Bot.*, N.S. **22**, 321–352. 13.18

WATSON D.J. & WATSON M.A. (1953). Comparative physiological studies on the growth of field crops. III. Effect of infection with beet yellows and beet mosaic virus on the growth and yield of the sugar beet crop. *Ann. appl. Biol.*, **40**, 1–37. 30.16

WATSON D.J. & WITTS K.J. (1959). The net assimilation rates of wild and cultivated beets. *Ann. Bot.*, N.S. **23**, 431–439. 13.18

WEBB J.A. & GORHAM P.R. (1964). Translocation of photosynthetically assimilated C^{14} in straight-necked squash. *Pl. Physiol., Lancaster*, **39**, 663–672. 23.6

WEBER C.A. (1879). Über specifische Assimilationsenergie. Inaugural-Dissertation der philosophischen Facultät der Kgl. Maximilians-Universität zu Würzburg. Stabel'schen Buchdruckerei, Würzburg, 27 pp. + Tab. 1–9. 13.8, 16.1, 29.1

WEBER C.A. (1882). Über specifische Assimilationsenergie. *Arbeiten bot. Inst. Würzburg*, **2**, 346–352. 13.8, 29.1

WELBANK P.J. (1957). Plant growth and root competition. Ph.D. Thesis, Univ. Cambridge. 12.4, 12.10

d WELBANK P.J. (1961). A study of the nitrogen and water factors in competition with *Agropyron repens* (L.) Beauv. *Ann. Bot.*, N.S. **25**, 116–137. **10.6**, 10.7, 15.6, 28.17

d WELBANK P.J. (1962). The effects of competition with *Agropyron repens* and of nitrogen- and water-supply on the nitrogen content of *Impatiens parviflora*. *Ann. Bot.*, N.S. **26**, 361–373. **13.14**, 13.20, **13.20**, 28.17, 32.5

WELBANK P.J. & WILLIAMS E.D. (1968). Root growth of a barley crop estimated by sampling with portable powered soil-coring equipment. *J. appl. Ecol.*, **5**, 477–481. 11.15

e WENT F.W. (1957). *The experimental control of plant growth*. Chronica Botanica, Waltham, pp. xvii + 343.

WEST C., BRIGGS G.E. & KIDD F. (1920). Methods and significant relations in a quantitative analysis of plant growth. *New Physol.*, **19**, 200–207. 13.5

WESTLAKE D.F. (1963). Comparisons of plant productivity. *Biol. Rev.*, **38**, 385–425. 31.1, 31.2, 31.3

WHITEHEAD F.H. & MYERSCOUGH P.J. (1962). Growth analysis of plants. The ratio of mean relative growth rate to mean relative rate of leaf area increase. *New Phytol.*, **61**, 314–321.   22.8, 23.4, 32.2

hilpq   WHITMORE T.C. (1966). A study of light conditions in forests in Ecuador with some suggestions for further studies in tropical forests. In *Light as an Ecological Factor* (*Symp. Brit. Ecol. Soc.* **6**, pp. 235–247) Blackwell Sci. Pubs, Oxford.

WHITTAKER R.H. & WOODWELL G.M. (1968). Dimension and production relations of trees and shrubs in the Brookhaven Forests, New York. *J. Ecol.*, **56**, 1–25.   31.8

WILKINS M.B. (ed.) (1969). *The physiology of plant growth and development.* McGraw-Hill, New York, pp. xxi + 695.   23.2, 26.1

WILLIAMS R.F. (1946). The physiology of plant growth with special reference to the concept of net assimilation rate. *Ann. Bot.*, N.S. **10**, 41–72.   13.14, 16.17, *16.17, 16.18*, 20.5

WILLIAMS R.F. (1948). The effects of phosphorus supply on the rates of intake of phosphorus and nitrogen and upon aspects of phosphorus metabolism in gramineous plants. *Aust. J. scient. Res.*, B **1**, 333–361.   13.20, 32.5

WILLIAMS R.F. (1960). The physiology of growth in the wheat plant. I. Seedling growth and pattern of growth at the shoot apex. *Aust. J. biol. Sci.*, **13**, 401–428.   28.19, 32.4

YODA K. (1967). Comparative ecological studies on three main types of forest vegetation in Thailand. III. Community respiration. *Nature and Life in S.E. Asia*, **5**, 83–148.   31.10, 31.15, 31.16

ZEHNI M.S. (1969). Photoperiod and flower bud development in *Phaseolus vulgaris* (L.) Savi. Ph.D. Thesis, Univ. Cambridge.   23.11

ZEHNI M.S., SAAD F.S. & MORGAN D.G. (1970). Photoperiod and flower bud development in *Phaseolus vulgaris. Nature, Lond.*, **227**, 628–629.   23.11

ZELAWSKI W. & FUCHS S. (1961). Verlauf der Atmungsänderungen im Holz nach dem Schnitt. Parallel-untersuchungen mit den CO_2 und Wasserdampf URAS. *Arch. Forstw.*, **10**, 1260–1268.   12.4

INDEX

Numbers in bold type refer to figures. Numbers in italic type refer to tables

24

25*

Unit root rate
> defined 13.20
> obstacles to determination of 13.20

Upper Teesdale (54°40′N 2°10′W), growth of *Viola rupestris* on sugar limestone of
> 10.5

Urtica dioica L.
> changes in root/top ratio of and lowered mineral nutrient supply 26.14
> extraction of root system during harvesting 11.15
> growth of, in the field as an indicator of phosphate supply 30.3
> mineral requirements of 5.6

Variability
> assessments of, of higher plant populations 4.4
> between germination and first harvest, control of 6.1–6.10
> of experimental plants, as affected by time from grading *8.3*, 8.9
> genetic structure of natural populations, in relation to 5.4
> of glasshouse-grown plants, control of 6.10, 9.13, 9.14
> higher plants, forms of 5.2
> plant, its causes 3.6, 4.1, 7.1, 17.1
> plant, in natural or semi-natural populations, investigation of 8.1
> of plant material, and physiological experiments on whole plants 28.15
> of plant size, due to non-simultaneous germination, means of reducing 5.10,
> > 5.11
> preliminary surveys of, as an aid in experimental planning 5.3
> reasons for, of plant material grown under natural conditions 3.6, 7.1, 17.1
> of unit leaf rate, and rate of progression along regular ontogenetic drifts 20.1,
> > 20.5

Variance
> comparison of relative growth rate for different sets of plants, reasons for and
> > against homogeneity of 15.2
> damage to plants by handling, effect on 11.20
> genetic variability, influence on 11.22, 11.23

Variance ratio (F test), example of use of *15.3*, 15.4, 15.5

Variances, comparison of as a means of detecting discrepancies in basic data 11.20

Variation
> in radiation conditions within a plant stand 6.9, 13.18, 13.19
> sources of, in higher plant growth 3.6, 4.1, 7.1, 17.1

Varietal differences in flowering behaviour, in relation to environment and ontogeny
> 23.10, 23.11

Vegetative propagation as a means of reducing variability, its dangers 5.4

Ventilation of glasshouses
> effects on changes in saturation deficit 9.8
> in relation to solar radiation 9.14

Vermiculite
> adjustment of alkalinity of 6.4
> aeration of 6.4
> as an artificial soil 6.4